Springer Proceedings in Mathematics & Statistics

This book series features volumes composed of selected contributions from workshops and conferences in all areas of current research in mathematics and statistics, including operation research and optimization. In addition to an overall evaluation of the interest, scientific quality, and timeliness of each proposal at the hands of the publisher, individual contributions are all refereed to the high quality standards of leading journals in the field. Thus, this series provides the research community with well-edited, authoritative reports on developments in the most exciting areas of mathematical and statistical research today.

More information about this series at http://www.springer.com/series/10533

Vladas Sidoravicius
Editor

Sojourns in Probability Theory and Statistical Physics - I

Spin Glasses and Statistical Mechanics, A Festschrift for Charles M. Newman

NYU SHANGHAI
上海纽约大学

Springer

Editor
Vladas Sidoravicius
NYU Shanghai
Shanghai, China

ISSN 2194-1009 ISSN 2194-1017 (electronic)
Springer Proceedings in Mathematics & Statistics
ISBN 978-981-15-0293-4 ISBN 978-981-15-0294-1 (eBook)
https://doi.org/10.1007/978-981-15-0294-1

Mathematics Subject Classification (2010): 60-XX, 82-XX, 82D30, 82C44

This Springer imprint is published by the registered company Springer Nature Singapore Pte Ltd.
The registered company address is: 152 Beach Road, #21-01/04 Gateway East, Singapore 189721, Singapore

Professor Charles M. Newman

Preface

This three-volume set, entitled Sojourns in Probability Theory and Statistical Physics, constitutes a Festschrift for Chuck Newman on the occasion of his 70th birthday. In these coordinated volumes, Chuck's closest colleagues and collaborators pay tribute to the immense impact he has had on these two deeply intertwined fields of research. The papers published here include original research articles and survey articles, on topics gathered by theme as follows:

Volume 1: Spin Glasses and Statistical Mechanics
Volume 2: Brownian Web and Percolation
Volume 3: Interacting Particle Systems and Random Walks

Our colleague Vladas Sidoravicius conceived the idea for this Festschrift during the conference on Probability Theory and Statistical Physics that was hosted on 25–27 March 2016 by the NYU-ECNU Institute of Mathematical Sciences at NYU Shanghai. This conference brought together more than 150 experts to discuss frontier research at the interface between these two fields, and it coincided with Chuck's 70th birthday. After the conference, Vladas approached various of Chuck's colleagues with invitations to contribute. Papers flowed in during the Fall of 2016 and the Spring of 2017. The Festschrift suffered delays in 2018, and then on 23 May 2019, Vladas passed away unexpectedly. Following discussions in June 2019 with NYU Shanghai and Springer Nature, we offered to assume editorial responsibility for bringing the volumes to completion.

We gratefully acknowledge Vladas's investment in these volumes, and we recognise that his presence in our community worldwide will be sorely missed. We offer our thanks to Julius Damarackas (NYU Shanghai) for his detailed preparation of the articles in these volumes.

Chuck has been one of the leaders in our profession for nearly 50 years. He has worked on a vast range of topics and has collaborated with and inspired at least three generations of mathematicians, sharing with them his deep insights into mathematics and statistical physics and his views on key developments, always leavened with his acute and captivating sense of humour. We wish him and his family many fruitful years to come.

July 2019

Federico Camia
Geoffrey Grimmett
Frank den Hollander
Daniel Stein

Contents

Probability Theory in Statistical Physics, Percolation, and Other Random Topics: The Work of C. Newman

Federico Camia[1,2] and Daniel L. Stein[3,4,5](✉)

[1] Division of Science, NYU Abu Dhabi, Saadiyat Island, Abu Dhabi, UAE
federico.camia@nyu.edu
[2] Department of Mathematics, VU University Amsterdam, De Boelelaan 1081A, 1081 HV Amsterdam, The Netherlands
[3] Department of Physics and Courant Institute of Mathematical Sciences, New York University, New York, NY 10012, USA
daniel.stein@nyu.edu
[4] NYU-ECNU Institutes of Physics and Mathematical Sciences at NYU Shanghai, 3663 Zhongshan Road North, Shanghai 200062, China
[5] SFI External Professor, Santa Fe Institute, 1399 Hyde Park Rd., Santa Fe, NM 87501, USA

Abstract. In the introduction to this volume, we discuss some of the highlights of the research career of Chuck Newman. This introduction is divided into two main sections, the first covering Chuck's work in statistical mechanics and the second his work in percolation theory, continuum scaling limits, and related topics.

Keywords: Spin glasses · Replica symmetry breaking · Pure states · Ground states · Metastate · Edwards–Anderson model · Fortuin–Kasteleyn · Random cluster representation · FK percolation · Nature vs. nurture · Deep quench · Riemann hypothesis · Lee–Yang theorem · deBruijn–Newman constant · Percolation · First passage percolation · Critical exponents · Continuum scaling limit · Normal fluctuations · SLE · CLE · Ising field theory · Brownian web

1 Equilibrium and Nonequilibrium Statistical Mechanics

Chuck has devoted a substantial portion of his research career—and made foundational advances—to the investigation of statistical mechanical systems, both homogeneous and inhomogeneous, elucidating both their thermal behavior in equilibrium and their nonequilibrium dynamical evolution following a deep quench. Here we briefly review only a few of his most important contributions, beginning with his papers on equilibrium thermodynamics.

V. Sidoravicius (Ed.): *Sojourns in Probability Theory and Statistical Physics - I*, PROMS 298, pp. 1–38, 2019.
https://doi.org/10.1007/978-981-15-0294-1_1

1.1 Thermodynamics of Disordered Systems

In this section we focus on three areas to which Chuck has contributed heavily, all largely focused on the thermodynamic structure of short-range spin glasses in finite dimensions. These are: general principles of pure state organization (using the metastate approach); multiplicity/non-multiplicity of pure and ground states in realistic and non-realistic models; and presence or absence of a thermodynamic phase transition in sufficiently high dimensions.

General Principles of Pure State Organization. In the mid-1990's there were (and as of this writing, still are) two leading (and very different) scenarios for the thermodynamic structure of short-range spin glasses: the many-state mean-field replica symmetry breaking (RSB) scenario put forward by Giorgio Parisi and co-workers [177,178,180,236,237] and the two-state droplet-scaling picture introduced by McMillan, Bray and Moore, and Fisher and Huse [41,42,94,97,175]. Given the analytical intractability of short-range spin glass models, studies relied (and continue to rely) largely on numerical simulations (for a sampling, see [21,32,33,35,36,48,63,71,80,125,139–141,151,168–173,181–183,189,233–235,245,274,284]). The actual pure state structure that would—or could—result from the application of RSB to short-range spin glasses had not been studied in depth; people were mostly looking for numerical evidence (often using spin overlap functions) of the signature features of RSB, in which (1) the thermodynamics is characterized by a mixed thermodynamic state decomposable into a countable infinity of pure states of varying weights within the mixed state; (2) the spin overlap distribution of these pure states is characterized by non-self averaging (over the coupling realizations) in the thermodynamic limit; and (3) the pure states are ordered hierarchically, in that their relative distances satisfy the ultrametric property (reviews can be found in [33,180,223,243,244,267]). This set of properties taken together at face value was called the "standard SK picture" in [214–216,223].

One of the distinguishing features of Chuck's work with one of us (DLS) on this problem was the use of rigorous mathematical techniques in an area where they had been seldom used before (and perhaps rarer still, where they were used to resolve—or at least sharpen understanding of—open physical questions). In the earliest foray on this problem [210], it was shown that the presence of distinct multiple pure states led to "chaotic size dependence": an infinite sequence of finite-volume Gibbs states, generated using boundary conditions chosen independently of the couplings (e.g., the standard periodic, free, or fixed boundary conditions), would not converge in the limit to a single thermodynamic state. However, one could still generate a thermodynamic state by appropriate averaging procedures, and if there were many pure states, this would be a mixture as required by RSB. A surprising conclusion then followed: no matter how the state was constructed (as long as it satisfied basic requirements, such as measurability), the main features of RSB were incompatible with each other. That is, short-range spin glasses could not support the standard RSB picture in any finite dimension.

This then led to the question, is there *any* scenario that is mean-field-like, and that *can* be supported in a mathematically consistent fashion in short-range spin glasses? In order to answer this question, the tool of the *metastate* was introduced in [213–215, 223]. (It was soon proved in [216] that this construction was essentially equivalent to an earlier construct introduced by Aizenman and Wehr [10].) Just as in spin models a thermodynamic state is a probability measure on spin configurations, the metastate is a probability measure on the thermodynamic states themselves. Although it can be used for any thermodynamic system, it is most useful (perhaps even essential) when dealing with systems with many competing thermodynamic states, providing an elegant means of connecting the behavior observed in finite volumes with the (infinite-volume) thermodynamics. (In homogeneous systems, the connection between the two is straightforward, but is anything but straightforward if many pure states exist.) Because of this, its significance and usefulness extend well beyond spin glasses alone (see, for example, [152, 153]).

Using the metastate, a "maximal" mean-field picture was constructed [213, 214, 223] that avoided the inconsistencies of the standard RSB model (leading to the moniker "nonstandard RSB model"). In a series of publications, Newman and Stein (NS) proved that if there are multiple spin glass pure/ground states in finite dimensions, then there must be an *uncountable* infinity of them [224]. In order to arrange them in an RSB-like fashion, there must then be (more formally, the metastate must be supported on) an uncountable set of distinct mixed thermodynamic states, *each one of which* is supported on a countable infinity of pure states with varying weights. Roughly speaking, in a given large finite volume, one would observe a single one of these mixed states, which appear with probabilites assigned by the metastate. Ultrametricity would hold among the pure states within a *single* mixed thermodynamic state, but not in general between any three pure states chosen arbitrarily from the metastate itself. Moreover, the meaning of non-self-averaging would change, from averaging over different coupling realizations to (roughly speaking) averaging over different volumes for *fixed* coupling realization. This picture, based on what remains logically possible based on self-consistency as determined by rigorous arguments, was recently shown by Read [244] to follow from a field-theoretic approach of direct application of RSB techniques to short-range spin glasses.

In very recent work [12] with Louis–Pierre Arguin, a new set of thermodynamic identities—connecting pure state weights, correlation functions, and overlaps—was derived, again using rigorous arguments. It was shown using these relations that pure state weights within a single mixed thermodynamic state consisting of infinitely many pure states (if such a thing exists) are distributed according to a Poisson–Dirichlet process—exactly as derived much earlier [76, 179, 249] for the RSB solution to the infinite-range Sherrington–Kirkpatrick [258] spin glass.

So, looking at the larger picture, this body of work used rigorous arguments to greatly narrow down the set of possibilities for the organization of pure states in spin glasses (if in fact there are multiple pure states). It is important to recognize, as this work emphasizes, that *a priori*, there are many possible many-state pictures; having many pure states is not synonymous with RSB. In fact, the RSB scenario is quite special, having an enormous amount of structure. One of the conclusions of the work briefly summarized above is that *if* there are many pure states in short-range spin glasses, then (regardless of dimension) there must be an uncountable number, and *if* they appear in nontrivial mixed thermodynamic states (each comprising a countable infinity with given weights), then the *only* possibility is the RSB one (as understood within the "nonstandard" picture). This is a strong conclusion, and is based on rigorous arguments.

Of course, none of this addresses the actual question of whether the RSB picture (at this point, it's no longer necessary to refer to the "nonstandard RSB picture") actually occurs in short-range spin glasses in any finite dimension. In [215, 223, 267], it was argued that this is unlikely. The arguments are based on a few rigorous results: the invariance of the metastate with respect to changes of boundary conditions [215], and important differences in the behavior of edge disorder chaos in infinite-range and short-range models, as pointed out by Chatterjee [65]. (We note that the argument claiming that metastate invariance is incompatible with RSB has been criticized in [244].) Perhaps more compelling is a nonrigorous argument based on pathologies arising from coupling strengths in infinite-range spin glasses scaling to zero with the number of spins, coupled with the randomness of coupling signs [222, 267]. However, the issue remains unresolved as of now.

While the body of work described above has led to the ruling out of an enormous number of possible scenarios, it has also led to the introduction of new ones. A very natural and (so far) viable picture, introduced and investigated in [213–215, 223] and called the "chaotic pairs" picture, arises naturally from the metastate approach. Chaotic pairs is a scenario that, like RSB, has uncountably many pure states in the (periodic boundary condition, for specificity) metastate, but they are organized much more simply than in RSB, with a simple, self-averaging overlap structure that is identical (in any finite volume) to that of droplet-scaling. In chaotic pairs, thermodynamic states are a *trivial* mixture of pure states, each consisting of a single pure state and its global flip, each with weight 1/2 in the mixed thermodynamic state. However, there are uncountably many of these thermodynamic states, each one comprising a pure state pair different from all the others.

For completeness, we should mention a fourth picture, introduced independently by Palassini and Young [235] and Krzakala and Martin [151] and sometimes referred to as the "TNT" picture. Here excitations above the ground state have boundaries with zero density in the edge set (like droplet-scaling, and unlike RSB and chaotic pairs) but whose energies do not scale as some increasing function of the volume (like RSB, and unlike droplet-scaling and chaotic pairs). While this last picture does not specify how many pure/ground states there are, it was argued in [221] that it is most naturally consistent with two-state pictures. If so, that leaves us with two potential many-state pictures for the short-range spin glass phase (RSB, chaotic pairs) and two two-state pictures (droplet-scaling, TNT). Which of these holds for short-range spin glasses, and in which dimensions, remains a fundamental open problem in the mathematics and physics of the statistical mechanics of disordered systems.

Multiplicity of Pure and Ground States in Short-Range Spin Glasses. The work described in the previous section clarified and restricted the ways in which pure states could be *organized* in short-range spin glasses; it did not address the actual *number* of pure or ground states of the spin glass phase. It should be noted that—from a logical standpoint at least—this question cannot be considered without answering several deeper ones: namely, is there a thermodynamical phase transition at nonzero temperature in *any* finite dimension, and if so, does the low-temperature phase break spin-flip symmetry (so that pure states come in spin-reversed pairs, as described in the previous section)? The first of these questions will be addressed in the following section; for now we will assume, in accordance with a good deal of numerical evidence [22, 33, 118, 124, 134, 232] that above some lower critical dimension there is a phase transition at a dimension-dependent positive temperature to a spin glass phase with broken spin-flip symmetry (for those readers interested in more offbeat possibilities, see [224]). For specificity we restrict the discussion in this section to spin glasses described by the nearest-neighbor Edwards–Anderson model [88] on the Euclidean lattice $\mathbb{Z}^d$ and with i.i.d. couplings taken from the normal (Gaussian) distribution with mean zero and variance one.

The main question, if pure states come in pairs, is whether there is only a single pair, as in droplet-scaling and (probably) TNT, or else uncountably many, as in RSB and chaotic pairs. (We will always refer in this section to infinite-volume pure or ground states. Definitions and constructions can be found in [223].) Aren't there other possibilities? Probably not: a countable number, either finite or infinite, was rigorously ruled out for RSB as noted earlier, and there exists strong evidence that it cannot occur for chaotic pairs either. The number of pure/ground states, of course, could in principle be dimension-dependent, or even temperature-dependent for fixed dimension. But we're a long way at this point from addressing these questions. (We confine our discussion to the so-called *incongruent* ground/pure states [95, 96], which differ from each other by positive-density interfaces, and are generated by coupling-independent boundary conditions [221, 223].)

Until recently the only dimensions in which the answer was fully known were one dimension (a single pure state at all positive temperatures and a single pair of ground states at zero temperature) and (strictly) infinite dimensions (where it is assumed that the state space structure is given by RSB). For the moment, let's assume the simplest (though not the most interesting) case that in each dimension there is a single transition temperature, above which there is a single pure state and below which there is a spin glass phase where the pure state cardinality is independent of temperature and equal to the ground state cardinality. This appears to be what most workers in the field believe. Many who fall into the RSB camp are inclined toward the possibility that at the lower critical dimension (possibly $d = 3$, probably no larger than $d = 4$) and above, the low-temperature phase is fully described by RSB. Others have argued that there is also an upper critical dimension at $d = 6$ [18], below which there is a single pair of pure/ground states and above which infinitely many. Still others [96] conjecture that there is only a single pair of pure/ground states in all finite dimensions (above the lower critical dimension), and the infinite-dimensional limit is singular in the sense of the structure of the spin glass phase.

All of this remains conjecture. Numerical experiments mostly (but not universally) agree that in $2D$ there is only a single pair of ground states (it is believed that $T_c = 0$ in $2D$), but in $3D$ and $4D$ different groups have arrived at different conclusions based on their numerical studies. The only rigorous results for any dimension between one and infinity appear in [11, 219, 220], all of which deal with two dimensions. In [219, 220], NS proved that in $2D$ any two distinct ground states can differ by only a *single*, positive density interface, providing evidence that there is only a single pair of ground states. Further evidence for this conclusion is provided in [11], where Arguin, Michael Damron, and NS proved essentially that the *half*-plane with a free boundary condition along its edge has only a single pair of ground states.

In a recent set of papers, Arguin, NS, and Janek Wehr attacked this problem from a different, dimension-independent, direction, by proving lower bounds on free energy fluctuations between pure states at positive temperature [13] and energy fluctuations between ground states at zero temperature [14]. An upper bound has existed for many years, where it was shown that in finite volumes free energy fluctuations scale no faster than the square root of the surface area of the volume under consideration [6, 209, 265]. Lower bounds have already been obtained for several cases, where it was shown that free energy fluctuations scale linearly with the square root of the volume under consideration. A (hopefully small) difficulty is that the quantity used to obtain the lower bound is slightly different from that used for obtaining the upper bound, so a corresponding upper bound (which is expected to behave similarly to the known upper bound just described) needs to be found for the quantity examined in [13, 14]. If it can, then for any cases where these results can be shown to apply, there must be no more than a single pair of pure/ground states. This work is in progress at the time of this writing.

While progress has been slow in determining numbers of pure/ground states in realistic spin glass models, there are other interesting, but unrealistic, models which can provide some interesting illumination. A very useful such model is the so-called *highly disordered* model [23, 211, 212], in which the Hamiltonian is Edwards–Anderson and couplings are independent random variables, but chosen from a volume-dependent distribution. The idea is that the distribution of coupling magnitudes depends on volume in such a way that, in sufficiently large volumes, each coupling magnitude is more than twice as large as the next smaller one, and no more than half as large as the next larger one. One interesting feature of this model is that the true ground state in any volume can be found by the greedy algorithm (in fact, this was how the model was originally arrived at). Moreover, the algorithm used to find the ground state can be mapped onto invasion percolation. Consequently, the number of ground states can be found in any dimension: there is only a single pair below $d = 6$ and uncountably many above. (The original papers [211, 212] used heuristic arguments to propose $d = 8$

as the crossover dimension, but a more careful analysis in [133] indicated the crossover dimension to be $d = 6$.)

Interestingly, the ground state multiplicity in this model is the same for both the spin glass and the random ferromagnet. The somewhat subtle distinction between the two comes from features of chaotic size dependence for the spin glass, and its absence in the ferromagnet. In any case, the highly disordered model remains a useful testing ground for new ideas.

Equilibrium Phase Transitions. Not much of the discussion so far is relevant to spin glasses if there is no equilibrium phase transition to a spin glass phase above some lower critical dimension d_c: at positive temperature there would simply be a unique Gibbs state. (Of course, even in that case the question of multiplicity of ground states at $T = 0$ would still exist.) Even with a phase transition, it could still be the case that there exists a unique Gibbs state above and below the transition temperature $T_c(d)$; the state would simply be qualitatively different in these two temperature regimes. Or there could be multiple transitions within a fixed dimension [224]. And even in the absence of a phase transition there could be some sort of dynamical transition with interesting features.

However, a great deal of mostly numerical work [22, 33, 118, 124, 134, 232] points toward a more conventional scenario in which in dimension $d > d_c$, there is a unique transition temperature $T_c(d)$, below which spin-flip symmetry is broken (equivalently, the Edwards–Anderson order parameter [88] $q_{EA} \neq 0$). Moreover, almost all (currently relevant) theoretical work addressing the low-temperature spin glass phase, whether using evidence from the infinite-range spin glass (i.e., the RSB scenario), or resulting from scaling arguments (droplet-scaling), or suggested from the structure of the metastate (chaotic pairs) or following numerical simulations (TNT) begin by assuming an equilibrium spin glass transition breaking spin-flip symmetry. So it's natural to ask whether such a transition can be proved to exist (in *any* finite dimension), and if so, what can one say about its structure?

Substantial progress toward this end was made by Jon Machta and NS in [165–167], where percolation-theoretic methods were used to uncover what is likely to be an underlying geometric structure for spin glass phase transitions. These random graph methods, in particular the Fortuin–Kastelyn (FK) random cluster (RC) representation [106, 138], provide a set of useful tools for studying phase transitions (more specifically, the presence of multiple Gibbs states arising from broken spin flip symmetry) in discrete spin models. These representations map Ising and Potts models onto a type of percolation problem, thereby allowing spin correlation functions to be expressed through the geometrical properties of associated random graphs. FK and related models are probably best known in the physics literature for providing the basis for powerful Monte Carlo methods, in particular the Swendsen–Wang algorithm for studying phase transitions [270]. At the same time they have proved important in obtaining rigorous results on phase transitions in discrete-spin ferromagnetic (including inhomogeneous and randomly diluted) models. Because of complications due to frustration, however, graphical representations have so far played a less important role in the study of spin glasses.

Some readers will already be familiar with the Fortuin–Kasteleyn random cluster representation. For those who are not and would like to read on, a brief one-page summary can be found in [167]. (For those who have both the interest and the time, the original references [106, 138] are the best place to go. Also see the article by Newman [201] and/or the lengthy and detailed review by Grimmett [114], as well as papers by Camia, Jianping Jiang, and Chuck containing recent applications of FK methods [54, 55].) In what follows, we assume a familiarity with the basic ideas of the FK random cluster representation for discrete spin models. For our purposes, it's sufficient to note that in the case of a ferromagnet, general theorems [47] ensure that when percolation of an FK cluster occurs, the percolating cluster is unique. We shall informally refer to such percolation as "FK percolation". It can then be shown that FK percolation in the ferromagnet corresponds to the presence of multiple Gibbs states (in the Ising ferromagnet, magnetization up and magnetization down) with broken spin-flip symmetry, and moreover that the onset of percolation occurs at the ferromagnetic critical temperature.

However, the situation is greatly complicated when couplings of both signs are allowed, as in the spin glass. In this case, FK percolation does *not* appear to be a sufficient condition for multiple Gibbs states (although it's undoubtedly necessary), and the numerical onset of FK percolation does not coincide with (what is believed to be) the transition to a spin glass state with broken spin-flip symmetry [270]. The essential problem is the following. If there is broken spin-flip symmetry, then in a finite-volume Gibbs state one should be able to change the orientation of the spin at the origin by changing the orientations of fixed spins at the boundary; or equivalently, the use of (at least some) fixed-spin boundary conditions should lead to a nonzero thermally averaged magnetization at the origin. As an example, choose the boundary condition where all boundary spins are fixed to be $+1$. At positive temperature, there are many possible realizations of the FK percolating cluster. But at least *a priori*, some of these will connect the origin to the boundary in a way that will lead to a positive magnetization at the origin, and others to a negative magnetization, and these could cancel leading to a net zero thermal average of the magnetization. Of course, we don't know for sure that this happens, but so far no one has been able to rule it out. Supporting evidence comes from the numerical studies described above, in which single FK percolation occurs well above what is believed to be the spin glass transition temperature, and proofs that FK percolation occurs [110] in $2D$ models where no broken spin-flip symmetry is expected. An interesting speculation, which remains to be studied numerically, is that FK percolation might indeed indicate a phase transition, but not one where spin-flip symmetry is broken. Rather, it could be the case that there is a change in behavior of two-point correlation functions with distance, but the EA order parameter remains zero [224].

So at first glance it would appear that FK percolation is the wrong tool to use in analyzing spin glass phase transitions. However, it turns out that FK percolation is indeed relevant, but that there is a considerably more complicated signature for the spin glass transition in both short-range and infinite-range models, as elucidated in [165–167]. These papers pointed out that understanding the percolation signature for the spin glass phase requires two ingredients beyond what is needed for ferromagnets. The first is the need to consider percolation within a two-replica representation, and the second is that spin glass ordering corresponds to a more subtle percolation phenomenon than simply the appearance of a percolating cluster—it involves a *pair* of percolating clusters, first proposed in [224].

The ferromagnetic phase transition corresponds to percolation of a unique infinite cluster in the FK representation, which will hereafter be referred to as "single" FK percolation. As noted above, single FK percolation is insufficient—so far as is currently known—to indicate the presence of multiple Gibbs states in spin glasses. However, if one switches to a two-replica formalism, one can study what might be called "double FK percolation", meaning the following: take two independent FK realizations and consider "doubly occupied" bonds—i.e., bonds occupied in both representations. If these percolate (as usual, with probability one), then one has double FK percolation. (In fact, there are two-replica representations other than FK than can be used equally well, as described in [165]. For the sake of brevity we focus here only on double FK percolation.)

Double FK percolation is much more difficult to study than single FK percolation, and a number of general theorems valid for the latter fail for the former. In particular, uniqueness of a percolating double cluster is no longer guaranteed. And not only is it not guaranteed, it doesn't happen. What was found instead was that the spin glass transition corresponds to the breaking of indistinguishability between *two* percolating networks of doubly FK-occupied bonds—in particular, by their having a nonzero difference in densities.

It's worth describing this in just a little more detail, starting with short-range spin glasses (in what follows, we note that results were obtained numerically for the EA spin glass in three dimensions and rigorously for the infinite-range SK model). For the EA model in $3D$ (and presumably in higher dimensions as well), there is a series of transitions as the temperature is lowered from infinity. At a temperature T_{c2} well above the putative spin glass transition temperature, there appear two doubly-infinite FK clusters of equal density, and presumably macroscopically indistinguishable. In terms of using boundary conditions to observe macroscopically distinguishable Gibbs states, this situation is no better than that of single FK percolation. Below a lower temperature T_{c1}, which is (within numerical error) equal to the spin glass transition temperature T_c observed using other means, the clusters separate in density—one grows and the other diminishes, leading to the presence of a macroscopic observable whose sign can be reversed using a flip of boundary conditions (and therefore corresponds to broken spin-flip symmetry).

The SK model behaves differently: above T_c, there is no double percolation at all, while below T_c two doubly infinite clusters of unequal density appear.

For more details we refer the reader to [165–167]. This work will hopefully prove useful in understanding better the differences between the nature of the phase transition in ferromagnets and in spin glasses. More importantly, the hope is that for short-range models, this work will be a significant step toward finally developing a rigorous proof for spin glass ordering and will eventually lead to a clean analysis of the differences between short- and infinite-range spin glass ordering. Finally, it can help to explain why there is no spin glass transition leading to broken spin-flip symmetry on simple planar lattices: two dimensions does not generally provide enough "room" for two disjoint infinite clusters to percolate. However, a system that is infinite in extent in two dimensions but finite in the third might be able to support two percolating clusters, with unequal densities at low temperature. These intriguing possibilities remain unexplored at the time of this writing.

1.2 Nonequilibrium Dynamics of Discrete Spin Systems

A second broad area of Chuck's research interests centers on the dynamical evolution of Ising and Potts-like systems, both ordered and disordered, in conditions far from thermodynamic equilibrium. His work in this general area addresses a wide variety of problems focusing on different aspects of nonequilibrium dynamics, but this section will touch on only two: the mostly unexplored relation between nonequilibrium dynamics and equilbrium thermodynamics, and the problem of predictability (or more colloquially, "nature vs. nurture") in spin systems following a deep quench. For brevity's sake, we unfortunately omit other areas of Chuck's notable contributions to topics as diverse as persistence [217], aging [100, 101, 103], broken ergodicity [266], biological evolution [148, 204], and food webs [70, 203]; we refer those interested to the references cited.

Nonequilibrium Dynamics and Equilbrium Thermodynamics. It is often noted in the literature that pure state multiplicity, as a purely equilibrium property of thermodynamic systems, plays no role in its dynamics, because (under conventional dynamics, such as one-spin-flip, to which we adhere throughout) a system in a pure state remains in that state forever. A natural conclusion to draw is that a system's pure state multiplicity and its dynamical behavior are largely distinct, and information about one says little about the other. However, a series of rigorous arguments in [218] showed a surprisingly deep connection between the two under what initially would seem the most unpromising conditions: a system undergoing a deep quench and thereafter evolving under conditions far from equilibrium.

The arguments are long and technical and won't be repeated here. The essence of the main result is the following. Consider a system at a temperature where, in equilibrium, there are (or are presumed to be) two or more pure states. Examples include the Ising ferromagnet and the EA spin glass above the critical dimension, among many others. Consider now a deep quench from high (well above T_c) to low (well below T_c) temperature. It was proved in [218] that as time $t \to \infty$, although the system is usually in some pure state locally (i.e., within a fixed volume), then either (a) it *never* settles permanently (within that volume) into a single pure state, or (b) it does but then the pure state depends on *both* the initial spin configuration *and* the realization of the stochastic dynamics.

It was further proved that the former case holds for deeply quenched $2D$ ferromagnets. But the latter case is particularly interesting, because it was shown in [218] that it can occur only if there exists an uncountable number of pure states with almost every pair having zero overlap. It was further shown that in both cases, almost *no* initial spin configuration is in the basin of attraction of a single pure state. That is, after a deep quench the resulting configuration space is almost all boundary; equivalently, the union of the basins of attraction of all pure states forms a set of measure zero in configuration space.

So in principle the nonequilibrium dynamics following a deep quench provides information about the multiplicity of pure states and vice-versa. However, this relation is likely to be more useful for future mathematical analysis than for numerical tests, which are confined to relatively small systems. Nevertheless, there may be experimental consequences as yet unexplored. Unlike the ferromagnet, it is not clear that spin glasses are easily prepared to lie within a single pure state, even under conditions as close to equilibrium as current technology allows. In particular, because of the possibility of chaotic temperature dependence [42,97], the conclusions of [218] could well apply to laboratory spin glasses prepared under conditions where small temperature changes are made slowly. Experimentally observed slow relaxation and long equilibration times in spin glasses could then be a consequence of small (relative to the system) domain size and slow (possibly due to pinning) motion of domain walls (a conclusion earlier reached by Fisher and Huse [98] using different considerations).

Nature vs. Nurture: Predictability in Discrete Spin Dynamics. Although the core issues raised in [218] remain open, the paper led to an unanticipated research direction that opened an entirely new area of inquiry: given a typical initial configuration, which then evolves under a specified dynamics, how much can one predict about the state of the system at later times? Chuck's papers with collaborators have colloquially referred to this as a "nature vs. nurture" problem, with "nature" representing the influence of the initial configuration (and disorder realization, if relevant) and "nurture" representing the influence of the random dynamics.

First one must determine under what conditions a system settles down [190, 218], in the sense of *local equilibration*: do domain walls cease to sweep across a fixed region after a (size-dependent) finite time? Local equilibration occurs in many quenched systems, for example, any disordered Ising model with continuous couplings [190], but has also been shown *not* to occur for others, such as the $2D$ homogeneous Ising ferromagnet [190]. When local *non*equilibration (LNE) occurs, one can still ask whether the *dynamically averaged* configuration has a limiting distribution. If so, this implies a complete lack of predictability, while the absence of a distributional limit implies that some amount of predictability remains [190, 218]. In [218], systems displaying LNE but having a limiting dynamically averaged distribution were said to exhibit "weak LNE", while those with no distributional limit were said to exhibit "chaotic time dependence".

So the distinction between these two types of local nonequilibration is important for the question of predictability. While the presence of LNE has been rigorously established for several systems [190, 218], determining which type of LNE occurs is considerably more difficult (when the initial state is chosen randomly, as discussed above). However, Renato Fontes, Marco Isopi, and Chuck were able to prove the presence of chaotic time dependence in the one-dimensional voter model with random rates at zero temperature (equivalently, the $1D$ zero-temperature stochastic Ising model) if the disorder distribution is heavy-tailed [100]. (This is a $1D$ version of Jean–Phillipe Bouchaud's trap model [40]). Otherwise, they showed that chaotic time dependence is absent. The latter conclusion was also shown to hold for the voter model with random rates in dimension greater than two. In further work [101], Fontes *et al.* extended the results of [100] by analyzing in detail the space-time scaling limit of the $1D$ voter model with random rates as a singular diffusion in a random environment. The limit object, now know as FIN (Fontes–Isopi–Newman) diffusion, has some striking scaling properties. Much subsequent work, in particular by Gerard Ben Arous, Jiri Cerny, and collaborators [15] has studied scaling limits beyond one dimension, which are rather different kinds of objects than the FIN diffusions.

Returning to the problem of nature vs. nurture, the questions raised by these considerations have led to a rich area of research that has brought in a variety of collaborators and students, and that touches on a number of related dynamical problems of interest, including phase-ordering kinetics, persistence, damage spreading, and aging. The history of these efforts, along with recent advances, is described in detail in a separate contribution to this volume by one of us (DLS), to which we refer the interested reader.

1.3 Lee–Yang Zeros and the Riemann Hypothesis

In 1950 T.D. Lee and C.N. Yang published a pair of papers [159, 279] that pioneered a new way of understanding phase transitions. They considered the ferromagnetic Ising model in an external magnetic field, and showed that the zeros of its partition function, as a function of the external magnetic field, all lie on the imaginary axis. Since that time it's been shown that a number of other statistical mechanical systems obey Lee–Yang type theorems of their own, and Chuck played a substantial part in these efforts. Given that Lee–Yang type theorems provide not only an understanding of the properties of the phase transition in a given system (e.g., existence of a mass gap [116, 239]), but can also lead to useful correlation inequalities [83, 194], such results are important in our overall understanding of phase transitions in condensed matter. For a general review and discussion of applications of Lee–Yang type theorems, see [108].

Among Chuck's contributions to this area was to extend the Lee–Yang theorem to ferromagnetic models with a very general class of single-spin distributions [193], to the classical XY model [84] (further generalized in [163]; see also [269]), and [109, 225, 226] to Villain models [273], which are closely related to XY models. In [226] it was further shown that complex Gaussian multiplicative chaos in general does *not* have the Lee–Yang property. These results have some interesting implications. In particular, the "spin wave conjecture" [85, 176] asserts that, below a critical temperature, the angular variable θ of the XY (and Villain) model at large scales behaves like a Gaussian free field (modulo 2π), suggesting in turn that the spin variables in these models could behave like a version of complex Gaussian multiplicative chaos. The spin wave conjecture might lead one to expect that complex Gaussian multiplicative chaos would display a Lee–Yang property; but the paper of Newman and Wu rules that out, at least for a range of inverse temperature β.

In a different direction, Chuck has long been interested in the possible connection between the Lee–Yang theorem and the famous Riemann Hypothesis. Recall that the Riemann Hypothesis states that the nontrivial zeroes of the Riemann zeta function $\zeta(s) = \sum_{n=1}^{\infty} \frac{1}{n^s}$ all have real part 1/2. While on the surface it sounds less than dramatic, many consider it to be the most important unsolved problem in mathematics. Its proof (or disproof) has important implications for the distribution of prime numbers, the behavior of various functions in number theory and combinatorics, eigenvalue distributions of random matrices, quantum chaos, and problems in many other areas. The similarity between the distribution of Lee–Yang zeros and those of the Riemann zeta function have long led to speculation that one path to a proof of the Riemann Hypothesis lies through the Lee–Yang theorem (for a review, see [149]). To date, however, attempts to carry out this program have been unsuccessful (otherwise, you surely would have heard).

However, Chuck made an important contribution to this problem in a slightly different and very interesting direction. In 1950 Nicolaas De Bruijn showed that a certain function $H(\lambda, z)$ (its precise form is unimportant for the purposes of this discussion) has only real zeros for $\lambda \geq 1/2$ [45]. In addition, if λ is such that $H(\lambda, z)$ has only real zeros, then for all $\lambda' > \lambda$, $H(\lambda', z)$ also has all real zeros. The Riemann Hypothesis is equivalent to $H(0, z)$ having only real zeros. Now if H has only real zeros for *all* real λ, then the Riemann Hypothesis would follow. This was a strategy attempted by Polya, among others. However, Chuck proved in [195] that there exist real λ for which H has a nonreal zero. This led to what is now known as the De Bruijn–Newman constant, usually denoted by Λ. It is defined to be the value such that if $\lambda \geq \Lambda$, H has only real zeros, while if $\lambda < \Lambda$, H has a nonreal zero.

The Riemann Hypothesis is true if and only if the De Bruijn–Newman constant $\Lambda \leq 0$. But in [195], Chuck conjectured that $\Lambda \geq 0$. If true, this immediately implies that the Riemann Hypothesis is true if and only if $\Lambda = 0$. Computer-aided rigorous calculations of a lower bound for the De Bruijn–Newman constant have been made over the years. Until 2018, the best lower bound was -1.1×10^{-11} [251], but recently Brad Rodgers and Terry Tao [247] posted a proof verifying Chuck's conjecture that $\Lambda \geq 0$.

There also exist upper bounds for $\mathit{\Lambda}$. The earliest was de Bruijn's $\mathit{\Lambda} \leq 1/2$ [45] in his original paper; an improvement to a strict inequality $\mathit{\Lambda} < 1/2$ was made in 2009 by Ki, Kim, and Lee [147]. An improved upper bound of 0.22 was very recently obtained by Tao and collaborators, as posted in [241]. In addition, a survey article by Chuck and Wei Wu on de Bruijn–Newman type constants has just been published [227].

While work is ongoing, we'll give the last word to Chuck. In [195], Chuck noted that his conjecture that $\Lambda \geq 0$ put a quantitative scaffolding behind the dictum that the "Riemann hypothesis, if true, is only barely so."

This introduction has barely described the extent and importance of Chuck's many contributions to mathematical physics and related areas. Our hope is that the reader will go on to take a look at some of Chuck's many papers, especially in areas that for lack of space, and the finite lifetimes of the authors, this introduction was not able to discuss.

In the following sections we turn to a different aspect of Chuck's work, involving seminal contributions to percolation theory, scaling limits, SLE, and related topics.

2 Percolation Theory and Continuum Scaling Limits

In this section, we briefly review some of the other fundamental advances, more probabilistic in nature, made by Chuck in his long and productive career. We will only briefly mention the Brownian Web and then discuss Chuck's contributions to percolation theory and to the study of continuum scaling limits. Like the results surveyed in the previous section, this side of Chuck's work was mainly motivated by questions arising in statistical physics, and sometimes quantum field theory, or a combination of the two. The breadth of Chuck's interests and his agile versatility as a mathematician and mathematical physicist are evident in the collection of problems discussed here and in the previous section.

2.1 The Brownian Web

Construction and relation to disordered systems. One of Chuck's strengths is his ability to move between fields and exploit fruitful connections. This is evident in his influential work on the Brownian web, which has its roots in the analysis of nonequilibrium dynamics in one-dimensional disordered systems, discussed in the previous section. Roughly speaking, the Brownian web is the scaling limit of the space-time graphical representation of an infinite collection of coalescing random walks. It originated from Arratia's Ph.D. thesis [16], where it is shown that a collection of coalescing random walks on $\mathbb{Z}$ starting from every vertex of $\mathbb{Z}$ converges to a collection of Brownian motions on $\mathbb{R}$ starting from every point of the real line at time zero. Arratia's attempt [17] to construct a process corresponding to a collection of coalescing Brownian motions starting at every point of the real line at every time $t \geq 0$ was never completed. However, building on Arratia's work, the process was constructed years later by Balint Tóth and Wendelin Werner [272], who discovered a surprising connection between Arratia's one-dimensional coalescing Brownian motions and a two-dimensional random process repelled by its own local time profile, which they called *true self-repelling motion.* A few more years later, Luiz Renato Fontes, Marco Isopi, Chuck and DLS realized that the same system of coalescing Brownian motions also arises from the scaling limits of one-dimensional spin systems [103]. The first three authors, together with Ravishankar, introduced [102, 104] a topology such that the system of coalescing Brownian motions starting from every space-time point can be realized as a random variable in a separable metric space, and they named this random variable the *Brownian web*. This remarkable object emerges in contexts as diverse as hydrology and the zero-temperature dynamics of the Ising model. The work of Chuck and coauthors on the Brownian web has inspired others to explore its properties and propose extensions, spurring a wealth of interesting papers. We will not discuss the topic further in this introduction; instead we refer the interested reader to [252, 268] for a survey of results and extensions.

2.2 Percolation

Percolation as a mathematical theory was introduced by Broadbent and Hammersley [43, 44] to model the spread of a "fluid" through a random "medium." Broadbent and Hammersley interpreted the terms fluid and medium broadly, having in mind situations such as a solute diffusing though a solvent, electrons moving through an atomic lattice, molecules penetrating a porous solid, or disease infecting a community. They were interested in situations where the randomness is associated with the medium rather than with the fluid. To mimic the randomness of the medium, one can think of the latter as a system of channels some of which are randomly declared closed to the passage of the fluid. This can be modeled by a d-dimensional cubic lattice, seen as an infinite graph, where the edges between nearest-neighbor vertices are independently declared *open* (to the passage or the fluid) with probability p or *closed* with probability $1-p$. Models of this type are called *bond percolation* models, and can be defined on any graph. Many other variants have been studied, attracting the interest of both mathematicians and physicists. One version in particular, site percolation on the triangular lattice, will be discussed in more detail in Sect. 2.3. In *site percolation*, the vertices of a graph rather than the edges are declared open or closed.

Mathematicians are interested in percolation because of its deceptive simplicity which hides difficult and elegant results. From the point of view of physicists, percolation is one of the simplest statistical mechanical models undergoing a continuous phase transition as the value of the parameter p is varied, with all the standard features typical of critical phenomena (scaling laws, conformal invariance, universality). On the applied side, percolation has been used to model the spread of a disease, a fire or a rumor, the displacement of oil by water, the behavior of random electrical circuits, and more recently the connectivity properties of communication networks.

Existence and Uniqueness of Infinite Clusters. One of the most interesting aspects of percolation, and a major reason for its popularity, is its phase transition, which is of a purely "geometric" nature. Defining open clusters to be the sets of vertices connected to each other by a path of open edges, one can ask whether there exists an infinite open cluster at a given value p of the density of open edges. This amounts to asking whether there is an open path from the origin of the lattice (any deterministic vertex) to infinity with strictly positive probability. Indeed, if the answer to this question is positive, then translation invariance and an application of Kolmogorov's zero-one law imply the existence of an infinite open cluster with probability one somewhere in the system. If the answer is negative, then with probability one, no infinite open cluster exists. This justifies the introduction of the *percolation function* $\theta(p)$ defined as the probability that the origin is connected to infinity by a path of open edges. In terms of θ, $\theta(p) > 0$ is equivalent to the existence of an infinite open cluster (with probability one), which naturally leads to a definition of the *critical probability* $p_c = \sup\{p : \theta(p) = 0\}$.

An argument analogous to that used by Peierls [238] to establish the existence of a phase transition in the Ising model can be used to show that $\theta(p) = 0$ when p is sufficiently small, and $\theta(p) > 0$ when p is sufficiently close to one, which implies that $0 < p_c < 1$. This fundamental result, proved early on by Broadbent and Hammersley [44,119,120], shows that the percolation model undergoes a phase transition and explains the subsequent interest in the subject (see, e.g., [37,115,144,264]).

When $\theta(p) > 0$, it is natural to ask about the multiplicity of infinite open clusters. A simple and elegant proof by Burton and Keane [46] shows that, under quite general conditions, there is almost surely a unique infinite cluster. The same conclusion had however already been reached two years earlier in a joint paper of Chuck's with Michael Aizenman and Harry Kesten [7]. Their analysis applies to both site and bond models in arbitrary dimension, including long range bond percolation where one considers edges (bonds) between vertices that are not nearest-neighbors. A particularly interesting example of such long-range models, covered in [7], is provided by one-dimensional $1/|x-y|^2$ models, where the probability that the bond between vertices x and y is open decays like $1/|x-y|^2$. We'll come back to such models shortly.

An important precursor to [7] is the paper by Chuck and Lawrence Shulman [207] which investigates the number and nature of infinite clusters in a large class of percolation models in general dimension. Using primarily methods from ergodic theory and measure theory, the paper shows that, under general conditions, the number of infinite clusters is either 0, 1 or ∞. The class of percolation models to which this result applies is characterized by translation invariance, translation ergodicity, and a "finite energy" condition which implies that the conditional probability of a local configuration, conditioned on the configuration in the rest of the system, is always strictly between 0 and 1. Intuitively, having strictly positive conditional probabilities for local configurations regardless of the rest of the system means that there are no "prohibited" (local) configurations. The proof by Burton and Keane [46] also uses this side of the finite energy condition in a crucial way, together with translation invariance and ergodicity. Their proof was extended in [110] to more general percolation models, including in particular long-range models. The existing proofs of uniqueness of the infinite cluster may, in fact, be adapted to all "periodic" graphs such that the number of vertices within distance n of the origin grows sub-exponentially in n. The situation is qualitatively different on trees where, although one still has two phases, one can show that above the critical density p_c there are *infinitely many* infinite open clusters instead of just one. Yet another situation arises when one considers the direct product of $\mathbb{Z}$ and a regular tree, as done in [205], where it is shown that percolation on such a graph has at least *three* distinct phases, with the number of infinite clusters being (almost surely) 0, ∞, and 1, respectively, as the density of open bonds increases. Coauthored by Chuck and Geoffrey Grimmett, [205] is the first paper to explore percolation on nonamenable graphs, and the subject has since then attracted much attention. (Roughly speaking, an infinite graph G is nonamenable if, for every finite subset W of G, the size of the boundary of W is of the same order as the size of W. A typical example of a nonamenable graph is a regular tree graph: as one "grows" the tree from the "root" adding more vertices to the graph, the number of "leaves" is always of the same order as the total number of vertices.) An introduction to percolation on nonamenable graphs and a list of problems (some of which have by now been solved) are contained in [30], and a somewhat more recent survey can be found in [164].

Another interesting result proved in [7], essentially as a corollary of the uniqueness of the infinite open cluster, is the continuity of the percolation function θ, except possibly at the critical density p_c. The continuity of the percolation function θ at p_c is still one of the major open problems in percolation theory. For bond percolation on the square lattice, it was established in a groundbreaking and very influential paper by Harry Kesten [143]. Ten years later, Takashi Hara and Gordon Slade [123] proved continuity in dimension $d \geq 19$ (a bound later improved to $d \geq 11$ [99]) and in more than six dimensions for sufficiently "spread-out" models where long-range bonds are allowed. A little later, Barsky, Grimmett and Newman [25] proved that the probability that there exists an infinite cluster in $\mathbb{N} \times \mathbb{Z}^{d-1}$ is zero for $p = p_c(\mathbb{Z}^d)$, but the continuity of θ at p_c remains a conjecture for percolation models in general dimensions.

The continuity results just described justify the characterization of the percolation phase transition as a *continuous* phase transition. Indeed, the percolation phase transition is often considered a prototypical example of a continuous phase transition. However, Chuck and Michael Aizenman [9] showed that the situation is different if one considers one-dimensional $1/|x-y|^2$ models, already mentioned earlier. In such models the percolation function $\theta(p)$ is *discontinuous* at $p = p_c$. Chuck and Larry Schulman [208] had already proved that one-dimensional $1/|x-y|^s$ percolation models have a phase transition for all $s \leq 2$, a result that is non-trivial for $s > 1$. This is analogous to the occurrence of a phase transition in long-range one-dimensional Ising models with interactions decaying like $1/|x-y|^s$ [86]. The results of [9] were extended to Ising and Potts models in [5], which provides a proof of a type of discontinuity originally predicted by Thouless [271].

Critical Exponents. Typically, for statistical mechanical models undergoing a continuous phase transition, close to the critical point, the correlation length and other thermodynamic quantities exhibit power-law behavior in the parameters of the model. The exponents in those power laws are called *critical exponents* and their values appear to be largely independent of the microscopic details of the model. Instead, they appear to depend only on global features such as the dimension and symmetries of the model. This phenomenon is called *universality* and has a natural explanation within the framework of the *renormalization group*, which will be briefly discussed in Sect. 2.3.

A first quantitative theory of critical phenomena was proposed by Landau [154], corresponding to the mean-field approximation that applies to systems on a Bethe lattice [31] or in sufficiently high dimensions or with long-range interactions. However, Onsager's exact solution [230] of the two-dimensional Ising model and Guggenheim's results [117] on the coexistence curve of simple fluids showed that critical exponents can take values different from those of Landau's mean-field theory. Indeed, critical exponents should only take their mean-field values above the *upper critical dimension* d_{uc}, already introduced in Section IA. Even today, only a small number of non-mean-field critical exponents have been derived rigorously (including, as we'll discuss later, a couple recently established by Chuck and co-authors in the case of the planar Ising model). But in 1963, a seminal paper by Rushbrooke [250] demonstrated how *inequalities* between critical exponents can be derived rigorously and exploited fruitfully to study the singularity of thermodynamic functions near the critical point. Since then, numerous such inequalities have been proved for various models of statistical mechanics.

Chuck's contributions to this line of inquiry appeared in several papers [8,198–200]. Papers [8,198,200] deal with the exponent γ for percolation, associated with the expected cluster size χ, namely $\chi(p) \sim (p_c - p)^{-\gamma}$ as $p \nearrow p_c$. In particular, among the results reported in [198,200] is the fact that in one-dimensional $1/|x-y|^2$ models, where a discontinuous phase transition occurs, $\gamma \geq 2$; conversely, for models such as standard site or bond percolation in dimension $d > 2$, where it is believed but not proved that the phase transition is continuous, it is shown that, in order to prove continuity of the percolation function at p_c, it would suffice to show that $\gamma < 2$. (Incidentally, for percolation in $d = 3$, γ is numerically estimated to be about 1.7—see, e.g., [263] and the references given there.) Another result discussed in [198,200] is the inequality $\gamma \geq 2(1 - 1/\delta)$, with $\delta \geq 2$, which improves on the bound $\gamma \geq 1$, proved in [8]. The critical exponents considered in [199] are the exponent β, which determines the divergence of the percolation function θ as $p \searrow p_c$, namely $\theta(p) \sim (p - p_c)^{\beta}$, and the exponent δ, which determines the behavior of the cluster size distribution at p_c, namely the probability $P_n(p_c)$ that the cluster of the origin contains exactly n vertices when $p = p_c$. Assuming that the percolation density vanishes at the critical point, the inequality proved in [199] is $\beta \geq 2/\delta$, improving on a previous result by Aizenman and Barsky [2] and on $\delta \geq 2$, since $\beta \leq 1$ [67].

Coming back to [8], besides the proof of the inequality $\delta \geq 1$, already mentioned and soon improved upon, and other results discussed there, the most remarkable and influential contribution of the paper is arguably the introduction of the *triangle diagram* ∇ and of the corresponding *triangle condition.*

The triangle diagram is defined as a sum of two-point functions τ, namely $\nabla = \sum_{x,y} \tau(0,x)\tau(x,y)\tau(y,0)$, and the triangle condition corresponds to the *finiteness* of the triangle diagram at p_c (or uniform boundness for $p < p_c$). A main result of [8] is a proof that, in finite-range percolation models, the triangle condition implies $\gamma = 1$. The relevance of this result stems from the fact that 1 is the mean-field value of the exponent γ. Therefore, if the triangle condition is satisfied for some dimension d, this suggests that d is greater than the upper critical dimension d_{uc} and that the mean-field approximation gives the correct prediction for the critical exponents in dimension d. Indeed, Barsky and Aizenman [24] extended the result of [8] to the critical exponents δ and β by showing that, in percolation models where the triangle condition is satisfied, the exponents δ and β exist and take their mean-field values: $\delta = 2$ and $\beta = 1$. In particular, the existence of the exponent β implies the continuity of the percolation function at p_c, i.e., $\theta(p_c) = 0$. Another result concerning the triangle condition is the proof by Nguyen [228] that, if the triangle condition is satisfied, then the *gap exponents*, characterizing the divergence of higher moments of the cluster size distribution, assume their mean-field value of 2. The triangle condition itself was proved in [122,123] for percolation in sufficiently high dimensions for nearest-neighbor models, and above six dimensions for a class of spread-out models. Similar conditions have been subsequently introduced in the literature [26,38,69,229].

First Passage Percolation. First passage percolation was introduced by Hammersley and Welsh [121] as a percolation-type model with a time dimension that makes it suitable for studying the spread of a disease or a fluid in a porous medium. In the standard version, one assigns to each edge e of $\mathbb{Z}^d$ a nonnegative random variable $t(e)$, which is usually interpreted as the passage time of the edge e. The *passage time* of a path r consisting of edges $e_1, \ldots, e_n$ is $T(r) = \sum_{i=1}^{n} t(e_i)$ and the *passage* or *travel time* between two vertices $u, v \in \mathbb{Z}^d$ is the infimum of $T(r)$ over all paths r from u to v. The stochastic region $\tilde{B}(t) = \{x \in \mathbb{Z}^d : T(0,x) \leq t\}$ is the set of vertices that can be reached from the origin by time t. The interested reader is referred to [20] for precise definitions and a comprehensive recent survey.

A main object of interest in first-passage percolation is the set $\tilde{B}(t)$, and in particular its asymptotic properties when t is large. The interface between $\tilde{B}(t)$ and its complement will, under natural hypotheses, grow linearly in t with a deterministic shape, while the magnitude of the fluctuations of this interface about its mean shape is believed to be typically of order t^{χ}, with a universal exponent χ (which should of course depend on the dimension d). The study of the fluctuations of growing interfaces is a subject that has attracted considerable attention in the physics literature (see [150] for a review). In the case of first-passage percolation, the first rigorous results were established by Kesten [145,146], who proved the first rigorous bounds on the variance of the passage time $T(0,x)$. Since then, there has been an extensive literature on both lower and upper bounds for the variance of passage times. A detailed review of that literature can be found in [20]. In two dimensions, the best lower bound to date was obtained in [206] by Chuck and Marcelo Piza, who showed that the variance of $T(0,x)$ must grow at least as fast as $\log|x|$ and provided the first proof of divergence of the fluctuations of the interface between $\tilde{B}(t)$ and its complement. Prior to this paper, there had been no proof that the shape fluctuations diverge for any model in any dimension $d > 1$.

In related work [161], Cristina Licea, Chuck and Marcelo Piza provided the best rigorous results on the *wandering exponent* ξ, which determines the traverse fluctuations of time-minimizing paths. Like χ, the exponent ξ is expected to depend on the dimension d but to be otherwise universal (e.g., independent of the distribution of the random variables $\tau(e)$). There are, a priori, many possible mathematical definitions of the exponent ξ, some based on point-to-plane and some based on point-to-point passage times, but it is believed that they yield the same exponent. Furthermore, it has been conjectured that $\xi(d) \geq 1/2$ for all dimensions $d \geq 2$, with a strict inequality (superdiffusivity) at least for low d, and with $\xi(2) = 2/3$. In [161], Licea, Newman and Piza, working primarily with definitions of ξ of the point-to-plane type, obtained the lower bounds $\xi(d) \geq 1/2$ for all d and $\xi(2) \geq 3/5$. It should be noted that the exponents χ and ξ had been conjectured in numerous physics papers to satisfy the scaling identity $\chi = 2\xi - 1$, irrespective of the dimension (see [150]). This relation was recently proved [66] by Sourav Chatterjee. Before Chatterjee's paper, the best result was due to Newman and Piza [206] who proved that $\chi' \geq 2\xi - 1$, where χ' is an exponent closely related (and perhaps equal) to χ.

Another line of inquiry pursued by Chuck concerns the study of infinite geodesics, motivated by the connection (in $d = 2$) between "bigeodesics" (i.e., doubly-infinite geodesics) in first-passage percolation and ground states of disordered ferromagnetic spin models [105,202]. Originating from the physics literature on disordered Ising models is a conjecture that, at least in two dimensions, bigeodesics should not exist. This derives from the conjecture that, in low dimensions (including $d = 2$), disordered Ising ferromagnets should have only two (constant) ground states. Indeed, the existence of nonconstant ground states for a disordered Ising model on $\mathbb{Z}^2$ with couplings J_e would imply the existence of bigeodesics on a dual square lattice $\mathbb{Z}^{2^*}$ with passage times $\tau(e^*) = J_e$, where the dual edge e^* is the perpendicular bisector of e.

A celebrated shape theorem [72,145,246] states, roughly speaking, that $\tilde{B}(t)$ behaves like $tB_0 + o(t)$ as $t \to \infty$, where B_0 is a convex subset of $\mathbb{R}^2$. Under the assumption that the boundary of B_0 has uniformly positive curvature, Cristina Licea and Chuck proved [162] that all infinite geodesics have an asymptotic direction, and that there is a set $D \subset [0, 2\pi)$ of full Lebesgue measure such that, for any $\theta \in D$, there are no bigeodesics with one end directed in direction θ. This result provides a partial verification of the conjecture mentioned above; however, a proof that the asymptotic shape B_0 has uniformly positive curvature seems to be still out of reach.

In [129], Doug Howard and Chuck introduced a Euclidean version of first-passage percolation where the vertex set of the lattice $\mathbb{Z}^d$ is replaced by the set of points of a homogeneous Poisson point process on $\mathbb{R}^d$. For this model, in [129] they proved a shape theorem, with asymptotic shape given by a Euclidean ball (due to the isotropy of the model), and the nonexistence of certain (geometrically non-realistic) doubly infinite geodesics. In [130], they continued their study of this new model, showing in particular that, for any dimension, with probability one, every semi-infinite geodesic has an asymptotic direction and every direction has at least one semi-infinite geodesic (starting from each Poisson point). In [131], they proved the bounds $\chi \leq 1/2$ and $\xi \leq 3/4$ for Euclidean first-passage percolation in all dimensions, together with other interesting results concerning spanning trees of semi-infinite geodesics and related random surfaces.

Chuck's work on first passage percolation has been and continues to be very influential. His results on the subject continue to be cited, and the ideas and methods developed by Chuck and coauthors have been used and extended in numerous papers (for a small selection, see [19, 64, 73, 74, 93, 112, 126, 242]).

2.3 Limit Theorems and Continuum Scaling Limits

Over the past two decades, besides working on disordered systems and nonequilibrium dynamics, Chuck has dedicated a lot of effort to the study of *continuum scaling limits*, particularly in the case of spanning trees, percolation and the Ising model. Chuck's interest in scaling limits, and in particular in the Euclidean fields emerging from such limits, has deep roots, going all the way back to his Ph.D. dissertation on Quantum Field Theory [191] (see also [192]).

A main goal of both probability theory and statistical physics is to understand and describe the behavior of random systems with a very large number of "degrees of freedom." In field theory, one deals with an infinite number of degrees of freedom, and there is indeed a deep connection between the theory of critical phenomena and field theory. This connection is particularly salient in the renormalization group approach to these theories in which, broadly speaking, one analyses the behavior of specific observable quantities at different scales (see, e.g., [276–278]). As more scales are included, the behavior of these observable quantities is random, and the quantities themselves need to be "renormalized" to "tame" their fluctuations. *Renormalizable* systems are such that this renormalization procedure is possible and one can take a limit over all the scales of the system. In field theory, a momentum cutoff introduced to reduce the number of degrees of freedom is sent to infinity as higher and higher energies are taken into consideration. In statistical mechanics, a version of the renormalization approach can be implemented with a continuum scaling limit in which some elementary scale of the system (e.g., the lattice spacing in lattice-based models such as percolation and the Ising model) is sent to zero. Renormalizable systems may possess a certain amount of *scale invariance* leading, in the limit of large scales or large momenta, to models that are (statistically) *self-similar*.

Thanks to the work of Polyakov [240] and others [28, 29], it was understood by physicists since the early seventies that critical statistical mechanical models should typically possess continuum scaling limits with a global conformal invariance that goes beyond simple scale invariance. Following this observation, in two dimensions, conformal methods were applied extensively to Ising and Potts models, Brownian motion, the self-avoiding walk, percolation, diffusion limited aggregation, and more. The large body of knowledge and techniques (mainly non-rigorous) that resulted from these efforts goes under the name of Conformal Field Theory (CFT)—see [107] for an extensive treatment.

Normal Fluctuations, the Gaussian Fixed Point, and Beyond. The renormalization group idea of gradually including the effect of more scales or degrees of freedom has a parallel in the theory of limit theorems in probability theory (see [34,135–137]).

In two influential papers [196,197], Chuck provided general conditions for the central limit theorem to apply to positively correlated random variables. Normal fluctuations are expected to occur in statistical mechanical systems away from a critical point, in which case, observables such as the renormalized magnetization in Ising ferromagnets are expected to converge to Gaussian random variables. In renormalization group terminology, the system converges to the *Gaussian fixed point.* Chuck's results have several important applications, including to magnetization and energy fluctuations in a large class of Ising ferromagnets, infinite cluster volume and surface density fluctuations in percolation models, and boson field fluctuations in (Euclidean) Yukawa quantum field theory models.

While Gaussian limits are very common, non-Gaussian limits can arise when dealing with random variables that are strongly positively correlated. This is believed to happen in various models studied in statistical mechanics, in particular in (ferromagnetic) Ising systems at the critical point. In the 1970s (extending earlier results of Simon and Griffiths [259]), Chuck and co-authors studied the emergence of non-Gaussian limits in Curie–Weiss (mean-field) models [89–91]. Part of the motivations for choosing a relatively simple class of systems such as mean-field models came from the fact that, at the time, the existence of non-Gaussian limits was an open problem, and seemed out of reach, for almost all non-trivial Ising systems. The situation has changed since then, and Chuck has contributed significantly to this change. As we will discuss at the end of this section, more than thirty years after his work on Curie–Weiss models, Chuck and a different set of co-authors were able to provide substantial contributions to the study of other non-Gaussian limits, this time in the context of the critical planar Ising model. Meanwhile, new interesting results on other mean-field models, some of which are analogous to those of [89,90], have also emerged—see the review article by Alberici, Contucci and Mingione in this volume, where the authors describe their recent work on mean-field monomer-dimer models.

Critical and Near-Critical Percolation. The renormalization group approach has greatly improved our understanding of critical phenomena, but from a mathematical perspective it remains to this day largely non-rigorous. In the late 1990s, Chuck contributed to the development of a new framework [1,3,4]. That framework and, in particular, the introduction of the Shramm–Loewner Evolution (SLE) by Oded Schramm [253] provided a new, mathematically rigorous, approach to study the geometry of critical systems on the plane. This new approach consists in viewing cluster boundaries in models such as Ising, Potts and percolation models, or loops in the $O(n)$ model, as random interfaces with a distribution that depends on the scale of the system under consideration (i.e., the lattice spacing), and in analyzing the continuum limit as the scale of the system is sent to zero. Schramm realized [253] that, at criticality, these interfaces become, in the continuum limit, random planar curves whose distributions can be uniquely identified thanks to their conformal invariance and a certain "Markovian" property. There is a one-parameter family of SLEs indexed by a positive real number κ: SLE_κ is a random growth process based on a Loewner chain driven by a one-dimensional Brownian motion with speed κ [248,253]. Lawler, Schramm and Werner [155–158], and Smirnov and Werner [262], among others, used SLE to confirm several results that had appeared in the physics literature, and to prove new ones.

In particular, substantial progress was made, thanks to SLE, in understanding the fractal and conformally invariant nature of (the scaling limit of) large percolation clusters, starting with the work of Schramm [253] and Smirnov [260], which identified the scaling limit of critical percolation interfaces with SLE_6. One of Chuck's early contributions to this area of research, in collaboration with one of us (FC), was the first complete and self-contained proof of the convergence of a percolation interface in critical site percolation on the triangular lattice to SLE_6 [60]. The proof, which follows roughly Smirnov's strategy [260] with some important differences, appeared first in an appendix of an unpublished preprint [58]. It was later deemed worth publishing as a separate paper [60] because of the importance of the result and of its many applications, which include the rigorous derivation of various percolation critical exponents [262], of Watt's crossing formula [81], and of Schramm's percolation formula [254].

The convergence result proved in [60] is also a crucial ingredient in the derivation of the *full scaling limit* of critical percolation [57,59], obtained by FC and Chuck by taking the continuum limit of the collection of all percolation interfaces at once. The resulting object, called the *continuum nonsimple loop process* in [57,59], is invariant under scale and conformal transformations, and inherits a spatial Markov property from the discrete percolation process. As shown in [61], the continuum nonsimple loop process constructed in [57,59] is a *Conformal Loop Ensemble* (CLE) [256,257]. The construction of the continuum scaling limit of planar critical percolation [57,59] has had several interesting applications, including [27,280–283].

Another notable application of [57,59] is in the construction of the *near-critical* scaling limit of planar percolation in which the percolation density p approaches the critical value p_c ($p_c = 1/2$ for site percolation on the triangular lattice) according to an appropriate power of the lattice spacing a, $p = p_c + \lambda a^\alpha$, as $a \searrow 0$. With an appropriate choice of α ($\alpha = 3/4$ for site percolation on the triangular lattice), one can show that the scaling limit leads to a one-parameter family, indexed by λ, of limits that are not scale invariant (except in the critical case, $\lambda = 0$) but retain some of the properties of the critical scaling limit (see, e.g., [56]). In [49,50], Camia, Fontes and Newman proposed an approach to construct the one-parameter family of near-critical scaling limits of planar percolation based on the critical full scaling limit and the "Poissonian marking" of some special ("macroscopically pivotal") points. This approach leads to a conceptual framework that can describe not only the scaling limit of near-critical percolation but also of related two-dimensional models such as dynamical percolation, the minimal spanning tree and invasion percolation. Partly inspired by [49,50], Garban, Pete and Schramm later constructed those scaling limits in [111]. Although Garban, Pete and Schramm use a notion of full scaling limit different from that of [57,59], they do take advantage of some of the results proved in [59], in particular to resolve issues of uniqueness, measurability, and conformal invariance of the limit.

Besides the applications mentioned above, a remarkable aspect of [57,59] is that they contain at once the first example of a nested CLE, including a description of some of its crucial properties, as well as the first rigorous construction of a CLE as a scaling limit, before the concept of CLE was formalized in full generality in [256]. Conformal Loop Ensembles have been extensively studied for their intrinsic interest (see, e.g., [142,184–188,255,257,275]) and for their applications to CFT (see, e.g., [77–79]), and we don't attempt to provide here a comprehensive list of references.

Critical and Near-Critical Ising Model. The Ising model [132], suggested by Lenz [160] and cast in its current form by Peierls [238], is one of the most studied models of statistical mechanics. Its two-dimensional version has played a special role since Peierls' proof of a phase transition [238] and Onsager's calculation of the free energy [230]. This phase transition has become a prototype for developing new techniques. Its analysis has helped test a fundamental tenet of critical phenomena, that near-critical physical systems are characterized by a *correlation length*, which provides the natural length scale for the system and diverges when the critical point is approached.

Substantial progress in the rigorous analysis of the two-dimensional Ising model at criticality was made by Smirnov [261] with the introduction and scaling limit analysis of *discrete holomorphic observables*. These have proved extremely useful in studying the Ising model in finite geometries with boundary conditions and in establishing conformal invariance of the scaling limit of various quantities, including the energy density [127,128] and spin correlation functions [68]. (An independent derivation of critical Ising correlation functions in the plane was obtained in [82].)

In [62], Chuck and one of us (FC) proposed a strategy to obtain the continuum scaling limit of the renormalized Ising magnetization field using the FK random cluster representation of the Ising model (see, e.g., [114]) and a new tool called, in later papers, a *conformal measure ensemble*. The strategy, which involves coupling conformal measure and conformal loop ensembles and leads to a geometric representation of the Ising magnetization in the continuum reminiscent of the Edwards–Sokal coupling [87] was not fully carried out in [62]. (In particular, no conformal measure ensemble is constructed in [62].) Nevertheless, the paper was the starting point for the subsequent work of Camia, Garban, Newman and of Camia, Jiang, Newman, which we now discuss.

In [52] FC, Christophe Garban and Chuck, partly following the strategy of [62] but without the use of conformal measure ensembles, showed that, in two dimensions at the critical point, when properly renormalized, the Ising magnetization field has a continuum scaling limit Φ^0 which satisfies the expected properties of conformal covariance. Φ^0 is a (generalized, Euclidean) random field on $\mathbb{R}^2$—i.e., for suitable test functions f on $\mathbb{R}^2$, one can construct random variables $\Phi^0(f)$, formally written as $\int_{\mathbb{R}^2} \Phi^0(x)f(x)\mathrm{d}x$. The tail behavior of the field Φ^0, obtained in [53], shows that Φ^0 is not a Gaussian field. (This follows also from the behavior of its correlations, which do not obey Wick's theorem—see [68]). Another significant contribution of [53] is the construction of the continuum scaling limit of the magnetization field for the near-critical Ising model with external magnetic field $ha^{15/8}$ on the rescaled lattice $a\mathbb{Z}^2$ as $a \searrow 0$. As stated in [52], it was expected that the truncated correlations of the resulting field Φ^h would decay exponentially whenever $h \neq 0$. A proof of that statement is provided in [54] together with a rigorous proof that the critical exponent for how the correlation length diverges as $h \searrow 0$ is 8/15, and related scaling properties of Φ^h. The related magnetization critical exponent δ, which determines how, at the critical temperature, the magnetization depends on the external magnetic field, was rigorously shown to be equal to 15 in a joint paper by FC, Garban and Chuck [51].

It is worth pointing out that, while the concept of conformal measure ensemble is not used in [52,53], it does play a crucial role in [54]. Indeed, a surprising contribution of [54] is the demonstration that conformal measure ensembles coupled with the corresponding conformal loop ensembles can be useful in analyzing near-critical scaling limits. The construction of conformal measure ensembles and their coupling to CLE_κ is carried out for $\kappa = 6$ and 16/3 in the article by FC, René Conijn and Demeter Kiss in Volume 2 of this Festschrift, where the authors also obtain the geometric representation of the Ising magnetization in the continuum mentioned earlier. We note that, as pointed out in [62], in addition to their utility for critical and near-critical two-dimensional models, measure ensembles may be more extendable than loop ensembles to scaling limits in dimensions $d > 2$.

It seems fitting to conclude this introduction with some comments on the relevance of [52,53] and [54] for (constructive) quantum field theory, one of Chuck's early scientific interests. Euclidean random fields, such as Φ^h, on the Euclidean "space-time" $\mathbb{R}^d := \{x = (x_0, w_1, \dots, w_{d-1})\}$ are related to quantum fields on relativistic space-time, $\{(t, w_1, \dots, w_{d-1})\}$, essentially by replacing x_0 with a complex variable and analytically continuing from the purely real x_0 to a pure imaginary $(-it)$—see [113,231]. In this context, non-Gaussian Euclidean fields such as those discussed in [52,53] are of particular interest since Gaussian Euclidean fields correspond to non-interacting (and therefore trivial) quantum fields—see, e.g., [92]. The construction of interacting field theories has been from the start one of the main goals of constructive field theory, albeit in the physically interesting case of four dimensions rather than in the two-dimensional setting of [52,53]. Notwithstanding, one major reason for interest in Φ^h is that the associated quantum field theory was predicted by Zamolodchikov [285] to have remarkable properties including a "particle content" of eight particles whose masses are related to the exceptional Lie algebra E_8—see [39,75,174]. One of the main contributions of [54] is to prove a strictly positive lower bound on all masses (i.e., a *mass gap*) predicted by Zamolodchikov's theory. This is a first natural step in the direction of a rigorous analysis of Zamolodchikov's theory, to which we hope Chuck will provide further insightful contributions.

The description of Chuck's contributions in this introduction is unavoidably biased and incomplete, and we apologize to all of his co-authors whose work has not been mentioned. The responsibility rests with Chuck for being so unreasonably productive.

References

1. Aizenman, M.: Scaling limit for the incipient spanning clusters. In: Mathematics of Multiscale Materials, pp. 1–24. Springer, New York (1998)
2. Aizenman, M., Barsky, D.J.: Sharpness of the phase transition in percolation models. Commun. Math. Phys. **108**(3), 489–526 (1987)
3. Aizenman, M., Burchard, A.: Hölder regularity and dimension bounds for random curves. Duke Math. J. **99**(3), 419–453 (1999)
4. Aizenman, M., Burchard, A., Newman, C.M., Wilson, D.B.: Scaling limits for minimal and random spanning trees in two dimensions. Random Struct. Algorithms **15**(3–4), 319–367 (1999)
5. Aizenman, M., Chayes, J.T., Chayes, L., Newman, C.M.: Discontinuity of the magnetization in one-dimensional $1/|x-y|^2$ Ising and Potts models. J. Stat. Phys. **50**, 1–40 (1988)
6. Aizenman, M., Fisher, D.S.: Unpublished
7. Aizenman, M., Kesten, H., Newman, C.M.: Uniqueness of the infinite cluster and continuity of connectivity functions for short and long range percolation. Commun. Math. Phys. **111**(4), 505–531 (1987)
8. Aizenman, M., Newman, C.M.: Tree graph inequalities and critical behavior in percolation models. J. Stat. Phys. **36**(1), 107–143 (1984)
9. Aizenman, M., Newman, C.M.: Discontinuity of the percolation density in one dimensional $1/|x-y|^2$ percolation models. Commun. Math. Phys. **107**(4), 611–647 (1986)
10. Aizenman, M., Wehr, J.: Rounding effects of quenched randomness on first-order phase transitions. Commun. Math. Phys. **130**, 489–528 (1990)
11. Arguin, L.P., Damron, M., Newman, C.M., Stein, D.L.: Uniqueness of ground states for short-range spin glasses in the half-plane. Commun. Math. Phys. **300**, 641–657 (2010)
12. Arguin, L.P., Newman, C.M., Stein, D.L.: Thermodynamic identities and symmetry breaking in short-range spin glasses. Phys. Rev. Lett. **115**, 187–202 (2015)
13. Arguin, L.P., Newman, C.M., Stein, D.L., Wehr, J.: Fluctuation bounds for interface free energies in spin glasses. J. Stat. Phys. **156**, 221–238 (2014)
14. Arguin, L.P., Newman, C.M., Stein, D.L., Wehr, J.: Fluctuation bounds in spin glasses at zero temperature. J. Stat. Phys. **165**, 1069–1078 (2016)
15. Arous, G.B., Cerný, J.: Dynamics of trap models. In: Mathematical Statistical Physics Lecture Notes, vol. LXXXIII, Les Houches Summer School, pp. 331–394. North-Holland, Amsterdam (2006)

16. Arratia, R.: Coalescing Brownian motions on the line. Ph.D. Thesis, Univ. Wisconsin, Madison (1979)
17. Arratia, R.: Coalescing Brownian motions and the voter model on F (1981). Unpiblished manuscript
18. Aspelmeier, T., Katzgraber, H.G., Larson, D., Moore, M.A., Wittmann, M., Yeo, J.: Finite-size critical scaling in Ising spin glasses in the mean-field regime. Phys. Rev. E **93**, 032123 (2016)
19. Auffinger, A., Damron, M.: Differentiability at the edge of the percolation cone and related results in first-passage percolation. Probab. Theory Relat. Fields **156**(1), 193–227 (2013)
20. Auffinger, A., Damron, M., Hanson, J.: 50 Years of First-Passage Percolation. American Mathematical Society, Providence (2017)
21. Badoni, D., Ciria, J., Parisi, G., Ritort, F., Pech, J., Ruiz-Lorenzo, J.: Numerical evidence of a critical line in the $4d$ Ising spin glass. Europhys. Lett. **21**, 495–499 (1993)
22. Ballesteros, H., Cruz, A., Fernández, L., Martin-Mayor, V., Pech, J., Ruiz-Lorenzo, J., Tarancón, A., Téllez, P., Ullod, C., Ungil, C.: Critical behavior of the three-dimensional Ising spin glass. Phys. Rev. B **62**, 14237 (2000)
23. Banavar, J.R., Cieplak, M., Maritan, A.: Optimal paths and domain walls in the strong disorder limit. Phys. Rev. Lett. **72**, 2320–2323 (1994)
24. Barsky, D., Aizenman, M.: Percolation critical exponents under the triangle condition. Ann. Probab. **19**, 1520–1536 (1991)
25. Barsky, D.J., Grimmett, G.R., Newman, C.M.: Percolation in half-spaces: equality of critical densities and continuity of the percolation probability. Probab. Theory Relat. Fields **90**(1), 111–148 (1991)
26. Barsky, D.J., Wu, C.C.: Critical exponents for the contact process under the triangle condition. J. Stat. Phys. **91**(1), 95–124 (1998)
27. Beffara, V., Nolin, P.: On monochromatic arm exponents for 2D critical percolation. Ann. Probab. **39**(4), 1286–1304 (2011)
28. Belavin, A.A., Polyakov, A.M., Zamolodchikov, A.B.: Infinite conformal symmetry in two-dimensional quantum field theory. Nucl. Phys. B **241**(2), 333–380 (1984)
29. Belavin, A.A., Polyakov, A.M., Zamolodchikov, A.B.: Infinite conformal symmetry of critical fluctuations in two dimensions. J. Stat. Phys. **34**(5), 763–774 (1984)
30. Benjamini, I., Schramm, O.: Percolation beyond F^d, many questions and a few answers. Electron. Commun. Probab. **1**, 71–82 (1996)
31. Bethe, H.: Statistical theory of superlattices. Proc. Roy. Soc. London A: Math. Phys. Eng. Sci. **150**(871), 552–575 (1935)
32. Billoire, A., Maiorano, A., Marinari, E., Martin-Mayor, V., Yllanes, D.: Cumulative overlap distribution function in realistic spin glasses. Phys. Rev. B **90**, 094201 (2014)
33. Binder, K., Young, A.P.: Spin glasses: experimental facts, theoretical concepts, and open questions. Rev. Mod. Phys. **58**, 801–976 (1986)
34. Bleher, P.M., Sinai, J.G.: Investigation of the critical point in models of the type of dyson's hierarchical models. Commun. Math. Phys. **33**(1), 23–42 (1973)
35. Bokil, H., Bray, A.J., Drossel, B., Moore, M.A.: Comment on 'General method to determine replica symmetry breaking transitions'. Phys. Rev. Lett. **82**, 5174 (1999)
36. Bokil, H., Drossel, B., Moore, M.A.: The influence of critical behavior on the spin glass phase (2000). Available as cond-mat/0002130

37. Bollobas, B., Riordan, O.: Percolation. Cambridge University Press, New York (2006)
38. Borgs, C., Chayes, J.T., van der Hofstad, R., Slade, G., Spencer, J.: Random subgraphs of finite graphs. II. The lace expansion and the triangle condition. Ann. Probab. **33**(5), 1886–1944 (2005)
39. Borthwick, D., Garibaldi, S.: Did a 1-dimensional magnet detect a 248-dimensional Lie algebra? Notices Amer. Math. Soc. **58**, 1055–1066 (2011)
40. Bouchaud, J.P.: Weak ergodicity breaking and aging in disordered systems. J. Phys. **I**(2), 1705–1713 (1992)
41. Bray, A.J., Moore, M.A.: Critical behavior of the three-dimensional Ising spin glass. Phys. Rev. B **31**, 631–633 (1985)
42. Bray, A.J., Moore, M.A.: Chaotic nature of the spin-glass phase. Phys. Rev. Lett. **58**, 57–60 (1987)
43. Broadbent, S.: Contribution to discussion on symposium on Monte Carlo methods. J. Roy. Statist. Soc. B **16**, 68 (1954)
44. Broadbent, S.R., Hammersley, J.M.: Percolation processes: I. Crystals and mazes. In: Mathematical Proceedings of the Cambridge Philosophical Society, **53**(3), pp. 629–641 (1957)
45. Bruijn, N.G.D.: The roots of trigonometric integrals. Duke Math. J. **17**, 197–226 (1950)
46. Burton, R.M., Keane, M.: Density and uniqueness in percolation. Commun. Math. Phys. **121**(3), 501–505 (1989)
47. Burton, R.M., Keane, M.: Topological and metric properties of infinite clusters in stationary two-dimensional site percolation. Isr. J. Math. **76**, 299–316 (1991)
48. Cacciuto, A., Marinari, E., Parisi, G.: A numerical study of ultrametricity in finite-dimensional spin glasses. J. Phys. A **30**, L263–L269 (1997)
49. Camia, F., Fontes, L.R.G., Newman, C.M.: The scaling limit geometry of near-critical 2D percolation. J. Stat. Phys. **125**, 57–69 (2006)
50. Camia, F., Fontes, L.R.G., Newman, C.M.: Two-dimensional scaling limits via marked nonsimple loops. Bull. Braz. Math. Soc. **37**(4), 537–559 (2006)
51. Camia, F., Garban, C., Newman, C.M.: The Ising magnetization exponent on $\mathbb{Z}^2$ is 1/15. Probab. Theory Rel. Fields **160**, 175–187 (2014)
52. Camia, F., Garban, C., Newman, C.M.: Planar Ising magnetization field I. Uniqueness of the critical scaling limits. Ann. Probab. **43**, 528–571 (2015)
53. Camia, F., Garban, C., Newman, C.M.: Planar Ising magnetization field II. Properties of the critical and near-critical scaling limits. Ann. Inst. H. Poincaré Probab. Statist. **52**, 146–161 (2016)
54. Camia, F., Jiang, J., Newman, C.M.: Exponential decay for the near-critical scaling limit of the planar Ising model. Commun. Pure Appl. Math. (2017, to appear). arXiv:1707.02668
55. Camia, F., Jiang, J., Newman, C.M.: New FK–Ising coupling applied to near-critical planar models. arXiv:1709.00582
56. Camia, F., Joosten, M., Meester, R.: Trivial, critical and near-critical scaling limits of two-dimensional percolation. J. Stat. Phys. **137**, 57–69 (2009)
57. Camia, F., Newman, C.M.: Continuum nonsimple loops and 2D critical percolation. J. Stat. Phys. **116**(1), 157–173 (2004)
58. Camia, F., Newman, C.M.: The full scaling limit of two-dimensional critical percolation (2005). arXiv:math.PR/0504036
59. Camia, F., Newman, C.M.: Two-dimensional critical percolation: the full scaling limit. Commun. Math. Phys. **268**(1), 1–38 (2006)

60. Camia, F., Newman, C.M.: Critical percolation exploration path and SLE_6: a proof of convergence. Probab. Theory Relat. Fields **139**(3), 473–519 (2007)
61. Camia, F., Newman, C.: SLE6 and CLE6 from critical percolation. In: Pinsky, M., Birnir, B. (eds.) Probability, Geometry and Integrable Systems, Mathematical Sciences Research Institute Publications 55, vol. 55, pp. 103–130. Cambridge University Press, Cambridge (2008)
62. Camia, F., Newman, C.M.: Ising (conformal) fields and cluster area measures. Proc. Natl. Acad. Sci. (USA) **14**, 5457–5463 (2009)
63. Caracciolo, S., Parisi, G., Patarnello, S., Sourlas, N.: Low temperature behaviour of 3-D spin glasses in a magnetic field. J. Phys. France **51**, 1877–1895 (1990)
64. Cator, E., Pimentel, L.P.R.: A shape theorem and semi-infinite geodesics for the Hammersley model with random weights. LEA: Lat. Am. J. Probab. Math. Stat. **8**, 163–175 (2011)
65. Chatterjee, S.: Disorder chaos and multiple valleys in spin glasses (2009). arXiv:0907.3381
66. Chatterjee, S.: The universal relation between scaling exponents in first-passage percolation. Ann. Math. **177**(2), 663–697 (2013)
67. Chayes, J., Chayes, L.: An inequality for the infinite cluster density in Bernoulli percolation. Phys. Rev. Lett. **56**, 1619–1622 (1986)
68. Chelkak, D., Hongler, C., Izyurov, K.: Conformal invariance of spin correlations in the planar Ising model. Ann. Math. **181**, 1087–1138 (2015)
69. Borgs, C., Chayes, J.T., van der Hofstad, R., Slade, G., Spencer, J.: Random subgraphs of finite graphs: I. The scaling window under the triangle condition. Random Struct. Algorithms **27**(2), 137–184 (2005)
70. Cohen, J.E., Newman, C.M., Briand, F.: A stochastic theory of community food webs: II. Individual webs. Proc. Royal Soc. London **B224**, 449–461 (1985)
71. Contucci, P., Giardiná, C., Giberti, C., Parisi, G., Vernia, C.: Ultrametricity in the Edwards-Anderson model. Phys. Rev. Lett. **99**, 057206 (2007)
72. Cox, J.T., Durrett, R.: Some limit theorems for percolation processes with necessary and sufficient conditions. Ann. Probab. **9**(4), 583–603 (1981)
73. Damron, M., Hanson, J.: Bigeodesics in first-passage percolation. Commun. Math. Phys. **349**(2), 753–776 (2017)
74. Damron, M., Hochman, M.: Examples of nonpolygonal limit shapes in I.I.D. First-passage percolation and infinite coexistence in spatial growth models. Ann. Appl. Probab. **23**(3), 1074–1085 (2013)
75. Delfino, G.: Integrable field theory and critical phenomena: the Ising model in a magnetic field. J. Phys. A: Math. Gen. **37**, R45–R78 (2004)
76. Derrida, B., Toulouse, G.: Sample to sample fluctuations in the random energy model. J. Phys. (Paris) Lett. **46**, L223–L228 (1985)
77. Doyon, B.: Calculus on manifolds of conformal maps and CFT. J. Phys. A: Math. Theor. **45**, 315202 (2012)
78. Doyon, B.: Conformal loop ensembles and the stress-energy tensor. Lett. Math. Phys. **103**, 233–284 (2013)
79. Doyon, B.: Random loops and conformal field theory. J. Stat. Mech. Ther. Expt. **46**, 46039207 (2014)
80. Drossel, B., Bokil, H., Moore, M.A., Bray, A.J.: The link overlap and finite size effects for the $3D$ Ising spin glass. Eur. Phys. J. B **13**, 369–375 (2000)
81. Dubédat, J.: Excursion decompositions for SLE and Watts' crossing formula. Probab. Theory Relat. Fields **134**(3), 453–488 (2006)
82. Dubédat, J.: Exact bosonization of the Ising model (2011). arXiv:1112.4399

83. Dunlop, F.: Zeros of the partition function and Gaussian inequalities for the plane rotator model. J. Stat. Phys. **21**, 561–572 (1979)
84. Dunlop, F., Newman, C.M.: Multicomponent field theories and classical rotators. Commun. Math. Phys. **44**, 223–235 (1975)
85. Dyson, F.: General theory of spin-wave interactions. Phys. Rev. **102**, 1217–1229 (1956)
86. Dyson, F.J.: Existence of a phase-transition in a one-dimensional Ising ferromagnet. Commun. Math. Phys. **12**(2), 91–107 (1969)
87. Edwards, R.G., Sokal, A.D.: Generalization of the Fortuin-Kasteleyn-Swendsen-Wang representation and Monte Carlo algorithm. Phys. Rev. D **38**, 2009–2012 (1988)
88. Edwards, S., Anderson, P.W.: Theory of spin glasses. J. Phys. F **5**, 965–974 (1975)
89. Ellis, R.S., Newman, C.M.: Limit theorems for sums of dependent random variables occurring in statistical mechanics. Zeitschrift für Wahrscheinlichkeitstheorie und Verwandte Gebiete **44**(2), 117–139 (1978)
90. Ellis, R.S., Newman, C.M.: The statistics of Curie-Weiss models. J. Stat. Phys. **19**(2), 149–161 (1978)
91. Ellis, R.S., Newman, C.M., Rosen, J.S.: Limit theorems for sums of dependent random variables occurring in statistical mechanics. Zeitschrift für Wahrscheinlichkeitstheorie und Verwandte Gebiete **51**(2), 153–169 (1980)
92. Fernandez, R., Fröhlich, J., Sokal, A.D.: Random Walks, Critical Phenomena, and Triviality in Quantum Field Theory. Springer, Berlin (1992)
93. Ferrari, P.A., Pimentel, L.P.R.: Competition interfaces and second class particles. Ann. Probab. **33**(4), 1235–1254 (2005)
94. Fisher, D.S., Huse, D.A.: Ordered phase of short-range Ising spin-glasses. Phys. Rev. Lett. **56**, 1601–1604 (1986)
95. Fisher, D.S., Huse, D.A.: Absence of many states in realistic spin glasses. J. Phys. A **20**, L1005–L1010 (1987)
96. Fisher, D.S., Huse, D.A.: Pure States in Spin Glasses. J. Phys. A **20**, L997–L1004 (1987)
97. Fisher, D.S., Huse, D.A.: Equilibrium behavior of the spin-glass ordered phase. Phys. Rev. B **38**, 386–411 (1988)
98. Fisher, D.S., Huse, D.A.: Nonequilibrium dynamics of spin glasses. Phys. Rev. B **38**, 373–385 (1988)
99. Fitzner, R., van den Hofstad, R.: Nearest-neighbor percolation function is continuous for $d>10$: extended version (2015). arXiv:1506.07977
100. Fontes, L.R.G., Isopi, M., Newman, C.M.: Chaotic time dependence in a disordered spin system. Prob. Theory Rel. Fields **115**, 417–433 (1999)
101. Fontes, L.R.G., Isopi, M., Newman, C.M.: Random walks with strongly inhomogeneous rates and singular diffusions: convergence, localization and aging in one dimension. Ann. Probab. **30**, 579–604 (2002)
102. Fontes, L.R.G., Isopi, M., Newman, C.M., Ravishankar, K.: The Brownian web: characterization and convergence. Ann. Probab. **32**(4), 2857–2883 (2004)
103. Fontes, L.R.G., Isopi, M., Newman, C.M., Stein, D.L.: Aging in $1D$ discrete spin models and equivalent systems. Phys. Rev. Lett. **87**, 1102011 (2001)
104. Fontes, L.R., Isopi, M., Newman, C.M., Ravishankar, K.: The Brownian web. Proc. Natl. Acad. Sci. **99**, 15888–15893 (2002)
105. Forgacs, G., Lipowsky, R., Nieuwenhuizen, T.M.: The behavior of interfaces in ordered and disordered systems. In: Domb, C., Lebowitz, J. (eds.) Phase Transitions and Critical Phenomena, vol. 14, pp. 135–363. Academic Press, London (1991)

106. Fortuin, C.M., Kasteleyn, P.W.: On the random-cluster model. I. Introduction and relation to other models. Physica **57**, 536–564 (1972)
107. Francesco, P., Mathieu, P., Senechal, D.: Conformal Field Theory. Graduate Texts in Contemporary Physics, corrected edn. Springer, New York (1997)
108. Fröhlich, J., Rodriguez, P.F.: Some applications of the Lee-Yang theorem. J. Math. Phys. **53**, 095218 (2012)
109. Fröhlich, J., Spencer, T.: On the statistical mechanics of classical Coulomb and dipole gases. J. Stat. Phys. **24**, 617–701 (1981)
110. Gandolfi, A., Keane, M., Newman, C.M.: Uniqueness of the infinite component in a random graph with applications to percolation and spin glasses. Probab. Theory Relat. Fields **92**, 511–527 (1992)
111. Garban, C., Pete, G., Schramm, O.: Pivotal, cluster, and interface measures for critical planar percolation. J. Amer. Math. Soc. **26**, 939–1024 (2013)
112. Garet, O., Marchand, R.: Moderate deviations for the chemical distance in Bernoulli percolation. ALEA: Lat. Am. J. Probab. Math. Stat. **7**, 171–191 (2010)
113. Glimm, J., Jaffe, A.: Quantum Physics: A Functional Integral Point of View, 2nd edn. Springer, Berlin (1987)
114. Grimmett, G.: The random-cluster model. Grundlehren der mathematischen Wissenschaften **333**, (2016)
115. Grimmett, G.R.: Percolation. Springer, Berlin (1999)
116. Guerra, F., Rosen, L., Simon, B.: Correlation inequalities and the mass gap in $p(\phi)_2$. Commun. Math. Phys. **41**, 19–32 (1975)
117. Guggenheim, E.A.: The principle of corresponding states. J. Chem. Phys. **13**(7), 253–261 (1945)
118. Gunnarson, K., Svedlindh, P., Nordblad, P., Lundgren, L., Aruga, H., Ito, A.: Static scaling in a short-range Ising spin glass. Phys. Rev. B **43**, 8199 (1991)
119. Hammersley, J.M.: Percolation processes: lower bounds for the critical probability. Ann. Math. Stat. **28**(3), 790–795 (1957)
120. Hammersley, J.M.: Bornes supérieures de la probabilité critique dans un processus de filtration. In: Le Calcul des Probabilités et ses Applications. Paris, 15-20 Juillet 1958, p. 790–795. Centre National de la Recherche Scientifique (1959)
121. Hammersley, J.M., Welsh, D.J.A.: First-Passage Percolation, Subadditive Processes, Stochastic Networks, and Generalized Renewal Theory, pp. 61–110. Springer, Heidelberg (1965)
122. Hara, T., Slade, G.: The triangle condition for percolation. Bull. Am. Math. Soc. (N.S.) **21**, 269–273 (1989)
123. Hara, T., Slade, G.: Mean-field critical behaviour for percolation in high dimensions. Commun. Math. Phys. **128**(2), 333–391 (1990)
124. Hasenbusch, M., Pelissetto, A., Vicari, E.: Critical behavior of three-dimensional Ising spin glass models. Phys. Rev. B **78**, 214205 (2008)
125. Hed, G., Young, A.P., Domany, E.: Lack of ultrametricity in the low-temperature phase of three-dimensional Ising spin glasses. Phys. Rev. Lett. **92**, 157201 (2004)
126. Hoffman, C.: Geodesics in first passage percolation. Ann. Appl. Probab. **18**(5), 1944–1969 (2008)
127. Hongler, C.: Conformal invariance of Ising model correlations. Ph.D. dissertation, Univ. Geneva (2010)
128. Hongler, C., Smirnov, S.: The energy density in the planar Ising model. Acta Math. **211**, 191–225 (2013)
129. Howard, C.D., Newman, C.M.: Euclidean models of first-passage percolation. Probab. Theory Relat. Fields **108**(2), 153–170 (1997)

130. Howard, C.D., Newman, C.M.: From Greedy Lattice Animals to Euclidean First-Passage Percolation, pp. 107–119. Birkhäuser Boston, Boston (1999)
131. Howard, C.D., Newman, C.M.: Geodesics and spanning trees for Euclidean first-passage percolation. Ann. Probab. **29**(2), 577–623 (2001)
132. Ising, E.: Beitrag zur theorie des ferromagnetismus. Z. Phys. **31**, 253–258 (1925)
133. Jackson, T.S., Read, N.: Theory of minimum spanning trees. I. Mean-field theory and strongly disordered spin-glass model. Phys. Rev. E **81**, 021130 (2010)
134. Baity-Jesi, M., et al.: (Janus Collaboration): Critical parameters of the three-dimensional Ising spin glass. Phys. Rev. B **88**, 224416 (2013)
135. Jona-Lasinio, G.: The renormalization group: a probabilistic view. Il Nuovo Cimento B (1971–1996) **26**(1), 99–119 (1975)
136. Jona-Lasinio, G.: Probabilistic Approach to Critical Behavior, pp. 419–446. Springer, Boston (1977)
137. Jona-Lasinio, G.: Renormalization group and probability theory. Phys. Rep. **352**(4), 439–458 (2001). Renormalization group theory in the new millennium. III
138. Kasteleyn, P.W., Fortuin, C.M.: Phase transitions in lattice systems with random local properties. J. Phys. Soc. Jpn. [Suppl.] **26**, 11–14 (1969)
139. Katzgraber, H.G., Krzakala, F.: Temperature and disorder chaos in three-dimensional Ising spin glasses. Phys. Rev. Lett. **98**, 017201 (2007)
140. Katzgraber, H.G., Palassini, M., Young, A.P.: Monte Carlo simulations of spin glasses at low temperatures. Phys. Rev. B **63**, 184422 (2001)
141. Katzgraber, H.G., Young, A.P.: Probing the Almeida-Thouless line away from the mean-field model. Phys. Rev. B **72**, 184416 (2005)
142. Kemppainen, A., Werner, W.: The nested simple conformal loop ensembles in the Riemann sphere. Probab. Theory Relat. Fields **165**, 835–866 (2016)
143. Kesten, H.: The critical probability of bond percolation on the square lattice equals 1/2. Commun. Math. Phys. **74**(1), 41–59 (1980)
144. Kesten, H.: Percolation Theory for Mathematicians. Birkhauser, Boston (1982)
145. Kesten, H.: Aspects of First-Passage Percolation, pp. 125–264. Springer, Berlin (1986)
146. Kesten, H.: On the speed of convergence in first-passage percolation. Ann. Appl. Probab. **3**, 296–338 (1993)
147. Ki, H., Kim, Y.O., Lee, J.: On the de Bruijn-Newman constant. Adv. Math. **222**, 281–306 (2009)
148. Kipnis, C., Newman, C.M.: The metastable behavior of infrequently observed, weakly random, one-dimensional diffusion processes. SIAM J. Appl. Math. **45**, 972–982 (1985)
149. Knauf, A.: Number theory, dynamical systems and statistical mechanics. Rev. Math. Phys. **11**, 1027–1060 (1999)
150. Krug, J., Spohn, H.: Kinetic Roughening of Growing Interfaces, pp. 479–582. Cambridge University Press, Cambridge (1991)
151. Krzakala, F., Martin, O.C.: Spin and link overlaps in three-dimensional spin glasses. Phys. Rev. Lett. **85**, 3013–3016 (2000)
152. Külske, C.: Limiting behavior in random Gibbs measures: metastates in some disordered mean field models. In: Bovier, A., Picco, P. (eds.) Mathematics of Spin Glasses and Neural Networks, pp. 151–160. Birkhauser, Boston (1998)
153. Külske, C.: Metastates in disordered mean-field models II: The superstates. J. Stat. Phys. **91**, 155–176 (1998)
154. Landau, L.D.: On the theory of phase transitions. Zh. Eksp. Teor. Fiz. **7**, 19–32 (1937). [Ukr. J. Phys.53,25(2008)]

155. Lawler, G.F., Schramm, O., Werner, W.: The dimension of the planar Brownian frontier is 4/3. Math. Res. Lett. **8**, 401–411 (2001)
156. Lawler, G.F., Schramm, O., Werner, W.: Values of Brownian intersection exponents, I: Half-plane exponents. Acta Math. **187**(2), 237–273 (2001)
157. Lawler, G.F., Schramm, O., Werner, W.: One-arm exponent for critical 2d percolation. Electron. J. Probab. **7**, 13 pp. (2002)
158. Lawler, G.F., Schramm, O., Werner, W.: Conformal invariance of planar loop-erased random walks and uniform spanning trees. Ann. Probab. **32**(1B), 939–995 (2004)
159. Lee, T.D., Yang, C.N.: Statistical theory of equations of state and phase transitions. II. Lattice gas and Ising model. Phys. Rev. **87**, 410–419 (1952)
160. Lenz, W.: Beitrag zum Verständnis der magnetischen Erscheinungen in festen Körpern. Physik Zeitschr. **21**, 613–615 (1920)
161. Licea, C., Newman, C., Piza, M.: Superdiffusivity in first-passage percolation. Probab. Theory Relat. Fields **106**(4), 559–591 (1996)
162. Licea, C., Newman, C.M.: Geodesics in two-dimensional first-passage percolation. Ann. Probab. **24**(1), 399–410 (1996)
163. Lieb, E.H., Sokal, A.D.: A general Lee-Yang theorem for one-component and multicomponent ferromagnets. Commun. Math. Phys. **80**, 153–179 (1981)
164. Lyons, R.: Phase transitions on nonamenable graphs. J. Math. Phys. **41**(3), 1099–1126 (2000)
165. Machta, J., Newman, C.M., Stein, D.L.: The percolation signature of the spin glass transition. J. Stat. Phys. **130**, 113–128 (2008)
166. Machta, J., Newman, C.M., Stein, D.L.: Percolation in the Sherrington–Kirkpatrick spin glass. In: Sidoravicious, V., Vares, M.E. (eds.) Progress in Probability, vol. 60: In and Out of Equilibrium II, pp. 527–542. Birkhauser, Boston (2009)
167. Machta, J., Newman, C.M., Stein, D.L.: A percolation-theoretic approach to spin glass phase transitions. In: de Monvel, A.B., Bovier, A. (eds.) Proceedings of the 2007 Paris Spin Glass Summer School, Progress in Probability Series, vol. 62, pp. 205–223. Birkhauser, Boston (2009)
168. Marinari, E., Parisi, G.: Effects of changing the boundary conditions on the ground state of Ising spin glasses. Phys. Rev. B **62**, 11677–11685 (2000)
169. Marinari, E., Parisi, G.: Effects of a bulk perturbation on the ground state of $3D$ Ising spin glasses. Phys. Rev. Lett. **86**, 3887–3890 (2001)
170. Marinari, E., Parisi, G., Ricci-Tersenghi, F., Ruiz-Lorenzo, J.J., Zuliani, F.: Replica symmetry breaking in spin glasses: theoretical foundations and numerical evidences. J. Stat. Phys. **98**, 973–1047 (2000)
171. Marinari, E., Parisi, G., Ritort, F.: On the $3D$ Ising spin glass. J. Phys. A **27**, 2687–2708 (1994)
172. Marinari, E., Parisi, G., Ruiz-Lorenzo, J.: Numerical simulations of spin glass systems. In: Young, A.P. (ed.) Spin Glasses and Random Fields, pp. 59–98. World Scientific, Singapore (1997)
173. Marinari, E., Parisi, G., Ruiz-Lorenzo, J.J., Ritort, F.: Numerical evidence for spontaneously broken replica symmetry in $3D$ spin glasses. Phys. Rev. Lett. **76**, 843–846 (1996)
174. McCoy, B., Mallard, J.M.: The importance of the Ising model. Program. Theor. Phys. **127**, 791–817 (2012)
175. McMillan, W.L.: Scaling theory of Ising spin glasses. J. Phys. C **17**, 3179–3187 (1984)

176. Mermin, N.D., Wagner, H.: Absence of ferromagnetism or antiferromagnetism in one-or two-dimensional isotropic Heisenberg models. Phys. Rev. Lett. **17**, 1133–1136 (1966)
177. Mézard, M., Parisi, G., Sourlas, N., Toulouse, G., Virasoro, M.: Nature of spin-glass phase. Phys. Rev. Lett. **52**, 1156–1159 (1984)
178. Mézard, M., Parisi, G., Sourlas, N., Toulouse, G., Virasoro, M.: Replica symmetry breaking and the nature of the spin-glass phase. J. Phys. (Paris) **45**, 843–854 (1984)
179. Mézard, M., Parisi, G., Virasoro, M.A.: Random free energies in spin glasses. J. Phys. (Paris) Lett. **46**, L217–L222 (1985)
180. Mézard, M., Parisi, G., Virasoro, M.A. (eds.): Spin Glass Theory and Beyond. World Scientific, Singapore (1987)
181. Middleton, A.A.: Numerical investigation of the thermodynamic limit for ground states in models with quenched disorder. Phys. Rev. Lett. **83**, 1672–1675 (1999)
182. Middleton, A.A.: Energetics and geometry of excitations in random systems. Phys. Rev. B **63**, 060202 (2000)
183. Middleton, A.A.: Extracting thermodynamic behavior of spin glasses from the overlap function. Phys. Rev. B **87**, 220201 (2013)
184. Miller, J., Sheffield, S., Werner, W.: CLE precolations. Forum Math. **5**, 102 pages (2017)
185. Miller, J., Sun, N., Wilson, D.B.: The Hausdorff dimension of the CLE gasket. Ann. Probab. **42**, 1644–1665 (2014)
186. Miller, J., Watson, S.S., Wilson, D.B.: The conformal loop ensemble nesting field. Probab. Theory Relat. Fields **163**, 769–801 (2015)
187. Miller, J., Watson, S.S., Wilson, D.B.: Extreme nesting in the conformal loop ensemble. Ann. Probab. **44**, 1013–1052 (2016)
188. Miller, J., Werner, W.: Connection probabilities for conformal loop ensembles. Commun. Math. Phys. **362**, 415–453 (2018)
189. Moore, M.A., Bokil, H., Drossel, B.: Evidence for the droplet picture of spin glasses. Phys. Rev. Lett. **81**, 4252–4255 (1998)
190. Nanda, S., Newman, C.M., Stein, D.L.: Dynamics of Ising spin systems at zero temperature. In: R. Minlos, S. Shlosman, Y. Suhov (eds.) On Dobrushin's Way (from Probability Theory to Statistical Physics), pp. 183–194. Amer. Math. Soc. Transl. (2) 198 (2000)
191. Newman, C.M.: Ultralocal quantum field theory in terms of currents. Commun. Math. Phys. **26**(3), 169–204 (1972)
192. Newman, C.M.: The construction of stationary two-dimensional Markoff fields with an application to quantum field theory. J. Funct. Anal. **14**(1), 44–61 (1973)
193. Newman, C.M.: Zeros of the partition function for generalized Ising systems. Commun. Pure Appl. Math. **27**, 143–159 (1974)
194. Newman, C.M.: Inequalities for Ising models and field theories which obey the Lee-Yang theorem. Commun. Math. Phys. **41**, 1–9 (1975)
195. Newman, C.M.: Fourier transforms with only real zeros. Proc. Am. Math. Soc. **61**, 245–251 (1976)
196. Newman, C.M.: Normal fluctuations and the FKG inequalities. Commun. Math. Phys. **74**(2), 119–128 (1980)
197. Newman, C.M.: A general central limit theorem for FKG systems. Commun. Math. Phys. **91**(1), 75–80 (1983)
198. Newman, C.M.: Some critical exponent inequalities for percolation. J. Stat. Phys. **45**(3), 359–368 (1986)

199. Newman, C.M.: Another critical exponent inequality for percolation: $\beta \geq 2/\delta$. J. Stat. Phys. **47**(5), 695–699 (1987)
200. Newman, C.M.: Inequalities for γ and related critical exponents in short and long range percolation. In: Percolation Theory and Ergodic Theory of Infinite Particle Systems, pp. 229–244. Springer, New York (1987)
201. Newman, C.M.: Disordered Ising systems and random cluster representations. In: Grimmett, G. (ed.) Probability and Phase Transition, pp. 247–260. Kluwer, Dordrecht (1994)
202. Newman, C.M.: Topics in Disordered Systems. Birkhauser, Basel (1997)
203. Newman, C.M., Cohen, J.E.: A stochastic theory of community food webs: I. Models and aggregated data. Proc. Royal. Soc. London **B224**, 421–448 (1985)
204. Newman, C.M., Cohen, J.E., Kipnis, C.: Neo-Darwinian evolution implies punctuated equilibria. Nature **315**, 400–401 (1985)
205. Newman, C.M., Grimmett, G.R.: Percolation in $\infty + 1$ Dimensions, pp. 167–190. Claredon Press, Oxford (1990)
206. Newman, C.M., Piza, M.S.T.: Divergence of shape fluctuations in two dimensions. Ann. Probab. **23**(3), 977–1005 (1995)
207. Newman, C.M., Schulman, L.S.: Infinite clusters in percolation models. J. Stat. Phys. **26**(3), 613–628 (1981)
208. Newman, C.M., Schulman, L.S.: One dimensional $1/|j-i|^s$ percolation models: the existence of a transition for $s \leq 2$. Commun. Math. Phys. **104**(4), 547–571 (1986)
209. Newman, C.M., Stein, D.L.: Unpublished
210. Newman, C.M., Stein, D.L.: Multiple states and thermodynamic limits in short-ranged Ising spin glass models. Phys. Rev. B **46**, 973–982 (1992)
211. Newman, C.M., Stein, D.L.: Spin-glass model with dimension-dependent ground state multiplicity. Phys. Rev. Lett. **72**, 2286–2289 (1994)
212. Newman, C.M., Stein, D.L.: Ground state structure in a highly disordered spin glass model. J. Stat. Phys. **82**, 1113–1132 (1996)
213. Newman, C.M., Stein, D.L.: Spatial inhomogeneity and thermodynamic chaos. Phys. Rev. Lett. **76**, 4821–4824 (1996)
214. Newman, C.M., Stein, D.L.: Metastate approach to thermodynamic chaos. Phys. Rev. E **55**, 5194–5211 (1997)
215. Newman, C.M., Stein, D.L.: Simplicity of state and overlap structure in finite-volume realistic spin glasses. Phys. Rev. E **57**, 1356–1366 (1998)
216. Newman, C.M., Stein, D.L.: Thermodynamic chaos and the structure of short-range spin glasses. In: Bovier, A., Picco, P. (eds.) Mathematics of Spin Glasses and Neural Networks, pp. 243–287. Birkhauser, Boston (1998)
217. Newman, C.M., Stein, D.L.: Blocking and persistence in the zero-temperature dynamics of homogeneous and disordered Ising models. Phys. Rev. Lett. **82**, 3944–3947 (1999)
218. Newman, C.M., Stein, D.L.: Equilibrium pure states and nonequilibrium chaos. J. Stat. Phys. **94**, 709–722 (1999)
219. Newman, C.M., Stein, D.L.: Nature of ground state incongruence in two-dimensional spin glasses. Phys. Rev. Lett. **84**, 3966–3969 (2000)
220. Newman, C.M., Stein, D.L.: Are there incongruent ground states in $2D$ Edwards-Anderson spin glasses? Commun. Math. Phys. **224**, 205–218 (2001)
221. Newman, C.M., Stein, D.L.: Interfaces and the question of regional congruence in spin glasses. Phys. Rev. Lett. **87**, 077201 (2001)
222. Newman, C.M., Stein, D.L.: Nonrealistic behavior of mean field spin glasses. Phys. Rev. Lett. **91**, 197205 (2003)

223. Newman, C.M., Stein, D.L.: Topical review: ordering and broken symmetry in short-ranged spin glasses. J. Phys. Condens. Matter **15**, R1319–R1364 (2003)
224. Newman, C.M., Stein, D.L.: Short-range spin glasses: results and speculations. In: Bolthausen, E., Bovier, A. (eds.) Spin Glass Theory, pp. 159–175. Springer, Berlin (2006)
225. Newman, C.M., Wu, W.: Gaussian fluctuations for the classical XY model. Ann. Inst. H. Poincare (B) **54**, 1759–1777 (2018)
226. Newman, C.M., Wu, W.: Lee-Yang property and Gaussian multiplicative chaos. Commun. Math. Phys. **369**, 153–170 (2019)
227. Newman, C.M., Wu, W.: Constants of deBruijn–Newman type in analytic number theory and statistical physics. Bull. Am. Math. Soc., in press. Published online April 19, 2019. https://doi.org/10.1090/bull/1668
228. Nguyen, B.G.: Gap exponents for percolation processes with triangle condition. J. Stat. Phys. **49**(1), 235–243 (1987)
229. Nguyen, B.G., Yang, W.S.: Triangle condition for oriented percolation in high dimensions. Ann. Probab. **21**(4), 1809–1844 (1993)
230. Onsager, L.: Crystal statisics. I. A two-dimensional model with an order-disorder transition. Phys. Rev. **65**, 117–149 (1944)
231. Osterwalder, K., Schrader, R.: Axioms for Euclidean Green's functions. Commun. Math. Phys. **31**, 83–112 (1973)
232. Palassini, M., Caracciolo, S.: Universal finite-size scaling functions in the $3d$ Ising spin glass. Phys. Rev. Lett. **82**, 5128–5131 (1999)
233. Palassini, M., Young, A.P.: Evidence for a trivial ground-state structure in the two-dimensional Ising spin glass. Phys. Rev. B **60**, R9919–R9922 (1999)
234. Palassini, M., Young, A.P.: Triviality of the ground state structure in Ising spin glasses. Phys. Rev. Lett. **83**, 5126–5129 (1999)
235. Palassini, M., Young, A.P.: Nature of the spin glass state. Phys. Rev. Lett. **85**, 3017–3020 (2000)
236. Parisi, G.: Infinite number of order parameters for spin-glasses. Phys. Rev. Lett. **43**, 1754–1756 (1979)
237. Parisi, G.: Order parameter for spin-glasses. Phys. Rev. Lett. **50**, 1946–1948 (1983)
238. Peierls, R.: On Ising's model of ferromagnetism. Proc. Cambridge Phil. Soc. **32**, 477–481 (1936)
239. Penrose, O., Lebowitz, J.L.: On the exponential decay of correlation functions. Commun. Math. Phys. **39**, 165–184 (1974)
240. Polyakov, A.M.: Conformal symmetry of critical fluctuations. JETP Lett. **12**, 381–383 (1970). [Pisma Zh. Eksp. Teor. Fiz. 12538(1970)]
241. Polymath, D.H.J.: Effective approximation of heat flow evolution of the Riemann ζ function, and a new upper bound for the deBruijn–Newman constant (2019). arXiv:1904.12438
242. R. Basu, S.S., Sly, A.: Coalescence of geodesics in exactly solvable models of last passage percolation (2017). arXiv:1704.05219
243. Rammal, R., Toulouse, G., Virasoro, M.A.: Ultrametricity for physicists. Rev. Mod. Phys. **58**, 765–788 (1986)
244. Read, N.: Short-range Ising spin glasses: the metastate interpretation of replica symmetry breaking. Phys. Rev. E **90**, 032142 (2014)
245. Reger, J.D., Bhatt, R.N., Young, A.P.: Monte Carlo study of the order-parameter distribution in the four-dimensional Ising spin glass. Phys. Rev. Lett. **64**, 1859–1862 (1990)

246. Richardson, D.: Random growth in a tessellation. Math. Proc. Cambridge Philos. Soc. **74**(3), 515–528 (1973)
247. Rodgers, B., Tao, T.: The De Bruijn–Newman constant is non-negative. arXiv:1801.05914
248. Rohde, S., Schramm, O.: Basic properties of SLE. Ann. Math. **161**(2), 883–924 (2005)
249. Ruelle, D.: A mathematical reformulation of derrida's REM and GREM. Commun. Math. Phys. **108**, 225–239 (1987)
250. Rushbrooke, G.S.: On the thermodynamics of the critical region for the Ising problem. J. Chem. Phys. **39**(3), 842–843 (1963)
251. Saouter, Y., Gourdon, X., Demichel, P.: An improved lower bound for the de Bruijn-Newman constant. Math. Comput. **80**, 2281–2287 (2011)
252. Schertzer, E., Sun, R., Swart, J.: The Brownian Web, The Brownian Net, and Their Universality, pp. 270–368. Cambridge University Press, Cambridge (2017)
253. Schramm, O.: Scaling limits of loop-erased random walks and uniform spanning trees. Israel J. Math. **118**(1), 221–288 (2000)
254. Schramm, O.: A percolation formula. Electron. Commun. Probab. **6**(12), 115–120 (2001)
255. Schramm, O., Sheffield, S., Wilson, D.: Conformal radii for conformal loop ensembles. Commun. Math. Phys. **288**, 43–53 (2009)
256. Sheffield, S.: Exploration trees and conformal loop ensembles. Duke Math. J. **147**(1), 79–129 (2009)
257. Sheffield, S., Werner, W.: Conformal loop ensembles: the Markovian characterization and the loop-soup construction. Ann. Math. **176**(3), 1827–1917 (2012)
258. Sherrington, D., Kirkpatrick, S.: Solvable model of a spin glass. Phys. Rev. Lett. **35**, 1792–1796 (1975)
259. Simon, B., Griffiths, R.B.: The $(\phi^4)_2$ field theory as a classical Ising model. Commun. Math. Phys. **33**(2), 145–164 (1973)
260. Smirnov, S.: Critical percolation in the plane: conformal invariance, Cardy's formula, scaling limits. Comptes Rendus de l'Académie des Sciences - Series I - Mathematics **333**(3), 239–244 (2001)
261. Smirnov, S.: Conformal invariance in random cluster models. I. Holomorphic fermions in the Ising model. Ann. Math. **172**, 1435–1467 (2010)
262. Smirnov, S., Werner, W.: Critical exponents for two-dimensional percolation. Math. Res. Lett. **8**, 729–744 (2001)
263. Stauffer, D.: Scaling properties of percolation clusters. In: Castellani, C., Di Castro, C., Peliti, L. (eds.) Disordered Systems and Localization, pp. 9–25. Springer, Heidelberg (1981)
264. Stauffer, D., Aharony, A.: Introduction to Percolation Theory. Taylor & Francis, London (1994). Revised Second Edition
265. Stein, D.L.: Frustration and fluctuations in systems with quenched disorder. In: Chandra, P., Coleman, P., Kotliar, G., Ong, P., Stein, D., Yu, C. (eds.) PWA90: A Lifetime of Emergence, pp. 169–186. World Scientific, Singapore (2016)
266. Stein, D.L., Newman, C.M.: Broken ergodicity and the geometry of rugged landscapes. Phys. Rev. E **51**, 5228–5238 (1995)
267. Stein, D.L., Newman, C.M.: Spin Glasses and Complexity. Princeton University Press, Princeton (2013)
268. Sun, R., Swart, J.M.: The Brownian net. Ann. Probab. **36**(3), 1153–1208 (2008)
269. Suzuki, M., Fisher, M.E.: Zeros of the partition function for the Heisenberg, ferroelectric, and general Ising models. J. Math. Phys. **12**, 235–246 (1971)

270. Swendsen, R.H., Wang, J.S.: Nonuniversal critical dynamics in Monte Carlo simulations. Phys. Rev. Lett. **58**, 86–88 (1987)
271. Thouless, D.J.: Long-range order in one-dimensional Ising systems. Phys. Rev. **187**, 732–733 (1969)
272. Tóth, B., Werner, W.: The true self-repelling motion. Probab. Theory Relat. Fields **111**(3), 375–452 (1998)
273. Villain, J.: Theory of one-and two-dimensional magnets with an easy magnetization plane. II. The planar, classical, two-dimensional magnet. J. Phys. **36**, 581–590 (1975)
274. Wang, W., Machta, J., Katzgraber, H.G.: Evidence against a mean field description of short-range spin glasses revealed through thermal boundary conditions. Phys. Rev. B **90**, 184412 (2014)
275. Werner, W., Wu, H.: From CLE(κ) to SLE(κ, ρ)'s. Electron. J. Probab. **18**, 1–20 (2013)
276. Wilson, K.G.: The renormalization group: critical phenomena and the Kondo problem. Rev. Mod. Phys. **47**, 773–840 (1975)
277. Wilson, K.G.: The renormalization group and critical phenomena. Rev. Mod. Phys. **55**, 583–600 (1983)
278. Wilson, K.G., Kogut, J.: The renormalization group and the ϵ expansion. Phys. Rep. **12**, 75–199 (1974)
279. Yang, C.N., Lee, T.D.: Statistical theory of equations of state and phase transitions. I. Theory of condensation. Phys. Rev. **87**, 404–409 (1952)
280. Yao, C.L.: Law of large numbers for critical first-passage percolation on the triangular lattice. Electron. Commun. Probab. **19**, 1–14 (2014)
281. Yao, C.L.: Multi-arm incipient infinite clusters in 2D: scaling limits and winding numbers. Ann. Inst. H. Poincaré Probab. Statist. **54**, 1848–1876 (2018)
282. Yao, C.L.: Asymptotics for 2D critical and near-critical first-passage percolation (2018). arXiv:1806.03737
283. Yao, C.L.: Limit theorems for critical first-passage percolation on the triangular lattice. Stoch. Proc. Appl. **128**, 445–460 (2018)
284. Yucesoy, B., Katzgraber, H.G., Machta, J.: Evidence of non-mean-field-like low-temperature behavior in the Edwards-Anderson spin-glass model. Phys. Rev. Lett. **109**, 177204 (2012)
285. Zamolodchikov, A.B.: Integrals of motion and S-matrix of the (scaled) $t = t_c$ Ising model with magnetic field. Int. J. Mod. Phys. **4**, 4235–4248 (1989)

Mean-Field Monomer-Dimer Models. A Review

Diego Alberici, Pierluigi Contucci(✉), and Emanuele Mingione

Department of Mathematics, University of Bologna,
Piazza di Porta San Donato 5, Bologna, Italy
{diego.alberici2,pierluigi.contucci,emanuele.mingione2}@unibo.it

To Chuck Newman, on his 70th birthday.

Abstract. A collection of rigorous results for a class of mean-field monomer-dimer models is presented. It includes a Gaussian representation for the partition function that is shown to considerably simplify the proofs. The solutions of the quenched diluted case and the random monomer case are explained. The presence of the attractive component of the Van der Waals potential is considered and phase transition analysed. In particular the breakdown of the central limit theorem is illustrated at the critical point where a non Gaussian, quartic exponential distribution is found for the number of monomers centered and rescaled with the volume to the power 3/4.

Keywords: Monomer-dimer models · Mean-field models · Central limit theorems

1 Introduction

The monomer-dimer models, an instance in the wide set of interacting particle systems, have a relevant role in equilibrium statistical mechanics. They were introduced to describe, in a simplified yet effective way, the process of absorption of monoatomic or diatomic molecules in condensed-matter physics [15,16,41] or the behaviour of liquid solutions composed by molecules of different sizes [26]. Exact solutions in specific cases (e.g. the perfect matching problem) have been obtained on planar lattices [25,27,36,38,43] and the problem on regular lattices is also interesting for the liquid crystals modelling [1,21,28,32,33,39]. The impact and the interest that monomer-dimer models have attracted has progressively grown beyond physics. Their thermodynamic behaviour has indeed proved to be useful in computer science for the matching problem [11,35] or for the applications of statistical physics methods to the social sciences [10].

From the physical point of view monomers and dimers cannot occupy the same site of a lattice due to the strong repulsion generated by the Pauli exclusion principle. Beside this interaction though, as already noticed by Peierls [40]

V. Sidoravicius (Ed.): *Sojourns in Probability Theory and Statistical Physics - I*, PROMS 298, pp. 39–62, 2019.
https://doi.org/10.1007/978-981-15-0294-1_2

in the first theoretical physics accounts, the attractive component of the Van der Waals potentials might influence the phase structure of the model and its thermodynamic behaviour. With the contemporary presence of those two interactions the global physical observables become particularly difficult to study. Generic Gaussian fluctuations on each ergodic component can still be expected but the nature of the critical point, if any, is a priori not obvious.

Here we focus on a set of monomer-dimer models in the mean field setting, i.e. on the complete graph where every site interacts in average with any other, and present a review of recent results. Section 2 introduces the general properties of the monomer-dimer systems that we approach with the help of a Gaussian representation for their partition function. This representation and its combinatorial features help to embed and ease part of the classical difficulties of their studies. The celebrated Heilmann and Lieb relation, so rich of rigorous consequences, emerges as the formula of integration by parts for Gaussian random vectors. The absence of phase transition for the pure hard-core case is therefore derived. Section 3 treats two quenched cases, namely the diluted complete graph of Erdős–Rényi type as well as the diluted random monomer field activity. For both cases we compute the exact solution. The diluted graph is treated with the help of correlation inequalities and the representation of the monomer density as the solution of an iterative distributional equation. The random monomer activity model is solved by reducing the computation of the equilibrium state to a standard variational problem, again with the help of the Gaussian representation. Section 4 introduces a genuine deterministic mean field model with and without the attractive interaction. It is shown how the model with attraction displays a phase space structure similar to the mean field ferromagnet but without the usual plus-minus symmetry. The model has a coexistence line bounded by a critical point with standard mean-field critical exponents. In Sect. 5 we show that while outside the critical point the central limit theorem holds, at criticality it breaks down and the limiting distribution is found at a scale of $N^{3/4}$ and turns out to be a quartic exponential, like in the well known results by Newman and Ellis [22, 23] for the ferromagnet.

2 Definition and General Properties

Let $G = (V, E)$ be a finite undirected graph with vertex set V and edge set $E \subseteq \{ij \equiv \{i,j\} \,|\, i \in V,\, j \in V,\, i \neq j\}$.

Definition 1 (Monomer-dimer configurations). *A set of edges $D \subseteq E$ is called a* monomer-dimer configuration, *or a* matching, *if the edges in D are pairwise non-incident (Fig. 1). The space of all possible monomer-dimer configurations on the graph G is denoted by $\mathscr{D}_G$.*

Given a monomer-dimer configuration D, we say that every edge in D is occupied by a *dimer*, while every vertex that does not appear in D is occupied by a *monomer*. The set of monomers associated to D is denoted by $M(D)$.

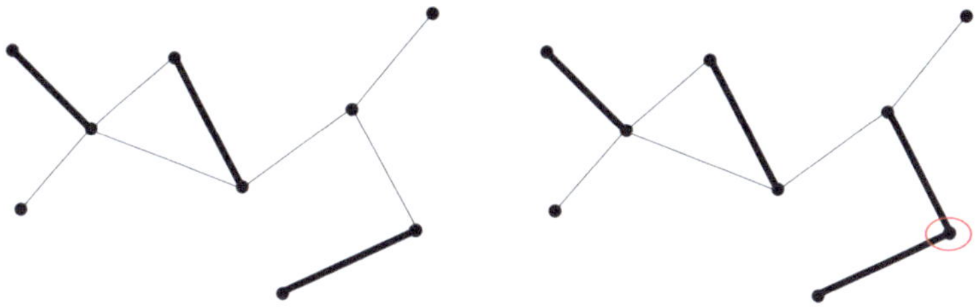

Fig. 1. The bold edges in the left figure form a monomer-dimer configuration on the graph, while those in the right figure do not because two of them share a vertex.

Remark 1. We can associate the dimer occupation variable $\alpha_{ij} \in \{0,1\}$ to each edge $ij \in E$: the edge ij is occupied by a dimer if and only if α_{ij} takes the value 1. It is clear that monomer-dimer configurations are in one-to-one correspondence with vectors $\alpha \in \{0,1\}^E$ satisfying the following constraint:

$$\sum_{j \sim i} \alpha_{ij} \leq 1\,, \quad \forall i \in V \tag{1}$$

where $j \sim i$ means that $ij \in E$. Therefore, with a slight abuse of notation, we denote by $\mathscr{D}_G$ also the set of $\alpha \in \{0,1\}^E$ that satisfy (1). The condition (1) guarantees that at most one dimer can be incident to a given vertex i, namely two dimers cannot be incident. This fact is usually referred as *hard-core interaction* or *hard-core constraint* or *monogamy constraint*. We also introduce an auxiliary variable, the monomer occupation variable,

$$\alpha_i := 1 - \sum_{j \sim i} \alpha_{ij} \in \{0,1\} \tag{2}$$

for each vertex $i \in V$: the vertex i is occupied by a monomer if and only if α_i takes the value 1.

The definition of monomer-dimer configurations already allows to raise non-trivial combinatorial questions as "How many monomer-dimer configurations, for a fixed number of dimers, exist on given a graph G?". This combinatorial problem is known to be NP-hard in general, but there are polynomial algorithms and exact solutions for specific cases [25,30,35,36,43]. In Statistical Mechanics a further structure is introduced and the previous problem becomes a specific limit case. We consider a Gibbs probability measure on the set of monomer-dimer configurations. There are several choices for the measure, depending on how we decide to model the interactions in the system.

2.1 Pure Hard-Core Interaction

This amounts to taking into account only the hard-core interaction among particles and assign a dimer activity $w_{ij} \geq 0$ to each edge $ij \in E$ and a monomer activity $x_i > 0$ to each vertex $i \in V$.

We notice that the representation (7) allows to express average values w.r.t. the measure (3) as Gaussian averages. For example, given a vertex $i \in V$, its monomer probability by (5) writes

$$\langle \alpha_i \rangle_N = \frac{\partial}{\partial h_i} \log Z_N . \tag{10}$$

Then, using the representation (7) in the r.h.s. of (10) together with the identity $\frac{\partial}{\partial h_i} \equiv x_i \frac{\partial}{\partial x_i}$, one obtains

$$\langle \alpha_i \rangle_N = x_i \, \mathbb{E}_\xi \left[\frac{1}{\xi_i + x_i} g_N(\xi, x) \right]$$

where $g_N(\xi, x) = \dfrac{1}{Z_N} \prod_{i=1}^N (\xi_i + x_i)$.

Heilmann and Lieb [30] provided a recursion for the partition functions of monomer-dimer models. As we will see this is a fundamental tool to obtain exact solutions and to prove general properties.

Proposition 2 (Heilmann–Lieb recursion *[30]***).** *Fixing any vertex $i \in V$ it holds:*

$$Z_G = x_i \, Z_{G-i} + \sum_{j \sim i} w_{ij} \, Z_{G-i-j} \, . \tag{11}$$

Here $G - i$ denotes the graph obtained from G deleting the vertex i and all its incident edges.

The Heilmann–Lieb recursion can be obtained directly from the definition (4), exploiting the hard-core constraint: the first term on the r.h.s. of (11) corresponds to a monomer on i, while the following terms correspond to a dimer on ij for some j neighbour of i. Here we show a different proof that uses Gaussian integration by parts.

Proof (see [5]*).* Set $N := |V|$. Introduce zero dimer weights $w_{hk} = 0$ for all the pairs $hk \notin E$, so that $Z_G \equiv Z_N$. Following Proposition 1, introduce an N-dimensional Gaussian vector ξ with mean 0 and covariance matrix W. Then write the identity (7) isolating the vertex i:

$$Z_G = \mathbb{E}_\xi \left[\prod_{k=1}^N (\xi_k + x_k) \right] = x_i \, \mathbb{E}_\xi \left[\prod_{k \neq i} (\xi_k + x_k) \right] + \mathbb{E}_\xi \left[\xi_i \prod_{k \neq i} (\xi_k + x_k) \right] . \tag{12}$$

Now apply the Gaussian integration by parts to the second term on the r.h.s. of (12):

$$\begin{aligned} \mathbb{E}_\xi \left[\xi_i \prod_{k \neq i} (\xi_k + x_k) \right] &= \sum_{j=1}^N \mathbb{E}_\xi [\xi_i \xi_j] \, \mathbb{E}_\xi \left[\frac{\partial}{\partial \xi_j} \prod_{k \neq i} (\xi_k + x_k) \right] \\ &= \sum_{j \neq i} w_{ij} \, \mathbb{E}_\xi \left[\prod_{k \neq i,j} (\xi_k + x_k) \right] . \end{aligned} \tag{13}$$

Notice that summing over $j \neq i$ in the r.h.s. of (13) is equivalent to summing over $j \sim i$, since by definition $w_{ij} = 0$ if $ij \notin E$. Substitute (13) into (12):

$$Z_G = x_i \, \mathbb{E}_\xi \Big[\prod_{k \neq i} (\xi_k + x_k) \Big] + \sum_{j \sim i} w_{ij} \, \mathbb{E}_\xi \Big[\prod_{k \neq i,j} (\xi_k + x_k) \Big] .$$

To conclude the proof observe that $(\xi_k)_{k \neq i}$ is an $(N-1)$-dimensional Gaussian vector with mean 0 and covariance $(w_{hk})_{h,k \neq i}$. Hence by Proposition 1

$$Z_{G-i} = \mathbb{E}_\xi \Big[\prod_{k \neq i} (\xi_k + x_k) \Big] .$$

And similarly

$$Z_{G-i-j} = \mathbb{E}_\xi \Big[\prod_{k \neq i,j} (\xi_k + x_k) \Big] .$$

□

The main general result about monomer-dimer models is the absence of phase transitions, proved by Heilmann and Lieb [30,31]. This result is obtained by localizing the complex zeros of the partition functions far from the positive real axis. A different probabilistic approach has been later proposed by van den Berg [12].

Theorem 1 (Zeros of the partition function [30]**).** *Consider uniform monomer activity x on the graph and arbitrary dimer activities w_{ij}. The partition function Z_G is a polynomial of degree N in x, where $N = |V|$. The complex zeros of Z_G are purely imaginary:*

$$\{x \in \mathbb{C} \,|\, Z_G(w_{ij}, x) = 0\} \subset \mathrm{i}\mathbb{R} .$$

Furthermore they interlace the zeros of Z_{G-i} for any given $i \in V$, that is:

$$a_1 \leq a'_1 \leq a_2 \leq a'_2 \leq \cdots \leq a'_{N-1} \leq a_N , \tag{14}$$

where $-\mathrm{i}a_1, \dots, -\mathrm{i}a_N$ are the zeros of Z_G and $-\mathrm{i}a'_1, \dots, -\mathrm{i}a'_{N-1}$ are the zeros of Z_{G-i}. The relation (14) holds with strict inequalities if $w_{ij} > 0$ for all $i, j \in V$.

Corollary 1 (Absence of phase transitions) *Consider dimer activities $w\, w_{ij}^{(N)}$ and monomer activities $x\, x_i^{(N)}$ and assume they are chosen in such a way that $p := \lim_{N\to\infty} \frac{1}{N} \log Z_N$ exists. Then the function p is analytic in the variables $(w, x) \in (0, \infty)^2$ and the derivatives $\frac{\partial^{h+k}}{\partial^h w \, \partial^k x}$ can be interchanged with the limit $N \to \infty$.*

2.2 Hard-Core and Imitative Interactions

Beyond the hard-core constraint it is possible to enrich monomer-dimer models with other kinds of interaction. For example in this work we consider, for a given $D \in \mathscr{D}_G$, the set of edges connecting particles of the same kind

$$I(D) = \{ij \in E \,|\, i, j \in M(D) \text{ or } i, j \notin M(D)\}$$

and we introduce an interaction between any pair of vertices $ij \in I(D)$ tuned by a coupling $J_{ij} \in \mathbb{R}$. More precisely

Definition 3 (Monomer-dimer models with imitative interactions) *An* imitative monomer-dimer model *on G is given by the following Gibbs probability measure on $\mathscr{D}_G$:*

$$\mu_G(D) := \frac{1}{Z_G} \prod_{ij \in D} w_{ij} \prod_{i \in M(D)} x_i \prod_{ij \in I(D)} \mathrm{e}^{J_{ij}} \tag{15}$$

for all $D \in \mathscr{D}_G$. The partition function is

$$Z_G := \sum_{D \in \mathscr{D}_G} \prod_{ij \in D} w_{ij} \prod_{i \in M(D)} x_i \prod_{ij \in I(D)} \mathrm{e}^{J_{ij}}$$

The dependence of the measure on the coefficients w_{ij}, x_i, J_{ij} is usually implicit in the notations.

When all the J_{ij}'s take the value zero this model is the pure hard-core model introduced in the previous section. Positive values of the J_{ij}'s favour the configurations with clusters of dimers and clusters of monomers.

Remark 5. The usual Gibbs form $\frac{1}{Z_G}\,\mathrm{e}^{-H_G(\alpha)}$ for the measure (15) is obtained by setting $x_i =: \mathrm{e}^{h_i}$, $w_{ij} =: \mathrm{e}^{h_{ij}}$ and taking as Hamiltonian function

$$H_G(\alpha) := -\sum_{ij \in E} h_{ij}\alpha_{ij} - \sum_{i \in V} h_i \alpha_i - \sum_{ij \in E} J_{ij}\big(\alpha_i\alpha_j + (1-\alpha_i)(1-\alpha_j)\big)$$

for all $\alpha \in \mathscr{D}_G$.

The Gaussian representation and the recursion relation found for the pure hard-core case can be extended to the imitative case.

Proposition 3. *The partition function of any monomer-dimer model over N vertices can be written as*

$$Z_N = \mathbb{E}_\xi\Big[\sum_{A \subset \{1,\dots,N\}} \prod_{i \in A} \xi_i \prod_{i \in A^c} x_i \prod_{\substack{i,j \in A \text{ or} \\ i,j \in A^c}} \mathrm{e}^{J_{ij}/2}\Big], \tag{16}$$

where $\xi = (\xi_1, \dots, \xi_N)$ is a Gaussian random vector with mean 0 and covariance matrix $W = (w_{ij})_{i,j=1,\dots,N}$ and $\mathbb{E}_\xi[\,\cdot\,]$ denotes the expectation with respect to ξ. The diagonal entries w_{ii} are arbitrary numbers, chosen in such a way that W is a positive semi-definite matrix. Moreover we set $J_{ii} = 0$.

The proof is the same as Proposition 1. It is interesting to observe that, when all the ξ_i's are positive, the sum inside the expectation on the r.h.s. of (16) is the partition function of an Ising model.

Proposition 4. *Fixing any vertex $i \in V$ it holds:*

$$Z_G = x_i \, \widetilde{Z}_{G-i} + \sum_{j \sim i} w_{ij} \, \widetilde{Z}_{G-i-j} \, , \tag{17}$$

where:

- *in the partition function $\widetilde{Z}_{G-i}$ the monomer activity x_k is replaced by $x_k \, e^{J_{ik}}$ for every vertex k (notice that only the neighbours of i actually change their activities);*
- *in the partition function $\widetilde{Z}_{G-i-j}$ the dimer activity $w_{kk'}$ is replaced by*

$$w_{kk'} \, e^{J_{ik}+J_{ik'}+J_{jk}+J_{jk'}}$$

for all vertices k, k' (notice that only the neighbours of i or j actually change their activities).

The relation (17) can be obtained directly from the definition: the first term on the r.h.s. corresponds to a monomer on i, while the following terms correspond to a dimer on ij for some j neighbour of i.

The hard-core interaction is not sufficient to cause a phase transition, but adding also the imitative interaction the system can have phase transitions [6, 15, 16, 32]: in Sect. 4 we will study this phase transition on the complete graph. The location of the zeros of the partition function in the complex plane in presence of imitation is an open problem.

3 Quenched Models: Erdős–Rényi and Random Field

In this section we consider monomer-dimer models with pure hard-core interactions in some random environment: the randomness is either in the structure of the graph or in the activities. In the first case we considered a class of random graphs that have locally tree-like structure and finite variance degree distribution [2]: this is the same for which the ferromagnetic Ising model was rigorously solved by Dembo and Montanari [18, 20], using the local weak convergence strategy developed in [9]. For the sake of clarity, in this review we have chosen to present the results on the Erdős–Rényi random graph, but they easily extend for example to random regular graphs and configuration models.

3.1 Self-averaging for Monomer-Dimer Models

One of the most important property describing the effects of randomness in statistical mechanics models is the *self-averaging* of physical quantities. Under quite general hypothesis a monomer-dimer model with independent random weights has a self-averaging pressure density [5].

Let $w_{ij}^{(N)} \geq 0$, $1 \leq i < j \leq N$, $N \in \mathbb{N}$, and $x_i > 0$, $i \in \mathbb{N}$, be *independent* random variables and consider the (random!) partition function

$$Z_N = \sum_{D \in \mathscr{D}_N} \prod_{ij \in D} w_{ij}^{(N)} \prod_{i \in M_N(D)} x_i \,.$$

Since the dimer weights may depend on N and may take the value zero, this framework is very general. Denote simply by $\mathbb{E}[\,\cdot\,]$ the expectation with respect to all the weights and assume

$$\sup_N \sup_{ij} \mathbb{E}[w_{ij}^{(N)}] =: C_1 < \infty\,, \quad \sup_i \mathbb{E}[x_i] =: C_2 < \infty\,, \quad \sup_i \mathbb{E}[x_i^{-1}] =: C_3 < \infty\,.$$

The pressure density $p_N := \frac{1}{N} \log Z_N$ is a random variable with finite expectation, indeed

$$N\, p_N \begin{cases} \geq \log \prod_{i=1}^N x_i \;=\; \sum_{i=1}^N \log x_i \;\geq\; \sum_{i=1}^N (1 + x_i^{-1}) \;\in L^1(\mathbb{P}) \\ \leq\; Z_N - 1 \;\in L^1(\mathbb{P}) \end{cases} .$$

The following theorem shows that in the limit $N \to \infty$ the pressure density p_N concentrates around its expectation.

Theorem 2 (see [5]). *For all $t > 0$, $N \in \mathbb{N}$, $q \geq 1$*

$$\mathbb{P}\big(\,|p_N - \mathbb{E}[p_N]| \geq t\big) \;\leq\; 2\, \exp\left(-\frac{t^2\, N}{4\, q^2\, \log^2 N}\right) + (a + b\, N)\, N^{1-q}\,, \qquad (18)$$

where $a := 4 + 2C_2C_3$, $b := 2C_1C_3^2$. As a consequence, choosing $q > 3$,

$$|p_N - \mathbb{E}[p_N]| \xrightarrow[N\to\infty]{} 0 \;\; \mathbb{P}\text{-almost surely}\,.$$

If the random variables $w_{ij}^{(N)}$, x_i, x_i^{-1} are bounded, then one can obtain an exponential rate of convergence instead of (18).

3.2 Erdős–Rényi Random Graph

Let G_N be a Erdős–Rényi random graph over N vertices: each pair of vertices is connected by an edge independently with probability $c/N > 0$. Denote by $Z_N(x)$ the partition function of a monomer-dimer model with monomer activity $x > 0$ and pure hard-core interaction on the graph G_N:

$$Z_N(x) \;=\; \sum_{D \in \mathscr{D}_{G_N}} x^{N-2|D|}\,,$$

$\langle\,\cdot\,\rangle_{G_N,x}$ will be the corresponding Gibbs expected value. The pressure density is

$$p_N(x) \;:=\; \frac{1}{N}\, \log Z_N(x)\,,$$

and the monomer density is

$$m_N(x) := \left\langle \frac{N - 2|D|}{N} \right\rangle_{N,x} = x \frac{\partial p_N}{\partial x}(x) .$$

Since the set of configurations $\mathscr{D}_{G_N}$ is random, the partition function, the pressure density and the monomer density are random variables and the Gibbs measure is a random measure. This randomness is treated as quenched with respect to the thermal fluctuations.

Theorem 3 (see [2,42]). *Almost surely and for all $x > 0$ the monomer density and the pressure density converge in the thermodynamical limit. Precisely:*

$$m_N(x) \xrightarrow[N\to\infty]{a.s.} \mathbb{E}[M(x)] \tag{19}$$

$$p_N(x) \xrightarrow[N\to\infty]{a.s.} \mathbb{E}\left[\log \frac{M(x)}{x}\right] - \frac{c}{2}\,\mathbb{E}\left[\log\left(1 + \frac{M_1(x)}{x}\,\frac{M_2(x)}{x}\right)\right] . \tag{20}$$

The law of the random variable $M(x)$ is the only solution supported in $[0,1]$ of the following fixed point distributional equation:

$$M \stackrel{d}{=} \frac{x^2}{x^2 + \sum_{i=1}^{\Delta} M_i} \tag{21}$$

where $(M_i)_{i\in\mathbb{N}}$ are i.i.d. copies of M and Δ is an independent Poisson(c)-distributed random variable. The limit monomer density and the limit pressure density are analytic functions of the activity $x > 0$ (Fig. 3).

The expression for the pressure on the right hand side of (20) was provided by Zdeborová and Mézard [46] via the theoretical physics method of cavity fields. This theorem provides a complete rigorous proof of their conjecture, partially studied in [13,42] (Fig. 2).

The proof of Theorem 3 relies on the locally tree-like structure of the Erdős–Rényi random graphs. Precisely fix a radius $r \in \mathbb{N}$ and for any vertex v denote by $[G_N, v]_r$ the ball of center v and radius r in the graph G_N; then consider a random tree T rooted at the vertex o and with independent Poisson(c)-distributed offspring sizes; it holds (see [19]):

$$\frac{1}{N} \sum_{v\in G_N} F\big([G_N, v]_r\big) \xrightarrow[n\to\infty]{a.s.} \mathbb{E}F\big([T, o]_r\big)$$

for every bounded real function F invariant under rooted graph isomorphisms.

Clearly the monomer density rewrites as an average over the vertices:

$$m_N(x) = \frac{1}{N} \sum_{v\in G_N} \mathcal{M}_x(G_N, v) , \quad \text{where}$$
$$\mathcal{M}_x(G_N, v) := \langle \mathbb{1}(v \text{ is a monomer}) \rangle_{G_N,x} .$$

A priori $\mathcal{M}_x(G_N, v)$ depends on the whole graph G_N, but it can be substituted by local quantities thanks to the following correlation inequalities:

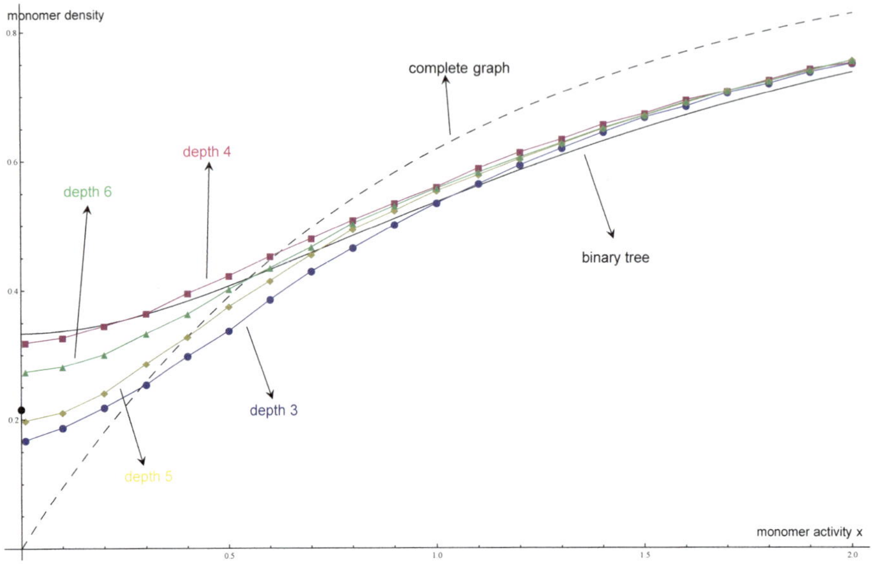

Fig. 2. Upper (even depths) and lower (odd depths) bounds for the limit monomer density $m(x) = \lim_{N\to\infty} m_N(x)$ versus the monomer activity x, in the Erdős–Rényi case with $c = 2$. The binary tree (continuous line) and the complete graph (dashed line) cases are also shown. The distributional recursion (21) is iterated a finite number r of times with initial values $M_i \equiv 1$: the obtained random variable $M_x(r)$ represents the root monomer probability of the random tree $[T, o]_r$. For values of $x = 0.01, 0.1, 0.2, \ldots, 2$, the random variables $M_x(r)$, $r - 3, 4, 5, 6$ are simulated numerically 10000 times and an empirical mean is taken in order to approximate $\mathbb{E}[M_x(r)]$. $\mathbb{E}[M_x(r)]$ provides an upper/low approximation of $m(x)$ when r is even/odd.

Lemma 1 **(Correlation inequalities)** *Let* (G, o) *be a rooted graph, let* $r \in \mathbb{N}$*. If* $[G, o]_{2r+1}$ *is a tree, then*

$$\mathcal{M}_x([G, o]_{2r+1}) \leq \mathcal{M}_x(G, o) \leq \mathcal{M}_x([G, o]_{2r}) \, .$$

Therefore one can deduce that

$$m_N(x) \xrightarrow[N\to\infty]{a.s.} \lim_{r\to\infty} \mathbb{E}\mathcal{M}_x\big([T, o]_r\big) \tag{22}$$

provided the existence of the $\lim_{r\to\infty}$. In this way the problem on random graphs is reduce to the study of the root monomer probability on a random tree. As usual in Statistical Mechanics working on trees is much easier since there are no loops in the interactions.

The problem is now approached by means of the Heilmann–Lieb recursion. By Lemma 1, the sequences of monomer probabilities respectively at even and odd depths of the tree are monotonic:

$$\mathcal{M}_x\big([T, o]_{2r}\big) \nearrow M_{\text{even}}(x) \, , \quad \mathcal{M}_x\big([T, o]_{2r+1}\big) \searrow M_{\text{odd}}(x) \quad \text{as } r \to \infty \, .$$

The relation (11) for partition functions gives the following relation for root monomer probabilities:

$$\begin{pmatrix} M_{\text{even}}(x) \\ M_{\text{odd}}(x) \end{pmatrix} \overset{d}{=} \begin{pmatrix} \frac{x^2}{x^2+\sum_{i=1}^{\Delta} M_{\text{odd}}^{(i)}(x)} \\ \frac{x^2}{x^2+\sum_{i=1}^{\Delta} M_{\text{even}}^{(i)}(x)} \end{pmatrix} \tag{23}$$

where $(M_{\text{even}}^{(i)}, M_{\text{odd}}^{(i)})$, $i \in \mathbb{N}$, are i.i.d. copies of $(M_{\text{even}}, M_{\text{odd}})$. A direct computation from Eq. (23) shows that

$$\mathbb{E}[|M_{\text{even}}(x) - M_{\text{odd}}(x)|] \leq \frac{c^2}{x^4}\, \mathbb{E}[|M_{\text{even}}(x) - M_{\text{odd}}(x)|]$$

therefore $M_{\text{even}}(x) = M_{\text{odd}}(x)$ almost surely for every $x > \sqrt{c}$. Now allow the monomer activity to take complex values in $\mathbb{H}_+ = \{z \in \mathbb{C} \,|\, \Re(z) > 0\}$. This has no physical or probabilistic meaning, but it is a technique to obtain powerful results by exploiting complex analysis. Using the Heilmann–Lieb recursion one can prove that for any rooted graph (G, o), the function $\mathcal{M}_z(G, o)$ is analytic in $z \in \mathbb{H}_+$ and is uniformly bounded as $|\mathcal{M}_z(G, o)| \leq |z|/\Re(z)$. It follows that the limit functions $M_{\text{even}}(z)$ and $M_{\text{odd}}(z)$ are analytic on $\mathbb{H}_+$. Therefore by uniqueness of the analytic continuation

$$M_{\text{even}}(x) = M_{\text{odd}}(x) =: M(x) \quad \text{almost surely for every } x > 0$$

and (19) follows by (22). $M(x)$ satisfies the distributional fixed point Eq. (21). The solution supported in $[0, 1]$ is unique, since for any random variable $M' \in [0, 1]$ that satisfies (21) it can be shown by induction on $r \in \mathbb{N}$ that

$$\mathcal{M}_x([T, o]_{2r+1}) \overset{d}{\leq} M' \overset{d}{\leq} \mathcal{M}_x([T, o]_{2r}) \,.$$

These are the ideas to prove the convergence of the monomer density. To complete the Theorem 3 it remains to prove the convergence of the pressure density. The convergence of $p_N(x)$ to some function $p(x)$ is guaranteed by the convergence of its derivative $m_N(x)/x$ together with the bounds 6. Call $\tilde{p}(x)$ the function defined by the right hand side of (20), which can be "guessed" by the heuristic method of energy shifts. Direct computations show that $x\,\tilde{p}'(x) = m(x) = x\,p'(x)$ for every $x > 0$ and $\lim_{x\to\infty} \tilde{p}(x) - \log x = 0 = \lim_{x\to\infty} p(x) - \log x$. Therefore $p = \tilde{p}$.

3.3 Random Field

For the class of models described above the randomness is in the graph structure. The model below instead introduces a randomness in the monomer activities and is useful to describe impurities. Consider the pure hard-core monomer-dimer model defined in Definition 2 and assume that $G = (V, E)$ is the complete graph with N vertices, the monomer activities $(x_i)_{i\in V}$ are *i.i.d.* positive random variables and the dimer activity is uniform $w_{ij} = w/N > 0$ $\forall i \neq j \in V$. The partition function is

$$Z_N = \sum_{D \in \mathscr{D}_N} \left(\frac{w}{N}\right)^{|D|} \prod_{i \in M(D)} x_i \,. \tag{24}$$

Notice that now the partition function and the pressure density $\frac{1}{N} \log Z_N$ are random variables. The first important fact is that under the assumptions of Theorem 2 the pressure density is self-averaging, namely it converges almost surely to its expectation usually called *quenched pressure density*. The Gaussian representation for the partition function (7) and a careful application of the Laplace method allows us to find its limiting value. More precisely the next theorem shows that thermodynamic limit the quenched pressure density exists and is given by a one-dimensional variational principle, which admits a unique solution.

Theorem 4 (see [5]). *Let $w > 0$. Let $x_i > 0$, $i \in \mathbb{N}$ be i.i.d. random variables with $\mathbb{E}_x[x] < \infty$ and $\mathbb{E}_x[(\log x)^2] < \infty$. Then:*

$$\exists \lim_{N \to \infty} \frac{1}{N} \mathbb{E}_{\boldsymbol{x}}[\log Z_N] = \sup_{\xi \geq 0} \Phi(\xi) \in \mathbb{R}$$

where

$$\Phi(\xi) := -\frac{\xi^2}{2w} + \mathbb{E}_x[\log(\xi + x)] \quad \forall \xi \geq 0 \,,$$

the function Φ reaches its maximum at a unique point ξ^* which is the only solution in $[0, \infty[$ of the fixed point equation

$$\xi = \mathbb{E}_x\left[\frac{w}{\xi + x}\right] . \tag{25}$$

Theorem 4 allows to compute the main macroscopic quantity of physical interest, that is the *dimer density*, in terms of the positive solution ξ^* of the fixed point Eq. (25).

Corollary 2. *In the hypothesis of Theorem 4 the limiting pressure per particle*

$$p(w) := \lim_{N \to \infty} \frac{1}{N} \mathbb{E}_{\boldsymbol{x}}\big[\log Z_N(w)\big]$$

exists and is a smooth function of $w > 0$. Moreover the limiting dimer density

$$d := \lim_{N \to \infty} \frac{1}{N} \mathbb{E}_{\boldsymbol{x}}[\langle\, |D|\, \rangle_N] = w \frac{\mathrm{d}p}{\mathrm{d}w} = \frac{(\xi^*)^2}{2w} \,.$$

A detailed proof of Theorem 4 can be found in [5]. Here we mention the main ideas. The first step is to use the Gaussian representation (7) for the partition function (24) that gives

$$Z_N = \mathbb{E}_\xi\left[\prod_{i=1}^N (\xi + x_i)\right] , \tag{26}$$

where ξ is a one-dimensional Gaussian random variable with mean 0 and variance w/N. Indeed by Proposition 1, $Z_N = \mathbb{E}_g\big[\prod_{i=1}^{N}(g_i+x_i)\big]$ where $g=(g_1,\dots,g_N)$ is an N-dimensional Gaussian random vector with mean 0 and constant covariance matrix $(w/N)_{i,j=1,\dots,N}$. It is easy to check that the vector g has the same joint distribution of the vector $(\xi,\dots,\xi)$ and the identity (26) follows. It is important to notice how easily, in this mean-field framework, the Gaussian representation reduces the degrees of freedom of the system. By explicitly rewriting (26) as

$$Z_N = \frac{\sqrt{N}}{\sqrt{2\pi w}} \int_{\mathbb{R}} \mathrm{e}^{-\frac{N}{2w}\,\xi^2} \prod_{i=1}^{N}(\xi+x_i)\,\mathrm{d}\xi\,. \tag{27}$$

and considering the function

$$f_N(\xi) := \mathrm{e}^{-\frac{N}{2w}\,\xi^2} \prod_{i=1}^{N}(\xi+x_i) \quad \forall\,\xi\in\mathbb{R}$$

one sees that Theorem 4 follows by approximating e^{Φ} in the integral (27) with the Laplace method.

4 The Mean-Field Case

Let $h\in\mathbb{R}$ and $J\geq 0$ and consider the imitative monomer-dimer model in Definition 3 within the following assumptions: $G=(V,E)$ is the complete graph with N vertices and $\forall i\neq j\in V$ we set $w_{ij}=1/N$, $x_i\equiv \mathrm{e}^h$ and $J_{ij}=J/N$. Since the number of edges is of order N^2, in order to keep the logarithm of the partition function of order N, a normalisation of the dimer activity as $1/N$ (see Remark 4) and the imitation coefficient as J/N are needed.

One can express the Hamiltonian in terms of occupancy variables as

$$H_N(\alpha) := -h\sum_{i=1}^{N}\alpha_i - \frac{J}{N}\sum_{1\leq i<j\leq N}\big(\alpha_i\,\alpha_j + (1-\alpha_i)\,(1-\alpha_j)\big) \tag{28}$$

for every monomer-dimer configuration on the complete graph $\alpha\in\mathscr{D}_N$. The partition function is

$$Z_N := \sum_{\alpha\in\mathscr{D}_N} N^{-D_N}\,\exp(-H_N)\,,$$

where $D_N := \sum_{1\leq i<j\leq N}\alpha_{ij}$ represents the total number of dimer for a given configuration $\alpha\in\mathscr{D}_N$. Observe that the only relevant quantity in this setting is actually the total number of monomers in a given monomer-dimer configuration

$$M_N := \sum_{i=1}^{N}\alpha_i$$

indeed the hardcore constraint (2) implies that $M_N + 2D_N = N$ and the Hamiltonian (28) is actually a function of M_N only. We denote the corresponding Gibbs measure as

$$\mu_N(\alpha) := \frac{N^{-D_N(\alpha)} \exp(-H_N(\alpha))}{Z_N} \quad \forall\, \alpha \in \mathscr{D}_N \tag{29}$$

and the expectation with respect to the measure μ_N is denoted by $\langle \cdot \rangle_N$. In particular, setting $m_N := \frac{1}{N}\sum_{i=1}^N \alpha_i$, the average monomer density is

$$\langle m_N \rangle_N = \sum_{\alpha \in \mathscr{D}_N} \frac{\sum_{i=1}^N \alpha_i}{N} \frac{\exp(-H_N(\alpha))}{Z_N} = \frac{\partial}{\partial h} \frac{\log Z_N}{N} \,.$$

This model has been initially studied in [6,7], where the behaviour of the pressure and monomer densities in the thermodynamic limit is analysed.

Theorem 5 (see [6]). *Let $h \in \mathbb{R}$, $J \geq 0$. Then there exists*

$$p := \lim_{N\to\infty} \frac{\log Z_N}{N} = \sup_m \psi(m) \tag{30}$$

the sup *can be taken indifferently over $m \in [0,1]$ or $m \in \mathbb{R}$ and*

$$\psi(m,h,J) := -Jm^2 + \frac{J}{2} + p^{(0)}(2Jm + h - J) \tag{31}$$

where for all $t \in \mathbb{R}$

$$p^{(0)}(t) := -\frac{1}{2}\big(1 - g(t)\big) - \frac{1}{2}\log\big(1 - g(t)\big) \,,$$

$$g(t) := \frac{1}{2}\big(\sqrt{\mathrm{e}^{4t} + 4\,\mathrm{e}^{2t}} - \mathrm{e}^{2t}\big) \,. \tag{32}$$

Furthermore the function $\psi(m)$ attains its maximum in (at least) one point $m^* = m^*(h,J) \in (0,1)$, which is a solution of the consistency equation

$$m = g\big((2m-1)\,J + h\big) \,.$$

At each value of the parameters (h,J) such that $h \mapsto m^*(h,J)$ is differentiable, the monomer density admits thermodynamic limit

$$\lim_{N\to\infty} \langle m_N \rangle_N = m^* \,.$$

In order to prove Theorem 5, first we need to deal with the case $J = 0$, then the limit (30) with $J > 0$ follows by a convexity argument introduced by Guerra [29] for the Curie–Weiss model. At $J = 0$ the model reduces to the pure monomer-dimer model of Definition 2 on the complete graph with $x_i = x = \mathrm{e}^h > 0$, $w_{ij} = 1/N > 0$ $\forall i \neq j \in V$. Let us denote by $Z_N^{(0)}$ and $\langle m_N \rangle_N^{(0)}$

respectively the partition function and the average monomer density at $J = 0$; it holds

$$\lim_{N\to\infty} \frac{1}{N} \log Z_N^{(0)} = p^{(0)}(h) \tag{33}$$
$$\lim_{N\to\infty} \langle m_N \rangle_N^{(0)} = g(h) \,.$$

The function $p^{(0)}$ is analytic thus at $J = 0$ there are no phase transition in agreement with the general result of Heilman–Lieb [30]. The limit (33) can be obtained in two different ways:

(1) by a combinatorial computation, since on the complete graph it is possible to compute explicitly the number of monomer-dimer configurations with a given number of monomers;
(2) by using the Gaussian representation (7) of the partition function and the Laplace method.

The latter method furnish a better estimation of the convergence (33)

$$Z_N^{(0)}(h) \underset{N\to\infty}{\sim} \frac{\exp\left(N p^{(0)}(h)\right)}{\sqrt{2-g(h)}} \,, \tag{34}$$

which will be fundamental in the study of the fluctuations of M_N (Sect. 5).

Remark 6. The limiting pressure density p can also be expressed as a different variational problem, equivalent to that of Theorem 5:

$$p = \sup_m \left(s(m) - \varepsilon(m)\right) \tag{35}$$

with

$$\begin{aligned} s(m) &:= -m \log m - \frac{1-m}{2} \log(1-m) + \frac{1+m}{2}, \\ \varepsilon(m) &:= -J\, m^2 - (h-J)\, m - \frac{J}{2} \,. \end{aligned} \tag{36}$$

The variational problem (35) can be obtained directly by the combinatorial computation mentioned before. The functions s and ε in (36) are the *entropy and energy densities* respectively.

The properties of the solution(s) of the one-dimensional variational problem (30) appearing in Theorem 5 determine the thermodynamic properties of the model. In particular we are interested in the value(s) of $m = m^*(h, J)$ where the maximum is reached, since it can be interpreted as the limiting value of the monomer density.

The function m^* (see [6]) is single-valued and smooth on the plane (h, J) with the exception of an implicitly defined open curve $h = \gamma(J)$ and its end-point (h_c, J_c). Instead on γ there are two global maximum points $m_1 < m_2$ that correspond to the *dimer* phase and the *monomer* phase respectively. Crossing the curve γ in the phase plane the order parameter m^* presents a jump discontinuity: in other words a *second order phase transition* occurs and γ is the coexistence curve. The point (h_c, J_c) is the *critical point* of the system, where m^* is continuous but not differentiable.

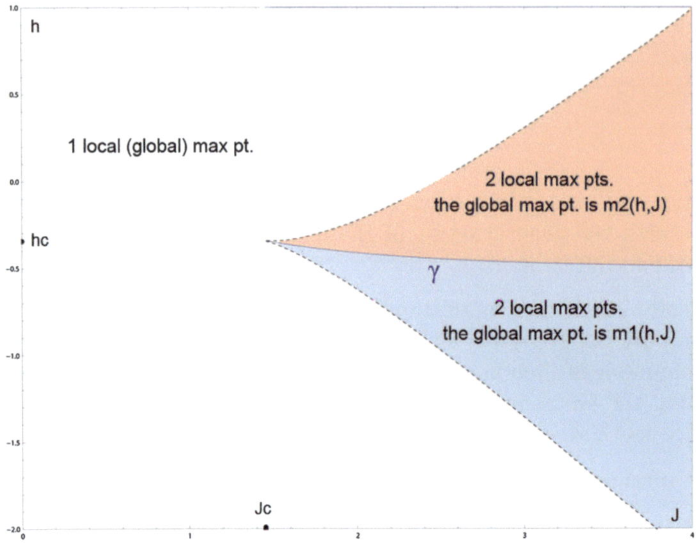

Fig. 3. Phase space (h, J). The curve γ separates the values (h, J) for which the global maximum point $m^*(h, J)$ of $m \mapsto \widetilde{p}\,(m, h, J)$ jumps between two values $m_1 < m_2$. This entails a discontinuity of $m^*(h, J)$ along the coexistence curve γ.

Remark 7. We notice that the techniques developed in [6] do not allow us to conclude the existence of the limiting monomer density on the coexistence curve γ. In the standard mean-field ferromagnetic model (Curie–Weiss) the existence of the magnetization on the coexistence curve ($h = 0$) follows directly by the global spin flip symmetry, a property that we do not have in the present case.

The non analytic behaviour of $m^*(h, J)$ near the critical point is described by its *critical exponents.*

Theorem 6 (see [6]). *Consider the global maximum point $m^*(h, J)$ of the function $m \mapsto \widetilde{p}\,(m, h, J)$ defined by (35). Set $m_c := m^*(h_c, J_c)$. The critical exponents of m^* at the critical point (h_c, J_c) are:*

$$\boldsymbol{\beta} = \lim_{J \to J_c +} \frac{\log |m^*(\delta(J), J) - m_c|}{\log(J - J_c)} = \frac{1}{2}$$

along any curve $h = \delta(J)$ such that $\delta \in C^2([J_c, \infty[)$, $\delta(J_c) = h_c$, $\delta'(J_c) = \gamma'(J_c)$ (i.e. if the curve δ has the same tangent of the coexistence curve γ at the critical point);

$$\frac{1}{\boldsymbol{\delta}} = \lim_{J \to J_c} \frac{\log |m^*(\delta(J), J) - m_c|}{\log |J - J_c|} = \frac{1}{3}$$

$$\frac{1}{\boldsymbol{\delta}} = \lim_{h \to h_c} \frac{\log |m^*(h, \delta(h)) - m_c|}{\log |h - h_c|} = \frac{1}{3}$$

along any curve $h = \delta(J)$ such that $\delta \in C^2(\mathbb{R}_+)$, $\delta(J_c) = h_c$, $\delta'(J_c) \neq \gamma'(J_c)$ or along a curve $J = \delta(h)$ such that $\delta \in C^2(\mathbb{R})$, $\delta(h_c) = J_c$, $\delta'(h_c) = 0$ (i.e. if the curve is not tangent to γ at the critical point).

Theorem 6 proves that the model belongs to the same *universality class* of the mean-field ferromagnet.

5 Distributional Limit Theorems at the Critical Point

In this section we study the distributional limit of the random variable *number of monomers* with respect to the Gibbs measure on the complete graph [3,4]. We show that a *law of large numbers* holds outside the coexistence curve γ, whereas on γ the limiting distribution is a convex combination of two Dirac deltas representing the two phases (Theorems 7 and 8). Moreover we show that a *central limit theorem* holds outside $\gamma \cup (h_c, J_c)$, while at the critical point a normalisation of order $N^{-3/4}$ for the fluctuations is required and the limiting distribution is $C\mathrm{e}^{-cx^4}\mathrm{d}x$ (Theorems 7 and 9).

In [3] we follow the Gaussian convolution method introduced by Ellis and Newman for the mean-field Ising model (Curie–Weiss) in [22–24] in order to deal with the imitative potential. An additional difficulty stems from the fact that even in the absence of imitation the system keeps an interacting nature due to the presence of the hard-core interaction: we use the Gaussian representation 1 to *decouple* the hard-core interaction.

We focus on the behaviour of the random variable $M_N = \sum_{i=1}^N \alpha_i$ (number of monomers) with respect to the Gibbs measure (29) with a suitable scaling when $N \to \infty$. From now on δ_x is the Dirac measure centred at x, $\mathcal{N}\left(x, \sigma^2\right)$ denotes the Gaussian distribution with mean x and variance σ^2 and $\xrightarrow{\mathcal{D}}$ denotes the convergence in distribution with respect to the Gibbs measure μ_N as $N \to \infty$.

At $J = 0$ the law of large numbers and the central limit theorem hold true:

Theorem 7 (see [3]). *At $J = 0$ the following results hold:*

$$\frac{M_N}{N} \xrightarrow{\mathcal{D}} \delta_{g(h)}$$

and

$$\frac{M_N - N\,g(h)}{\sqrt{N}} \xrightarrow{\mathcal{D}} \mathcal{N}\left(0, \frac{\partial g}{\partial h}(h)\right) \tag{37}$$

where g is the function defined by (32).

Notice that, even if $J = 0$, (37) is not a consequence of the standard central limit theorem, indeed M_N is not a sum of i.i.d. random variables because of the presence of the hard-core interaction. Theorem 7 follows from the recent results of Lebowitz–Pittel–Ruelle–Speer [37]. A different proof is presented here which includes also the general value of $J > 0$. We should mention that a slightly improvement of the result presented has been obtained with different methods in [17].

like voip conference calls and messaging [8]. At each single time every user cannot be in more than a call, i.e. the occupation number fullfills a hard-core constraint. While the old style phone calls were well described by a monomer-dimer system, novel technological devices need a wider space of higher dimensional polymers that allow for the presence of multiple individuals in the same virtual room: the monomer corresponds to a silent user, the dimer is a two-body conversation, the trimer a three-body and so on. The models to be investigated in this case are therefore polymer models with hard-core interaction on hypergraphs with no physical dimension, i.e. better described as some form of dilution of the complete hypergraph. The mean-field case and its diluted versions are therefore at the heart of the problem and not only mere approximations.

References

1. Alberici, D.: A cluster expansion approach to the Heilmann-Lieb liquid crystal model. J. Stat. Phys. **162**(3), 761–791 (2016)
2. Alberici, D., Contucci, P.: Solution of the monomer-dimer model on locally tree-like graphs. Rigorous results. Commun. Math. Phys. **331**, 975–1003 (2014)
3. Alberici, D., Contucci, P., Fedele, M., Mingione, E.: Limit theorems for monomer-dimer mean-field models with attractive potential. Commun. Math. Phys. **346**, 781–799 (2016)
4. Alberici, D., Contucci, P., Fedele, M., Mingione, E.: Limit theorems for monomer-dimer mean-field models with attractive potential. Europhys. Lett. **114**, 10006 (2016)
5. Alberici, D., Contucci, P., Mingione, E.: A mean-field monomer-dimer model with randomness. Exact solution and rigorous results. J. Stat. Phys. **160**, 1721–1732 (2015)
6. Alberici, D., Contucci, P., Mingione, E.: A mean-field monomer-dimer model with attractive interaction. Exact solution and rigorous results. J. Math. Phys. **55**, 1–27 (2014)
7. Alberici, D., Contucci, P., Mingione, E.: The exact solution of a mean-field monomer-dimer model with attractive potential. Europhys. Lett. **106**, 1–5 (2014)
8. Alberici, D., Contucci, P., Mingione, E., Molari, M.: Aggregation models on hypergraphs. Ann. Phys. **376**, 412–424 (2017)
9. Aldous, D., Steele, J.M.: The objective method: probabilistic combinatorial optimization and local weak convergence. Encyclopaedia Math. Sci. **110**, 1–72 (2004)
10. Barra, A., Contucci, P., Sandell, R., Vernia, C.: An analysis of a large dataset on immigrant integration in Spain. The statistical mechanics perspective on social action. Sci. Rep. **4**, 4174 (2014)
11. Bayati, M., Nair, C.: A rigorous proof of the cavity method for counting matchings. In: Proceedings of the 44th Annual Allerton Conference on Communication, Control and Computing (2006)
12. van den Berg, J.: On the absence of phase transition in the monomer-dimer model. In: Perplexing Problems in Probability, pp. 185–195 (1999)
13. Bordenave, C., Lelarge, M., Salez, J.: Matchings on infinite graphs. Probab. Theory Relat. Fields **157**, 183–208 (2013)
14. Burioni, R., Contucci, P., Fedele, M., Vernia, C., Vezzani, A.: Enhancing participation to health screening campaigns by group interactions. Sci. Rep. **5**, 9904 (2015)

15. Chang, T.S.: Statistical theory of the adsorption of double molecules. Proc. Roy. Soc. London A **169**, 512–531 (1939)
16. Chang, T.S.: The number of configurations in an assembly and cooperative phenomena. Proc. Camb. Philos. Soc. **38**, 256–292 (1939)
17. Chen, W.-K.: Limit theorems in the imitative monomer-dimer mean-field model via Stein's method. J. Math. Phys. **57**, 083302 (2016)
18. Dembo, A., Montanari, A.: Ising models on locally tree-like graphs. Ann. Appl. Probab. **20**, 565–592 (2010)
19. Dembo, A., Montanari, A.: Gibbs measures and phase transitions on sparse random graphs. Brazil. J. Prob. Stat. **24**, 137–211 (2010)
20. Dembo, A., Montanari, A., Sun, N.: Factor models on locally tree-like graphs. Ann. Prob. **41**(6), 4162–4213 (2013)
21. Disertori, M., Giuliani, A.: The nematic phase of a system of long hard rods. Commun. Math. Phys. **323**(1), 143–175 (2013)
22. Ellis, R.S., Newman, C.M.: The statistics of curie-weiss models. J. Stat. Phys. **19**, 149–161 (1978)
23. Ellis, R.S., Newman, C.M.: Limit theorems for sums of dependent random variables occurring in statistical mechanics. Probab. Theory Relat. Fields **44**, 117–139 (1978)
24. Ellis, R.S., Newman, C.M., Rosen, J.S.: Limit theorems for sums of dependent random variables occurring in statistical mechanics. Probab. Theory Relat. Fields **51**, 153–169 (1980)
25. Fisher, M.E.: Statistical mechanics of dimers on a plane lattice. Phys. Rev. **124**(6), 1664–1672 (1961)
26. Fowler, R.H., Rushbrooke, G.S.: An attempt to extend the statistical theory of perfect solutions. Trans. Faraday Soc. **33**, 1272–1294 (1937)
27. Giuliani, A., Jauslin, I., Lieb, E.H.: A Pfaffian formula for monomer-dimer partition functions. J. Stat. Phys. **163**(2), 211–238 (2016)
28. Giuliani, A., Lieb, E.H.: Columnar phase in quantum dimer models. J. Phys. A: Math. Theor. **48**(23), 235203 (2015)
29. Guerra, F.: Mathematical aspects of mean field spin glass theory. In: 4th European Congress of Mathematics, Stockholm June 27-July 2 2004, ed. Laptev, A., European Mathematical Society (2005)
30. Heilmann, O.J., Lieb, E.H.: Theory of monomer-dimer systems. Commun. Math. Phys. **25**, 190–232 (1972)
31. Heilmann, O.J., Lieb, E.H.: Monomers and dimers. Phys. Rev. Lett. **24**, 1412–1414 (1970)
32. Heilmann, O.J., Lieb, E.H.: Lattice models for liquid crystals. J. Stat. Phys. **20**, 680–693 (1979)
33. Jauslin, I., Lieb, E.H.: Nematic liquid crystal phase in a system of interacting dimers and monomers. Commun. Math. Phys. **363**, 955–1002 (2018)
34. Kac, M., Thompson, C.J.: Critical behavior of several lattice models with long-range interaction. J. Math. Phys. **10**, 1373 (1969)
35. Karp, R., Sipser, M.: Maximum matchings in sparse random graphs. In: Proceedings of the 22nd Annual Symposium on Foundations of Computer Science, pp. 364–375. IEEE Computer Society Press (1981)
36. Kasteleyn, P.W.: The statistics of dimers on a lattice. I. The number of dimer arrangements on a quadratic lattice. Physica **27**(12), 1209–1225 (1961)
37. Lebowitz, J.L., Pittel, B., Ruelle, D., Speer, E.R.: Central limit theorems, Lee–Yang zeros, and graph-counting polynomials. Preprint arxiv:1408.4153 (2014)
38. Lieb, E.H.: The solution of the dimer problems by the transfer matrix method. J. Math. Phys. **8**, 2339–2341 (1967)

39. Onsager, L.: The effects of shape on the interaction of colloidal particles. Ann. N. Y. Acad. Sci. **51**, 627–659 (1949)
40. Peierls, R.: On Ising's model of ferromagnetism. Math. Proc. Cambridge Philos. Soc. **32**(3), 477–481 (1936)
41. Roberts, J.K.: Some properties of mobile and immobile adsorbed films. In: Proceedings of the Cambridge Philosophical Society, vol. 34, pp. 399–411 (1938)
42. Salez, J.: Weighted enumeration of spanning subgraphs in locally tree-like graphs. Random Struct. Algorithms **43**, 377–397 (2013)
43. Temperley, H.N.V., Fisher, M.E.: Dimer problem in statistical mechanics - an exact result. Phil. Mag. **6**(68), 1061–1063 (1961)
44. Thompson, C.J.: Ising model in the high density limit. Commun. Math. Phys. **36**, 255–262 (1974)
45. Vladimirov, I.G.: The monomer-dimer model and Lyapunov exponents of homogeneous Gaussian random fields. Discrete Continuous Dyn. Syst. B **18**, 575–600 (2013)
46. Zdeborová, L., Mézard, M.: The number of matchings in random graphs. J. Stat. Mech. **5**, P05003 (2006)

Is the Riemann Zeta Function in a Short Interval a 1-RSB Spin Glass?

Louis-Pierre Arguin[1(✉)] and Warren Tai[2]

[1] Department of Mathematics, Baruch College and Graduate Center, City University of New York, New York, NY 10010, USA
louis-pierre.arguin@baruch.cuny.edu

[2] Department of Mathematics, Graduate Center, City University of New York, New York, NY 10010, USA
wtai@gradcenter.cuny.edu

Abstract. Fyodorov, Hiary & Keating established an intriguing connection between the maxima of log-correlated processes and the ones of the Riemann zeta function on a short interval of the critical line. In particular, they suggest that the analogue of the free energy of the Riemann zeta function is identical to the one of the Random Energy Model in spin glasses. In this paper, the connection between spin glasses and the Riemann zeta function is explored further. We study a random model of the Riemann zeta function and show that its two-overlap distribution corresponds to the one of a one-step replica symmetry breaking (1-RSB) spin glass. This provides evidence that the local maxima of the zeta function are strongly clustered.

Keywords: Riemann zeta function · Disordered systems · Spin glasses

1 Introduction and Main Result

1.1 Background

The Riemann zeta function is defined on $\mathbb{C}$ by

$$\zeta(s) = \sum_{n\geq 1} \frac{1}{n^s} = \prod_{p \text{ primes}} \left(1-p^{-s}\right)^{-1} \text{ if } \mathrm{Re}\ s > 1, \tag{1}$$

and can be analytically continued to the whole complex plane by the functional equation

$$\zeta(s) = \chi(s)\zeta(1-s), \qquad \chi(s) = 2^s\pi^{s-1}\sin\left(\frac{\pi}{2}s\right)\Gamma(1-s).$$

L.-P. Arguin—Supported by NSF CAREER 1653602, NSF grant DMS-1513441, and a Eugene M. Lang Junior Faculty Research Fellowship.
W. Tai—Partially supported by NSF grant DMS-1513441.

V. Sidoravicius (Ed.): *Sojourns in Probability Theory and Statistical Physics - I*, PROMS 298, pp. 63–88, 2019.
https://doi.org/10.1007/978-981-15-0294-1_3

Trivial zeros are located at negative even integers where $\chi(s) = 0$. The non-trivial zeros are restricted to the critical strip $0 \leq \text{Re } s \leq 1$. The Riemann hypothesis states that they all lie on the critical line Re $s = 1/2$. A weaker statement, yet with deep implications on the distribution of the primes, is the Lindelöf hypothesis which stipulates that the maximum of ζ on a large interval $[0, T]$ of the critical line grows slower than any power of T, i.e. $\zeta(1/2 + \mathrm{i}T)$ is $\mathrm{O}(T^{\varepsilon})$ for any $\varepsilon > 0$, see e.g. [41].

Mathematical physics has provided several important insights in the study of the Riemann zeta function over the years. We refer the reader to [39] for a broad discussion on this topic. We briefly highlight three contributions from statistical mechanics and probability. First, there are deep connections between the statistics of eigenvalues of random matrices and the zeros of zeta as exemplified by the Montgomery's pair correlation conjecture, see for example [13]. Second, the Riemann hypothesis can be recast in the framework of Ising models of statistical mechanics where it bears a resemblance to the Lee–Yang theorem. This perspective was investigated in details by Newman [28–30]. It led to an equivalent reformulation of the Riemann hypothesis in terms of the exact value of the de Bruijn–Newman constant [31], see [38] for a numerical estimate of the constant and [35] for a proof that the constant is non-negative. Third, Fyodorov et al. [21] and Fyodorov and Keating [22] recently unraveled a striking connection between the local statistics of the large values of the Riemann zeta function on the critical line and the extremes of a class of disordered systems, the *log-correlated processes*, that includes among others branching Brownian motion and the two-dimensional Gaussian free field. This connection has also been extended recently to the theory of Gaussian multiplicative chaos by Saksman and Webb [36,37].

The Fyodorov–Hiary–Keating conjecture is as follows [21,22]: if τ is sampled uniformly on a large interval $[T, 2T]$, then the maximum on a short interval, say $[0, 1]$, around τ is

$$\max_{h \in [0,1]} \log |\zeta(1/2 + \mathrm{i}(\tau + h)| = \log\log T - \frac{3}{4} \log\log\log T + \mathcal{M}_T, \tag{2}$$

where $(\mathcal{M}_T)$ is a sequence of random variables converging in distribution. The deterministic order of the maximum corresponds exactly to the one of a log-correlated process, such as a branching random walk and the two-dimensional Gaussian free field, see for example [1,25] for more background on this class of processes. The precise value of the leading order can be predicted heuristically since the process for $\log \zeta$ has effectively $\log T$ distinct values on $[0, 1]$ (because there are on average $\log T$ zeros on $[0, 1]$, see for example [41]), and the marginal distribution of $\log |\zeta(1/2 + i(\tau + h))|$ should be close to Gaussian with variance $\frac{1}{2} \log\log T$ as predicted by Selberg's Central Limit Theorem [34]. The log-correlations already appear at the level of the typical values from the multivariate CLT proved in [12]. The first order of the Conjecture (2) was proved recently in parallel: conditionally on the Riemann hypothesis in [27], and unconditionally in [3]. The evidence in favor of the conjecture laid out by Fyodorov and Keating [22] suggests that the large values of the Riemann zeta function

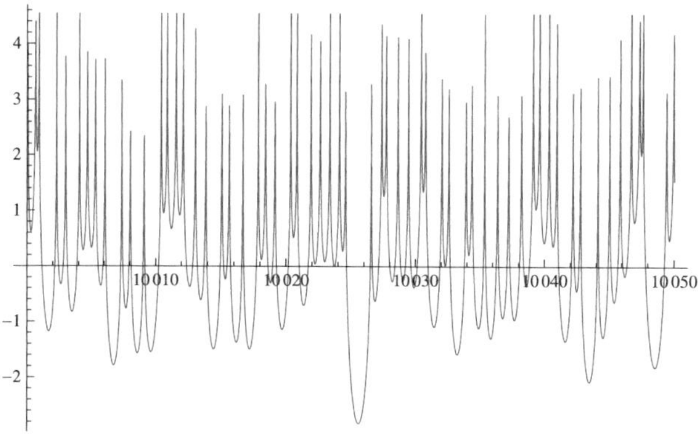

Fig. 1. The value of $-\log|\zeta(1/2 + \mathrm{i}(T+h))|$ for $T = 10000$ and $h \in [0, 50]$.

locally behaves like a disordered system of the spin-glass type characterized by an energy landscape with multiple minima, see Fig. 1. In particular, by considering $-\log|\zeta(1/2 + \mathrm{i}(\tau+h))|$ as the energy of a disordered system on the state space $[0, 1]$, they predict that the analogue of the free energy is in the limit

$$\lim_{T\to\infty} \frac{1}{\log\log T} \log\left(\log T \cdot \int_0^1 \left|\zeta\left(\frac{1}{2} + \mathrm{i}(\tau+h)\right)\right|^\beta \mathrm{d}h\right) = \begin{cases} 1 + \frac{\beta^2}{4} & \text{if } \beta < 2, \\ \beta & \text{if } \beta \geq 2, \end{cases}$$

similarly to a Random Energy Model (REM) with $\log T$ independent Gaussian variables of variance $\frac{1}{2}\log\log T$.

In this paper, we explore the connection with spin glasses further by providing evidence that $\log|\zeta|$ behaves locally like a spin glass with one-step replica symmetry breaking (1-RSB), cf. Theorem 1. More precisely, we study a simple random model introduced by Harper [23] for the large values of $\log|\zeta|$. We show that two points sampled from the Gibbs measure at low temperature have correlation coefficients (or *overlap*) 0 or 1 in the limit, similarly to a 1-RSB spin glass. We expect that part of our approach could be extended to prove a similar result for the Riemann zeta function itself as stated in Conjecture 1 below.

1.2 The Model and Main Result

Let $(U_p, p \text{ primes})$ be IID uniform random variables on the unit circle in $\mathbb{C}$. We write $\mathbb{E}$ for the expectation over the U_p's. We study the stochastic process

$$X_h = \sum_{p\leq T} \frac{\mathrm{Re}(U_p p^{-ih})}{p^{1/2}}, \ h \in [0, 1]. \tag{3}$$

We drop the dependence on T in the notation for simplicity. The process $(X_h, h \in [0, 1])$ is a good model for the large values of $\log|\zeta(1/2 + \mathrm{i}(\tau+h))|$, $h \in [0, 1]$, see [2,23,40] for more details. For example, it is known that the deterministic order of $\max_{h\in[0,1]} X_h$ corresponds to the one in (2), as proved in [2]. Roughly speaking, the process X_h corresponds to the leading order of the logarithm of the Euler product (1) with the identification

$$(p^{-\mathrm{i}\tau}, p \text{ primes}) \longleftrightarrow (U_p, p \text{ primes}) .$$

It is easily checked by computing the joint moments that the above identification is exact as $T \to \infty$ in the sense of finite-dimensional distribution.

The covariance can be calculated using the explicit distribution of the U_p's:

$$\begin{aligned}\mathbb{E}[X_h X_{h'}] &= \sum_{p\le T} \int_0^{2\pi} \frac{1}{2}\left(e^{\mathrm{i}(\theta - h\log p)} + e^{-\mathrm{i}(\theta-h\log p)}\right) \\ &\quad \times \frac{1}{2}\left(e^{\mathrm{i}(\theta-h'\log p)} + e^{-\mathrm{i}(\theta-h'\log p)}\right)\frac{\mathrm{d}\theta}{2\pi} \\ &= \frac{1}{2}\sum_{p\le T}\frac{\cos(|h-h'|\log p)}{p}. \end{aligned} \tag{4}$$

We are interested in the correlation coefficient or *overlap* (in the spin glass terminology):

$$\rho(h,h') = \frac{\mathbb{E}[X_h X_{h'}]}{\sqrt{\mathbb{E}[X_h^2]\ \mathbb{E}[X_{h'}^2]}}, \quad \text{for a given pair } (h,h'). \tag{5}$$

Any sum over primes can be estimated using the Prime Number Theorem [26], which gives the density of the primes up to very good errors,

$$\#\{p \le x : p \text{ prime}\} = \int_2^x \frac{1}{\log y}\mathrm{d}y + \mathrm{O}(xe^{-c\sqrt{\log x}}). \tag{6}$$

(The error term, which is already more than sufficient for our purpose, is improved under the Riemann hypothesis.) In particular, this can be used to rewrite the covariances as (see Lemma 1 below for details),

$$\mathbb{E}[X_h^2] = \frac{1}{2}\sum_{p\le T} p^{-1} = \frac{1}{2}\log\log T + \mathrm{O}(1) \qquad \mathbb{E}[X_h X_{h'}] = \frac{1}{2}\log|h-h'|^{-1} + \mathrm{O}(1). \tag{7}$$

The process (X_h) is said to be *log-correlated*, since the covariance decays approximately like the logarithm of the distance. The correlation coefficients as a function of the distance become

$$\rho(h,h') = \frac{\log|h-h'|^{-1}}{\log\log T} + \mathrm{o}(1)\ , \ \text{ for } |h-h'| \ge (\log T)^{-1}. \tag{8}$$

Throughout the paper, we will use the notation $f(T) = \mathrm{o}(g(T))$ if $f(T)/g(T) \to 0$ and $f(T) = \mathrm{O}(g(T))$ if $f(T)/g(T)$ is bounded. We will sometimes use $f(T) \ll g(T)$ for short if $f(T) = \mathrm{O}(g(T))$ (the Vinogradov notation).

The main result of this paper is the limiting distribution of the correlation coefficient when h and h' are sampled from the Gibbs measure. This is referred to as the *two-overlap distribution* in the spin-glass terminology. We denote the Gibbs measure by

$$G_{\beta,T}(A) = \int_A \frac{e^{\beta X_h}}{Z_{\beta,T}}\,\mathrm{d}h \qquad Z_{\beta,T} = \int_0^1 e^{\beta X_h}\,\mathrm{d}h. \tag{9}$$

Theorem 1. *For every $\beta > 2$ and for any interval $I \subseteq [0,1]$,*

$$\lim_{T\to\infty} \mathbb{E}\left[G_{\beta,T}^{\times 2}\{(h,h') : \rho(h,h') \in I\}\right] = \frac{2}{\beta}\mathbb{1}_I(0) + \left(1 - \frac{2}{\beta}\right)\mathbb{1}_I(1).$$

where $\mathbb{1}_I$ is the indicator function of the set I. In other words, when h, h' are sampled independently from the Gibbs measure $G_{\beta,T}$, the random variable $\rho(h,h')$ is Bernoulli-distributed with parameter $2/\beta$ in the limit $T \to \infty$.

The limit is exactly the two-overlap distribution of a 1-RSB spin glass. In view of the relation (8) between the correlation coefficient and the distance $|h - h'|$, the result means that the large values of X_h must lie at a distance $\mathrm{O}(1)$ or $\mathrm{O}((\log T)^{-1})$. The mesoscopic distances $(\log T)^{-\alpha}$, $0 < \alpha < 1$ are effectively ruled out. Similar results were obtained for the REM model [18], and log-correlated processes [4, 7, 8, 14, 15, 19, 24, 32].

In the spirit of the Fyodorov–Hiary–Keating conjecture, Theorem 1 suggests that $\log|\zeta|$ exhibits 1-RSB for β large enough.

Conjecture 1. Consider

$$\mathcal{G}_\beta(t) = |\zeta(1/2 + \mathrm{i}t)|^\beta \qquad \mathcal{Z}_\beta(t) = \int_0^1 \mathcal{G}_\beta(t+h)\mathrm{d}h.$$

For $\beta > 2$, and any interval $I \subseteq [0,1]$, if τ is sampled uniformly on $[T, 2T]$:

$$\lim_{T\to\infty} \mathbb{E}\left[\int_{\{(h,h'):\rho(h,h')\in I\}} \frac{\mathcal{G}_\beta(\tau+h)\cdot\mathcal{G}_\beta(\tau+h')}{\mathcal{Z}_\beta(\tau)^2}\,\mathrm{d}h\mathrm{d}h'\right] = \frac{2}{\beta}\mathbb{1}_I(0) + (1 - \frac{2}{\beta})\mathbb{1}_I(1)\ .$$

In other words, points h, h' whose ζ-value is of the order of $\log\log T$ are at a distance of $\mathrm{O}(1)$ or $\mathrm{O}((\log T)^{-1})$.

The above conjecture implies a strong clustering of the high values of ζ at a scale $(\log T)^{-1}$ akin to the one observed in log-correlated process [4]. In turns, this phenomenon has important consequences for the joint statistics of high values which should be Poissonian at a suitable scale as for log-correlated processes [5, 10]. In particular, it is expected that the statistics of the Gibbs weights is Poisson–Dirichlet [7, 8], and that the Gibbs measure converges to an atomic measure on [0, 1], see [20]. This perspective is studied in [33], and will be discussed further in a forthcoming paper.

1.3 Main Propositions and Proof of the Theorem 1

The proof of Theorem 1 is based on a method developed for log-correlated Gaussian processes by Arguin and Zindy [7,8]. It was adapted from a method of Bovier and Kurkova [14,15] for Generalized Random Energy Models (GREM's). The main idea is to relate the distribution of the overlaps with the free energy of a perturbed process. In the present case, the process is not Gaussian and the method has to be modified. To this aim, consider the process at *scale* α, for $0 < \alpha < 1$, where the sum over primes is truncated at $\exp((\log T)^\alpha)$,

$$X_h(\alpha) = \sum_{\log p \leq (\log T)^\alpha} \frac{\mathrm{Re}(U_p\, p^{-ih})}{p^{1/2}}, \qquad h \in [0,1]. \tag{10}$$

Note that $X_h(1) = X_h$. For a small parameter $|u| < 1$, we consider the *free energy* of the perturbed process $X_h + uX_h(\alpha)$ at scale α:

$$F_T(\beta;\alpha,u) = \mathbb{E}\left[\log \int_0^1 \exp\left(\beta(X_h + uX_h(\alpha)\right)\mathrm{d}h\right]. \tag{11}$$

The connection between the free energy (11) and the distribution of the correlation coefficients is through Gaussian integration by parts. Of course, for the process X_h, this step is only approximate. It follows closely the work of Carmona and Hu [16] and Auffinger and Chen [9] on the universality of the free energy and overlap distributions in the Sherrington–Kirkpatrick model.

Proposition 1. *For any* $0 < \alpha < 1$,

$$\left|\int_0^\alpha \mathbb{E}\left[G_{\beta,T}^{\times 2}\{(h,h') : \rho(h,h') \leq y\}\right]\,\mathrm{d}y - \frac{2}{\beta^2 \log\log T}\frac{\partial F_T}{\partial u}(\beta;\alpha,0)\right| = \mathrm{o}(1).$$

The free energy of the perturbed process is calculated using Kistler's multiscale second moment method [25]. The treatment is similar to the one of Arguin and Ouimet [6] for the perturbed Gaussian free field. The same result can be obtained by adapting the method of Bolthausen et al. [11] and Daviaud [17] to the model as was done in [7,8]. Kistler's method is simpler and more flexible. The result is better stated by first defining

$$f(\beta,\sigma^2) = \begin{cases} \beta^2\sigma^2/4 & \text{if } \beta \leq 2/\sigma, \\ \beta\sigma - 1 & \text{if } \beta \geq 2/\sigma. \end{cases} \tag{12}$$

Proposition 2. *For every* $\beta > 0$ *and* $|u| < 1$, *the following limit holds*

$$\lim_{T\to\infty} \frac{1}{\log\log T} F_T(\beta;\alpha,u) = \begin{cases} f(\beta,(1+u)^2\alpha + (1-\alpha)) & \text{if } u < 0, \\ \alpha f(\beta,(1+u)^2) + (1-\alpha) f(\beta,1) & \text{if } u \geq 0. \end{cases}$$

The theorem follows from the above two propositions. They are proved in Sects. 3 and 4 respectively. Estimates on the model needed for the proofs are given in Sect. 2.

Proof of Theorem 1. We need to show that the distribution of $\rho(h,h')$ converges weakly to $\frac{2}{\beta}\delta_0 + (1-\frac{2}{\beta})\delta_1$ where δ_a stands for the Dirac measure at a. Write $x_{\beta,T}(s)$ for $\mathbb{E}[G_{\beta,T}^{\times 2}\{(h,h') : \rho(h,h') \le s\}]$. By compactness of the space of probability measures on $[0,1]$, we can find a subsequence of $(x_{\beta,T})$ that converges weakly to x_β as $T \to \infty$. We show that the limit x_β is unique and equals $x_\beta(s) = 2/\beta$ for $0 \le s < 1$, thereby proving the claimed convergence.

By definition of weak convergence, $x_{\beta,T}(s)$ converges to $x_\beta(s)$ at all points of continuity of s. Since x_β is non-decreasing, this implies convergence almost everywhere. Thus, the dominated convergence theorem implies

$$\lim_{T\to\infty}\int_0^\alpha x_{\beta,T}(s)\,\mathrm{d}s = \int_0^\alpha x_\beta(s)\,\mathrm{d}s \quad , \text{for } 0<\alpha<1. \tag{13}$$

The left-hand side can be rewritten using Proposition 1 as

$$\lim_{T\to\infty}\int_0^\alpha x_{\beta,T}(s)\,\mathrm{d}s = \lim_{T\to\infty}\frac{2}{\beta^2\log\log T}\frac{\partial F_T}{\partial u}(\beta;\alpha,0). \tag{14}$$

Since $\big(((\log\log T)^{-1}F_T(\beta;\alpha,u)\big)_T$ is a sequence of convex functions of u, the limit of the derivatives is the derivative of the limit at any point of differentiability. Here the limit of the expectation of the free energy is given by Proposition 2, for u small enough so that $\beta > 2/\sigma$ whenever $\beta > 2$,

$$\lim_{T\to\infty}\frac{1}{\log\log T}F_T(\beta;\alpha,u) = \begin{cases}\beta\Big((1+u)^2\alpha + (1-\alpha)\Big)^{1/2} - 1 & \text{if } u<0,\\ \alpha\beta(1+u) + (1-\alpha)\beta - 1 & \text{if } u \ge 0.\end{cases} \tag{15}$$

In particular, the expected free energy is differentiable at $u=0$. Therefore, Eqs. (13), (14) and (15) altogether imply

$$\int_0^\alpha x_\beta(s)\mathrm{d}s = \alpha\frac{2}{\beta}, \text{ for } 0<\alpha<1.$$

This means that for any $0<\alpha<\alpha'<1$ we have

$$\frac{1}{\alpha'-\alpha}\int_\alpha^{\alpha'} x_\beta(s)\,\mathrm{d}s = \frac{2}{\beta}.$$

By taking $\alpha'-\alpha\to 0$, we conclude from the Lebesgue differentiation theorem that $x_\beta(s) = 2/\beta$ almost everywhere. Since x_β is non-decreasing and right-continuous, this implies that $x_\beta(s) = 2/\beta$ for every $0 \le s < 1$ as claimed.

2 Estimates on the Model of Zeta

In this section, we gather the estimates on the model of zeta needed for the proof of Propositions 1 and 2. Most of these results are contained in [2]. We include them for completeness since we will need to deal with a perturbed version of the process (X_h). It is important to point out that most (but not all!) of these estimates can be obtained for zeta itself with some more work, see [3].

The essential input from number theory for the model is the Prime Number Theorem (6). It shows that the density of the primes is approximately $1/\log p$. This implies, for example, that $\sum_p p^{-a} < \infty$ for $a > 1$. The Eq. (7) expressing the log-correlations for $h \neq h'$ is straightforward from the following lemma by taking $\Delta = |h-h'|$ and by splitting the sum (4) into the ranges $\log p \leq |h-h'|^{-1}$ and $|h-h'|^{-1} < \log p \leq \log T$.

Lemma 1. *Let* $2 \leq P < Q < \infty$. *Then for* $\Delta > 0$, *we have*

$$\begin{aligned}\sum_{P \leq p \text{ primes} \leq Q} & \frac{\cos(\Delta \cdot \log p)}{p} \\ &= \int_P^Q \frac{\cos(\Delta \cdot \log v)}{v \log v} \mathrm{d}v + \mathrm{O}(e^{-c\sqrt{\log P}}) \\ &= \begin{cases} \log\log Q - \log\log P + \mathrm{O}(1) & \textit{for}\, \Delta \cdot \log Q \leq 1, \\ \mathrm{O}(\frac{1}{\Delta \cdot \log P}) + \mathrm{O}(e^{-c\sqrt{\log P}}) & \textit{for}\, \Delta \cdot \log P \geq 1. \end{cases}\end{aligned} \tag{16}$$

Proof. Denote by $\mathrm{Li}(x) = \int_2^x \frac{1}{\log y} \mathrm{d}y$ the logarithmic integal. Write $\mathcal{E}(x)$ for the function of bounded variation $\pi(x) - \mathrm{Li}(x)$ giving the error, and $f(x)$ for $\frac{\cos(\Delta \cdot \log x)}{x}$. Clearly, we have

$$\sum_{P \leq p \leq Q} f(p) = \int_P^Q f(x) \pi(\mathrm{d}x) = \int_P^Q \frac{f(x)}{\log x} \mathrm{d}x + \int_P^Q f(x) \mathcal{E}(\mathrm{d}x).$$

It remains to estimate the error term. By integration by parts,

$$\int_P^Q f(x) \mathcal{E}(\mathrm{d}x) = \mathcal{E}(Q) f(Q) - \mathcal{E}(P) f(P) - \int_P^Q \mathcal{E}(x) f'(x) \mathrm{d}x.$$

Note that $f(x)$ is of the order of $1/x$ and $f'(x)$ is of the order of $1/x^2$. Since $\mathcal{E}(x) = O(xe^{-c\sqrt{\log x}})$, the first claimed equality follows. For the dichotomy in the second equality, in the case $\Delta \cdot \log Q \leq 1$, we expand the cosine to get after the change of variable $y = \log x$

$$\int_P^Q \frac{f(x)}{\log x} \mathrm{d}x = \int_{\log P}^{\log Q} \frac{\cos(\Delta \cdot y)}{y} \mathrm{d}y = \int_{\log P}^{\log Q} \left(\frac{1}{y} + \mathrm{O}(\Delta^2 \cdot y) \right) \mathrm{d}y.$$

The result follows by integration. In the case $\Delta \cdot \log P \geq 1$, we integrate by parts to get

$$\int_P^Q \frac{f(x)}{\log x} \mathrm{d}x = \frac{\sin(\Delta \cdot y)}{\Delta \cdot y} \Big|_{\log P}^{\log Q} + \int_{\log P}^{\log Q} \frac{\sin(\Delta \cdot y)}{\Delta \cdot y^2} \mathrm{d}y.$$

Both terms are $\mathrm{O}(\frac{1}{\Delta \cdot \log P})$ as claimed.

Proposition 2 gives an expression for the free energy (11) of the perturbed process at scale α. For simplicity, we denote this process by

$$\widetilde{X}_h = (1+u)X_h(\alpha) + X_h(\alpha,1) \quad \text{for } X_h(\alpha,1) = X_h - X_h(\alpha), h \in [0,1]. \tag{17}$$

Note that we recover X_h at $u=0$. The finite-dimensional distributions of $(\widetilde{X}_h)$ can be explicitly computed. In fact, it is not hard to compute explicitly the moment generating function for any increment of (X_h). We will only need the two-dimensional case.

Proposition 3. *Let $0 \le \alpha_1 < \alpha_2 \le 1$. Consider $X_h(\alpha_1,\alpha_2) = X_h(\alpha_2) - X_h(\alpha_1)$. We have for $\lambda, \lambda' \in \mathbb{R}$ and $h,h' \in [0,1]$,*

$$\mathbb{E}\left[\exp\left(\lambda X_h(\alpha_1,\alpha_2) + \lambda' X_{h'}(\alpha_1,\alpha_2)\right]\right.$$

$$= C(\lambda,\lambda') \cdot \exp\left(\frac{1}{2}\sum_{\substack{\log p \le (\log T)^{\alpha_2} \\ \log p > (\log T)^{\alpha_1}}} \frac{1}{2p}\left(\lambda^2 + \lambda'^2 + 2\lambda\lambda' \cos(|h-h'|\log p)\right)\right),$$

where $C = C(\lambda,\lambda')$ is bounded if λ and λ' are bounded uniformly in T.

Proof. The expression can be evaluated explicitly as follows. Since the U_p's are independent, we can first restrict the computation to a single p. Straightforward manipulations yield

$$\mathbb{E}\left[\exp\left(p^{-1/2}\lambda \cdot \mathrm{Re}(U_p p^{-ih}) + p^{-1/2}\lambda' \cdot \mathrm{Re}(U_p p^{-ih'})\right)\right] = \mathbb{E}\left[\exp(aU_p + \bar{a}\overline{U}_p)\right]$$

for $a = (2p^{1/2})^{-1}(\lambda p^{-ih} + \lambda' p^{-ih'})$. By expanding the exponentials and using the fact that U_p is uniform on the unit circle, we get

$$\begin{aligned}
\mathbb{E}\left[\exp(aU_p + \bar{a}\overline{U}_p)\right] &= \sum_{n=0}^{\infty}\sum_{k=0}^{n} \frac{a^k \bar{a}^{n-k}}{n!}\binom{n}{k}\mathbb{E}[U_p^k \overline{U}_p^{n-k}] \\
&= \sum_{m=0}^{\infty} \frac{1}{(m!)^2}\left(\frac{\lambda^2 + \lambda'^2 + 2\lambda\lambda'\cos(|h-h'|\log p)}{4p}\right)^m \\
&= 1 + \left(\frac{\lambda^2 + \lambda'^2 + 2\lambda\lambda'\cos(|h-h'|\log p)}{4p}\right) + \mathrm{O}(p^{-2}),
\end{aligned} \tag{18}$$

where the O-term depends on λ, λ'. The second equality follows from the fact that the expectation is non-zero only if $k = n/2$. It remains to take the product over the range of p. The claim then follows from the fact that the sum of p^{-2} is finite by (6).

Proposition 3 yields Gaussian bounds in the large deviation regime we are interested in. Indeed, by Chernoff's bound (optimizing over λ), it implies that, for $\gamma > 0$,

$$\mathbb{P}\left(X_h(\alpha_1,\alpha_2) > \gamma \log\log T\right) \ll \exp\left(-\frac{\gamma^2 \log\log T}{(\alpha_2 - \alpha_1)}\right) = (\log T)^{\frac{-\gamma^2}{\alpha_2-\alpha_1}}, \tag{19}$$

where we used Lemma 1 to estimate the sum over primes. This supports the heuristic that $X_h(\alpha_1, \alpha_2)$ is approximately Gaussian of variance $\frac{\alpha_2 - \alpha_1}{2} \log\log T$. This implies for $\widetilde{X}_h$ in (17)

$$\mathbb{P}\left(\widetilde{X}_h > \gamma \log\log T\right) \ll \exp\left(-\frac{\gamma^2 \log\log T}{(1+u)^2\alpha + (1-\alpha)}\right) = (\log T)^{\frac{-\gamma^2}{(1+u)^2\alpha+(1-\alpha)}}. \tag{20}$$

The same can be done for two points h, h'. Using Lemma 1 again, we get

$$\begin{aligned}\mathbb{P}\left(X_h(\alpha_1,\alpha_2) > \gamma \log\log T, X_{h'}(\alpha_1,\alpha_2) > \gamma \log\log T\right)\\ \ll \begin{cases} \exp\left(-\frac{\gamma^2 \log\log T}{(\alpha_2-\alpha_1)}\right) & \text{if } |h-h'| \le (\log T)^{-\alpha_2}, \\ \exp\left(-2\frac{\gamma^2 \log\log T}{(\alpha_2-\alpha_1)}\right) & \text{if } |h-h'| \ge (\log T)^{-\alpha_1}. \end{cases}\end{aligned} \tag{21}$$

This can be interpreted as follows. The increments are (almost) independent if the distance between the points is larger than the relevant scales of the increments, and are (almost) perfectly correlated if the distance is smaller than the scales.

It is important to note that if $\alpha_1 > 0$, then a stronger estimate than the one of Proposition 3 holds. This is because the sum over primes in (18) is then negligible since it is the tail of a summable series. This means that the constant $C(\lambda, \lambda')$ is then $1 + \mathrm{O}(1)$. This gives a precise Gaussian estimate by inverting the moment generating function (or the Fourier transform if we pick $\lambda, \lambda' \in \mathbb{C}$). We omit the proof for conciseness and we refer to [2] where this is done using a general version of the Berry Esseen theorem.

Proposition 4 (see Propositions 2.9, 2.10, 2.11 in [2]**).** *For* $0 < \alpha_1 < \alpha_2 \le 1$ *and* $0 < \gamma < 1$, *we have for* $h \in [0,1]$,

$$\begin{aligned}\mathbb{P}(X_h(\alpha_1,\alpha_2) > \gamma \log\log T) &\gg \frac{1}{\sqrt{\log\log T}} \exp\left(-\frac{\gamma^2 \log\log T}{(\alpha_2-\alpha_1)}\right)\\ &= (\log T)^{\frac{-\gamma^2}{\alpha_2-\alpha_1}+\mathrm{o}(1)}.\end{aligned}$$

Moreover, if $|h-h'| > (\log T)^{-\alpha_1}$, *then*

$$\begin{aligned}\mathbb{P}(X_h(\alpha_1,\alpha_2) > \gamma \log\log T, X_{h'}(\alpha_1,\alpha_2) > \gamma \log\log T)\\ = (1+\mathrm{o}(1))\mathbb{P}(X_h(\alpha_1,\alpha_2) > \gamma \log\log T)^2.\end{aligned}$$

Since the process $(X_h, h \in [0,1])$ is continuous and not discrete, we need a last estimate to control all values in an interval of length corresponding to the relevant scale. This is needed when proving rough bound on the maximum in Lemma 4. Heuristically, it says that the maximum of $X_h(\alpha_1, \alpha_2)$ over an interval of width smaller than $(\log T)^{-\alpha_2}$ behaves like a single value $X_h(\alpha_1, \alpha_2)$. This is done in [2] by a chaining argument and we omit the proof for conciseness.

Lemma 2 (Corollary 2.6 in [2]). *Let $0 \leq \alpha_1 < \alpha_2 \leq 1$. For every $h \in [0,1]$ and $\gamma > 0$, we have*

$$\mathbb{P}\left(\max_{|h-h'|\leq(\log T)^{-\alpha_2}} X_{h'}(\alpha_1,\alpha_2) > \gamma \log\log T\right) \ll (\log T)^{-\frac{\gamma^2}{\alpha_2-\alpha_1}}.$$

In particular, we have

$$\mathbb{P}\left(\max_{|h-h'|\leq(\log T)^{-1}} \widetilde{X}_{h'} > \gamma \log\log T\right) \ll (\log T)^{-\frac{\gamma^2}{\alpha(1+u)^2+(1-\alpha)}}.$$

3 Proof of Proposition 1

As mentioned in Sect. 1.3, the proof of Proposition 1 is based on an approximate Gaussian integration by parts as in [9,16]. The following lemma is an adaptation for complex random variables of Lemma 4 in [16] .

Lemma 3. *Let ξ be a complex random variable such that $\mathbb{E}[|\xi|^3] < \infty$, and $\mathbb{E}[\xi^2] = \mathbb{E}[\xi] = 0$. Let $F : \mathbb{C} \to \mathbb{C}$ be a twice continuously differentiable function such that for some $M > 0$,*

$$\left\|\partial_z^2 F\right\|_\infty, \left\|\partial_{\bar{z}}^2 F\right\|_\infty < M\ ,$$

where $\|f\|_\infty = \sup_{z\in\mathbb{C}} |f(z,\bar{z})|$. Then

$$\left|\mathbb{E}\left[\xi F(\xi,\bar{\xi})\right] - \mathbb{E}[|\xi|^2]\,\mathbb{E}\left[\partial_{\bar{z}} F(\xi,\bar{\xi})\right]\right| \ll M\ \mathbb{E}[|\xi|^3].$$

Proof. Since $\mathbb{E}[\xi^2] = \mathbb{E}[\xi] = 0$, the left-hand side can be written as

$$\mathbb{E}\left[\xi\big(F(\xi,\bar{\xi}) - F(0,0) - \xi\partial_z F(0,0) - \bar{\xi}\partial_{\bar{z}} F(0,0)\big)\right] \\ - \mathbb{E}[|\xi|^2]\,\mathbb{E}\left[\Big(\partial_{\bar{z}} F(\xi,\bar{\xi}) - \partial_{\bar{z}} F(0,0)\Big)\right]. \quad (22)$$

By Taylor's theorem in several variables and the assumptions, the following estimates hold

$$\left|F(\xi,\bar{\xi}) - F(0,0) - \xi\partial_z F(0,0) - \bar{\xi}\partial_{\bar{z}} F(0,0)\right| \ll M|\xi|^2$$
$$\left|\partial_{\bar{z}} F(\xi,\bar{\xi}) - \partial_{\bar{z}} F(0,0)\right| \ll M|\xi|.$$

Therefore the norm of (22) gives

$$\left|\mathbb{E}\left[\xi F(\xi,\bar{\xi})\right] - \mathbb{E}[|\xi|^2]\,\mathbb{E}\left[\partial_{\bar{z}} F(\xi,\bar{\xi})\right]\right| \ll M(\mathbb{E}[|\xi|^3] + \mathbb{E}[|\xi|^2]\mathbb{E}[|\xi|]).$$

The claim then follows by Hölder's inequality.

As in [16], the lemma can be applied to relate the derivative of the free energy to the two-point correlations of the process.

Proposition 5. *For any $p \leq T$, we have*

$$\frac{\partial}{\partial u}\mathbb{E}\left[\log \int_0^1 \exp\left(\beta(X_h(T) + u\ \mathrm{Re}(U_p p^{-ih-1/2})\right)\mathrm{d}h\right]\Big|_{u=0}$$
$$= \frac{\beta}{2}\mathbb{E}\left[\int_{[0,1]^2} \frac{1-\cos(|h-h'|\log p)}{p}\ \mathrm{d}G_{\beta,T}^{\times 2}(h,h')\right] + \mathrm{O}(p^{-3/2}).$$

Proof. Write for short $\omega_p(h) = (2p^{1/2})^{-1}p^{-ih}$. Direct differentiation yields at $u = 0$

$$\int_0^1 U_p\omega_p(h)\ \mathrm{d}G_{\beta,T}(h) + \int_0^1 \overline{U_p}\overline{\omega_p}(h)\ \mathrm{d}G_{\beta,T}(h). \tag{23}$$

We make the dependence on U_p in the measure $G_{\beta,T}$ explicit. For this, define

$$Y_p(h) = \beta \sum_{\substack{q \leq T \\ p \neq q}} \mathrm{Re}\left(\frac{U_q q^{-ih}}{q^{1/2}}\right).$$

Clearly, $Y_p(h)$ is independent of U_p by definition. Consider

$$F_p(z,\overline{z}) = \frac{\int_0^1 \omega_p(h)\ \exp(\beta\omega_p(h)z + \beta\overline{\omega_p}(h)\overline{z} + Y_p(h))\ \mathrm{d}h}{\int_0^1 \exp(\beta\omega_p(h')z + \beta\overline{\omega_p}(h')\overline{z} + Y_p(h'))\ \mathrm{d}h'}.$$

Note that with this definition, the first integral in (23) is $U_pF_p(U_p,\overline{U}_p)$ and the second is its complex conjugate. This shows that the derivative of the expectation at $u = 0$ is

$$\mathbb{E}\left[U_p \cdot F_p(U_p,\overline{U}_p) + \overline{U_p}\cdot\overline{F_p}(U_p,\overline{U}_p)\right].$$

It remains to apply Lemma 3 with the function $F_p(z,\overline{z})$ and $\xi = U_p$. Write for short for a function H on $[0,1]$

$$\langle H\rangle^{(z,\overline{z})} = \frac{\int_0^1 H(h)\ e^{\beta(z\ \omega_p(h)+\overline{z}\ \overline{\omega_p}(h))+Y_p(h)}\mathrm{d}h}{\int_0^1 e^{\beta(z\ \omega_p(h)+\overline{z}\ \overline{\omega_p}(h))+Y_p(h)}\mathrm{d}h}.$$

Direct differentiation of the above yields

$$\partial_{\overline{z}}\langle H\rangle^{(z,\overline{z})} = \beta\left(\langle H\overline{\omega_p}\rangle^{(z,\overline{z})} - \langle H\rangle^{(z,\overline{z})}\langle\overline{\omega_p}\rangle^{(z,\overline{z})}\right). \tag{24}$$

In particular, for $H = \omega_p$, we get

$$\partial_{\overline{z}}F_p(z,\overline{z}) = \beta\left(\langle|\omega_p|^2\rangle^{(z,\overline{z})} - |\langle\omega_p\rangle^{(z,\overline{z})}|^2\right). \tag{25}$$

When evaluated at $z = U_p$, this is by definition of ω_p

$$\partial_{\overline{z}}F_p(U_p,\overline{U}_p) = \frac{\beta}{4}\int_{[0,1]^2}(p^{-1} - p^{-1}\cos(|h-h'|\log p))\ \mathrm{d}G_{\beta,T}^{\times 2}(h,h'). \tag{26}$$

Clearly, $|\omega_p| \leq p^{-1/2}$. Therefore the second derivatives are easily checked to be bounded by $\mathrm{O}(p^{-3/2})$ by applying the formula (24) to each term of (25). The statement of the lemma then follows from Lemma 3 and (26), after noticing that the second term of (23) is the conjugate of the first.

The proof of Proposition 1 is an application of Proposition 5 to a range of primes.

Proof of Proposition 1. Recall the definition of $\rho(h,h')$ in Eqs. (5) and (8). On one hand, Fubini's theorem directly implies that

$$\begin{aligned}\int_0^\alpha & G_{\beta,T}^{\times 2}\{(h,h') : \rho(h,h') \le y\}\, \mathrm{d}y \\ &= \int_{[0,1]^2} \left(\int_0^\alpha \mathbb{1}_{\{\rho(h,h')\le r\}}\, \mathrm{d}r \right) \mathrm{d}G_{\beta,T}(h,h') \\ &= \int_{[0,1]^2} (\alpha - \rho(h,h'))\, \mathbb{1}_{\{\rho(h,h')\le\alpha\}} \mathrm{d}G_{\beta,T}(h,h'). \qquad (27)\end{aligned}$$

It remains to check on the other hand that the derivative in the proposition is close to the expectation of the above. Direct differentiation of (11) at $u=0$ yields by Proposition 5

$$\begin{aligned}\frac{\partial F_T}{\partial u}(\beta;\alpha,0) = \frac{\beta^2}{2} \int_{[0,1]^2} & \sum_{\log p \le (\log T)^\alpha} \mathbb{E}\big[p^{-1}(1-\cos(|h-h'|\log p))\mathrm{d}G_{\beta,T}^{\times 2}(h,h')\big] \\ & + \mathrm{O}\left(\sum_{p \le e^{(\log T)^\alpha}} p^{-3/2} \right). \qquad (28)\end{aligned}$$

The error term is of order one by (6). Similarly, if $|h-h'| \le (\log T)^{-\alpha}$, the sum in the integral is by (16)

$$\sum_{\log p \le (\log T)^\alpha} p^{-1}(1-\cos(|h-h'|\log p)) = \alpha \log\log T - \alpha \log\log T + \mathrm{O}(1) = \mathrm{O}(1).$$

On the other hand, if $|h-h'| > (\log T)^{-\alpha}$, the sum can be divided into three parts

$$\begin{aligned}\sum_{\log p \le (\log T)^\alpha} p^{-1} - \sum_{\log p \le |h-h'|^{-1}} & p^{-1} \cos(|h-h'|\log p) \\ & - \sum_{|h-h'|^{-1} < \log p \le (\log T)^\alpha} p^{-1} \cos(|h-h'|\log p).\end{aligned}$$

When Eq. (16) is applied to each of the parts, this equals

$$\alpha \log\log T - \log|h-h'|^{-1} + \mathrm{O}(1).$$

Furthermore, recall from (8) that $\rho(h,h')\log\log T$ differs from $\log|h-h'|^{-1}$ by $\mathrm{o}(\log\log T)$. This implies that the conditions on $\log|h-h'|^{-1}$ can be replaced by $\rho(h,h')\log\log T$ at a cost of a term $\mathrm{o}(\log\log T)$ (since the sum would differ by a range of $\log p$ of at most $\mathrm{o}(\log T)$ primes). All these observations together imply

the interval. See Remark 1 for more on this.) Putting this back in (31) and summing over j, k, and q finally give the estimate:

$$\begin{aligned}
&\mathbb{P}\left(\max_{h\in[0,1]} \widetilde{X}_h > (1+\varepsilon)\gamma^\star \log\log T\right) \\
&\ll (\log\log T)\cdot(\log T)^{\alpha}\cdot e^{-\frac{\left((1+\varepsilon)\alpha\log\log T\right)^2}{\alpha\log\log T}}\cdot(\log T)^{1-\alpha}\cdot e^{-\frac{\left((1+\varepsilon)(1-\alpha)\log\log T\right)^2}{(1-\alpha)\log\log T}} \\
&\ll (\log\log T)\cdot \log T^{1-(1+\varepsilon)^2} = \mathrm{o}(1).
\end{aligned}$$

This concludes the proof of the lemma.

Consider for $0 < \alpha \leq 1$ and $|u| < 1$, the (normalized) log-measure of γ-high points

$$\mathcal{E}_{\alpha,u}(\gamma;T) = \frac{1}{\log\log T}\log \mathrm{Leb}\{h\in[0,1] : \widetilde{X}_h > \gamma\log\log T\}\ ,0<\gamma<\gamma^\star. \tag{32}$$

The limit of these quantities in probability can be computed following [6].

Lemma 5. *The limit* $\mathcal{E}_{\alpha,u}(\gamma) = \lim_{T\to\infty}\mathcal{E}_{\alpha,u}(\gamma;T)$ *exists in probability. We have for* $u<0$,

$$\mathcal{E}_{\alpha,u}(\gamma) = -\frac{\gamma^2}{(1+u)^2\alpha+(1-\alpha)},$$

and for $u \geq 0$,

$$\mathcal{E}_{\alpha,u}(\gamma) = \begin{cases} -\frac{\gamma^2}{(1+u)^2\alpha+(1-\alpha)} & \text{if } \gamma<\gamma_c \\ -\alpha-\frac{(\gamma-(1+u)\alpha)^2}{(1-\alpha)} & \text{if } \gamma\geq\gamma_c. \end{cases} \quad \text{for } \gamma_c = \frac{(1+u)^2\alpha+(1-\alpha)}{(1+u)}.$$

Remark 1. The dichotomy in the log-measure is due to the fact that for h with values beyond $\gamma_c\log\log T$, the intermediate values $X_h(\alpha)$ is restricted by the maximal level $\alpha\log\log T$. More precisely, consider

$$\begin{aligned}
\mathcal{M}_T &= \mathrm{Leb}\{h\in[0,1] : \widetilde{X}_h > \gamma\log\log T\} \\
\mathcal{M}'_T &= \mathrm{Leb}\{h\in[0,1] : (1+u)X_h(\alpha)\geq\lambda\log\log T\} \\
\mathcal{M}''_T &= \mathrm{Leb}\{h\in[0,1] : (1+u)X_h(\alpha)\geq\lambda\log\log T, X_h(\alpha,1)\geq(\gamma-\lambda)\log\log T\}.
\end{aligned} \tag{33}$$

Clearly, we must have $\mathcal{M}''_T \leq \mathcal{M}_T$. It turns out that $\mathcal{M}''_T$ and $\mathcal{M}_T$ are comparable for an optimal choice of λ given by, when $u<0$,

$$\lambda^\star = \frac{\gamma(1+u)^2\alpha}{(1+u)^2\alpha+1-\alpha}, \qquad \gamma<\gamma^\star, \tag{34}$$

and when $u>0$,

$$\lambda^\star = \begin{cases} \frac{\gamma(1+u)^2\alpha}{(1+u)^2\alpha+1-\alpha} & \text{if } 0<\gamma<\gamma_c\ , \\ (1+u)\alpha & \text{if } \gamma_c\leq\gamma<\gamma^\star. \end{cases} \tag{35}$$

One can see this at a heuristic level by considering first moments. Since the maximum of X_h is well approximated by the maximum over lattice points spaced $(\log T)^{-1}$ apart, there should be γ-high points only if

$$(\log T) \cdot \mathcal{M}''_T \geq 1. \tag{36}$$

Moreover, we have that if $\mathcal{M}'_T = 0$, then $\mathcal{M}''_T = 0$. And the maximum of $X_h(\alpha)$ is well approximated by the maximum over lattice points spaced $(\log T)^{-\alpha}$ apart, so there should be γ-high points only if

$$(\log T)^{\alpha} \cdot \mathcal{M}'_T \geq 1. \tag{37}$$

Since $X_h(\alpha)$ and $X_h(\alpha, 1)$ are approximately Gaussian with variance $\frac{1}{2}\alpha \log\log T$ and $\frac{1}{2}(1-\alpha)\log\log T$, the following should hold approximately

$$\frac{\log \mathbb{E}[(\log T)^{\alpha} \cdot \mathcal{M}'_T]}{\log\log T} = \alpha - \frac{\lambda^2}{(1+u)^2\alpha} + \mathrm{o}(1)$$
$$\frac{\log \mathbb{E}[(\log T) \cdot \mathcal{M}''_T]}{\log\log T} = 1 - \frac{\lambda^2}{(1+u)^2\alpha} - \frac{(\gamma-\lambda)^2}{1-\alpha} + \mathrm{o}(1)$$

Together with conditions (36) and (37), we obtain constraints on the value of λ:

$$\alpha - \frac{\lambda^2}{(1+u)^2\alpha} \geq 0\ , \tag{38}$$

$$1 - \frac{\lambda^2}{(1+u)^2\alpha} - \frac{(\gamma-\lambda)^2}{1-\alpha} \geq 0. \tag{39}$$

By maximizing $\mathcal{M}''_T$, under the constraints (38) and (39), one gets the values (34) and (35) for λ.

With Remark 1 in mind, we are ready to bound the log-measure.

Proof of Lemma 5. Upper bound on the log-measure. For $0 < \gamma < \gamma^\star$, consider $\mathcal{M}_T$ as in (33). We need to show that for $\varepsilon > 0$

$$\lim_{T\to\infty} \mathbb{P}\left(\mathcal{M}_T > (\log T)^{\mathcal{E}_{\alpha,u}(\gamma)+\varepsilon}\right) = 0. \tag{40}$$

We first prove the easiest cases where $u \geq 0$ and $\gamma < \gamma_c$, as well as $u \leq 0$. Let $\varepsilon > 0$. And write $V = 1 - \alpha + (1+u)^2\alpha$ for short. Observe that by Markov's inequality and Fubini's theorem

$$\begin{aligned}\mathbb{P}(\mathcal{M}_T > (\log T)^{-\frac{\gamma^2}{V}+\varepsilon}) &\leq (\log T)^{\frac{\gamma^2}{V}-\varepsilon} \int_0^1 \mathbb{P}(\widetilde{X}_h > \gamma \log\log T)\, \mathrm{d}h \\ &= (\log T)^{\frac{\gamma^2}{V}-\varepsilon} \mathbb{P}(\widetilde{X}_h > \gamma\log\log T)\ ,\end{aligned} \tag{41}$$

where we used the fact that the variables $\widetilde{X}_h$, $h \in [0,1]$, are identically distributed. Since $\mathbb{P}(\widetilde{X}_h > \gamma \log\log T) \ll \exp(-\gamma^2 \log\log T/V)$ by Eq. (20), the claim (40) follows.

The case $u > 0$, $\gamma > \gamma_c$ is more delicate as we need to control the values at scale α. For $\varepsilon' > 0$ to be fixed later, note that the same argument as for Eq. (30) gives

$$\begin{aligned}\mathbb{P}\left(\text{Leb}\{h \in [0,1] : X_h(\alpha) > (\alpha+\varepsilon')\log\log T\} > 0\right) \\ \le \mathbb{P}\left(\exists h \in [0,1] : X_h(\alpha) > (\alpha+\varepsilon')\log\log T\right) \to 0.\end{aligned} \tag{42}$$

The same hold by symmetry for $-X_h(\alpha)$. This implies

$$\begin{aligned}&\mathbb{P}\left(\mathcal{M}_T > (\log T)^{\mathcal{E}_{\alpha,u}(\gamma)+\varepsilon}\right) \\ &= \mathbb{P}\left(\text{Leb}\{h : \widetilde{X}_h > \gamma\log\log T, |X_h(\alpha)| \le (\alpha+\varepsilon')\log\log T\} > (\log T)^{\mathcal{E}_{\alpha,u}(\gamma)+\varepsilon}\right) \\ &\quad + \text{o}(1).\end{aligned}$$

It remains to prove that the first term is o(1). As in the proof of Lemma 4, we consider the partition of $[0,1]$ by intervals J_j, $j \le (\log T)^\alpha$, and the sub-partition I_{jk}, $k \le (\log T)^{1-\alpha}$. We also divide the interval $[-(\alpha+\varepsilon')\log\log T, (\alpha+\varepsilon')\log\log T]$ into intervals $[q, q+1]$. Then by Markov's inequality and the additivity of the Lebesgue measure

$$\begin{aligned}&\mathbb{P}\left(\text{Leb}\{h : \widetilde{X}_h > \gamma\log\log T, |X_h(\alpha)| \le (\alpha+\varepsilon')\log\log T\} > (\log T)^{\mathcal{E}_{\alpha,u}(\gamma)+\varepsilon}\right) \\ &\le (\log T)^{-\mathcal{E}_{\alpha,u}(\gamma)-\varepsilon} \sum_{j,k} \sum_{|q|\le(\alpha+\varepsilon')\log\log T} \\ &\qquad \mathbb{E}\left[\text{Leb}\{h \in I_{jk} : \widetilde{X}_h > \gamma\log\log T, X_h(\alpha) \in [q, q+1]\}\right] \\ &\le (\log T)^{-\mathcal{E}_{\alpha,u}(\gamma)-\varepsilon} \sum_{j,k} \sum_{|q|\le(\alpha+\varepsilon')\log\log T} \\ &\qquad (\log T)^{-1}\mathbb{P}\left(X_h(\alpha,1) > \gamma\log\log T - (1+u)(q+1), X_h(\alpha) \ge q\right).\end{aligned} \tag{43}$$

The last line follows from Fubini's theorem and the fact that

$$\text{Leb}(I_{jk}) = (\log T)^{-1}.$$

The probabilities can be bounded by the Gaussian bound (19)

$$\begin{aligned}&\mathbb{P}\left(X_h(\alpha,1) > \gamma\log\log T - (1+u)(q+1), X_h(\alpha) \ge q\}\right) \\ &\ll \exp\left(\frac{-q^2}{\alpha\log\log T}\right) \cdot \exp\left(\frac{-\left(\gamma\log\log T - (1+u)(q+1)\right)^2}{(1-\alpha)\log\log T}\right).\end{aligned}$$

It is easily checked that the expression is maximized at $q > (\alpha+\varepsilon')\log\log T$ for ε'. Moreover, at the optimal $q = (\alpha+\varepsilon')\log\log T$ in the considered range, the probability equals $(1+\text{o}(1))(\log T)^{\mathcal{E}_{\alpha,u}(\gamma)}$. Using this observation to bound the probability for each q in (43), we get

$$\begin{aligned}&\mathbb{P}\left(\text{Leb}\{h : \widetilde{X}_h > \gamma\log\log T, |X_h(\alpha)| \le (\alpha+\varepsilon')\log\log T\} > (\log T)^{\mathcal{E}_{\alpha,u}(\gamma)+\varepsilon}\right) \\ &\ll (\log T)^{-\varepsilon}\log\log T = \text{o}(1).\end{aligned}$$

This finishes the proof of the upper bound.
Lower bound on the log-measure. For $\varepsilon > 0$, the goal is to show

$$\mathbb{P}\left(\mathcal{M}_T > (\log T)^{\mathcal{E}_{\alpha,u}(\gamma)-\varepsilon}\right) \to 1 \text{ as } T \to \infty. \tag{44}$$

This is done using the Paley–Zygmund inequality, which states that for a random variable $\mathcal{M} \geq 0$ and $0 \leq \eta_T \leq 1$,

$$\mathbb{P}\left(\mathcal{M} > \eta_T E\left[\mathcal{M}\right]\right) \geq \left(1-\eta_T\right)^2 \frac{\mathbb{E}\left[\mathcal{M}\right]^2}{\mathbb{E}\left[\mathcal{M}^2\right]}. \tag{45}$$

We will have $\eta_T \to 0$, so the main task will be to demonstrate

$$\mathbb{E}\left[\mathcal{M}^2\right] = (1+\mathrm{o}(1))\,\mathbb{E}\left[\mathcal{M}\right]^2. \tag{46}$$

This cannot be achieved when $\mathcal{M} = \mathcal{M}_T$ because of the correlations in X_h. To overcome this problem, we define a modified version of $\mathcal{M}_T$ by *coarse graining* the field as described in [25].

For $K \in \mathbb{N}$ (that will depend eventually on ε), assume without loss of generality that $\{0, \frac{1}{K}, \frac{2}{K}, \ldots, \frac{K-1}{K}, 1\}$ is a partition of $[0,1]$ that is a refinement of $\{0, \alpha, 1\}$. Consider $\lambda < \lambda^\star$ as defined in (34) and (35), and $\delta > 0$ (that will depend on ε). Define the events for the K-level coarse increments:

$$\mathcal{J}_h(m) = \begin{cases} \left\{(1+u)X_h\left(\frac{m-1}{K}, \frac{m}{K}\right) \geq (1+\delta)\frac{\lambda \log\log T}{\alpha K}\right\} & \text{if } m = 1, \ldots, \alpha K, \\ \left\{X_h\left(\frac{m-1}{K}, \frac{m}{K}\right) \geq (1+\delta)\frac{(\gamma-\lambda)\log\log T}{(1-\alpha)K}\right\} & \text{if } m = \alpha K+1, \ldots, K. \end{cases} \tag{47}$$

Moreover, define the sets

$$\begin{aligned} \mathcal{A} &= \{h : \mathcal{J}_h(m) \text{ occurs } \forall m = 2, \ldots, K\}, \\ \mathcal{B} &= \left\{h : (1+u)X_h\left(\frac{1}{K}\right) \leq -\frac{\delta}{2}\log\log T\right\}. \end{aligned} \tag{48}$$

Note that if $h \in \mathcal{A}$, by adding up the inequalities in $\mathcal{J}_h(m)$, we have for K large enough,

$$\begin{aligned} \widetilde{X}_h - (1+u)X_h\left(\frac{1}{K}\right) &\geq (1+\delta)\left(\gamma - \frac{\lambda}{\alpha K}\right)\log\log T \\ &\geq \left(\gamma + \frac{\delta}{2}\right)\log\log T. \end{aligned}$$

Therefore this implies the inclusion

$$\mathcal{A} \subset \left\{h \in [0,1] : \widetilde{X}_h \geq \gamma \log\log T\right\} \cup \mathcal{B},$$

so that $\mathcal{M}_T \geq \mathrm{Leb}(\mathcal{A}) - \mathrm{Leb}(\mathcal{B})$. Equation (19) and Fubini's theorem shows that $\mathbb{E}[\mathrm{Leb}(\mathcal{B})] \ll (\log T)^{-\frac{\delta^2 K}{4(1+u)^2}}$. For K large enough, Markov's inequality then implies

$$\mathbb{P}\left(\mathrm{Leb}\left\{h \in [0,1] : (1+u)X_h\left(\frac{1}{K}\right) \leq -\frac{\delta}{2}\log\log T\right\} \leq (\log T)^{\mathcal{E}_{\alpha,u}(\gamma)-\varepsilon}\right) \to 1.$$

The proof of (44) is then reduced to show

$$\mathbb{P}(\mathrm{Leb}(\mathcal{A}) > 2(\log T)^{\mathcal{E}_{\alpha,u}(\gamma)-\varepsilon}) = \mathbb{P}(\mathrm{Leb}(\mathcal{A}) > \eta_T \mathbb{E}[\mathrm{Leb}(\mathcal{A})]) \to 1, \tag{49}$$

where η_T is defined by $2(\log T)^{\mathcal{E}_{\alpha,u}(\gamma)-\varepsilon} = \eta_T \mathbb{E}[\mathrm{Leb}(\mathcal{A})]$.

Following (45), we first show $\eta_T \to 0$. By (48), Fubini's theorem, and independence,

$$\mathbb{E}[\mathrm{Leb}(\mathcal{A})] = \int_0^1 \prod_{m=2}^{K} \mathbb{P}\left(\mathcal{J}_h(m)\right) \mathrm{d}h = \prod_{m=2}^{K} \mathbb{P}\left(\mathcal{J}_h(m)\right), \tag{50}$$

since the X_h's are identically distributed. By Proposition 4,

$$\mathbb{P}(\mathcal{J}_h(m)) \gg \begin{cases} (\log T)^{-\frac{(1+\delta)^2\lambda^2}{\alpha^2 K(1+u)^2}+\mathrm{o}(1)} & \text{when } m = 1, \ldots, \alpha K, \\ (\log T)^{-\frac{(1+\delta)^2(\gamma-\lambda)^2}{(1-\alpha)^2 K}+\mathrm{o}(1)} & \text{when } m = \alpha K + 1, \ldots, K. \end{cases} \tag{51}$$

Thus, by (50) and (51), we have

$$\mathbb{E}[\mathrm{Leb}(\mathcal{A})] \gg (\log T)^{-\frac{(1+\delta)^2\lambda^2}{\alpha(1+u)^2} - \frac{(1+\delta)^2(\gamma-\lambda)^2}{(1-\alpha)}} (\log T)^{\frac{(1+\delta)^2\lambda^2}{\alpha^2 K(1+u)^2}+\mathrm{o}(1)}. \tag{52}$$

We can take λ close enough to $\lambda^\star$, δ small enough, and K large enough so that

$$\mathbb{E}[\mathrm{Leb}(\mathcal{A})] \gg (\log T)^{-\frac{\lambda^{\star 2}}{\alpha(1+u)^2} - \frac{(\gamma-\lambda^\star)^2}{(1-\alpha)} - \frac{\varepsilon}{2}} = (\log T)^{\mathcal{E}_{\alpha,u}(\gamma)-\frac{\varepsilon}{2}},$$

where we replace the value of $\lambda^\star$ of (34) and (35). This shows that $\eta_T \to 0$. Observe that, we also have the reverse inequality

$$\mathbb{E}[\mathrm{Leb}(\mathcal{A})] \ll (\log T)^{\mathcal{E}_{\alpha,u}(\gamma)+\frac{\varepsilon}{2}}, \tag{53}$$

using (19) instead of Proposition 4.

It remains to show (46). By independence of increments and Fubini's theorem, we have

$$\mathbb{E}[\mathrm{Leb}(\mathcal{A})^2] = \int_0^1 \int_0^1 \prod_{m=2}^{K} \mathbb{P}(\mathcal{J}_h(m) \cap \mathcal{J}_{h'}(m))\ \mathrm{d}h\mathrm{d}h'. \tag{54}$$

We split the integral into four integrals: I for $|h - h'| > (\log T)^{\frac{-1}{2K}}$, II for $(\log T)^{\frac{-1}{K}} \leq |h - h'| \leq (\log T)^{\frac{-1}{2K}}$, III for $(\log T)^{\frac{-r}{K}} < |h - h'| \leq (\log T)^{\frac{-(r-1)}{K}}$, $r = 2, \ldots K$, and IV for $|h - h'| \leq (\log T)^{-1}$. We will show that

$$\mathrm{I} = \mathbb{E}\left[\mathrm{Leb}(\mathcal{A})\right]^2 (1 + \mathrm{o}(1))$$

and the others $\mathrm{o}(\mathbb{E}[\mathrm{Leb}(\mathcal{A})]^2)$.

- For II, note that $\text{Leb}^{\times 2}\{(h,h') : (\log T)^{\frac{-1}{K}} \leq |h-h'| \leq (\log T)^{\frac{-1}{2K}}\} \ll (\log T)^{\frac{-1}{2K}}$. Moreover, by (21) and Proposition 4, we have

$$\mathbb{P}(\mathcal{J}_h(m) \cap \mathcal{J}_{h'}(m)) \ll \mathbb{P}(\mathcal{J}_h(m))^2.$$

  This implies $\text{II} = \text{o}(\mathbb{E}[\text{Leb}(\mathcal{A})]^2)$.
- For IV, note that clearly $\mathbb{P}(\mathcal{J}_h(m) \cap \mathcal{J}_{h'}(m)) \leq \mathbb{P}(\mathcal{J}_h(m))$. Thus, $\text{IV} \ll (\log T)^{-1}\mathbb{E}[\text{Leb}(\mathcal{A})]$. Using (53) and the fact that $1+\mathcal{E}_{\alpha,u}(\gamma) > 0$ for $\gamma < \gamma^\star$, one gets $\text{IV} = \text{o}(\mathbb{E}[\text{Leb}(\mathcal{A})]^2)$.
- For I, note that $\text{Leb}^{\times 2}\{(h,h') : |h-h'| > (\log T)^{\frac{-1}{2K}}\} = 1+\text{o}(1)$. Moreover, by Proposition 4, $\mathbb{P}(\mathcal{J}_h(m) \cap \mathcal{J}_{h'}(m)) = (1+\text{o}(1))\mathbb{P}(\mathcal{J}_h(m))^2$. This implies $\text{I} = (1+\text{o}(1))\mathbb{E}[\text{Leb}(\mathcal{A})]^2)$.
- For III, the integral is a sum over $r = 2, \ldots, K$ of integrals of pairs with

$$(\log T)^{\frac{-r}{K}} < |h-h'| \leq (\log T)^{\frac{-(r-1)}{K}}.$$

  The measure of this set is $\ll (\log T)^{\frac{-(r-1)}{K}}$. For fix r, the integrand is

$$\begin{aligned}\prod_{m=2}^{K} \mathbb{P}(\mathcal{J}_h(m) \cap \mathcal{J}_{h'}(m)) &\leq \prod_{m=2}^{r} \mathbb{P}(\mathcal{J}_h(m)) \prod_{m=r+1}^{K} \mathbb{P}(\mathcal{J}_h(m) \cap \mathcal{J}_{h'}(m)) \\ &\ll \prod_{m=2}^{r} \mathbb{P}(\mathcal{J}_h(m)) \prod_{m=r+1}^{K} \mathbb{P}(\mathcal{J}_h(m))^2 ,\end{aligned}$$

  where the last line follows by (21) and Proposition 4. Putting all this together and factoring the square of the one-point probabilities, one gets

$$\text{III} \ll \mathbb{E}[\text{Leb}(\mathcal{A})]^2 \sum_{r=2}^{K} (\log T)^{\frac{-(r-1)}{K}} \prod_{m=2}^{r} \Big(\mathbb{P}(\mathcal{J}_h(m))\Big)^{-1}.$$

  We show $\prod_{m=2}^{r} \Big(\mathbb{P}(\mathcal{J}_h(m))\Big)^{-1} < (\log T)^{\frac{(r-1)}{K}}$ uniformly in T. This finishes the proof since the sum is then the tail of a convergent geometric series. In the case $u < 0$, since $\lambda < \lambda^\star$, and $(1+\delta)\gamma < \gamma^\star$ for δ small, we have by (51),

$$\mathbb{P}(\mathcal{J}_h(m))^{-1} \ll \begin{cases} (\log T)^{\frac{\lambda^{\star 2}}{\alpha^2 K(1+u)^2}} & \text{if } m \leq \alpha K \\ (\log T)^{\frac{(\gamma^\star - \lambda^\star)^2}{(1-\alpha)^2 K}} & \text{if } m = \alpha K+1, \ldots, K. \end{cases}$$

  By the definition of $\lambda^\star$ and $\gamma^\star = V^{1/2}$, this implies

$$\prod_{m=2}^{r} \Big(\mathbb{P}(\mathcal{J}_h(m))\Big)^{-1} \ll \begin{cases} (\log T)^{\frac{(1+u)^2}{V}\frac{r-1}{K}} & \text{if } r \leq \alpha K \\ (\log T)^{\frac{(\alpha-1/K)(1+u)^2}{V} + \frac{1}{V}\frac{r-\alpha K}{K}} & \text{if } r = \alpha K+1, \ldots, K. \end{cases}$$

  Since $u < 0$, it is straightforward to check that the exponent is smaller than $\frac{r-1}{K}$ as claimed. The case $u \geq 0$ is done similarly by splitting into two cases $\gamma_c \leq \gamma < \gamma^\star$ and $0 < \gamma < \gamma_c$. We omit the proof for conciseness.

We now have all the results to finish the proof of Proposition 2 using Laplace's method.

Proof of Proposition 2. We first prove the limit in probability. The convergence in L^1, and in particular the convergence of the expectation, will be a consequence of Lemma 6 below. For fixed $\varepsilon > 0$ and $M \in \mathbb{N}$, consider

$$\gamma_j = \frac{j(1+\varepsilon)}{M}\gamma^\star \qquad 0 \le j \le M,$$

and the event

$$A = \bigcap_{j=1}^{M} \Big\{ (\log T)^{\mathcal{E}_{\alpha,u}(\gamma_j)-\varepsilon} \le \mathrm{Leb}\{h : \widetilde{X}_h > \gamma_j \log\log T\} \le (\log T)^{\mathcal{E}_{\alpha,u}(\gamma_j)+\varepsilon} \Big\}$$
$$\bigcap \Big\{ \mathrm{Leb}\{h : \widetilde{X}_h > \gamma_M \log\log T\} = 0 \Big\}. \quad (55)$$

By Lemmas 4 and 5, we have that $\mathbb{P}(A^c) \to 0$ as $T \to \infty$. It remains to prove that the free energy is close to the claimed expression on the event A. On one hand, the following upper bound holds on A:

$$\begin{aligned}\int_0^1 \exp \beta \widetilde{X}_h \, \mathrm{d}h &\le \sum_{j=1}^{M} \int_0^1 \exp \beta \widetilde{X}_h \, \mathbb{1}_{\{(\log T)^{\gamma_{j-1}} < e^{\widetilde{X}_h} \le (\log T)^{\gamma_j}\}} \, \mathrm{d}h \\ &\quad + \int_0^1 \exp \beta \widetilde{X}_h \, \mathbb{1}_{\{e^{\widetilde{X}_h} \le 1\}} \, \mathrm{d}h \\ &\le \sum_{j=1}^{M} (\log T)^{\beta\gamma_j + \mathcal{E}_{\alpha,u}(\gamma_{j-1}) + \varepsilon} + 1.\end{aligned}$$

On the other hand, we have the lower bound

$$\begin{aligned}\int_0^1 \exp \beta \widetilde{X}_h \, \mathrm{d}h &\ge \sum_{j=1}^{M} \int_0^1 \exp \beta \widetilde{X}_h \, \mathbb{1}_{\{(\log T)^{\gamma_{j-1}} < e^{\widetilde{X}_h} \le (\log T)^{\gamma_j}\}} \, \mathrm{d}h \\ &\ge \sum_{j=1}^{M} (\log T)^{\beta\gamma_{j-1} + \mathcal{E}_{\alpha,u}(\gamma_j) - \varepsilon}.\end{aligned}$$

Altogether, this implies

$$\begin{aligned}\max_{1 \le j \le M} \{\beta\gamma_{j-1} + \mathcal{E}_{\alpha,u}(\gamma_j) - \varepsilon\} &\le \frac{\log \int_0^1 \exp \beta \widetilde{X}_h \, \mathrm{d}h}{\log\log T} \\ &\le \max_{1 \le j \le M} \{\beta\gamma_j + \mathcal{E}_{\alpha,u}(\gamma_{j-1}) + \varepsilon\} + \mathrm{o}(1).\end{aligned}$$

In particular, by continuity of $\mathcal{E}_{\alpha,u}(\gamma)$, we can pick M large enough depending on ε and T large enough so that

$$\left| \frac{\log \int_0^1 \exp \beta \widetilde{X}_h \, \mathrm{d}h}{\log\log T} - \max_{\gamma \in [0,\gamma^\star]} \{\beta\gamma + \mathcal{E}_{\alpha,u}(\gamma)\} \right| \le 2\varepsilon.$$

As mentioned above, since $\mathbb{P}(A^c) \to 0$ as $T \to \infty$, this proves the convergence in probability

$$\lim_{T\to\infty} \frac{\log \int_0^1 \exp \beta \widetilde{X}_h \, \mathrm{d}h}{\log\log T} = \max_{\gamma\in[0,\gamma^\star]} \{\beta\gamma + \mathcal{E}_{\alpha,u}(\gamma)\} .$$

It remains to check that the right side has the desired form. Let $V = (1+u)^2\alpha + (1-\alpha)$. If $u < 0$, the optimal γ is $\beta V/2$ whenever $\beta V/2 < \gamma^\star$, i.e., $\beta < 2/V^{1/2}$. If $\beta \geq 2/V^{1/2}$, then the optimal γ is simply $\gamma^\star$. Therefore, we have

$$\max_{\gamma\in[0,\gamma^\star]} \{\beta\gamma + \mathcal{E}_{\alpha,u}(\gamma)\} = \begin{cases} \frac{\beta^2 V}{4} & \text{if } \beta < 2/V^{1/2} \\ \beta V^{1/2} - 1 & \text{if } \beta \geq 2/V^{1/2}. \end{cases}$$

If $u \geq 0$, the optimal γ is $\beta V/2$ if $\gamma < \gamma_c$, i.e., $\beta < 2/(1+u)$. If $\gamma > \gamma_c$, then the optimal γ is $(1+u)\alpha + \frac{\beta(1-\alpha)}{2}$ until it equals $\gamma^\star$. This happens at $\beta \geq 2$. Putting all this together, we obtain that

$$\max_{\gamma\in[0,\gamma^\star]} \{\beta\gamma + \mathcal{E}_{\alpha,u}(\gamma)\} = \begin{cases} \frac{\beta^2\left((1+u)^2\alpha + (1-\alpha)\right)}{4} & \text{if } \beta < \frac{2}{(1+u)} \\ \beta(1+u)\alpha - \alpha + \frac{\beta^2(1-\alpha)}{4} & \text{if } \frac{2}{(1+u)} \leq \beta < 2 \\ \beta\Big((1+u)\alpha + (1-\alpha)\Big) - 1 & \text{if } \beta \geq 2. \end{cases}$$

This corresponds to the expression in Proposition 2 expressed in terms of (12).

Lemma 6. *The sequence of random variables*

$$\Big(\frac{1}{\log\log T} \log \int_0^1 \exp\big(\beta(X_h + uX_h(\alpha)\big)\mathrm{d}h\Big)_{T>1}$$

is uniformly integrable. In particular, the convergence in probability of the sequence is equivalent to the convergence in L^1.

Proof. Write for short

$$f_T = (\log\log T)^{-1} \log \int_0^1 \exp \beta \widetilde{X}_h \, \mathrm{d}h.$$

We need to show that for any $\varepsilon > 0$, there exists C large enough so that uniformly in T,

$$\mathbb{E}[|f_T| \mathbb{1}_{\{|f_T)|>C\}}] < \varepsilon.$$

It is easy to check that

$$\begin{aligned} \mathbb{E}[|f_T| \mathbb{1}_{\{|f_T|>C\}}] &= \int_C^\infty \mathbb{P}(f_T > y)\mathrm{d}y + C\mathbb{P}(f_T > C) \\ &+ \int_{-\infty}^{-C} \mathbb{P}(f_T < y)\,\mathrm{d}y + C\mathbb{P}(f_T < -C). \end{aligned} \tag{56}$$

Therefore, it remains to get a good control on the right and left tail of f_T. For the right tail, observe that by Markov's inequality

$$\mathbb{P}(f_T > y) = \mathbb{P}\Big(\int \exp \beta \widetilde{X}_h \, \mathrm{d}h > (\log T)^y\Big) \leq (\log T)^{-y} \, \mathbb{E}\left[\int \exp \beta \widetilde{X}_h \, \mathrm{d}h\right].$$

Using Proposition 3 and Fubini's theorem, we get

$$\mathbb{P}(f_T > y) \ll (\log T)^{((1+u)^2\alpha + (1-\alpha))\frac{\beta^2}{4} - y}.$$

This implies

$$\int_C^\infty \mathbb{P}(f_T > y) \, \mathrm{d}y + C\mathbb{P}(f_T > C) \ll \frac{(\log T)^{((1+u)^2\alpha + (1-\alpha))\frac{\beta^2}{4} - C}}{\log\log T} + C(\log T)^{((1+u)^2\alpha + (1-\alpha))\frac{\beta^2}{4} - C}.$$

It suffices to take $C > ((1+u)^2\alpha + (1-\alpha))\frac{\beta^2}{4}$ for this to be uniformly small in T. The left tail is bounded the same way after noticing that by Markov's and Jensen's inequalities,

$$\begin{aligned}
\mathbb{P}(f_T < -y) &= \mathbb{P}\Big(\int \exp \beta \widetilde{X}_h \, \mathrm{d}h < (\log T)^{-y}\Big) \\
&\leq (\log T)^{-y} \, \mathbb{E}\left[\left(\int \exp \beta \widetilde{X}_h \, \mathrm{d}h\right)^{-1}\right] \\
&\leq (\log T)^{-y} \, \mathbb{E}\left[\int \exp -\beta \widetilde{X}_h \, \mathrm{d}h\right] \\
&\ll (\log T)^{((1+u)^2\alpha + (1-\alpha))\frac{\beta^2}{4} - y}.
\end{aligned}$$

These estimates imply that $\mathbb{E}[|f_T| \mathbb{1}_{\{|f_T)| > C\}}]$ can be made arbitrarily small in (56) by taking C larger than $((1+u)^2\alpha + (1-\alpha))\frac{\beta^2}{4}$.

Acknowledgements. L.-P. A. is supported by NSF CAREER 1653602, NSF grant DMS-1513441, and a Eugene M. Lang Junior Faculty Research Fellowship. W. T. is partially supported by NSF grant DMS-1513441. Both authors would like to thank Frédéric Ouimet for useful comments on a first version of the paper. L.-P. A. is indebted to Chuck Newman for his constant support and his scientific insights throughout the years.

References

1. Arguin, L.P.: Extrema of log-correlated random variables principles and examples. In: Advances in Disordered Systems. Random Processes and Some Applications, pp. 166–204. Cambridge Univ. Press, Cambridge (2017)
2. Arguin, L.P., Belius, D., Harper, A.J.: Maxima of a randomized Riemann zeta function, and branching random walks. Ann. Appl. Probab. **27**(1), 178–215 (2017)

3. Arguin, L.P., Belius, D., Bourgade, P., Radziwiłł, M., Soundararajan, K.: Maximum of the Riemann zeta function on a short interval of the critical line. Commun. Pure Appl. Math. **72**(3), 500–535 (2019)
4. Arguin, L.P., Bovier, A., Kistler, N.: Genealogy of extremal particles of branching Brownian motion. Commun. Pure Appl. Math. **64**(12), 1647–1676 (2011)
5. Arguin, L.P., Bovier, A., Kistler, N.: Poissonian statistics in the extremal process of branching Brownian motion. Ann. Appl. Probab. **22**(4), 1693–1711 (2012)
6. Arguin, L.P., Ouimet, F.: Extremes of the two-dimensional Gaussian free field with scale-dependent variance. ALEA Lat. Am. J. Probab. Math. Stat. **13**(2), 779–808 (2016)
7. Arguin, L.P., Zindy, O.: Poisson-Dirichlet statistics for the extremes of a log-correlated Gaussian field. Ann. Appl. Probab. **24**(4), 1446–1481 (2014)
8. Arguin, L.P., Zindy, O.: Poisson–Dirichlet statistics for the extremes of the two-dimensional discrete Gaussian free field. Electron. J. Probab. **20**(59), 19 pp. (2015)
9. Auffinger, A., Chen, W.K.: Universality of chaos and ultrametricity in mixed p-spin models. Commun. Pure Appl. Math. **69**(11), 2107–2130 (2016)
10. Biskup, M., Louidor, O.: Extreme local extrema of two-dimensional discrete Gaussian free field. Commun. Math. Phys. **345**(1), 271–304 (2016)
11. Bolthausen, E., Deuschel, J.D., Giacomin, G.: Entropic repulsion and the maximum of the two-dimensional harmonic crystal. Ann. Probab. **29**(4), 1670–1692 (2001)
12. Bourgade, P.: Mesoscopic fluctuations of the zeta zeros. Probab. Theor. Relat. Fields **148**(3–4), 479–500 (2010)
13. Bourgade, P., Keating, J.: Quantum chaos, random matrix theory, and the Riemann ζ-function. In: Chaos: Poincaré Seminar 2010, Birkhauser, Boston, vol. 66, pp. 125–168 (2013)
14. Bovier, A., Kurkova, I.: Derrida's generalised random energy models. I. Models with finitely many hierarchies. Ann. Inst. H. Poincaré Probab. Statist. **40**(4), 439–480 (2004)
15. Bovier, A., Kurkova, I.: Derrida's generalized random energy models. II. Models with continuous hierarchies. Ann. Inst. H. Poincaré Probab. Statist. **40**(4), 481–495 (2004)
16. Carmona, P., Hu, Y.: Universality in Sherrington-Kirkpatrick's spin glass model. Ann. Inst. H. Poincaré Probab. Statist. **42**(2), 215–222 (2006)
17. Daviaud, O.: Extremes of the discrete two-dimensional Gaussian free field. Ann. Probab. **34**(3), 962–986 (2006)
18. Derrida, B.: Random-energy model: an exactly solvable model of disordered systems. Phys. Rev. B **24**, 2613–2626 (1981)
19. Derrida, B., Spohn, H.: Polymers on disordered trees, spin glasses, and traveling waves. J. Statist. Phys. **51**(5–6), 817–840 (1988)
20. Duplantier, B., Rhodes, R., Sheffield, S., Vargas, V.: Critical Gaussian multiplicative chaos: convergence of the derivative martingale. Ann. Probab. **42**(5), 1769–1808 (2014)
21. Fyodorov, Y.V., Hiary, G.A., Keating, J.P.: Freezing transition, characteristic polynomials of random matrices, and the Riemann zeta function. Phys. Rev. Lett. **108**, 170601 (2012)
22. Fyodorov, Y.V., Keating, J.P.: Freezing transitions and extreme values: random matrix theory, and disordered landscapes. Philos. Trans. R. Soc. Lond. Ser. A Math. Phys. Eng. Sci. **372**(2007), 20120503, 32 (2014)
23. Harper, A.J.: A note on the maximum of the Riemann zeta function, and log-correlated random variables. Preprint arxiv: 1304.0677 (2013)

24. Jagannath, A.: On the overlap distribution of branching random walks. Electron. J. Probab. **21**, 16 pp. (2016)
25. Kistler, N.: Derrida's random energy models. From spin glasses to the extremes of correlated random fields. In: Correlated Random Systems: Five Different Methods, Lecture Notes in Math., vol. 2143, pp. 71–120. Springer, Cham (2015)
26. Montgomery, H.L., Vaughan, R.C.: Multiplicative Number Theory. I. Classical Theory, Cambridge Studies in Advanced Mathematics, vol. 97. Cambridge University Press, Cambridge (2007)
27. Najnudel, J.: On the extreme values of the Riemann zeta function on random intervals of the critical line. Probab. Theory Related Fields, pp. 1–66 (2017)
28. Newman, C.M.: Zeros of the partition function for generalized Ising systems. Commun. Pure Appl. Math. **27**(2), 143–159 (1974)
29. Newman, C.M.: Inequalities for Ising models and field theories which obey the Lee-Yang theorem. Commun. Math. Phys. **41**(1), 1–9 (1975)
30. Newman, C.M.: Classifying general Ising models. In: Les Méthodes Mathématiques de la Théorie Quantique des Champs, C.N.R.S., Paris, pp. 273–288 (1976)
31. Newman, C.M.: Fourier transforms with only real zeros. Proc. Amer. Math. Soc. **61**, 245–251 (1976)
32. Ouimet, F.: Geometry of the Gibbs measure for the discrete 2D Gaussian free field with scale-dependent variance. ALEA Lat. Am. J. Probab. Math. Stat. **14**(2), 851–902 (2017)
33. Ouimet, F.: Poisson-Dirichlet statistics for the extremes of a randomized Riemann zeta function. Electron. Commun. Probab. **23**(46), 15 (2018)
34. Radizwiłł, M., Soundararajan, K.: Selberg's central limit theorem for $\log|\zeta(1/2+it)|$. Enseign. Math. **63**, 1–19 (2017). https://doi.org/10.4171/LEM/63-1/2-1
35. Rodgers, B., Tao, T.: The de Bruijn–Newman constant is non-negative. Preprint arXiv:1801.05914 (2018)
36. Saksman, E., Webb, C.: Multiplicative chaos measures for a random model of the Riemann zeta function. Preprint arxiv:1604.08378 (2016)
37. Saksman, E., Webb, C.: The Riemann zeta function and gaussian multiplicative chaos: statistics on the critical line. Preprint arxiv:1609.00027 (2017)
38. Saouter, Y., Gourdon, X., Demichel, P.: An improved lower bound for the de Bruijn-Newman constant. Math. Comput. **80**, 2281–2287 (2011)
39. Schumayer, D., Hutchinson, D.A.W.: Colloquium: physics of the Riemann hypothesis. Rev. Mod. Phys. **83**, 307–330 (2011)
40. Soundararajan, K.: Moments of the Riemann zeta function. Ann. Math. (2) **170**(2), 981–993 (2009)
41. Titchmarsh, E.: The Theory of the Riemann Zeta-Function, 2nd edn. Oxford Univ. Press, New York (1986)

A Homogenization Theorem for Langevin Systems with an Application to Hamiltonian Dynamics

Jeremiah Birrell[1,2] and Jan Wehr[3(✉)]

[1] Department of Mathematics, University of Arizona, Tucson, AZ 85721, USA
[2] Department of Mathematics and Statistics, University of Massachusetts Amherst, Amherst, MA 01003, USA
birrell@math.umass.edu
[3] Department of Mathematics and Program in Applied Mathematics, University of Arizona, Tucson, AZ 85721, USA
wehr@math.arizona.edu

We dedicate this work to Chuck Newman on the occasion of his 70th birthday

Abstract. This paper studies homogenization of stochastic differential systems. The standard example of this phenomenon is the small mass limit of Hamiltonian systems. We consider this case first from the heuristic point of view, stressing the role of detailed balance and presenting the heuristics based on a multiscale expansion. This is used to propose a physical interpretation of recent results by the authors, as well as to motivate a new theorem proven here. Its main content is a sufficient condition, expressed in terms of solvability of an associated partial differential equation ("the cell problem"), under which the homogenization limit of an SDE is calculated explicitly. The general theorem is applied to a class of systems, satisfying a generalized detailed balance condition with a position-dependent temperature.

Keywords: Homogenization · Stochastic differential equation · Hamiltonian system · Small mass limit · Noise-induced drift

1 Introduction and Background

This paper studies the small mass limit of a general class of Langevin equations. Langevin dynamics is defined in terms of canonical variables—positions and momenta—by adding damping and (Itô) noise terms to Hamiltonian equations. In the limit when the mass, or masses, of the system's particles, go to zero, the momenta homogenize, and one obtains a limiting equation for the position variables only. This is a great simplification which often allows one to see the

V. Sidoravicius (Ed.): *Sojourns in Probability Theory and Statistical Physics - I*, PROMS 298, pp. 89–122, 2019.
https://doi.org/10.1007/978-981-15-0294-1_4

nature of the dynamics more clearly. If the damping matrix of the original system depends on its state, a *noise-induced drift* arises in the limit. We analyze and interpret this term from several points of view. The paper consists of four parts. The first part contains general background on stochastic differential equations. In the second part, the small-mass limit of Langevin equations is studied using a multiscale expansion. This method requires making additional assumptions, but it leads to correct results in all cases in which rigorous proofs are known. The third part presents a new rigorous result about homogenization of a general class of singularly perturbed SDEs. The final part applies this result to prove a theorem about the homogenization of a large class of Langevin systems.

1.1 Stochastic Differential Equations

Let us start from a general background on Langevin equations. The material presented here is not new, and its various versions can be found in many textbooks, see for example [1]. We do not strive for complete precision or a listing of all necessary assumptions in our discussions here. The aim of the first two sections is to motivate and facilitate reading the remainder of the paper. Detailed technical considerations will be reserved for Sects. 3 and 4, where we present our new results.

Consider the stochastic differential equation

$$dy_t = b(y_t)\,dt + \sigma(y_t)\,dW_t. \tag{1}$$

The process y_t takes values in $\mathbb{R}^m$, b is a vector field in $\mathbb{R}^m$, W is an n-dimensional Wiener process and σ is an $m \times n$-matrix-valued function. Define an $m \times m$ matrix Σ by $\Sigma = \sigma\sigma^T$. The Eq. (1) defines a Markov process with the infinitesimal operator

$$(Lf)(y) = \frac{1}{2}\Sigma_{ij}\partial_i\partial_j f + b_i\partial_i f$$

where we are writing ∂_i for ∂_{y_i} and suppressing the dependence of Σ, b_i and f on y from the notation. Summation over repeating indices is implied.

We assume that this process has a unique stationary probability measure with a C^2-density $h(y)$. Under this assumption h satisfies the equation

$$L^*h = 0$$

where L^* denotes the formal adjoint of L,

$$L^*f = \frac{1}{2}\partial_i\partial_j\left(\Sigma_{ij}f\right) - \partial_i\left(b_i f\right).$$

That is, we have

$$\partial_i\left(\frac{1}{2}\partial_j\left(\Sigma_{ij}h\right) - b_i h\right) = 0. \tag{2}$$

Consider the special case when h solves the equation

$$\frac{1}{2}\partial_j\left(\Sigma_{ij}h\right) - b_i h = 0. \tag{3}$$

In this case the operator L is symmetric on the space $L^2\left(\mathbb{R}^m, h\right) \equiv L_h^2$ of square-integrable functions with the weight h, as the following calculation shows. Using product formula, we have

$$\int (Lf)\,gh = \int fL^*(gh) = \int f\partial_i\left(\frac{1}{2}\partial_j\left(\Sigma_{ij}gh\right) - b_i gh\right).$$

The expression in parentheses equals

$$\frac{1}{2}\partial_j g\left(\Sigma_{ij}h\right) + g\frac{1}{2}\partial_j\left(\Sigma_{ij}h\right) - gb_i h = \frac{1}{2}\partial_j g\left(\Sigma_{ij}h\right)$$

by (3). Applying product formula again, we obtain

$$\int (Lf)\,gh = \int f\left(\frac{1}{2}\Sigma_{ij}\left(\partial_i\partial_j g\right)h + \frac{1}{2}\partial_i\left(\Sigma_{ij}h\right)\partial_j g\right)$$

which, by another application of (3), equals

$$\int f\left(\frac{1}{2}\Sigma_{ij}\partial_i\partial_j g - b_j\partial_j g\right)h = \int f\,(Lg)\,h.$$

Here is a more complete discussion.

1.2 Detailed Balance Condition and Symmetry of the Infinitesimal Operator

We have

$$\begin{aligned}
\int (Lf)\,gh &= \int\left(\frac{1}{2}\Sigma_{ij}\partial_i\partial_j f + b_i\partial_i f\right)gh \\
&= -\frac{1}{2}\int \partial_i f\partial_j\left(\Sigma_{ij}gh\right) + \int\left(\partial_i f\right)b_i gh \\
&= -\frac{1}{2}\int \partial_i f\left[\partial_j\left(\Sigma_{ij}h\right)g + \Sigma_{ij}h\partial_j g\right] + \partial\left(\partial_i f\right)b_i gh \\
&= \int \partial_i f\left[-\frac{1}{2}\partial_j\left(\Sigma_{ij}h\right) + b_i h\right]g - \frac{1}{2}\int \Sigma_{ij}\partial_i f\partial_j gh.
\end{aligned}$$

Interchanging the roles of f and g and canceling the term symmetric in f and g, we obtain

$$\int (Lf)\,gh - \int f\,(Lg)\,h = \int\left[\left(\partial_i f\right)g - \left(\partial_i g\right)f\right]\left(-\frac{1}{2}\partial_j\left(\Sigma_{ij}h\right) + b_i h\right).$$

If h is a solution to the equation

$$-\frac{1}{2}\partial_j\left(\Sigma_{ij}h\right)+b_ih=0$$

then the above expression is zero, showing that the operator L is symmetric on the space L^2_h. Conversely, for this symmetry to hold, the $\mathbb{R}^m$-valued function $-\frac{1}{2}\partial_j\left(\Sigma_{ij}h\right)+b_ih$ has to be orthogonal to all elements of the space L^2 (of functions with values in $\mathbb{R}^m$) of the form $\left(\partial_i f\right)g-\left(\partial_i g\right)f$. It is not hard to prove that every C^1 function with this property must vanish, and thus, that $\frac{1}{2}\partial_j\left(\Sigma_{ij}h\right)-b_ih=0$. Here is a sketch of a proof: suppose ϕ is C^1 and orthogonal to all such functions. That is, for every f and g,

$$\int\left[\phi_i\left(\partial_i f\right)g-\phi_i\left(\partial_i g\right)f\right]=0.$$

Integrating the first term by parts we obtain

$$\int\left[-\left(\partial_i\phi_i\right)g-2\phi_i\partial_i g\right]f=0.$$

Since this holds for all f, it follows that

$$-\left(\partial_i\phi_i\right)g-2\phi_i\partial_i g=0$$

and thus also

$$\int\left[-\left(\partial_i\phi_i\right)g-2\phi_i\partial_i g\right]=0.$$

Integrating the second term by parts, we get

$$\int\left(\partial_i\phi_i\right)g=0$$

and, since this is true for every g, it follows that $\partial_i\phi_i$ vanishes. We thus have, for every g

$$\phi_i\partial_i g=0$$

and this implies that ϕ vanishes. In summary:

Proposition: If the density h of the stationary probability measure is C^2, then h satisfies the stationary Fokker–Planck equation

$$\partial_i\left[-\frac{1}{2}\partial_j\left(\Sigma_{ij}h\right)+b_ih\right]=0.$$

The stronger statement

$$-\frac{1}{2}\partial_j\left(\Sigma_{ij}h\right)+b_ih=0$$

is equivalent to symmetry of the operator L on the space L^2_h.

We are now going to relate the above symmetry statement to the detailed balance property of the stationary dynamics. First, it is clearly equivalent to the analogous property for the backward Kolmogorov semigroup:

$$\int (P_t f)\, gh = \int f\,(P_t g)\, h$$

since $P_t = \exp(tL)$. Now, $(P_t f)(x)$ is the expected value of $f(x_t)$ for the process, starting at x at time 0. In particular, for $f = \delta_y$, we obtain $P_t f(x) = p_t(x, y)$—the density of the transition probability from x to y in time t. Using the above symmetry of P_t with $f = \delta_y$ and $g = \delta_x$, we obtain the detailed balance condition:

$$h(x) p_t(x, y) = h(y) p_t(y, x)$$

which, conversely, implies the symmetry statement for arbitrary f and g.

1.3 The Case of a Linear Drift and Constant Noise

When both $b(y)$ and $\sigma(y)$ are constant or depend linearly on y, (1) can be solved explicitly [2] and an explicit formula for its stationary distribution can be found, when it exists. We consider the special case $b(y) = -\gamma y$ and $\sigma(y) \equiv \sigma$, where γ and σ are constant matrices and the eigenvalues of γ have positive real parts. The stationary Fokker–Planck equation, (2), reads

$$\nabla \cdot \left(\frac{1}{2} \Sigma \nabla h + (\gamma y) h \right) = 0$$

where $\Sigma = \sigma \sigma^T$. It has a Gaussian solution

$$h(y) = (2\pi)^{-\frac{m}{2}} (\det M)^{-\frac{1}{2}} \exp \left(-\frac{1}{2} \left(M^{-1} y, y \right) \right)$$

with the covariance matrix M which is the unique solution of the Lyapunov equation

$$\gamma M + M \gamma^T = \Sigma$$

and can be written as (see, for example, [3])

$$M = \int_0^\infty \exp(-t\gamma)\, \Sigma \exp\left(-t\gamma^T\right) dt.$$

This result can be verified by a direct calculation. We emphasize that it holds without assuming the detailed balance condition. The latter condition is satisfied if and only if the above h solves the equation

$$\frac{1}{2} \Sigma \nabla h + (\gamma y) h = 0$$

which is equivalent to $M = \frac{1}{2} \gamma^{-1} \Sigma$ or, in terms of the coefficients of the system, to

$$\Sigma \gamma^T = \gamma \Sigma \tag{4}$$

To see the physical significance of this condition, let us go back to the general case and write (adapting the discussion in [4] to our notation)

$$\gamma = \frac{1}{2}\Sigma M^{-1} - i\Omega.$$

Ω represents the "oscillatory degrees of freedom" of the diffusive system. The above calculations show that the detailed balance condition is equivalent to $\Omega = 0$, in agreement with the physical intuition that there are no macroscopic currents in the stationary state.

2 Small Mass Limit—A Perturbative Approach

We are now going to apply the general facts about Langevin equations to a model of a mechanical system, interacting with a noisy environment. The dynamical variables of this system are positions and momenta, and, in general, the Langevin equations which describe its time evolution, are not linear. However, when investigating the small mass limit of the system by a perturbative method, we will encounter equations closely related to those studied above. This will be explained later, when we interpret the limiting equations.

Consider a mechanical system with the Hamiltonian $\mathcal{H}(q,p)$ where $q, p \in \mathbb{R}^n$. We want to study a small mass limit of this system, coupled to a damping force and the noise. Therefore, we introduce the variable $z = \frac{p}{\sqrt{m}}$ and assume the Hamiltonian can be written $\mathcal{H}(q,p) = H(q,z)$ where H is independent of m. We thus have

$$\begin{aligned} dq_t &= \frac{1}{\sqrt{m}}\nabla_z H(q_t, z_t)\,dt \\ dz_t &= -\frac{1}{\sqrt{m}}\nabla_q H(q_t, z_t)\,dt - \frac{1}{m}\gamma(q_t)\nabla_z H(q_t, z_t)\,dt + \frac{1}{\sqrt{m}}\sigma(q_t)\,dW_t. \end{aligned}$$

γ is $n \times n$-matrix-valued, σ is $n \times k$-matrix-valued and W is a k-dimensional Wiener process. We emphasize that σ does not play here the same role that it played in our discussion of the general Langevin equation, since the noise term enters only the equation for dz_t. The number k of the components of the driving noise does not have to be related to the dimension of the system in any particular way. The corresponding backward Kolmogorov equation for a function $\rho(q, z, t)$ is

$$\partial_t \rho = L\rho$$

where the differential operator L equals

$$L = \frac{1}{m}L_1 + \frac{1}{\sqrt{m}}L_2$$

with

$$\begin{aligned} L_1 &= \frac{1}{2}\Sigma\nabla_z \cdot \nabla_z - \gamma\nabla_z H\nabla_z \\ L_2 &= \nabla_z H \cdot \nabla_q - \nabla_q H \cdot \nabla_z \end{aligned}$$

where $\Sigma(q) = \sigma(q)\sigma(q)^T$. We represent the solution of the Kolmogorov equation as a formal series

$$\rho = \rho_0 + \sqrt{m}\rho_1 + m\rho_2 + \dots$$

Equating the expressions, proportional to m^{-1}, $m^{-\frac{1}{2}}$ and m^0, we obtain the equations:

$$\begin{aligned} L_1\rho_0 &= 0, \\ L_1\rho_1 &= -L_2\rho_0, \\ \partial_t\rho_0 &= L_1\rho_2 + L_2\rho_1. \end{aligned}$$

To satisfy the first equation it is sufficient to choose ρ_0 which does not depend on z:

$$\rho_0 = \rho_0(q,t).$$

If we now search for ρ_1 which is linear in z, the second equation simplifies to

$$\gamma\nabla_z H \cdot \nabla_z\rho_1 = \nabla_z H \cdot \nabla_q\rho_0$$

which has a solution

$$\rho_1(q,z) = \left(\gamma^{-1}\right)^T \nabla_q\rho_0 \cdot z = \nabla_q\rho_0 \cdot \gamma^{-1}z.$$

Writing the third equation as

$$\partial_t\rho_0 - L_2\rho_1 = L_1\rho_2$$

and applying the identity

$$RanL_1 = (KerL_1^*)^\perp$$

to the space L^2 with respect to the z variable, we see that $\partial_t\rho_0 - L_2\rho_1 = L_1\rho_2$ must be orthogonal in this space to any function h in the null space of L_1^*. We have

$$L_1^*h = \nabla_z \cdot \left(\frac{1}{2}\Sigma\nabla_z h + (\gamma\nabla_z H)\,h\right)$$

where $\Sigma = \sigma\sigma^T$.

It is impossible to continue the analysis without further, simplifying assumptions. We are first going to study the case of a general H, assuming a form of the detailed balance condition in the variable z, at fixed q.

Assumption 1. for every q there exists a nonnegative solution of the equation

$$\frac{1}{2}\Sigma\nabla_z h + (\gamma\nabla_z H)\,h = 0 \tag{5}$$

of finite $L^1(dz)$-norm. We can thus choose

$$\int h(q,z)\,dz = 1.$$

We will say in this case that the system satisfies the *conditional detailed balance property* in the variable z. Since ρ_0 does not depend on z, the orthogonality condition can be written as

$$\partial_t \rho_0 = \int L_2 \rho_1(q,z) h(q,z)\, dz.$$

We have the following explicit formula for $L_2\rho_1$ (summation over repeated indices is implied):

$$\begin{aligned} L_2\rho_1 =& \partial_{z_i} H \left(\partial_{q_i}\partial_{q_j}\rho_0\right)\left(\gamma^{-1}\right)_{jk} z_k + \partial_{z_i} H \left(\partial_{q_j}\rho_0\right)\partial_{q_i}\left(\left(\gamma^{-1}\right)_{jk}\right) z_k \\ &- \partial_{q_i} H \left(\partial_{q_j}\rho_0\right)\left(\gamma^{-1}\right)_{ji}. \end{aligned}$$

To integrate it against $h(q,z)$, we will use the following consequence of (5)

$$\begin{aligned} &\int \left(\partial_{z_i} H\right) z_k h(q,z)\, dz = -\frac{1}{2}\int \left(\gamma^{-1}\Sigma\nabla_z h\right)_i z_k\, dz \qquad (6) \\ =& -\int \left(\gamma^{-1}\Sigma\right)_{ij}\left(\partial_{z_j} h\right) z_k\, dz = -\frac{1}{2}\left(\gamma^{-1}\Sigma\right)_{ij}\int \left(\partial_{z_j} h\right) z_k\, dz \\ =& \frac{1}{2}\left(\gamma^{-1}\Sigma\right)_{ij}\int h\delta_{jk}\, dz = \frac{1}{2}\left(\gamma^{-1}\Sigma\right)_{ik}. \end{aligned}$$

The orthogonality condition is thus

$$\begin{aligned} \partial_t\rho_0 =& -\left(\gamma^{-1}\right)_{ji}\langle\partial_{q_i} H\rangle\partial_{q_j}\rho_0 + \frac{1}{2}\left(\gamma^{-1}\Sigma\right)_{ik}\partial_{q_i}\left(\left(\gamma^{-1}\right)_{jk}\right)\partial_{q_j}\rho_0 \\ &+ \frac{1}{2}\left(\gamma^{-1}\Sigma\right)_{ik}\left(\gamma^{-1}\right)_{jk}\left(\partial_{q_i}\partial_{q_j}\rho_0\right). \end{aligned}$$

In this formula, which is more general than the detailed-balance case of the rigorous result of [5], $\langle - \rangle$ denotes the average (i.e. the integral over z with the density $h(q,z)$). This notation is used only in the term in which the average has not been calculated explicitly. Passing from the Kolmogorov equation to the corresponding SDE, we obtain the effective Langevin equation in the $m \to 0$ limit:

$$dq_t = -\gamma(q_t)^{-1}\left(\langle\nabla_q H\rangle(q_t) + S(q_t)\right) dt + \gamma^{-1}(q_t)\sigma(q_t)\, dW_t \qquad (7)$$

where the components of the noise-induced drift, $S(q)$, are given by

$$S_i(q) = \frac{1}{2}\left(\gamma^{-1}\Sigma\right)_{jk}\partial_{q_j}\left(\left(\gamma^{-1}\right)_{ik}\right)$$

and we have used

$$\gamma^{-1}\sigma\left(\gamma^{-1}\sigma\right)^T = \gamma^{-1}\Sigma\left(\gamma^{-1}\right)^T.$$

We are now going to interpret the limiting Eq. (7), using the stationary probability measure $h(q,z)\, dz$, as follows: from the original equations for q_t and z_t we obtain

$$dq_t = -\gamma(q_t)^{-1}\nabla_q H\, dt + \gamma(q_t)^{-1}\sigma(q_t)\, dW_t - \sqrt{m}\gamma(q_t)^{-1}\, dz_t.$$

Integrating the last term by parts, we obtain

$$\sqrt{m}\left(\gamma^{-1}(q_t)\right)_{ij} dz_t^j = d\left(\sqrt{m}\left(\gamma_{ij}^{-1}(q_t)\right) z_t^j\right) - \sqrt{m}\, d\left(\left(\gamma^{-1}\right)_{ij}\right) z_t^j.$$

We leave the first term out, since, under fairly general natural assumptions, it is of order $m^{\frac{1}{2}}$ [5]. The second term equals

$$-\partial_{q_k}\left(\left(\gamma^{-1}\right)_{ij}\right)\left(\partial_{z_k} H\right) z_j\, dt.$$

We substitute this into the equation for dq_t and average, multiplying by $h(q, z)$ and integrating over z. The calculation is as in (6) and the result is thus the same as the equation obtained by the multiscale expansion (7). This provides the following heuristic physical interpretation of the perturbative result: the smaller m is, the faster the variation of z becomes, and in the limit $m \to 0$, z homogenizes instantaneously, with q changing only infinitesimally.

Let us now discuss conditions, under which one may expect our conditional detailed balance assumption to hold. As seen above, at fixed q this assumption is equivalent to existence of a non-negative, integrable solution of the equation

$$\frac{1}{2}\Sigma\nabla_z h + \gamma\left(\nabla_z H\right) h = 0. \tag{8}$$

This equation can be rewritten as

$$\frac{\nabla_z h}{h} = -2\Sigma^{-1}\gamma\nabla_z H.$$

The left-hand side equals $\nabla_z \log h$. Letting $B = -2\Sigma^{-1}\gamma$ to simplify notation, we see that a necessary condition for existence of a solution is that $B\nabla_z H$ be a gradient. This requires

$$\partial_{z_k}\left(b_{ij}\partial_{z_j} H\right) = \partial_{z_i}\left(b_{kj}\partial_{z_j} H\right)$$

for all i, k, where b_{ij} are matrix elements of B. Introducing the matrix $R = (r_{ij})$ of second derivatives of H,

$$r_{ij} = \partial_{z_i}\partial_{z_j} H$$

we see that solvability of (8) is equivalent to symmetry of the product BR:

$$BR = RB^T.$$

For the system to satisfy the conditional detailed balance property, this relation has to be satisfied for all q and z. When H is a quadratic function of z, the matrix R is constant. Even though in this case we will derive the limiting equation without assuming conditional detailed balance, let us remark that the above approach provides a method of determining when that condition holds, different from that used earlier. Namely, let

$$H(q, z) = V(q) + \frac{1}{2}Q(q)z \cdot z$$

where $Q(q)$ is a symmetric matrix. We then have $R = Q$ and the solvability condition becomes

$$BQ = QB^T.$$

In a still more special—but the most frequently considered—case when Q is a multiple of identity, this reduces to

$$B = B^T$$

which is easily seen to be equivalent to the relation

$$\gamma\Sigma = \Sigma\gamma^T.$$

We have derived this condition earlier by a different argument (4).

If γ is symmetric, this becomes the commutation relation

$$\gamma\Sigma = \Sigma\gamma.$$

Note that if $\gamma\Sigma = \Sigma\gamma^T$, the solution of the Lyapunov equation

$$J\gamma^T + \gamma J = \Sigma$$

is given by $J = \frac{1}{2}\gamma^{-1}\Sigma$. In this the case the linear Langevin equation in the z variable, whose conditional equilibrium at fixed value of q we are studying, has no "oscillatory degrees of freedom", as discussed earlier (see also [4]).

In the case when H is not a quadratic function of z, the matrix $BR(q, z)$ has to be symmetric for all q and z, which means satisfying a continuum of conditions for every fixed q. It is interesting to ask whether there exist physically natural examples in which this happens, without each $B(q)$ being a multiple of identity. We are not going to pursue this question here.

In the case when $B(q)$ is a multiple of identity, we can write

$$A = 2\beta(q)^{-1}\gamma$$

with $\beta(q)^{-1} = k_B T(q)$ and call the scalar function $T(q)$ *generalized temperature.* The limiting Kolmogorov equation reads then

$$\partial_t \rho_0 = -\left(\gamma^{-1}\right)_{ji} \langle \partial_{q_i} H \rangle \partial_{q_j} \rho_0 + k_B T \partial_{q_k} \left(\gamma^{-1}\right)_{jk} \partial_{q_j} \rho_0 + k_B T \left(\gamma^{-1}\right)_{ij} \left(\partial_{q_i} \partial_{q_j} \rho_0\right)$$

and the components of the noise-induced drift are thus

$$S_j(q) = k_B T \partial_{q_k} \left(\gamma^{-1}\right)_{jk}.$$

The above applies in particular in the one-dimensional case, in which $\sigma(q)^2$ and $\gamma(q)$ are scalars and hence one is always an (q-dependent) multiple of the other:

$$k_B T(q) = \frac{\sigma(q)^2}{2\gamma(q)}.$$

The limiting Langevin equation is in this case

$$dq_t = -\frac{\langle \nabla_q H \rangle}{\gamma}\, dt - \frac{1}{2}\frac{\nabla_q \gamma}{\gamma^3}\sigma^2\, dt + \frac{\sigma}{\gamma}\, dW_t.$$

For a Hamiltonian equal to a sum of potential and quadratic kinetic energy, $H = V(q) + \frac{z^2}{2}$, the first term equals $\frac{F}{\gamma}$, where $F = -\nabla_q V\, dt$, in agreement with earlier results.

The second situation, in which the perturbative treatment of the original system can be carried out explicitly is the general quadratic kinetic energy case.

Assumption 2. $H = V(q) + \frac{z^2}{2}$

If we follow the singular perturbation method used above, we again need to find the integral (6), where $\partial_{z_i} H = z_i$. In this case we know the solution of $L_1^* h = 0$ explicitly:

$$h(q, z) = (2\pi)^{-\frac{n}{2}} (\det M)^{-\frac{1}{2}} \exp\left(-\frac{1}{2}M^{-1}z \cdot z\right)$$

so the integral in (6) is the mean of $z_i z_k$ in the Gaussian distribution with the covariance $M = (m_{ik})$, that is, m_{ik}. The second-order term in the Kolmogorov equation is thus $m_{ik}\left(\gamma^{-1}\right)_{jk} \partial_{q_i}\partial_{q_j}\rho_0$. The corresponding Langevin equation, which has been derived rigorously in [5] is in this case

$$dq_t = -\gamma^{-1}(q_t)\nabla_q V(q_t)\, dt + S(q_t)\, dt + \gamma^{-1}(q_t)\sigma(q_t)\, dW_t.$$

The homogenization heuristics proposed under Assumption 1 applies here as well: the limiting Langevin equation can be interpreted as a result of averaging over the conditional stationary distribution of the z variable. A rigorous result, corroborating this picture has recently been proven in [6].

3 A Rigorous Homogenization Theorem

We now develop a framework for the homogenization of Langevin equations that is able to make many of the heuristic results from the previous two sections rigorous. Our results will concern Hamiltonians of the form

$$H(t, x) = K(t, q, p - \psi(t, q)) + V(t, q) \tag{9}$$

where $x = (q, p) \in \mathbb{R}^n \times \mathbb{R}^n$, $K = K(t, q, z)$ and $V = V(t, q)$ are C^2, $\mathbb{R}$-valued functions, K is non-negative, and ψ is a C^2, $\mathbb{R}^n$-valued function. The splitting of H into K and V does not have to correspond physically to any notion of kinetic and potential energy, although we will use those terms for convenience. The splitting is not unique; it will be constrained further as we continue. We now define the family of scaled Hamiltonians, parameterized by $\epsilon > 0$ (generalizing the above mass parameter):

$$H^\epsilon(t,q,p) \equiv K^\epsilon(t,q,p) + V(t,q) \equiv K(t,q,(p-\psi(t,q))/\sqrt{\epsilon}) + V(t,q).$$

Consider the following family of SDEs:

$$dq_t^\epsilon = \nabla_p H^\epsilon(t,x_t^\epsilon)dt, \tag{10}$$
$$dp_t^\epsilon = (-\gamma(t,x_t^\epsilon)\nabla_p H^\epsilon(t,x_t^\epsilon) - \nabla_q H^\epsilon(t,x_t^\epsilon) + F(t,x_t^\epsilon))dt + \sigma(t,x_t^\epsilon)dW_t, \tag{11}$$

where $\gamma : [0,\infty)\times\mathbb{R}^{2n} \to \mathbb{R}^{n\times n}$ and $\sigma : [0,\infty)\times\mathbb{R}^{2n} \to \mathbb{R}^{n\times k}$ are continuous, γ is positive definite, and W_t is a $\mathbb{R}^k$-valued Brownian motion on a filtered probability space $(\Omega,\mathcal{F},\mathcal{F}_t,P)$ satisfying the usual conditions [7].

Our objective in this section is to develop a method for investigating the behavior of x_t^ϵ in the limit $\epsilon \to 0^+$; more precisely, we wish to prove the existence of a limiting "position" process q_t and derive a homogenized SDE that it satisfies. In fact, the method we develop is applicable to a more general class SDEs that share certain properties with (10)–(11). In the following subsection, we discuss some prior results concerning (10)–(11). This will help motivate the assumptions made in the development of our general homogenization method, starting in Subsect. 3.2.

3.1 Summary of Prior Results

Let x_t^ϵ be a family of solutions to the SDE (10)–(11) with initial condition $x_0^\epsilon = (q_0^\epsilon, p_0^\epsilon)$. We assume that a solution exists for all $t \geq 0$ (i.e. there are no explosions). See Appendix B in [8] for assumptions that guarantee this. Under Assumptions 1–3 in [8] (repeated as Assumptions A.1–A.3 in Appendix A, we showed that for any $T > 0$, $p > 0$, $0 < \beta < p/2$ we have

$$\sup_{t\in[0,T]} E\left[\|p_t^\epsilon - \psi(t,q_t^\epsilon)\|^p\right] = O(\epsilon^{p/2}) \text{ and } E\left[\sup_{t\in[0,T]} \|p_t^\epsilon - \psi(t,q_t^\epsilon)\|^p\right] = O(\epsilon^\beta) \tag{12}$$

as $\epsilon \to 0^+$ i.e. the point (p,q) is attracted to the surface defined by $p = \psi(t,q)$.

Adding Assumption 4 (Assumption (A.4) in the appendix) we also showed that

$$\begin{aligned} d(q_t^\epsilon)^i =& (\tilde{\gamma}^{-1})^{ij}(t,q_t^\epsilon)(-\partial_t\psi_j(t,q_t^\epsilon) - \partial_{q^j}V(t,q_t^\epsilon) + F_j(t,x_t^\epsilon))dt \\ &+ (\tilde{\gamma}^{-1})^{ij}(t,q_t^\epsilon)\sigma_{j\rho}(t,x_t^\epsilon)dW_t^\rho - (\tilde{\gamma}^{-1})^{ij}(t,q_t^\epsilon)\partial_{q^j}K(t,q_t^\epsilon,z_t^\epsilon)dt \\ &+ (z_t^\epsilon)_j\partial_{q^l}(\tilde{\gamma}^{-1})^{ij}(t,q_t^\epsilon)\partial_{z_l}K(t,q_t^\epsilon,z_t^\epsilon)dt - d((\tilde{\gamma}^{-1})^{ij}(t,q_t^\epsilon)(u_t^\epsilon)_j) \\ &+ (u_t^\epsilon)_j\partial_t(\tilde{\gamma}^{-1})^{ij}(t,q_t^\epsilon)dt, \end{aligned} \tag{13}$$

where $u_t^\epsilon \equiv p_t^\epsilon - \psi(t,q_t^\epsilon)$, $z_t^\epsilon \equiv u_t^\epsilon/\sqrt{\epsilon}$, and

$$\tilde{\gamma}_{ik}(t,q) \equiv \gamma_{ik}(t,q) + \partial_{q^k}\psi_i(t,q) - \partial_{q^i}\psi_k(t,q). \tag{14}$$

We define the components of $\tilde{\gamma}^{-1}$ such that

$$(\tilde{\gamma}^{-1})^{ij}\tilde{\gamma}_{jk} = \delta^i_k,$$

and for any v_i we define $(\tilde{\gamma}^{-1}v)^i = (\tilde{\gamma}^{-1})^{ij}v_j$.

Under the additional Assumptions 5–7 in [8], which include further restrictions on the form of the Hamiltonian, we were then able to show that q_t^ϵ converges in an L^p-norm as $\epsilon \to 0^+$ to the solution of a lower dimensional SDE,

$$\begin{aligned} dq_t =& \tilde{\gamma}^{-1}(t,q_t)(-\partial_t\psi(t,q_t) - \nabla_q V(t,q_t) + F(t,q_t,\psi(t,q_t)))dt + S(t,q_t)dt \\ &+ \tilde{\gamma}^{-1}(t,q_t)\sigma(t,q_t,\psi(t,q_t))dW_t. \end{aligned} \tag{15}$$

The *noise-induced drift* term, $S(t,q)$, that arises in the limit is the term of greatest interest here. Its form is given in Eq. (3.26) in [8].

The homogenization technique used in [8] to arrive at (15) relies heavily on the specific structural assumptions on the form of the Hamiltonian. Those assumptions cover a wide variety of important systems, such as a particle in an electromagnetic field, and motion on a Riemannian manifold, but it is desirable to search for a more generally applicable homogenization method. In this paper, we develop a significantly more general technique, adapted from the methods presented in [9], that is capable of homogenizing terms of the form $G(t,q_t^\epsilon,(p_t^\epsilon - \psi(t,q_t^\epsilon))/\sqrt{\epsilon})dt$ for a general class of SDEs that satisfy the property (12), as well as prove convergence of q_t^ϵ to the solution of a limiting, homogenized SDE. In particular, it will be capable of homogenizing q_t^ϵ from the Hamiltonian system (10)–(11) under less restrictive assumptions on the form of the Hamiltonian, than those made in [8]. We emphasize that the convergence statements are proven in the strong sense, see Sect. 3.2.

3.2 General Homogenization Framework

Here we describe our homogenization technique in a more general context than the Hamiltonian setting from the previous section. This method is related to the cell problem method from [9], our proof applies to a larger class of SDEs and demonstrates L^p-convergence rather that weak convergence.

We will denote an element of $\mathbb{R}^n \times \mathbb{R}^m$ by $x = (q,p)$, where we no longer require the q and p degrees of freedom to have the same dimensionality, though we still employ the convention of writing q indices with superscripts and p indices with subscripts. We let W_t be an $\mathbb{R}^k$-valued Wiener process, $\psi : [0,\infty)\times\mathbb{R}^n \to \mathbb{R}^m$ be C^2 and $G_1, F_1 : [0,\infty)\times\mathbb{R}^{n+m}\times\mathbb{R}^m \to \mathbb{R}^n$, $G_2, F_2 : [0,\infty)\times\mathbb{R}^{n+m}\times\mathbb{R}^m \to \mathbb{R}^m$, $\sigma_1 : [0,\infty)\times\mathbb{R}^{n+m} \to \mathbb{R}^{n\times k}$, and $\sigma_2 : [0,\infty)\times\mathbb{R}^{n+m} \to \mathbb{R}^{m\times k}$ be continuous. With these definitions, we consider the following family of SDEs, depending on a parameter $\epsilon > 0$:

$$dq_t^\epsilon = \left(\frac{1}{\sqrt{\epsilon}}G_1(t,x_t^\epsilon,z_t^\epsilon) + F_1(t,x_t^\epsilon,z_t^\epsilon)\right)dt + \sigma_1(t,x_t^\epsilon)dW_t, \tag{16}$$

$$dp_t^\epsilon = \left(\frac{1}{\sqrt{\epsilon}}G_2(t,x_t^\epsilon,z_t^\epsilon) + F_2(t,x_t^\epsilon,z_t^\epsilon)\right)dt + \sigma_2(t,x_t^\epsilon)dW_t, \tag{17}$$

where we define $z_t^\epsilon = (p_t^\epsilon - \psi(t, q_t^\epsilon))/\sqrt{\epsilon}$. We will assume, in analogy with (12), that:

Assumption 3.1. For any $T > 0$, $p > 0$, $0 < \beta < p/2$ we have

$$\sup_{t\in[0,T]} E\left[\|p_t^\epsilon - \psi(t, q_t^\epsilon)\|^p\right] = O(\epsilon^{p/2}) \text{ and } E\left[\sup_{t\in[0,T]} \|p_t^\epsilon - \psi(t, q_t^\epsilon)\|^p\right] = O(\epsilon^\beta)$$

as $\epsilon \to 0^+$.

In words, we assume that the p degrees of freedom are attracted to the values defined by $p = \psi(t, q)$. This is an appropriate setting to expect some form of homogenization, as it suggests that the dynamics in the limit $\epsilon \to 0^+$ can be characterized by fewer degrees of freedom—the q-variables.

Homogenization of Integral Processes. In this section we derive a method capable of homogenizing processes of the form

$$M_t^\epsilon \equiv \int_0^t G(s, x_s^\epsilon, z_s^\epsilon) ds \tag{18}$$

in the limit $\epsilon \to 0^+$. More specifically, our aim is to find conditions under which there exists some function, $S(t, q)$, such that

$$\int_0^t G(s, x_s^\epsilon, z_s^\epsilon) ds - \int_0^t S(s, q_s^\epsilon) ds \to 0 \tag{19}$$

in some norm, as $\epsilon \to 0^+$, i.e. only the q-degrees of freedom are needed to characterize M_t^ϵ in the limit. We will call a family of processes, $S(t, q_t^\epsilon)dt$, that satisfies such a limit, a homogenization of $G(t, x_t^\epsilon, z_t^\epsilon)dt$. The technique we develop will also be useful for proving existence of a limiting process q_s (i.e. $q_s^\epsilon \to q_s$), and showing that

$$\int_0^t G(s, x_s^\epsilon, z_s^\epsilon) ds \to \int_0^t S(s, q_s) ds.$$

as $\epsilon \to 0^+$. We will consider this second question in Sect. 3.2. Here, our focus is on (19).

As a starting point, let $\chi(t, x, z) : [0, \infty) \times \mathbb{R}^{n+m} \times \mathbb{R}^m \to \mathbb{R}$ be $C^{1,2}$, where $C^{1,2}$ is defined as follows:

- If $\sigma_1 \neq 0$ then we take this to mean χ is C^1 and, for each t, $\chi(t, x, z)$ is C^2 in (x, z) with second derivatives continuous jointly in all variables.
- If $\sigma_1 = 0$ then we take this to mean χ is C^1 and, for each t, q, $\chi(t, q, p, z)$ is C^2 in (p, z) with second derivatives continuous jointly in all variables.

Eventually, we will need to carefully choose χ so that we achieve our aim, but for now we simply use Itô's formula to compute $\chi(t, x_t^\epsilon, z_t^\epsilon)$. We defined $C^{1,2}$ precisely so that Itô's formula is justified. For this computation, we will define $\chi^\epsilon(t,x) = \chi(t, x, (p - \psi(t,q))/\sqrt{\epsilon})$, and

$$\Sigma_{11}^{ij} = \sum_\rho (\sigma_1)^i_\rho (\sigma_1)^j_\rho,\ (\Sigma_{12})^i_j = \sum_\rho (\sigma_1)^i_\rho (\sigma_2)_{j\rho},\ (\Sigma_{22})_{ij} = \sum_\rho (\sigma_2)_{i\rho}(\sigma_2)_{j\rho}. \tag{20}$$

Itô's formula gives

$$\begin{aligned}
\chi(t, x_t^\epsilon, z_t^\epsilon) = \chi(0, x_0^\epsilon, z_0^\epsilon) &+ \int_0^t \partial_s \chi^\epsilon(s, x_s^\epsilon) ds \\
&+ \int_0^t \nabla_q \chi^\epsilon(s, x_s^\epsilon) \cdot dq_s^\epsilon + \int_0^t \nabla_p \chi^\epsilon(s, x_s^\epsilon) \cdot dp_s^\epsilon \\
&+ \frac{1}{2}\int_0^t \partial_{q^i}\partial_{q^j}\chi^\epsilon(s, x_s^\epsilon)\Sigma_{11}^{ij}(s, x_s^\epsilon) ds \\
&+ \frac{1}{2}\int_0^t \partial_{q^i}\partial_{p_j}\chi^\epsilon(s, x_s^\epsilon)(\Sigma_{12})^i_j(s, x_s^\epsilon) ds \\
&+ \frac{1}{2}\int_0^t \partial_{p_i}\partial_{q^j}\chi^\epsilon(s, x_s^\epsilon)(\Sigma_{12})^j_i(s, x_s^\epsilon) ds \\
&+ \frac{1}{2}\int_0^t \partial_{p_i}\partial_{p_j}\chi^\epsilon(s, x_s^\epsilon)(\Sigma_{22})_{ij}(s, x_s^\epsilon) ds.
\end{aligned}$$

Note that if $\sigma_1 = 0$ then only the second derivatives that we have assumed exist are involved in this computation.

We can compute these terms as follows:

$$\begin{aligned}
\partial_t \chi^\epsilon(t,x) =& \partial_t \chi(t,x,z) - \partial_{z_i}\chi(t,x,z)\partial_t \psi_i(t,q)/\sqrt{\epsilon}, \\
\partial_{q^i}\chi^\epsilon(t,x) =& (\partial_{q^i}\chi)(t,x,z) - \epsilon^{-1/2}\partial_{q^i}\psi_k(t,q)(\partial_{z_k}\chi)(t,x,z), \\
\partial_{p_i}\chi^\epsilon(t,x) =& (\partial_{p_i}\chi)(t,x,z) + \epsilon^{-1/2}(\partial_{z_i}\chi)(t,x,z), \\
\partial_{q^i}\partial_{q^j}\chi^\epsilon(t,x) =& (\partial_{q^i}\partial_{q^j}\chi)(t,x,z) + \epsilon^{-1/2}\left(-\partial_{q^j}\psi_k(t,q)(\partial_{q^i}\partial_{z_k}\chi)(t,x,z)\right. \\
&\left. - \partial_{q^i}\partial_{q^j}\psi_k(t,q)(\partial_{z_k}\chi)(t,x,z) - \partial_{q^i}\psi_k(t,q)(\partial_{q^j}\partial_{z_k}\chi)(t,x,z)\right) \\
&+ \epsilon^{-1}\partial_{q^i}\psi_k(t,q)\partial_{q^j}\psi_l(t,q)(\partial_{z_k}\partial_{z_l}\chi)(t,x,z). \\
\partial_{p_i}\partial_{p_j}\chi^\epsilon(t,x) =& (\partial_{p_i}\partial_{p_j}\chi)(t,x,z) + \epsilon^{-1/2}\left((\partial_{z_j}\partial_{p_i}\chi)(t,x,z)\right. \\
&\left. + (\partial_{p_j}\partial_{z_i}\chi)(t,x,z)\right) + \epsilon^{-1}(\partial_{z_i}\partial_{z_j}\chi)(t,x,z), \\
\partial_{q^i}\partial_{p_j}\chi^\epsilon(t,x) =& (\partial_{q^i}\partial_{p_j}\chi)(t,x,z) + \epsilon^{-1/2}\left((\partial_{q^i}\partial_{z_j}\chi)(t,x,z)\right. \\
&\left. - \partial_{q^i}\psi_k(t,q)(\partial_{p_j}\partial_{z_k}\chi)(t,x,z)\right) \\
&- \epsilon^{-1}\partial_{q^i}\psi_k(t,q)(\partial_{z_j}\partial_{z_k}\chi)(t,x,z),
\end{aligned}$$

where z is evaluated at $z(t,x,\epsilon) = (p - \psi(t,q))/\sqrt{\epsilon}$ in each of the above formulae.

Using these expressions, together with the SDE (16)–(17) we find

$$
\begin{aligned}
\chi(t,x_t^\epsilon,z_t^\epsilon) = \chi(0,x_0^\epsilon,z_0^\epsilon)&\\
&+\int_0^t \partial_s\chi(s,x_s^\epsilon,z_s^\epsilon)-\epsilon^{-1/2}(\partial_{z_i}\chi)(s,x_s^\epsilon,z_s^\epsilon)\partial_s\psi_i(s,q_s^\epsilon)ds\\
&+\int_0^t \Big((\partial_{q^i}\chi)(s,x_s^\epsilon,z_s^\epsilon)-\epsilon^{-1/2}\partial_{q^i}\psi_k(s,q_s^\epsilon)(\partial_{z_k}\chi)(s,x_s^\epsilon,z_s^\epsilon)\Big)\\
&\quad\times\left[\left(\frac{1}{\sqrt{\epsilon}}G_1(s,x_s^\epsilon,z_s^\epsilon)+F_1(s,x_s^\epsilon,z_s^\epsilon)\right)ds+\sigma_1(s,x_s^\epsilon)dW_s\right]^i\\
&+\int_0^t \Big((\partial_{p_i}\chi)(s,x_s^\epsilon,z_s^\epsilon)+\epsilon^{-1/2}(\partial_{z_i}\chi)(s,x_s^\epsilon,z_s^\epsilon)\Big)\\
&\quad\times\left[\left(\frac{1}{\sqrt{\epsilon}}G_2(s,x_s^\epsilon,z_s^\epsilon)+F_2(s,x_s^\epsilon,z_s^\epsilon)\right)ds+\sigma_2(s,x_s^\epsilon)dW_s\right]_i\\
&+\frac{1}{2}\int_0^t \Sigma_{11}^{ij}(s,x_s^\epsilon)\Big[(\partial_{q^i}\partial_{q^j}\chi)(s,x_s^\epsilon,z_s^\epsilon)\\
&\quad+\epsilon^{-1/2}\big(-\partial_{q^j}\psi_k(s,q_s^\epsilon)(\partial_{q^i}\partial_{z_k}\chi)(s,x_s^\epsilon,z_s^\epsilon)\\
&\qquad-\partial_{q^i}\partial_{q^j}\psi_k(s,q_s^\epsilon)(\partial_{z_k}\chi)(s,x_s^\epsilon,z_s^\epsilon)\\
&\qquad-\partial_{q^i}\psi_k(s,q_s^\epsilon)(\partial_{q^j}\partial_{z_k}\chi)(s,x_s^\epsilon,z_s^\epsilon)\big)\\
&\quad+\epsilon^{-1}\partial_{q^i}\psi_k(s,q_s^\epsilon)\partial_{q^j}\psi_l(s,q_s^\epsilon)(\partial_{z_k}\partial_{z_l}\chi)(s,x_s^\epsilon,z_s^\epsilon)\Big]ds\\
&+\int_0^t (\Sigma_{12})^i_j(s,x_s^\epsilon)\Big[(\partial_{q^i}\partial_{p_j}\chi)(s,x_s^\epsilon,z_s^\epsilon)+\epsilon^{-1/2}\left((\partial_{q^i}\partial_{z_j}\chi)(s,x_s^\epsilon,z_s^\epsilon)\right.\\
&\quad\left.-\partial_{q^i}\psi_k(s,q_s^\epsilon)(\partial_{p_j}\partial_{z_k}\chi)(s,x_s^\epsilon,z_s^\epsilon)\right)\\
&\quad-\epsilon^{-1}\partial_{q^i}\psi_k(s,q_s^\epsilon)(\partial_{z_j}\partial_{z_k}\chi)(s,x_s^\epsilon,z_s^\epsilon)\Big]ds\\
&+\frac{1}{2}\int_0^t (\Sigma_{22})_{ij}(s,x_s^\epsilon)\Big[(\partial_{p_i}\partial_{p_j}\chi)(s,x_s^\epsilon,z_s^\epsilon)\\
&\quad+\epsilon^{-1/2}\left((\partial_{z_j}\partial_{p_i}\chi)(s,x_s^\epsilon,z_s^\epsilon)\right.\\
&\quad\left.+(\partial_{p_j}\partial_{z_i}\chi)(s,x_s^\epsilon,z_s^\epsilon)\right)+\epsilon^{-1}(\partial_{z_i}\partial_{z_j}\chi)(s,x_s^\epsilon,z_s^\epsilon)\Big]ds.
\end{aligned}
$$

Multiplying by ϵ and collecting powers, we arrive at

$$
\int_0^t (L\chi)(s,x_s^\epsilon,z_s^\epsilon)ds=\epsilon^{1/2}(R_1^\epsilon)_t+\epsilon\left(\chi(t,x_t^\epsilon,z_t^\epsilon)-\chi(0,x_0^\epsilon,z_0^\epsilon)+(R_2^\epsilon)_t\right), \tag{21}
$$

where we define

$$
\begin{aligned}
(L\chi)(t,x,z)=&\bigg(\frac{1}{2}\Sigma_{11}^{ij}(t,x)\partial_{q^i}\psi_k(t,q)\partial_{q^j}\psi_l(t,q) \qquad (22)\\
&-(\Sigma_{12})^i_l(t,x)\partial_{q^i}\psi_k(t,q)+\frac{1}{2}(\Sigma_{22})_{kl}(t,x)\bigg)(\partial_{z_k}\partial_{z_l}\chi)(t,x,z)\\
&+\left((G_2)_k(t,x,z)-\partial_{q^i}\psi_k(t,q)G_1^i(t,x,z)\right)(\partial_{z_k}\chi)(t,x,z),
\end{aligned}
$$

$$
\begin{aligned}
&(R_1^\epsilon)_t \qquad (23)\\
&= \int_0^t (\partial_{z_i}\chi)(s,x_s^\epsilon,z_s^\epsilon)\partial_s\psi_i(s,q_s^\epsilon)ds - \int_0^t (\partial_{q^i}\chi)(s,x_s^\epsilon,z_s^\epsilon)G_1^i(s,x_s^\epsilon,z_s^\epsilon)ds\\
&+ \int_0^t \partial_{q^i}\psi_k(s,q_s^\epsilon)(\partial_{z_k}\chi)(s,x_s^\epsilon,z_s^\epsilon)\left[F_1(s,x_s^\epsilon,z_s^\epsilon)ds + \sigma_1(s,x_s^\epsilon)dW_s\right]^i\\
&- \int_0^t (\partial_{z_i}\chi)(s,x_s^\epsilon,z_s^\epsilon)\left[F_2(s,x_s^\epsilon,z_s^\epsilon)ds + \sigma_2(s,x_s^\epsilon)dW_s\right]_i\\
&- \int_0^t (\partial_{p_i}\chi)(s,x_s^\epsilon,z_s^\epsilon)(G_2)_i(s,x_s^\epsilon,z_s^\epsilon)ds\\
&- \frac{1}{2}\int_0^t \Sigma_{11}^{ij}(s,x_s^\epsilon)\big(-\partial_{q^j}\psi_k(s,q_s^\epsilon)(\partial_{q^i}\partial_{z_k}\chi)(s,x_s^\epsilon,z_s^\epsilon)\\
&\qquad - \partial_{q^i}\partial_{q^j}\psi_k(s,q_s^\epsilon)(\partial_{z_k}\chi)(s,x_s^\epsilon,z_s^\epsilon) - \partial_{q^i}\psi_k(s,q_s^\epsilon)(\partial_{q^j}\partial_{z_k}\chi)(s,x_s^\epsilon,z_s^\epsilon)\big)\,ds\\
&- \int_0^t (\Sigma_{12})^i_j(s,x_s^\epsilon)\big((\partial_{q^i}\partial_{z_j}\chi)(s,x_s^\epsilon,z_s^\epsilon) - \partial_{q^i}\psi_k(s,q_s^\epsilon)(\partial_{p_j}\partial_{z_k}\chi)(s,x_s^\epsilon,z_s^\epsilon)\big)\,ds\\
&- \int_0^t (\Sigma_{22})_{ij}(s,x_s^\epsilon)(\partial_{z_j}\partial_{p_i}\chi)(s,x_s^\epsilon,z_s^\epsilon)ds,
\end{aligned}
$$

and

$$
\begin{aligned}
&(R_2^\epsilon)_t \qquad (24)\\
&= -\int_0^t \partial_s\chi(s,x_s^\epsilon,z_s^\epsilon)ds\\
&- \int_0^t (\partial_{q^i}\chi)(s,x_s^\epsilon,z_s^\epsilon)\left[F_1(s,x_s^\epsilon,z_s^\epsilon)ds + \sigma_1(s,x_s^\epsilon)dW_s\right]^i\\
&- \int_0^t (\partial_{p_i}\chi)(s,x_s^\epsilon,z_s^\epsilon)\left[F_2(s,x_s^\epsilon,z_s^\epsilon)ds + \sigma_2(s,x_s^\epsilon)dW_s\right]_i\\
&- \frac{1}{2}\int_0^t \Sigma_{11}^{ij}(s,x_s^\epsilon)(\partial_{q^i}\partial_{q^j}\chi)(s,x_s^\epsilon,z_s^\epsilon)ds\\
&- \int_0^t (\Sigma_{12})^i_j(s,x_s^\epsilon)(\partial_{q^i}\partial_{p_j}\chi)(s,x_s^\epsilon,z_s^\epsilon)ds\\
&- \frac{1}{2}\int_0^t (\Sigma_{22})_{ij}(s,x_s^\epsilon)(\partial_{p_i}\partial_{p_j}\chi)(s,x_s^\epsilon,z_s^\epsilon)ds.
\end{aligned}
$$

First, think of simply homogenizing (18) to a quantity of the form

$$\int_0^t \tilde{G}(s,x_s^\epsilon)ds.$$

Suppose we have a candidate for $\tilde{G}$. If we can find a $C^{1,2}$ solution, χ, to the PDE

$$(L\chi)(t,x,z) = G(t,x,z) - \tilde{G}(t,x)$$

then substituting this into (21) gives

$$\begin{aligned} &\int_0^t G(s, x_s^\epsilon, z_s^\epsilon)ds - \int_0^t \tilde{G}(s, x_s^\epsilon)ds \\ =&\epsilon^{1/2}(R_1^\epsilon)_t + \epsilon\left(\chi(t, x_t^\epsilon, z_t^\epsilon) - \chi(0, x_0^\epsilon, z_0^\epsilon) + (R_2^\epsilon)_t\right). \end{aligned} \tag{25}$$

Given sufficient growth bounds for χ and its derivatives, one anticipates that the right hand side of (25) vanishes in the limit. If in addition, $\tilde{G}$ is Lipschitz in p, uniformly in (t, q), then, based on Assumption 3.1, one expects

$$\int_0^t G(s, x_s^\epsilon, z_s^\epsilon)ds - \int_0^t \tilde{G}(s, q_s^\epsilon, \psi(s, q_s^\epsilon))ds \to 0$$

as $\epsilon \to 0^+$.

We make this informal discussion precise in Theorem 1, below. For this, we will need the following assumptions:

Assumption 3.2. For all $T > 0$, the following quantities are polynomially bounded in z, with the bounds uniform on $[0, T] \times \mathbb{R}^{n+m}$:
G_1, F_1, G_2, F_2, σ_1, σ_2, $\partial_t\psi$, $\partial_{q^i}\psi$, $\partial_{q^i}\partial_{q^j}\psi$. If $\sigma_1 = 0$ then we can remove the requirement on $\partial_{q^i}\partial_{q^j}\psi$.

Recall that an $\mathbb{R}^l$-valued function, $\phi(t, x, z)$, is called *polynomially bounded* in z, uniformly on $[0, T] \times \mathbb{R}^{n+m}$ if there exists $q, C > 0$ such that

$$\|\phi(t, x, z)\| \leq C(1 + \|z\|^q)$$

for all $(t, x, z) \in [0, T] \times \mathbb{R}^{n+m} \times \mathbb{R}^m$. In particular, if ϕ is independent of z, this just means it is bounded on $[0, T] \times \mathbb{R}^{n+m}$. Applying this to ψ, we note that Assumption 3.2 implies ψ is Lipschitz in q, uniformly in $t \in [0, T]$.

Assumption 3.3. Given a continuous $G : [0, \infty) \times \mathbb{R}^{n+m} \times \mathbb{R}^m \to \mathbb{R}$, assume that there exists a $C^{1,2}$ function $\chi : [0, \infty) \times \mathbb{R}^{n+m} \times \mathbb{R}^m \to \mathbb{R}$ and a continuous function $\tilde{G}(t, x) : [0, \infty) \times \mathbb{R}^{n+m} \to \mathbb{R}$ that together satisfy the PDE

$$(L\chi)(t, x, z) = G(t, x, z) - \tilde{G}(t, x), \tag{26}$$

where the differential operator, L, is defined in (22).

Assume that, for a given $T > 0$, $\tilde{G}$ is Lipschitz in p, uniformly for $(t, q) \in [0, T] \times \mathbb{R}^n$. Also suppose that χ, its first derivatives, and the second derivatives $\partial_{q^i}\partial_{q^j}\chi$, $\partial_{q^i}\partial_{p_j}\chi$, $\partial_{q^i}\partial_{z_j}\chi$, $\partial_{p_i}\partial_{p_j}\chi$, and $\partial_{p_i}\partial_{z_j}\chi$ are polynomially bounded in z, uniformly for $(t, x) \in [0, T] \times \mathbb{R}^{n+m}$. If $\sigma_1 = 0$ then the only second derivatives that we require to be polynomially bounded are $\partial_{p_i}\partial_{p_j}\chi$ and $\partial_{p_i}\partial_{z_j}\chi$.

Theorem 1. *Fix $T > 0$. Let Assumptions 3.1–3.3 hold and $x_t^\epsilon = (q_t^\epsilon, p_t^\epsilon)$ satisfy the SDE* (16)–(17). *Then for any $p > 0$ we have*

$$E\left[\sup_{t\in[0,T]}\left|\int_0^t G(s, x_s^\epsilon, z_s^\epsilon)ds - \int_0^t \tilde{G}(s, x_s^\epsilon)ds\right|^p\right] = O(\epsilon^{p/2}).$$

and

$$E\left[\sup_{t\in[0,T]}\left|\int_0^t G(s,x_s^\epsilon,z_s^\epsilon)ds-\int_0^t \tilde{G}(s,q_s^\epsilon,\psi(s,q_s^\epsilon))ds\right|^p\right]=O(\epsilon^{p/2})$$

as $\epsilon\to 0^+$.

Proof. Fix $T>0$. First let $p\geq 2$. (25) gives

$$\begin{aligned}&E\left[\sup_{t\in[0,T]}\left|\int_0^t G(s,x_s^\epsilon,z_s^\epsilon)ds-\int_0^t \tilde{G}(s,x_s^\epsilon)ds\right|^p\right]\\ \leq& 3^{p-1}\left(\epsilon^{p/2}E\left[\sup_{t\in[0,T]}|(R_1^\epsilon)_t|^p\right]+2^p\epsilon^pE\left[\sup_{t\in[0,T]}|\chi(t,x_t^\epsilon,z_t^\epsilon)|^p\right]\right.\\ &\left.+\epsilon^pE\left[\sup_{t\in[0,T]}|(R_2^\epsilon)_t|^p\right]\right).\end{aligned}$$

From (23) and (24) we see that R_1^ϵ and R_2^ϵ have the forms

$$(R_i^\epsilon)_t=\int_0^t V_i(s,x_s^\epsilon,z_s^\epsilon)ds+\int_0^t Q_{ij}(s,x_s^\epsilon,z_s^\epsilon)dW_s^j,$$

where V_i and Q_{ij} are linear combinations of products of (components of) one or more terms from the following list:
$G_1, F_1, G_2, F_2, \sigma_1, \sigma_2, \partial_t\psi, \partial_{q^i}\psi, \partial_{q^i}\partial_{q^j}\psi, \partial_t\chi, \partial_{q^i}\chi, \partial_{z_i}\chi, \partial_{p_i}\chi, \partial_{q^i}\partial_{q^j}\chi, \partial_{q^i}\partial_{p_j}\chi, \partial_{q^i}\partial_{z_j}\chi, \partial_{p_i}\partial_{p_j}\chi, \partial_{p_i}\partial_{z_j}\chi$. Also note that if $\sigma_1=0$ then the only second derivatives terms that are involved are $\partial_{p_i}\partial_{p_j}\chi$ and $\partial_{p_i}\partial_{z_j}\chi$.

By assumption, these are all polynomially bounded in z, uniformly on $[0,T]\times\mathbb{R}^{n+m}$, as is χ. Therefore, letting $\tilde{C}$ denote a constant that potentially varies line to line, there exists $r>0$ such that

$$\begin{aligned}&E\left[\sup_{t\in[0,T]}\left|\int_0^t G(s,x_s^\epsilon,z_s^\epsilon)ds-\int_0^t \tilde{G}(s,x_s^\epsilon)ds\right|^p\right]\\ \leq&\tilde{C}\epsilon^{p/2}\left(E\left[\left(\int_0^T|V_1(s,x_s^\epsilon,z_s^\epsilon)|ds\right)^p\right]+E\left[\sup_{t\in[0,T]}\left|\int_0^t Q_{1j}(s,x_s^\epsilon,z_s^\epsilon)dW_s^j\right|^p\right]\right)\\ &+\tilde{C}\epsilon^p\left(E\left[\left(\int_0^T|V_2(s,x_s^\epsilon,z_s^\epsilon)|ds\right)^p\right]+E\left[\sup_{t\in[0,T]}\left|\int_0^t Q_{2j}(s,x_s^\epsilon,z_s^\epsilon)dW_s^j\right|^p\right]\right.\\ &\left.+1+E\left[\sup_{t\in[0,T]}\|z_t^\epsilon\|^{rp}\right]\right).\end{aligned}$$

Hölder's inequality and polynomial boundedness yields

$$E\left[\left(\int_0^T |V_i(s,x_s^\epsilon,z_s^\epsilon)|ds\right)^p\right] \leq T^{p-1}E\left[\int_0^T |V_i(s,x_s^\epsilon,z_s^\epsilon)|^p ds\right]$$
$$\leq \tilde{C}T^p\left(1+\sup_{t\in[0,T]} E\left[\|z_t^\epsilon\|^{rp}\right]\right).$$

Applying the Burkholder–Davis–Gundy inequality to the terms involving Q_{ij}, (as found in, for example, Theorem 3.28 in [7]), and then Hölder's inequality, we obtain

$$E\left[\sup_{t\in[0,T]}\left|\int_0^t Q_{ij}(s,x_s^\epsilon,z_s^\epsilon)dW_s^j\right|^p\right]$$
$$\leq \tilde{C}E\left[\left(\int_0^T \|Q_i(s,x_s^\epsilon,z_s^\epsilon)\|^2 ds\right)^{p/2}\right]$$
$$\leq \tilde{C}T^{p/2-1}E\left[\int_0^T \|Q_i(s,x_s^\epsilon,z_s^\epsilon)\|^p ds\right]$$
$$\leq \tilde{C}T^{p/2}\left(1+\sup_{t\in[0,T]} E[\|z_t^\epsilon\|^{rp}]\right).$$

Combining these bounds, and using Assumption 3.1, we find

$$E\left[\sup_{t\in[0,T]}\left|\int_0^t G(s,x_s^\epsilon,z_s^\epsilon)ds - \int_0^t \tilde{G}(s,x_s^\epsilon)ds\right|^p\right]$$
$$\leq \tilde{C}\epsilon^{p/2}\left(1+\sup_{t\in[0,T]} E[\|z_t^\epsilon\|^{rp}]\right)$$
$$+\tilde{C}\epsilon^p\left(1+\sup_{t\in[0,T]} E[\|z_t^\epsilon\|^{rp}] + E\left[\sup_{t\in[0,T]}\|z_t^\epsilon\|^{rp}\right]\right)$$
$$\leq \tilde{C}\epsilon^{p/2}(1+O(1)) + \tilde{C}\epsilon^p\left(1+O(1)+O(\epsilon^{-\delta})\right)$$

for any $\delta > 0$. Letting $\delta = p/2$ we find

$$E\left[\sup_{t\in[0,T]}\left|\int_0^t G(s,x_s^\epsilon,z_s^\epsilon)ds - \int_0^t \tilde{G}(s,x_s^\epsilon)ds\right|^p\right] = O(\epsilon^{p/2}).$$

Now use Hölder's inequality, the uniform Lipschitz property of $\tilde{G}$, and Assumption 3.1 again to compute

$$\begin{aligned}
&E\left[\sup_{t\in[0,T]}\left|\int_0^t G(s,x_s^\epsilon,z_s^\epsilon)ds-\int_0^t \tilde{G}(s,q_s^\epsilon,\psi(s,q_s^\epsilon))ds\right|^p\right]\\
&\leq O(\epsilon^{p/2})+\tilde{C}E\left[\sup_{t\in[0,T]}\left|\int_0^t \tilde{G}(s,x_s^\epsilon)-\tilde{G}(s,q_s^\epsilon,\psi(s,q_s^\epsilon))ds\right|^p\right]\\
&\leq O(\epsilon^{p/2})+\tilde{C}T^{p-1}E\left[\int_0^T |\tilde{G}(s,x_s^\epsilon)-\tilde{G}(s,q_s^\epsilon,\psi(s,q_s^\epsilon))|^p ds\right]\\
&\leq O(\epsilon^{p/2})+\tilde{C}T^p\sup_{t\in[0,T]}E\left[\|p_t^\epsilon-\psi(t,q_t^\epsilon)\|^p\right]\\
&=O(\epsilon^{p/2}).
\end{aligned}$$

This proves the claim for $p\geq 2$. The result for arbitrary $p>0$ then follows from an application of Hölder's inequality. □

Formal Derivation of $\tilde{G}$. Formally applying the Fredholm alternative to (26) motivates the form that $\tilde{G}$ must have in order for χ and its derivatives to possess the growth bounds required by Theorem 1. The formal calculation is simple enough that we repeat it here:
Let L^* be the formal adjoint to L and suppose we have a solution, $h(t,x,z)$, to

$$L^*h=0,\quad \int h(t,x,z)dz=1. \tag{27}$$

If χ and its derivatives grow slowly enough and h and its derivatives decay quickly enough, then $\int hL\chi dz$ will exist, the boundary terms from integration by parts will vanish at infinity, and we find

$$0=\int (L^*h)\chi dz=\int hL(\chi)dz=\int h(G-\tilde{G})dz=\int hGdz-\tilde{G}.$$

Therefore we must have

$$\tilde{G}(t,x)=\int h(t,x,z)G(t,x,z)dz.$$

In essence, the homogenized quantity is obtained by averaging over h, the instantaneous equilibrium distribution for the fast variables, z. This corroborates the heuristic discussion in Sect. 2.

Limiting Equation. We now apply the above framework to prove existence of a limiting process $q_t^\epsilon\to q_s$ and deriving an SDE satisfied by q_s. Specifically, we have:

Theorem 2. *Let $T > 0$, $p \geq 2$, $0 < \beta \leq p/2$, $x_t^\epsilon = (q_t^\epsilon, p_t^\epsilon)$ satisfy the SDE (16)–(17), suppose Assumptions 3.1–3.3 hold, and that the SDE for q_t^ϵ, (16), can be rewritten in the form*

$$q_t^\epsilon = q_0^\epsilon + \int_0^t \tilde{F}(s, x_s^\epsilon)ds + \int_0^t G(s, x_s^\epsilon, z_s^\epsilon)ds + \int_0^t \tilde{\sigma}(s, x_s^\epsilon)dW_s + R_t^\epsilon \tag{28}$$

where the components of G have the properties described in Assumption 3.3,

$$\tilde{F}(t,x) : [0,\infty) \times \mathbb{R}^{n+m} \to \mathbb{R}^n, \quad \tilde{\sigma}(t,x) : [0,\infty) \times \mathbb{R}^{n+m} \to \mathbb{R}^{n\times k}$$

are continuous, Lipschitz in x, uniformly in $t \in [0,T]$, and R_t^ϵ are continuous semimartingales that satisfy

$$E\left[\sup_{t\in[0,T]} \|R_t^\epsilon\|^p\right] = O(\epsilon^\beta) \text{ as } \epsilon \to 0^+.$$

Suppose $\tilde{G}$ (from Assumption 3.3) is Lipschitz in x, uniformly in $t \in [0,T]$, and we have initial conditions $E[\|q_0^\epsilon\|^p] < \infty$, $E[\|q_0\|^p] < \infty$, and $E[\|q_0^\epsilon - q_0\|^p] = O(\epsilon^{p/2})$. Then

$$E\left[\sup_{t\in[0,T]} \|q_t^\epsilon - q_t\|^p\right] = O(\epsilon^\beta) \text{ as } \epsilon \to 0^+$$

where q_t satisfies the SDE

$$\begin{aligned} q_t = q_0 + &\int_0^t \tilde{F}(s, q_s, \psi(s, q_s))ds + \int_0^t \tilde{G}(s, q_s, \psi(s, q_s))ds \\ &+ \int_0^t \tilde{\sigma}(s, q_s, \psi(s, q_s))dW_s. \end{aligned} \tag{29}$$

Proof. We will prove this theorem by verifying all the hypotheses of Lemma A.3. Define

$$\tilde{R}_t^\epsilon = R_t^\epsilon + \int_0^t G(s, x_s^\epsilon, z_s^\epsilon)ds - \int_0^t \tilde{G}(s, x_s^\epsilon)ds.$$

Then

$$q_t^\epsilon = q_0^\epsilon + \int_0^t \tilde{F}(s, x_s^\epsilon)ds + \int_0^t \tilde{G}(s, x_s^\epsilon)ds + \int_0^t \tilde{\sigma}(s, x_s^\epsilon)dW_s + \tilde{R}_t^\epsilon$$

where $\tilde{F} + \tilde{G}$ and $\tilde{\sigma}$ are Lipschitz in x, uniformly for $t \in [0,T]$ and

$$E\left[\sup_{t\in[0,T]} \|\tilde{R}_t^\epsilon\|^p\right] = O(\epsilon^\beta)$$

by Theorem 1. $E[\|q_0^\epsilon - q_0\|^p] = O(\epsilon^\beta)$ as $\epsilon \to 0^+$ by assumption and

$$\sup_{t\in[0,T]} E[\|p_t^\epsilon - \psi(t,q_t^\epsilon)\|^p] = O(\epsilon^{p/2}) \text{ as } \epsilon \to 0^+$$

by Assumption 3.1. Note that the assumptions also imply that a solution q_t to (29) exists for all $t \geq 0$ [1].

For any $\epsilon > 0$, using the Burkholder–Davis–Gundy inequality and Hölder's inequality we obtain the bound

$$\begin{aligned}
&E\left[\sup_{t\in[0,T]} \|q_t^\epsilon\|^p\right] \\
&\leq 4^{p-1}\Bigg(E\left[\|q_0^\epsilon\|^p\right] + \epsilon^{-p/2} E\left[\left(\int_0^T \|G_1(s,x_s^\epsilon,z_s^\epsilon)\|ds\right)^p\right] \\
&\quad + E\left[\left(\int_0^T \|F_1(s,x_s^\epsilon,z_s^\epsilon)\|ds\right)^p\right] + E\left[\sup_{t\in[0,T]}\left\|\int_0^t \sigma_1(s,x_s^\epsilon)dW_s\right\|^p\right]\Bigg) \\
&\leq 4^{p-1}\Bigg(E\left[\|q_0^\epsilon\|^p\right] + \epsilon^{-p/2}T^{p-1}\int_0^T E\left[\|G_1(s,x_s^\epsilon,z_s^\epsilon)\|^p\right]ds \\
&\quad + T^{p-1}\int_0^T E\left[\|F_1(s,x_s^\epsilon,z_s^\epsilon)\|^p\right]ds + \tilde{C}T^{p/2-1}\int_0^T E\left[\|\sigma_1(s,x_s^\epsilon)\|_F^p\right]ds\Bigg).
\end{aligned}$$

Polynomial boundedness (see Assumption 3.2) gives

$$E\left[\sup_{t\in[0,T]} \|q_t^\epsilon\|^p\right] \leq 4^{p-1}\left(E\left[\|q_0^\epsilon\|^p\right] + \tilde{C}\int_0^T E\left[(1+\|z_s^\epsilon\|^q)^p\right]ds\right),$$

where we absorbed all factors of T and ϵ into the constant $\tilde{C}$. Using Assumption 3.1 then gives

$$E\left[\sup_{t\in[0,T]} \|q_t^\epsilon\|^p\right] < \infty$$

for all ϵ sufficiently small.

Finally, for $n > 0$ define the stopping time $\tau_n = \inf\{t \geq 0 : \|q_t\| \geq n\}$. Then for $0 \leq t \leq T$ the Lipschitz properties together with the Burkholder–Davis–Gundy and Hölder's inequalities imply

$$\begin{aligned}
&E\left[\sup_{s\in[0,t]} \|q_s^{\tau_n}\|^p\right] \\
&\leq 3^{p-1}\Bigg(E[\|q_0\|^p] + E\left[\left(\int_0^{t\wedge\tau_n} \|(\tilde{F}+\tilde{G})(s,q_s^{\tau_n},\psi(s,q_s^{\tau_n}))\|ds\right)^p\right] \\
&\quad + E\left[\sup_{t\in[0,t]}\left\|\int_0^{t\wedge\tau_n} \tilde{\sigma}(s,q_s^{\tau_n},\psi(s,q_s^{\tau_n}))dW_s\right\|^p\right]\Bigg)
\end{aligned}$$

$$
\begin{aligned}
&\leq 3^{p-1}E[\|q_0\|^p] + \tilde{C}\int_0^t E\left[\|q_s^{\tau_n}\|^p\right] ds \\
&\quad + \tilde{C}\int_0^t \|(\tilde{F}+\tilde{G})(s,0,\psi(s,0))\|^p + \|\tilde{\sigma}(s,0,\psi(s,0))\|_F^p ds \\
&\leq \tilde{C}\left(1 + \int_0^t E\left[\sup_{r\in[0,s]} \|q_r^{\tau_n}\|^p\right] ds\right),
\end{aligned}
$$

where $\tilde{C}$ changes line to line, and is independent of t.

The definition of τ_n, together with $E[\|q_0\|^p] < \infty$, implies that

$$
\sup_{s\geq 0} E\left[\sup_{r\in[0,s]} \|q_r^{\tau_n}\|^p\right] < \infty.
$$

Therefore we can apply Gronwall's inequality to get

$$
E\left[\sup_{t\in[0,T]} \|q_t^{\tau_n}\|^p\right] \leq \tilde{C}e^{\tilde{C}T},
$$

where the constant $\tilde{C}$ is independent of n. Hence, the monotone convergence theorem yields

$$
E\left[\sup_{t\in[0,T]} \|q_t\|^p\right] \leq \tilde{C}e^{\tilde{C}T} < \infty.
$$

This completes the verification that the hypotheses of Lemma A.3 hold, allowing us to conclude that

$$
E\left[\sup_{t\in[0,T]} \|q_t^\epsilon - q_t\|^p\right] = O(\epsilon^\beta) \text{ as } \epsilon \to 0^+.
$$

□

4 Homogenization of Hamiltonian Systems

In this final section, we apply the above framework to our original Hamiltonian system, (10)–(11) (in particular, $m = n$ in this section) in order to prove the existence of a limiting process $q_t^\epsilon \to q_t$ and derive a homogenized SDE for q_t. Specifically, in Sects. 4.1 and 4.2 we will study a class of Hamiltonian systems for which the PDEs (27) for h and (26) for χ that are needed to derive the limiting equation are explicitly solvable and the required bounds can be verified by elementary means.

The SDE (10)–(11) can be rewritten in the general form (16)–(17):

$$
\begin{aligned}
dq_t^\epsilon &= \frac{1}{\sqrt{\epsilon}} \nabla_z K(t, q_t^\epsilon, z_t^\epsilon) dt, \\
dp_t^\epsilon &= \left(-\frac{1}{\sqrt{\epsilon}} \left(\gamma_l(t, x_t^\epsilon) - \nabla_q \psi_l(t, q_t^\epsilon) \right) \partial_{z_l} K(t, q_t^\epsilon, z_t^\epsilon) - \nabla_q K(t, q_t^\epsilon, z_t^\epsilon) \right. \\
&\quad \left. - \nabla_q V(t, q_t^\epsilon) + F(t, x_t^\epsilon) \right) dt + \sigma(t, x_t^\epsilon) dW_t,
\end{aligned}
$$

where γ_l denotes the vector obtained by taking the lth column of γ. Specifically,

$$
\begin{aligned}
&F_1 = 0, \;\; \sigma_1 = 0, \;\; \sigma_2 = \sigma, \;\; G_1(t, x, z) = \nabla_z K(t, q, z), \\
&F_2(t, x, z) = -\nabla_q K(t, q, z) - \nabla_q V(t, q) + F(t, x), \\
&G_2(t, x, z) = -\left(\gamma_l(t, x) - \nabla_q \psi_l(t, q) \right) \partial_{z_l} K(t, q, z).
\end{aligned}
$$

In particular, $\sigma_1 = 0$, so below we use the definition of $C^{1,2}$ applicable to this case.

The operator L, (22), and its formal adjoint have the following form:

$$
\begin{aligned}
(L\chi)(t, x, z) &= \frac{1}{2} \Sigma_{kl}(t, x) (\partial_{z_k} \partial_{z_l} \chi)(t, x, z) \\
&\quad - \tilde{\gamma}_{kl}(t, x) \partial_{z_l} K(t, q, z) (\partial_{z_k} \chi)(t, x, z), \\
(L^* h)(t, x, z) &= \partial_{z_k} \left(\frac{1}{2} \Sigma_{kl}(t, x) \partial_{z_l} h(t, x, z) \right. \\
&\quad \left. + \tilde{\gamma}_{kl}(t, x) \partial_{z_l} K(t, q, z) h(t, x, z) \right),
\end{aligned}
\tag{30}
$$

where

$$
\Sigma_{ij} = \sum_\rho \sigma_{i\rho} \sigma_{j\rho} \tag{31}
$$

and $\tilde{\gamma}$ was defined in (14). Here σ and Σ denote what were σ_2 and Σ_{22} respectively in (20) and $\sigma_1 = 0$. In particular, the indices on Σ have the meaning $\Sigma_{ij} \equiv (\Sigma_{22})_{ij}$.

4.1 Computing the Noise Induced Drift

In general, an explicit solution to $L^* h = 0$ is not available, and so the homogenized equation can only be defined implicitly, as in Theorem 2. However, there are certain classes of systems where we can explicitly derive the form of the additional vector field, $\tilde{G}$, appearing in the homogenized equation. In [8], one such class was studied by a different method. Here, we explore the case where the noise and dissipation satisfy the fluctuation dissipation relation pointwise for a time and state dependent generalized temperature $T(t, q)$,

$$
\Sigma_{ij}(t, q) = 2 k_B T(t, q) \gamma_{ij}(t, q). \tag{32}
$$

where Σ was defined in (31). We will make Assumptions A.1–A.4, but make no further constraints on the form of the Hamiltonian here.

As can be verified by a direct calculation, under the assumption (32), the adjoint equation (30) is solved by

$$h(t,q,z) = \frac{1}{Z(t,q)} \exp[-\beta(t,q)K(t,q,z)], \tag{33}$$

where we define $\beta(t,q) = 1/(k_B T(t,q))$ and Z, the "partition function", is chosen so that $\int h dz = 1$. Note that Assumption A.3 ensures such a normalization exists. We also point out that in this case, the antisymmetric part of $\tilde{\gamma}$ does not contribute to the right hand side of (30).

An interesting point to note is that when the antisymmetric part of $\tilde{\gamma}$ vanishes (physically, for K quadratic in z this means a vanishing magnetic field), the vector field that we are taking the divergence of in (30) vanishes identically. When $\tilde{\gamma}$ has a non-vanishing antisymmetric part, only once we take the divergence does the expression in (30) vanish.

From (13), we see that the terms that require homogenization are

$$\begin{aligned} G(t,q_t^\epsilon,z_t^\epsilon) = &- (\tilde{\gamma}^{-1})^{ij}(t,q_t^\epsilon)\partial_{q^j}K(t,q_t^\epsilon,z_t^\epsilon)dt \\ &+ (z_t^\epsilon)_j\partial_{q^l}(\tilde{\gamma}^{-1})^{ij}(t,q_t^\epsilon)\partial_{z_l}K(t,q_t^\epsilon,z_t^\epsilon)dt. \end{aligned} \tag{34}$$

Using (33), the formal calculation of Sect. 3.2 gives

$$\tilde{G}(t,q) = -(\tilde{\gamma}^{-1})^{ij}(t,q)\langle\partial_{q^j}K(t,q,z)\rangle + \frac{\partial_{q^l}(\tilde{\gamma}^{-1})^{il}(t,q)}{\beta(t,q)},$$

where we define

$$\langle\partial_{q^j}K(t,q,z)\rangle = \frac{1}{Z(t,q)}\int \partial_{q^j}K(t,q,z)\exp[-\beta(t,q)K(t,q,z)]dz. \tag{35}$$

Of course, this calculation is only formal. In the next section, we study a particular case where everything can be made rigorous.

4.2 Rigorous Homogenization of a Class of Hamiltonian Systems

In this section we explore a class of Hamiltonian systems for which Assumption 3.3 can be rigorously verified via an explicit solution to the PDE for χ. We will work with Hamiltonian systems that satisfy Assumptions A.1–A.5, A.7. In particular, we are restricting to the class of Hamiltonians with

$$K(t,q,z) = \tilde{K}(t,q,A^{ij}(t,q)z_i z_j), \tag{36}$$

where $A(t,q)$ is valued in the space of positive definite $n\times n$-matrices. We will write $\tilde{K} \equiv \tilde{K}(t,q,\zeta)$ and $\tilde{K}' \equiv \partial_\zeta \tilde{K}$.

We will also need the following relations between Σ, γ, and A to hold:

Assumption 4.1. σ is independent of p and

$$\Sigma(t,q) = b_1(t,q)A^{-1}(t,q), \quad \gamma(t,q) = b_2(t,q)A^{-1}(t,q) \tag{37}$$

where, for every $T > 0$, the b_i are bounded, C^2 functions that have positive lower bounds and bounded first derivatives, both on $[0,T] \times \mathbb{R}^n$.

Note that these relations imply a fluctuation-dissipation relation with a time and state dependent generalized temperature $T = \frac{b_1}{2k_B b_2}$.

In [8], we showed that Assumptions A.1–A.5 imply:

$$d(q_t^\epsilon)^i = \tilde{F}^i(t,x)dt + \tilde{\sigma}^i_\rho(t,x_t^\epsilon)dW_t^\rho + G^i(t,x_t^\epsilon,z_t^\epsilon)dt + d(R_t^\epsilon)^i,$$

where

$$\begin{aligned}
\tilde{F}^i(t,x) &= (\tilde{\gamma}^{-1})^{ij}(t,q)(-\partial_t\psi_j(t,q) - \partial_{q^j}V(t,q) + F_j(t,x)) + S^i(t,q),\\
G^i(t,q,z) &= -(\tilde{\gamma}^{-1})^{ij}(t,q)(\partial_{q^j}\tilde{K})(t,q,A^{ij}(t,q)z_i z_j),\\
\tilde{\sigma}^i_\rho(t,x) &= (\tilde{\gamma}^{-1})^{ij}(t,q)\sigma_{j\rho}(t,x),\\
S^i(t,q) &= k_B T(t,q)\left(\partial_{q^j}(\tilde{\gamma}^{-1})^{ij}(t,q) - \frac{1}{2}(\tilde{\gamma}^{-1})^{ik}(t,q)A^{-1}_{jl}(t,q)\partial_{q^k}A^{jl}(t,q)\right),
\end{aligned} \tag{38}$$

and R_t^ϵ is a family of continuous semimartingales. $S(t,q)$ is called the *noise induced drift* (see Eq. 3.26 in [8]).

Note, that with $K(t,q,z)$ defined by (36), the first term in (34) consists of two contributions—one coming from the q-dependence of $\tilde{K}$ and one coming from the q-dependence of A. The G defined here comprises only the first contribution. The method of [8] is able to homogenize the second term in (34), as well the first contribution of the first term, leading to the noise induced drift, S, but fails when $\tilde{K}$ depends explicitly on q. However, under certain circumstances, the method developed in Sect. 3.2 is succeeds in homogenizing the system when $\tilde{K}$ has q dependence, as we now show.

We will need one final assumption:

Assumption 4.2. For every $T > 0$:

1. There exists $\zeta_0 > 0$ and $C > 0$ such that $\tilde{K}'(t,q,\zeta) \geq C$ for all $(t,q,\zeta) \in [0,T] \times \mathbb{R}^n \times [\zeta_0,\infty)$.
2. $\tilde{K}(t,q,\zeta)$, $\partial_t\partial_{q^i}\tilde{K}(t,q,\zeta)$, $\partial_{q^i}\partial_\zeta\tilde{K}(t,q,\zeta)$, and $\partial_{q^i}\partial_{q^j}\tilde{K}(t,q,\zeta)$ are polynomially bounded in ζ, uniformly in $(t,q) \in [0,T] \times \mathbb{R}^n$.

We are now prepared to prove the following homogenization result:

Theorem 3. *Let $x_t^\epsilon = (q_t^\epsilon, p_t^\epsilon)$ satisfy the Hamiltonian SDE (10)–(11) and suppose Assumptions A.1–A.5, A.7, 4.1, and 4.2 hold. Let $p \geq 2$ and suppose we have initial conditions that satisfy $E[\|q_0^\epsilon\|^p] < \infty$, $E[\|q_0\|^p] < \infty$, and $E[\|q_0^\epsilon - q_0\|^p] = O(\epsilon^{p/2})$. Then for any $T > 0$, $0 < \beta < p/2$ we have*

$$E\left[\sup_{t\in[0,T]}\|q_t^\epsilon - q_t\|^p\right] = O(\epsilon^\beta) \text{ as } \epsilon \to 0^+$$

where q_t is the solution to the SDE

$$dq_t^i = (\tilde{\gamma}^{-1})^{ij}(t,q_t)(-\partial_t\psi_j(t,q_t) - \partial_{q^j}V(t,q_t) + F_j(t,q_t,\psi(t,q_t)))dt \qquad (39)$$
$$+ S^i(t,q_t)dt + \tilde{G}^i(t,q_t)dt + (\tilde{\gamma}^{-1})^{ij}(t,q_t)\sigma_{j\rho}(t,q_t)dW_t^\rho$$

with initial condition q_0. See (14), (38), *and* (40) *for the definitions of $\tilde{\gamma}$, S, and $\tilde{G}$, respectively.*

Proof. From [8], Assumptions A.1–A.5, A.7 imply:

1. q_t^ϵ satisifies an equation of the form (28), where, for every $T > 0$, $\tilde{F}$, $\tilde{\sigma}$ are bounded, continuous, and Lipschitz in x, on $[0,T] \times \mathbb{R}^{2n}$, with Lipschitz constant uniform in t.
2. Assumptions 3.1 holds.
3. For any $p > 0$, $T > 0$, $0 < \beta < p/2$ we have

$$E\left[\sup_{t\in[0,T]} \|R_t^\epsilon\|^p\right] = O(\epsilon^\beta) \text{ as } \epsilon \to 0^+.$$

Combined with polynomial boundedness of $\tilde{K}$ (Assumption 4.2) we see that Assumption 3.2 also holds. Therefore, to apply Theorem 2, we have to verify Assumption 3.3 and that the $\tilde{G}$ referenced therein is Lipschitz in x, uniformly in $t \in [0,T]$.

From Sect. 3.2, we expect that

$$\tilde{G}^i(t,q) = -(\tilde{\gamma}^{-1})^{ij}(t,q)\langle\partial_{q^j}\tilde{K}(t,q,A^{ij}(t,q)z_iz_j)\rangle \qquad (40)$$

where, similarly to (35),

$$\langle\partial_{q^j}\tilde{K}(t,q,\|z\|_A^2)\rangle = \frac{1}{Z(t,q)}\int \partial_{q^j}\tilde{K}(t,q,\|z\|_A^2)\exp[-\beta(t,q)\tilde{K}(t,q,\|z\|_A^2)]dz.$$

Here we use the shorthand $\|z\|_A^2 \equiv A^{ij}(t,q)z_iz_j$ when the implied values of t,q are apparent from the context.

Using our assumptions, along with several applications of the DCT, one can see that $\tilde{G}$ is C^1 and, for every $T > 0$, is bounded with bounded first derivatives on $[0,T] \times \mathbb{R}^n$. In particular, it is Lipschitz in q, uniformly in $t \in [0,T]$.

We now turn to solving the equation

$$L\chi = G - \tilde{G}. \qquad (41)$$

Since $G - \tilde{G}$ is independent of p and depends on z only through $\|z\|_A^2$, we look for χ with the same behavior. Using the ansatz $\chi(t,q,z) = \tilde{\chi}(t,q,\|z\|_A^2)$, and defining $G^i(t,q,\zeta) = -(\tilde{\gamma}^{-1})^{ij}(t,q)(\partial_{q^j}\tilde{K})(t,q,\zeta)$, leads (on account of the antisymmetry of the matrix $\tilde{\gamma} - \gamma$) to the ODE in the variable ζ:

$$\zeta\tilde{\chi}''(t,q,\zeta) + \left(\frac{n}{2} - \beta(t,q)\zeta\tilde{K}'(t,q,\zeta)\right)\tilde{\chi}'(t,q,\zeta)$$
$$= \frac{1}{2b_1(t,q)}(G(t,q,\zeta) - \tilde{G}(t,q)).$$

This has the solution

$$\tilde{\chi}(t,q,\zeta) = \frac{1}{2b_1(t,q)} \int_0^{\zeta} \zeta_1^{-n/2} \exp[\beta(t,q)\tilde{K}(t,q,\zeta_1)] \tag{42}$$
$$\times \int_0^{\zeta_1} \zeta_2^{(n-2)/2} \exp[-\beta(t,q)\tilde{K}(t,q,\zeta_2)] \left(G(t,q,\zeta_2) - \tilde{G}(t,q)\right) d\zeta_2 d\zeta_1$$

Therefore $\chi(t,q,z) \equiv \tilde{\chi}(t,q,\|z\|_A^2)$ solves the PDE (41). One can show that it is $C^{1,2}$ and that χ and its first derivatives are polynomially bounded in z, uniformly for $(t,q) \in [0,T] \times \mathbb{R}^n$. As a representative example, in Appendix B we outline the proof that $\tilde{\chi}(t,q,\zeta)$ is polynomially bounded in ζ, uniformly in $(t,q) \in [0,T] \times \mathbb{R}^n$. The remainder of the computations are similar and we leave them to the reader.

χ is independent of p, so $\partial_{p_i}\partial_{p_j}\chi = 0$ and $\partial_{p_i}\partial_{z_j}\chi = 0$. Therefore, this completes the verification of Assumption 3.3 and we are justified in using Theorem 2 to conclude

$$E\left[\sup_{t\in[0,T]} \|q_t^{\epsilon} - q_t\|^p\right] = O(\epsilon^{\beta}) \text{ as } \epsilon \to 0^+,$$

where q_t satisfies the SDE

$$q_t = q_0 + \int_0^t \tilde{F}(s,q_s,\psi(s,q_s))ds + \int_0^t \tilde{G}(s,q_s)ds + \int_0^t \tilde{\sigma}(s,q_s)dW_s$$

as claimed. □

Lastly, we give an example of a general class of Hamiltonians that satisfy the hypotheses of Theorem 3. The proof of this corollary is straightforward, so we leave it to the reader.

Corollary 1. *Consider the class of Hamiltonians of the form*

$$H(t,q,p) = \sum_{l=k_1}^{k_2} d_l(t,q) \left[A^{ij}(t,q)(p-\psi(t,q))_i(p-\psi(t,q))_j\right]^l + V(t,q)$$

where $1 \le k_1 \le k_2$ are integers and the following properties hold on $[0,T] \times \mathbb{R}^n$ for every $T > 0$:

1. *V is C^2 and $\nabla_q V$ is bounded and Lipschitz in q, uniformly in $t \in [0,T]$.*
2. *ψ is C^3 and $\partial_t\psi$, $\partial_{q^i}\psi$, $\partial_t\partial_{q^i}\psi$, $\partial_{q^i}\partial_{q^j}\psi$, $\partial_t\partial_{q^j}\partial_{q^i}\psi$, and $\partial_{q^l}\partial_{q^j}\partial_{q^i}\psi$ are bounded.*
3. *d_l are C^2, non-negative, bounded, and have bounded first and second derivatives.*
4. *d_{k_1} and d_{k_2} are uniformly bounded below by a positive constant.*
5. *A is C^2, positive-definite, and A, $\partial_t A$, $\partial_{q_i} A$, $\partial_t\partial_{q^i} A$, and $\partial_{q^i}\partial_{q^j} A$ are bounded.*
6. *The eigenvalues of A are uniformly bounded below by a positive constant.*

Also suppose that

1. *σ is independent of p and*

$$\Sigma(t,q) = b_1(t,q)A^{-1}(t,q), \;\; \gamma(t,q) = b_2(t,q)A^{-1}(t,q)$$

   where, for every $T > 0$, the b_i are bounded, C^2 functions with positive lower bounds and bounded first derivatives.
2. *γ is C^2, is independent of p, and $\partial_t\gamma$, $\partial_{q^i}\gamma$, $\partial_t\partial_{q^j}\gamma$, $\partial_{q^i}\partial_{q^j}\gamma$ are bounded on $[0,T] \times \mathbb{R}^n$.*
3. *The eigenvalues of γ are bounded below by some $\lambda > 0$.*
4. *γ, F, and σ are bounded.*
5. *F and σ are Lipschitz in x uniformly in $t \in [0,T]$.*
6. *There exists $C > 0$ such that the (random) initial conditions satisfy*

$$K^\epsilon(0, x_0^\epsilon) \leq C$$

   for all $\epsilon > 0$ and all $\omega \in \Omega$.
7. *There is a $p \geq 2$ such that*

$$E[\|q_0^\epsilon\|^p] < \infty, \;\; E[\|q_0\|^p] < \infty, \;\; and\; E[\|q_0^\epsilon - q_0\|^p] = O(\epsilon^{p/2}).$$

Then all the hypotheses of Theorem 3 hold, in particular Assumptions A.1–A.5, A.7, 4.1, and 4.2 hold, and hence, for any $\beta \in \left(0, \frac{p}{2}\right)$,

$$E\left[\sup_{t\in[0,T]} \|q_t^\epsilon - q_t\|^p\right] = O(\epsilon^\beta) \;\; as\; \epsilon \to 0^+,$$

where $x_t^\epsilon = (q_t^\epsilon, p_t^\epsilon)$ satisfy the Hamiltonian SDE (10)–(11) *and q_t satisfies the homogenized SDE,* (39).

Acknowledgments. J. W. was partially supported by NSF grants DMS 131271 and DMS 1615045.

A Material from [8]

In this appendix, we collect several useful assumptions and results from [8]. The assumptions listed here are *not* used in the entirety of this current work. When they are needed for a particular result we explicitly reference them.

Assumption A.1. We assume that the Hamiltonian has the form given in (9) where K and ψ are C^2 and K is non-negative. For every $T > 0$, we assume the following bounds hold on $[0,T] \times \mathbb{R}^{2n}$:

1. There exist $C > 0$ and $M > 0$ such that

$$\max\{|\partial_t K(t,q,z)|, \|\nabla_q K(t,q,z)\|\} \leq M + CK(t,q,z).$$

2. There exist $c > 0$ and $M \geq 0$ such that

$$\|\nabla_z K(t,q,z)\|^2 + M \geq cK(t,q,z).$$

3. For every $\delta > 0$ there exists an $M > 0$ such that

$$\max\left\{\|\nabla_z K(t,q,z)\|, \left(\sum_{ij} |\partial_{z_i}\partial_{z_j} K(t,q,z)|^2\right)^{1/2}\right\} \leq M + \delta K(t,q,z).$$

Assumption A.2. For every $T > 0$, we assume that the following hold uniformly on $[0,T] \times \mathbb{R}^n$:

1. V is C^2 and $\nabla_q V$ is bounded
2. γ is symmetric with eigenvalues bounded below by some $\lambda > 0$.
3. γ, F, $\partial_t \psi$, and σ are bounded.
4. There exists $C > 0$ such that the (random) initial conditions satisfy

$$K^\epsilon(0, x_0^\epsilon) \leq C$$

for all $\epsilon > 0$ and all $\omega \in \Omega$.

Assumption A.3. We assume that for every $T > 0$ there exists $c > 0$, $\eta > 0$ such that

$$K(t,q,z) \geq c\|z\|^{2\eta}$$

on $[0,T] \times \mathbb{R}^{2n}$.

Assumption A.4. We assume that γ is C^1 and is independent of p.

Assumption A.5. We assume that K has the form

$$K(t,q,z) = \tilde{K}(t,q,A^{ij}(t,q)z_i z_j)$$

where $\tilde{K}(t,q,\zeta)$ is C^2 and non-negative on $[0,\infty) \times \mathbb{R}^n \times [0,\infty)$ and $A(t,q)$ is a C^2 function whose values are symmetric $n \times n$-matrices. We also assume that for every $T > 0$, the eigenvalues of A are bounded above and below by some constants $C > 0$ and $c > 0$ respectively, uniformly on $[0,T] \times \mathbb{R}^n$.

We will write $\tilde{K}'$ for $\partial_\zeta \tilde{K}$ and will use the abbreviation $\|z\|_A^2$ for $A^{ij}(t,q)z_i z_j$ when the implied values of t and q are apparent from the context.

Assumption A.7. We assume that, for every $T > 0$, $\nabla_q V$, F, and σ are Lipschitz in x uniformly in $t \in [0,T]$. We also assume that A and γ are C^2, ψ is C^3, and $\partial_t\psi$, $\partial_{q^i}\psi$, $\partial_{q^i}\partial_{q_j}\psi$, $\partial_t\partial_{q^i}\psi$, $\partial_t\partial_{q^j}\partial_{q^i}\psi$, $\partial_{q^l}\partial_{q^j}\partial_{q^i}\psi$, $\partial_t\gamma$, $\partial_{q^i}\gamma$, $\partial_t\partial_{q^j}\gamma$, $\partial_{q^i}\partial_{q^j}\gamma$, $\partial_t A$, $\partial_{q^i} A$, $\partial_t\partial_{q^i} A$, and $\partial_{q^i}\partial_{q^j} A$ are bounded on $[0,T] \times \mathbb{R}^{2n}$ for every $T > 0$.

Note that, combined with Assumptions A.1–A.4, this implies $\tilde{\gamma}$, $\tilde{\gamma}^{-1}$, $\partial_t\tilde{\gamma}^{-1}$, $\partial_{q^i}\tilde{\gamma}^{-1}$, $\partial_t\partial_{q^j}\tilde{\gamma}^{-1}$, and $\partial_{q^i}\partial_{q^j}\tilde{\gamma}^{-1}$ are bounded on compact t-intervals.

Lemma A.1. *Under Assumptions A.1 and A.2, for any $T>0$, $p>0$ we have*

$$E\left[\sup_{t\in[0,T]}\|q_t^\epsilon\|^p\right]<\infty.$$

Lemma A.2. *Under Assumptions A.1–A.3, for any $T>0$, $p>0$ we have*

$$\sup_{t\in[0,T]}E[\|p_t^\epsilon-\psi(t,q_t^\epsilon)\|^p]=O(\epsilon^{p/2})\text{ as }\epsilon\to 0^+$$

and for any $p>0$, $T>0$, $0<\beta<p/2$ we have

$$E\left[\sup_{t\in[0,T]}\|p_t^\epsilon-\psi(t,q_t^\epsilon)\|^p\right]=O(\epsilon^\beta)\text{ as }\epsilon\to 0^+.$$

The following is a slight variant of the result from [8], but the proof is identical.

Lemma A.3. *Let $T>0$ and suppose we have continuous functions $\tilde{F}(t,x):[0,\infty)\times\mathbb{R}^{n+m}\to\mathbb{R}^n$, $\tilde{\sigma}(t,x):[0,\infty)\times\mathbb{R}^{n+m}\to\mathbb{R}^{n\times k}$, and $\psi:[0,\infty)\times\mathbb{R}^n\to\mathbb{R}^m$ that are Lipschitz in x, uniformly in $t\in[0,T]$.*

Let W_t be a k-dimensional Wiener process, $p\geq 2$ and $\beta>0$ and suppose that we have continuous semimartingales q_t and, for each $0<\epsilon\leq\epsilon_0$, $\tilde{R}_t^\epsilon$, $x_t^\epsilon=(q_t^\epsilon,p_t^\epsilon)$ that satisfy the following properties:

1. $q_t^\epsilon=q_0^\epsilon+\int_0^t\tilde{F}(s,x_s^\epsilon)ds+\int_0^t\tilde{\sigma}(s,x_s^\epsilon)dW_s+\tilde{R}_t^\epsilon$.
2. $q_t=q_0+\int_0^t\tilde{F}(s,q_s,\psi(s,q_s))ds+\int_0^t\tilde{\sigma}(s,q_s^\epsilon,\psi(s,q_s))dW_s$.
3. $E[\|q_0^\epsilon-q_0\|^p]=O(\epsilon^\beta)$ *as* $\epsilon\to 0^+$.
4. $E\left[\sup_{t\in[0,T]}\|\tilde{R}_t^\epsilon\|^p\right]=O(\epsilon^\beta)$ *as* $\epsilon\to 0^+$.
5. $\sup_{t\in[0,T]}E[\|p_t^\epsilon-\psi(t,q_t^\epsilon)\|^p]=O(\epsilon^\beta)$ *as* $\epsilon\to 0^+$.
6. $E\left[\sup_{t\in[0,T]}\|q_t^\epsilon\|^p\right]<\infty$ *for all* $\epsilon>0$ *sufficiently small.*
7. $E\left[\sup_{t\in[0,T]}\|q_t\|^p\right]<\infty$.

Then

$$E\left[\sup_{t\in[0,T]}\|q_t^\epsilon-q_t\|^p\right]=O(\epsilon^\beta)\text{ as }\epsilon\to 0^+.$$

B Polynomial Boundedness of $\tilde{\chi}$

Changing variables, $\tilde{\chi}$ can be expressed as

$$\tilde{\chi}(t,q,\zeta)=\frac{1}{2b_1(t,q)}\zeta\int_0^1\exp[\beta(t,q)\tilde{K}(t,q,s\zeta)]\\ \times\int_0^1 r^{(m-2)/2}\exp[-\beta(t,q)\tilde{K}(t,q,rs\zeta)]\left(G(t,q,rs\zeta)-\tilde{G}(t,q)\right)drds.$$

Applying the DCT to this expression several times, one can prove that $\tilde{\chi}$ is $C^{1,2}$. Using the fact that $\tilde{K}$ and $\partial_{q^i}\tilde{K}$ are polynomially bounded in ζ, uniformly in $(t,q) \in [0,T] \times \mathbb{R}^n$, one can see that $\tilde{\chi}(t,q,\zeta)$ is bounded on $[0,T] \times \mathbb{R}^n \times [0,\zeta_0]$ for any $\zeta_0 > 0$. From Assumption 4.2, there exists ζ_0 and $C > 0$ such that $\tilde{K}'(t,q,\zeta) \geq C$ for all $(t,q,\zeta) \in [0,T] \times \mathbb{R}^n \times [\zeta_0,\infty)$.

By combining (42) with (40), one finds that for $\zeta \geq \zeta_0$, $\tilde{\chi}$ can alternatively be written as

$$\tilde{\chi}(t,q,\zeta) = \tilde{\chi}(t,q,\zeta_0) + \frac{1}{2b_1(t,q)} \int_{\zeta_0}^{\zeta} \zeta_1^{-m/2} \exp[\beta(t,q)\tilde{K}(t,q,\zeta_1)] \times \int_{\zeta_1}^{\infty} \exp[-\beta(t,q)\tilde{K}(t,q,\zeta_2)]\zeta_2^{(m-2)/2}(\tilde{G}(t,q) - G(t,q,\zeta_2))d\zeta_2 d\zeta_1.$$

Therefore, if we can show that the second term has the polynomial boundedness property then so does $\tilde{\chi}$, and hence χ.

Letting $\tilde{C}$ denote a constant that potentially changes in each line and choosing ζ_0 as in Assumption 4.2, we have

$$\begin{aligned}
&\left\| \frac{1}{2b_1(t,q)} \int_{\zeta_0}^{\zeta} \zeta_1^{-m/2} \exp[\beta(t,q)\tilde{K}(t,q,\zeta_1)] \right.\\
&\quad \left. \times \int_{\zeta_1}^{\infty} \exp[-\beta(t,q)\tilde{K}(t,q,\zeta_2)]\zeta_2^{(m-2)/2}(\tilde{G}(t,q) - G(t,q,\zeta_2))d\zeta_2 d\zeta_1 \right\| \\
&\leq \tilde{C} \int_{\zeta_0}^{\zeta} \zeta_1^{-m/2} \exp[\beta(t,q)\tilde{K}(t,q,\zeta_1)] \\
&\quad \times \int_{\zeta_1}^{\infty} \exp[-\beta(t,q)\tilde{K}(t,q,\zeta_2)]\zeta_2^{(m-2)/2+q} d\zeta_2 d\zeta_1 \\
&\leq \tilde{C}\zeta_0^{-m/2} \int_{\zeta_0}^{\zeta} \exp[\beta(t,q)\tilde{K}(t,q,\zeta_1)] \\
&\quad \times \int_{\zeta_1}^{\infty} \exp[-\beta(t,q)\tilde{K}(t,q,\zeta_2)]\tilde{K}(t,q,\zeta_2)^{((m-2)/2+q)/\eta}\tilde{K}'(t,q,\zeta_2) d\zeta_2 d\zeta_1 \\
&= \tilde{C}\zeta_0^{-m/2} \int_{\zeta_0}^{\zeta} \exp[\beta(t,q)\tilde{K}(t,q,\zeta_1)] \\
&\qquad \times \int_{\tilde{K}(t,q,\zeta_1)}^{\infty} \exp[-\beta(t,q)u]u^{((m-2)/2+q)/\eta} du d\zeta_1
\end{aligned}$$

for some $q > 0$. To obtain the first inequality, we use polynomial boundedness of $\partial_{q^i}\tilde{K}$. For the second, we used Assumption (A.3) together with the fact that $\tilde{K}' \geq C > 0$ on $[0,T] \times \mathbb{R}^n \times [\zeta_0,\infty)$.

Therefore we obtain

$$\|\tilde{\chi}(t,q,\zeta)\| \leq \tilde{C}\left(1 + \int_{\zeta_0}^{\zeta} P(\tilde{K}(t,q,\zeta_1))d\zeta_1\right)$$

for some polynomial $P(x)$ with positive coefficients that are independant of t and q. Polynomial boundedness of $\tilde{K}$ then implies

$$\|\tilde{\chi}(t,q,\zeta)\| \leq \tilde{C}\left(1+\int_{\zeta_0}^{\zeta} Q(\zeta_1)d\zeta_1\right)$$

for some polynomial $Q(\zeta)$. This proves the desired polynomial boundedness property for $\tilde{\chi}$.

References

1. Khasminskii, R.: Stochastic Stability of Differential Equations, vol. 66. Springer, Heidelberg (2011)
2. Arnold, L.: Stochastic Differential Equations: Theory and Applications. Krieger Publishing Company (1992)
3. Ortega, J.M.: Matrix Theory: A Second Course. Springer, Berlin (2013)
4. Zwanzig, R.: Nonequilibrium Statistical Mechanics. Oxford University Press, Oxford (2001)
5. Hottovy, S., McDaniel, A., Volpe, G., Wehr, J.: The Smoluchowski-Kramers limit of stochastic differential equations with arbitrary state-dependent friction. Commun. Math. Phys. **336**, 1259–1283 (2015)
6. Birrell, J., Wehr, J.: Phase space homogenization of noisy Hamiltonian systems. Annales Henri Poincaré **19**(4), 1081–1114 (2018)
7. Karatzas, I., Shreve, S.: Brownian Motion and Stochastic Calculus. Graduate Texts in Mathematics. Springer, New York (2014)
8. Birrell, J., Wehr, J.: Homogenization of dissipative, noisy, Hamiltonian dynamics. Stoch. Process. Appl. **128**(7), 2367–2403 (2018)
9. Pavliotis, G., Stuart, A.: Multiscale Methods: Averaging and Homogenization. Texts in Applied Mathematics. Springer, New York (2008)

Dyson Models Under Renormalization and in Weak Fields

Rodrigo Bissacot[1], Eric O. Endo[1,2,3], Aernout C. D. van Enter[2(✉)], Bruno Kimura[4], Arnaud Le Ny[5], and Wioletta M. Ruszel[4]

[1] Institute for Mathematics and Statistics – IME USP– Universidade de São Paulo, São Paulo, Brazil
rodrigo.bissacot@gmail.com, ericossamiendo@gmail.com

[2] Johann Bernoulli Institute, University of Groningen, Nijenborgh 9, 9747AG Groningen, The Netherlands
aenter@phys.rug.nl

[3] NYU-ECNU Institute of Mathematical Sciences, NYU Shanghai, 1555 Century Avenue, Pudong New District, Shanghai, China

[4] Delft Institute for Applied Mathematics, Technical University Delft, Van Mourik Broekmanweg 6, 2628 Delft, XE, The Netherlands
bruno.hfkimura@gmail.com, W.M.Ruszel@tudelft.nl

[5] LAMA UMR CNRS 8050, UPEC, Université Paris-Est, 94010 Créteil, France
arnaud.le-ny@u-pec.fr

Abstract. We consider one-dimensional long-range spin models (usually called *Dyson models*), consisting of Ising ferromagnets with slowly decaying long-range pair potentials of the form $1/|i-j|^\alpha$, mainly focusing on the range of slow decays $1 < \alpha \leq 2$. We describe two recent results, one about renormalization and one about the effect of external fields at low temperature.

The first result states that a decimated long-range Gibbs measure in one dimension becomes non-Gibbsian, in the same vein as comparable results in higher dimensions for short-range models.

The second result addresses the behaviour of such models under inhomogeneous fields, in particular external fields which decay to zero polynomially as $1/(|i|+1)^\gamma$. We study how the critical decay power of the field, γ, for which the phase transition persists and the decay power α of the Dyson model compare, extending recent results for short-range models on lattices and on trees. We also briefly point out some analogies between these results.

Keywords: Long-range Ising models · Hidden phase transitions · Generalized Gibbs measures · Slowly decaying correlated external fields

1 Introduction

In this short review we investigate some properties of one-dimensional long-range spin models, also known as Dyson models. In his original work, Dyson [13] considered an Ising spin system with formal Hamiltonian given by

V. Sidoravicius (Ed.): *Sojourns in Probability Theory and Statistical Physics - I*, PROMS 298, pp. 123–137, 2019.
https://doi.org/10.1007/978-981-15-0294-1_5

We denote by $C(\Omega)$ the set of continuous (quasilocal) functions on Ω, characterized by

$$f \in C(\Omega) \iff \lim_{\Lambda \uparrow \mathcal{S}} \sup_{\sigma,\omega:\sigma_\Lambda=\omega_\Lambda} \mid f(\omega) - f(\sigma) \mid = 0. \tag{1}$$

Monotonicity for functions and measures concerns the natural partial (FKG) order "$\leq$", which we have on our Ising spin systems: $\sigma \leq \omega$ if and only if $\sigma_i \leq \omega_i$ for all $i \in \mathbb{Z}$. Its maximal and minimal elements are the configurations $+$ and $-$, and this order extends to functions: $f : \Omega \longrightarrow \mathbb{R}$ is called *monotone increasing* when $\sigma \leq \omega$ implies $f(\sigma) \leq f(\omega)$. For measures, we write $\mu < \nu$ if and only if $\mu[f] \leq \nu[f]$ for all f monotone increasing[1].

Macroscopic states are represented by probability measures on $(\Omega, \mathcal{F}, \rho)$, whose main description – at least in mathematical statistical mechanics – is in terms of consistent systems of regular versions of finite-volume conditional probabilities with prescribed boundary conditions, within the so-called *DLR formalism* [11,37,51]. To do so, one introduces families of probability kernels that are natural candidates to represent such versions of conditional probabilities.

Definition 1 (Specification). *A specification $\gamma = (\gamma_\Lambda)_{\Lambda \in \mathcal{S}}$ on $(\Omega, \mathcal{F})$ is a family of probability kernels $\gamma_\Lambda : \Omega_\Lambda \times \mathcal{F}_{\Lambda^c} \longrightarrow [0,1]$; $(\omega, A) \longmapsto \gamma_\Lambda(A \mid \omega)$ s.t. for all $\Lambda \in \mathcal{S}$:*

1. *(Properness) For all $\omega \in \Omega$, $\gamma_\Lambda(B|\omega) = \mathbf{1}_B(\omega)$ when $B \in \mathcal{F}_{\Lambda^c}$.*
2. *(Finite-volume consistency) For all $\Lambda \subset \Lambda' \in \mathcal{S}$, $\gamma_{\Lambda'}\gamma_\Lambda = \gamma_{\Lambda'}$ where*

$$\forall A \in \mathcal{F},\ \forall \omega \in \Omega,\ (\gamma_{\Lambda'}\gamma_\Lambda)(A|\omega) = \int_\Omega \gamma_\Lambda(A|\sigma)\gamma_{\Lambda'}(\mathrm{d}\sigma|\omega). \tag{2}$$

These kernels also act on functions and on measures: for all $f \in C(\Omega)$ or $\mu \in \mathcal{M}_1^+$,

$$\gamma_\Lambda f(\omega) := \int_\Omega f(\sigma)\gamma_\Lambda(\mathrm{d}\sigma|\omega) = \gamma_\Lambda[f|\omega]$$

and

$$\mu\gamma_\Lambda[f] := \int_\Omega (\gamma_\Lambda f)(\omega)\mathrm{d}\mu(\omega) = \int_\Omega \gamma_\Lambda[f|\omega]\mu(\mathrm{d}\omega).$$

Definition 2 (DLR measures). *A probability measure μ on $(\Omega, \mathcal{F})$ is said to be consistent with a specification γ (or specified by γ) when for all $A \in \mathcal{F}$ and $\Lambda \in \mathcal{S}$.*

$$\mu[A|\mathcal{F}_{\Lambda^c}](\omega) = \gamma_\Lambda(A|\omega),\ \mu\text{–a.e. } \omega. \tag{3}$$

We denote by $\mathcal{G}(\gamma)$ the set of measures consistent with γ which forms a Choquet simplex [12,20].

A *specification* γ is said to be **quasilocal** when for any local function f, the image $\gamma_\Lambda f$ should be a continuous function of the boundary condition:

$$\gamma \text{ quasilocal} \iff \gamma_\Lambda f \in C(\Omega) \text{ for any } f \text{ local (or any } f \text{ in } C(\Omega)). \tag{4}$$

[1] We denote $\mu[f]$ for the expectation $\mathbb{E}_\mu[f]$ under a measure μ.

A *measure* is said to be quasilocal when it is specified by a quasilocal specification.

A particularly important subclass of quasilocal measures consists of the *Gibbs measures* with (formal) Hamiltonian H defined via a potential Φ, a family $\Phi = (\Phi_A)_{A\in\mathcal{S}}$ of local functions $\Phi_A \in \mathcal{F}_A$. The contributions of spins in finite sets A to the total energy define the *finite-volume Hamiltonians with free boundary conditions*

$$\forall \Lambda \in \mathcal{S},\ H_\Lambda(\omega) = \sum_{A\subset\Lambda} \Phi_A(\omega),\ \forall\omega \in \Omega. \tag{5}$$

To define Gibbs measures, we require for Φ that it is *Uniformly Absolutely Convergent* (UAC), i.e. that $\sum_{A\ni i} \sup_\omega |\Phi_A(\omega)| < \infty, \forall i \in \mathbb{Z}$. One then can give sense to the *Hamiltonian at volume* $\Lambda \in \mathcal{S}$ *with boundary condition* ω defined for all $\sigma, \omega \in \Omega$ as $H_\Lambda^\Phi(\sigma|\omega) := \sum_{A\cap\Lambda\neq\emptyset} \Phi_A(\sigma_\Lambda\omega_{\Lambda^c})(<\infty)$. The *Gibbs specification at inverse temperature* $\beta > 0$ is then defined by

$$\gamma_\Lambda^{\beta\Phi}(\sigma \mid \omega) = \frac{1}{Z_\Lambda^{\beta\Phi}(\omega)}\ \mathrm{e}^{-\beta H_\Lambda^\Phi(\sigma|\omega)}(\rho_\Lambda \otimes \delta_{\omega_{\Lambda^c}})(d\sigma) \tag{6}$$

where the partition function $Z_\Lambda^{\beta\Phi}(\omega)$ is an important normalizing constant. Due to the, in fact rather strong, UAC condition, these specifications are quasilocal. It appears that the converse is also true up to a non-nullness condition[2] (see e.g. [18,26,36,42,52]) and one can take:

Definition 3 (Gibbs measures). *$\mu \in \mathcal{M}_1^+$ is a Gibbs measure iff $\mu \in \mathcal{G}(\gamma)$, where γ is a non-null and quasilocal specification.*

Quasilocality, called *Almost Markovianness* in [52], is a natural way to extend the global (two-sided) Markov property. When $\mu \in \mathcal{G}(\gamma)$ is quasilocal, then for any f local and $\Lambda \in \mathcal{S}$, the conditional expectations of f w.r.t. the outside of Λ are μ-a.s. given by $\gamma_\Lambda f$, by (2), and each conditional probability has a version which itself is a continuous function of the boundary condition, so one gets for any ω

$$\lim_{\Delta\uparrow\mathbb{Z}} \sup_{\omega^1,\omega^2\in\Omega} \Big|\mu\big[f|\mathcal{F}_{\Lambda^c}\big](\omega_\Delta\omega^1_{\Delta^c}) - \mu\big[f|\mathcal{F}_{\Lambda^c}\big](\omega_\Delta\omega^2_{\Delta^c})\Big| = 0 \tag{7}$$

Thus, for Gibbs measures the conditional probabilities always have continuous versions, or equivalently there is no point of essential discontinuity. Those are configurations which are points of discontinuity for ALL versions of the conditional probability. In particular one cannot make conditional probabilities continuous by redefining them on a measure-zero set if such points exist. In the generalized Gibbsian framework, one also says that such a configuration is a *bad configuration* for the considered measure, see e.g. [42]. The existence of such bad configurations implies non-Gibbsianness of the associated measures.

---

[2] expressing that $\forall\Lambda \in \mathcal{S},\ \forall A \in \mathcal{F}_\Lambda$, $\rho(A) > 0$ implies that $\gamma_\Lambda(A|\omega) > 0$ for any $\omega \in \Omega$.

2.2 Dyson Models: Ferromagnets in One Dimension

Definition 4 (Dyson models). *Let $\beta > 0$ be the inverse temperature and consider $1 < \alpha \leq 2$. We call a* Dyson model with decay parameter α *the Gibbs specification (6) with pair-potential Φ^D defined for all $\omega \in \Omega$ by*

$$\Phi^D_A(\omega) = -\frac{1}{|i-j|^\alpha}\omega_i\omega_j \text{ when } A = \{i,j\} \subset \mathbb{Z}, \text{ and } \Phi^D_A \equiv 0 \text{ otherwise.} \tag{8}$$

We shall also consider Dyson models with non-zero magnetic fields $h = (h_i)_{i\in\mathbb{Z}}$ acting as an extra self-interaction part $\Phi^D_A(\omega) = -h_i\omega_i$
when $A = \{i\} \subset \mathbb{Z}$

We first use that as a consequence of the FKG property [22,30], the Dyson specification is *monotonicity-preserving*[3], which implies that the weak limits obtained by using as boundary conditions the maximal and minimal elements of the order $\leq$ are well defined and are the extremal elements of $\mathcal{G}(\gamma^D)$.

Proposition 1 ([19,30,38]). *For $\alpha > 1$ (and not only for $\alpha \in (1,2]$), the weak limits*

$$\mu^-(\cdot) := \lim_{\Lambda\uparrow\mathbb{Z}} \gamma^D_\Lambda(\cdot|-) \quad \text{and} \quad \mu^+(\cdot) := \lim_{\Lambda\uparrow\mathbb{Z}} \gamma^D_\Lambda(\cdot|+) \tag{9}$$

are well-defined, translation-invariant and extremal elements of $\mathcal{G}(\gamma^D)$. For any f bounded increasing, any other measure $\mu \in \mathcal{G}(\gamma^D)$ satisfies

$$\mu^-[f] \leq \mu[f] \leq \mu^+[f]. \tag{10}$$

Moreover, μ^- and μ^+ are respectively left-continuous and right-continuous.

While μ^- and μ^+ coincide at high temperatures, and at all temperatures when there is fast decay, $\alpha > 2$, one main peculiarity of this one-dimensional model is thus that when the range is long enough ($1 < \alpha \leq 2$), it is possible to recover low-temperature behaviours usually associated to higher dimensions for the standard Ising model, in particular phase transitions can occur. For more details on the history of the proofs, one can consult [17] and references therein or below.

Proposition 2 ([1,7,14,23–26,33,41,44,48]). *The Dyson model with potential (8), for $1 < \alpha \leq 2$, exhibits a* phase transition at low temperature*:*

$$\exists \beta^D_c > 0, \text{ such that } \beta > \beta^D_c \implies \mu^- \neq \mu^+ \text{ and } \mathcal{G}(\gamma^D) = [\mu^-, \mu^+]$$

where the extremal measures μ^+ and μ^- are translation-invariant. They have in particular opposite magnetisations $\mu^+[\sigma_0] = -\mu^-[\sigma_0] = M_0(\beta,\alpha) > 0$ at low temperature. Moreover, the Dyson model in a non-zero homogeneous field h has a unique Gibbs measure.

---

[3] in the sense that for all bounded increasing functions f, and $\Lambda \in \mathcal{S}$, the function $\gamma^D_\Lambda f$ is increasing.

We remark that the infinite-volume limit of a state (or a magnetisation) in which there is a + (resp. −)-measure or a Dyson model in a field $h > 0$ (resp. $h < 0$) outside some interval is the same as that obtained from + (resp. −)-boundary conditions (independent of the magnitude of h). This can be e.g. seen by an extension of the arguments of [40], see also [39]. Notice that taking the +-measure of the zero-field Dyson model outside a finite volume enforces this same measure inside (even before taking the limit); adding a field makes it more positive, and taking the thermodynamic limit then recovers the same measure again.

The case of $\alpha = 2$ is more complicated to analyse, and richer in its behaviour, than the other ones. There exists a hybrid transition (the "Thouless effect"), as the magnetisation is discontinuous while the energy density is continuous at the transition point. Moreover, there is second transition below this transition temperature. In the intermediate phase there is a positive magnetisation with non-summable covariance, while at very low temperatures the covariance decays at the same rate as the interaction, which is summable. For these results, see [1,31,32,50], and also the more recent description in [44].

3 Decimation

We first apply a decimation transformation to the lattice $2\mathbb{Z}$. Similarly to what was discussed in [15], to analyse whether the transformed measure is a Gibbs measure, and in particular to show that it is non-Gibbsian, we have to show that conditioned on a particular configuration of the transformed spins, the "hidden spins" display a phase transition. If we choose this particular configuration to be the alternating one, each hidden spin feels opposite terms from the left and the right side, coming from all odd distances. Thus the conditioned model is a Dyson model in zero field, at a reduced temperature. As such it has phase transition.

To translate this hidden phase transition into nonlocality of the "visible" transformed spins follows straightforwardly the arguments of [15]. See [17] for the details. We make use of the fact that one can define global specifications, so there are no measurability problems due to global conditioning.

We start from μ^+ as defined in (9), the +-phase of a Dyson model without external field in the phase transition region, and apply the *decimation transformation*

$$T\colon (\Omega, \mathcal{F}) \longrightarrow (\Omega', \mathcal{F}') = (\Omega, \mathcal{F});\ \omega \ \longmapsto \omega' = (\omega'_i)_{i\in\mathbb{Z}},\ \text{with}\ \omega'_i = \omega_{2i} \qquad (11)$$

Denote $\nu^+ := T\mu^+$ the decimated +-phase, formally defined as an image measure via

$$\forall A' \in \mathcal{F}',\ \nu^+(A') = \mu^+(T^{-1}A') = \mu^+(A)$$

where

$$A = T^{-1}A' = \{\omega : \omega' = T(\omega) \in A'\}.$$

We study the continuity of conditional expectations under decimated Dyson Gibbs measures of the spin at the origin when the outside is fixed in some special configuration ω'_{alt}. By definition,

$$\nu^+[\sigma'_0|\mathcal{F}_{\{0\}^c}](\omega') = \mu^+[\sigma_0|\mathcal{F}_{S^c}](\omega),\ \nu^+\text{–a.s.} \tag{12}$$

where $S^c = (2\mathbb{Z}) \cap \{0\}^c$, i.e. with $S = (2\mathbb{Z})^c \cup \{0\}$ is not finite: *the conditioning is not on the complement of a finite set*, and although the extension of the DLR equation to infinite sets is direct in case of uniqueness of the DLR-measure for a given specification [19,21,27], it can be more problematic otherwise: it is valid for finite sets only and measurability problems might arise in case of phase transitions when one wants to extend them to infinite sets. Nevertheless, beyond the uniqueness case, such an extension was made possible by Fernández and Pfister [19] in the case of attractive models. As we will make essential use of it, we describe it now in our particular case. The concept they introduced is that of a *global specification*, and this is in fact a central tool in some of our arguments.

Definition 5 (Global specification, [19]**).** *A* global specification Γ *on* $\mathbb{Z}$ *is a family of probability kernels* $\Gamma = (\Gamma_S)_{S\subset\mathbb{Z}}$ *on* $(\Omega_S, \mathcal{F}_{S^c})$ *such that for* any S *subset of* $\mathbb{Z}$*:*

1. *$\Gamma_S(\cdot|\omega)$ is a probability measure on $(\Omega, \mathcal{F})$ for all $\omega \in \Omega$.*
2. *$\Gamma_S(A|\cdot)$ is $\mathcal{F}_{S^c}$-measurable for all $A \in \mathcal{F}$.*
3. *$\Gamma_S(B|\omega) = \mathbf{1}_B(\omega)$ when $B \in \mathcal{F}_{S^c}$.*
4. *For all $S_1 \subset S_2 \subset \mathbb{Z}$, $\Gamma_{S_2}\Gamma_{S_1} = \Gamma_{S_2}$ where the product of kernels is made as in (2).*

Similarly to the consistency with a (local) specification, one introduces the *compatibility of measures with a global specification.*

Definition 6. *Let Γ be a global specification. We write $\mu \in \mathcal{G}(\Gamma)$, or say that $\mu \in \mathcal{M}_1^+$ is* Γ-compatible, *if for all $A \in \mathcal{F}$ and* any $S \subset \mathbb{Z}$,

$$\mu[A|\mathcal{F}_{S^c}](\omega) = \Gamma_S(A|\omega),\ \mu\text{–a.e. } \omega. \tag{13}$$

Note, by considering $S = \mathbb{Z}$, that $\mathcal{G}(\Gamma)$ contains at most one element.
In the case considered here, we get a global specification Γ^+ such that $\mu^+ \in \mathcal{G}(\Gamma^+)$, with $S = (2\mathbb{Z})^c \cup \{0\}$ consisting of the *odd integers plus the origin.* Hence $S = (2\mathbb{Z})^c \cup \{0\}$ and (12) yields for ν^+-a.e. all $\omega' \in \mathcal{N}_{\Lambda'}(\omega'_{\text{alt}})$ and $\omega \in T^{-1}\{\omega'\}$:

$$\nu^+[\sigma'_0|\mathcal{F}_{\{0\}^c}](\omega') = \Gamma_S^+[\sigma_0|\omega]\ \ \mu^+\text{–a.e.}(\omega). \tag{14}$$

to eventually get (see [17,19]) an expression of the latter in terms of a constrained measure $\mu^{+,\omega}_{(2\mathbb{Z})^c\cup\{0\}}$, with $\omega \in T^{-1}\{\omega'\}$ so that we get for any $\omega' \in \mathcal{N}_{\Lambda'}(\omega'_{\text{alt}})$,

$$\nu^+[\sigma'_0|\mathcal{F}_{\{0\}^c}](\omega') = \mu^{+,\omega}_{(2\mathbb{Z})^c\cup\{0\}} \otimes \delta_{\omega_{2\mathbb{Z}\cap\{0\}^c}}[\sigma_0].$$

Thanks to monotonicity-preservation, the constrained measure is explicitly built as the weak limit obtained by +-boundary conditions fixed after a freezing the constraint to be ω on the even sites:

$$\forall \omega' \in \mathcal{N}_{\Lambda'}(\omega'_{\text{alt}}),\ \forall \omega \in T^{-1}\{\omega'\},$$
$$\mu^{+,\omega}_{(2\mathbb{Z})^c\cup\{0\}}(\cdot) = \lim_{I\in\mathcal{S}, I\uparrow(2\mathbb{Z})^c\cup\{0\}} \gamma_I^D(\cdot \mid +_{(2\mathbb{Z})^c\cup\{0\})}\omega_{2\mathbb{Z}\cap\{0\}^c}). \tag{15}$$

Observe that when a phase transition holds for the Dyson specification – at low enough T for $1 < \alpha \leq 2$ – the same is true for the constrained specification *with alternating constraint* (although at a lower T). This phase transition then implies non-Gibbsianness of ν^+ (and for all other Gibbs measures of the model, see [17]).

Theorem 1 **([17]).** *Let $\alpha \in (1, 2]$, let μ be a Gibbs measure for the interaction given by* (8) *and let the transformation T be defined by* (11)*. Then for low temperatures, $\beta > 2^\alpha \beta_c^D$, the decimated measure $\nu = T \circ \mu$ is non-Gibbs.*

Sketch of Proof. The main idea is to prove that the alternating configuration is an essential point of discontinuity for the decimated conditional expectations. As already observed, because any non-fixed site at all odd distances has a positive and a negative spin whose influences cancel, conditioning by this alternating configuration yields a constrained model that is again a Dyson model at zero field, but at a temperature which is higher by 2^α. This again has a low-temperature transition in our range of decays $1 < \alpha \leq 2$. The coupling constants are indeed multiplied by a factor $2^{-\alpha}$, due to only even distances occurring between interacting (hidden) spins.

To prove non-Gibbsianness in [17], we essentially follow the proof strategy sketched in [15], by showing that within a neighborhood $\mathcal{N}_L(\omega'_{\text{alt}})$, there exists two subneighborhoods $\mathcal{N}^{\pm}_{N,L}$ of positive measure on which the conditional magnetizations defined on $\mathcal{N}_N, L^{\pm}$

$$M^+ = M^+(\omega) = \mu^{+,\omega^+}_{(2\mathbb{Z})^c \cup \{0\}}[\sigma_0] \text{ and } M^- = M^-(\omega) = \mu^{+,\omega^-}_{(2\mathbb{Z})^c \cup \{0\}}[\sigma_0] \qquad (16)$$

differ significantly.

The role of the "annulus" where configurations are constrained to be either + or − is played by two large intervals $[-N, -L-1]$ and $[L+1, N]$. Due to the long range of the interaction, their might be a direct influence from the boundary beyond the annulus, to the central interval. To avoid effects from this influence, we take N *much* larger than L. An argument based on "equivalence of boundary conditions" as in e.g. [6], under a choice $N = L^{1/(\alpha-1)}$ then implies that (16) does hardly depend on ω.

Once this choice of big annulus is made, observe that if we constrain the spins in these two intervals to be either + or −, within these two intervals the measures on the unfixed spins are close to those of the Dyson-type model in a positive, c.q. negative, magnetic field. As those measures are unique Gibbs measures, no influence from the boundary can be transmitted.

Indeed, in contrast to the case of the purely alternating configuration, in the case when we condition on all primed spins to be + (resp. −) in these large annuli, there is no phase transition and the system of unprimed spins has a unique Gibbs measure. It is a Dyson model, again at a heightened temperature, but now in a homogeneous external field, with positive (resp. negative) magnetisation $+M_0(\beta, \alpha) > 0$ (resp. $-M_0(\beta, \alpha) < 0$), stochastically larger (resp. smaller) than the zero-field + (resp. −)-measure.

In the −-case, in the annulus the magnetisation of the -even-distance- Dyson–Ising model is essentially that of the model with a negative homogeneous external field $-h$ everywhere, which at low enough temperature and for L large enough is close to (and in fact smaller than) the magnetisation of the Dyson–Ising model under the zero-field −-measure, i.e to $-M_0(\beta,\alpha) < 0$. Thus the inner interval where the constraint is alternating feels a −-like condition from outside its boundary. On the other hand, the magnetisation with the constraint ω^+ will be close to or bigger than $+M_0(\beta,\alpha) > 0$ so that a non-zero difference is created at low enough temperature. One needs again to adjust the sizes of L and N to be sure that boundary effects from outside the annulus are negligible in the inner interval.

Thus, for a given $\delta > 0$, e.g. $\delta = M_0(\beta,\alpha)/2$, for arbitrary L one can find $N(L)$ large enough, such that the expectation of the spin at the origin differs by more than δ. One can therefore feel the influence from the decimated spins in the far-away annulus, however large the central interval of decimated alternating spins is chosen. Thus, indeed it holds that $M^+ - M^- > \delta$, uniformly in L. □

In our choice of decimated lattice we made use of the fact that the constrained system, due to cancellations, again formed a zero-field Dyson-like model. This does not work for decimations to more dilute lattices, but although the original proofs of Dyson [13] and of Fröhlich and Spencer [24], or the Reflection Positivity proof of [23] do no longer apply to such periodic-field cases, the contour-like arguments of [7] and [33] could presumably still be modified to include such cases. Compare also [16,35].

The analysis of [9] which proves existence of a phase transition for Dyson models in random magnetic fields for a certain interval of α-values should imply that in that case there are many more, random, configurations which all are points of discontinuity. We note that choosing independent spins as a constraint provides a random field which is correlated. However, these correlations decay enough that this need actually not spoil the argument. Similarly, one should be able to prove that decimation of Dyson models in a weak external field will result in a non-Gibbsian measure. An interesting question would be to perform the analysis of [45]) or [43] to get a.s. configuration-dependent correlation decays.

On the other side of the Gibbs-non-Gibbs analysis, when the range of the interaction is lower, i.e. for $\alpha > 2$, or the temperature is too high, uniqueness holds, for all possible constraints and the transformed measures should be Gibbsian. Some standard high-temperature results apply, which were already discussed in [15].

About these shorter-range models, (i.e. long-range models with faster polynomial decay), Redig and Wang [46] have proved that Gibbsianness was conserved, providing in some cases ($\alpha > 3$) a decay of correlation for the transformed potential. In our longer-range models, for intermediate temperatures (below the transition temperature but above the transition temperature of the alternating-configuration-constrained model) decimating, both +- and −-measures, should imply Gibbsianness, essentially due to the arguments as proposed for short-range models in [29].

4 Dyson Models in Decaying Fields

In this section we consider one-dimensional Dyson models in a decaying field with decay parameter γ. The corresponding interaction $\Phi_A^D(\omega)$ is defined by

$$\Phi_A^D(\omega) = \begin{cases} -\frac{J}{|i-j|^\alpha}\omega_i\omega_j & \text{if } A = \{i,j\} \\ -\frac{h}{(|i|+1)^\gamma}\omega_i & \text{if } A = \{i\} \end{cases} \tag{17}$$

for some $J, h > 0$. The question raised in [5] is whether it is possible to extend results from e.g. [7,9], and in particular to investigate whether and under which conditions the existence of two distinct phases prevails in the presence of an external field. Let us mention that one-dimensional Dyson models in a field were considered before, for example in [35], where uniqueness was proven for fields which are either strong enough ($\Phi_{\{i\}}^D = h_i\omega_i$, where there exists $h_0 > 0$ such that $|h_i| > h_0$) or periodic in large enough blocks.

The main tool we will use are the one-dimensional contours of [7]. Recall that in [7] the authors prove that under the technical constraints $\alpha \in (\alpha^*, 2]$, where $\alpha^* := 3 - \log(3)/\log(2)$, $J(1) >> 1$ and $h = 0$, there exists $\beta_{c,0}^D > 0$ such that for all $\beta > \beta_{c,0}^D$

$$M_0(\beta, \alpha) = \mu^+[\sigma_0] = -\mu^-[\sigma_0] > 0$$

i.e. there is spontaneous magnetization yielding non-uniqueness of the Gibbs measures, $\mu^+ \neq \mu^-$. This result was generalized to all values of $\alpha \in (1, 2]$ in [44], again assuming the technical condition $J(1) >> 1$.

Phase coexistence in a positive external field is an unusual phenomenon, since typically Gibbs measures for models in a field are unique. It was previously observed in nearest-neighbour pair potentials with polynomially decaying fields in $d \geq 2$, see [2,3,10] or for sufficiently fast decaying (but not necessarily summable) fields on trees [4]. In [3] it is proven that in nearest-neighbour models for $\gamma > 1$ and low enough temperatures, there are multiple Gibbs states, whereas for $\gamma < 1$ there is a unique one.

Pirogov-Sinai is a robust and often applicable version of the Peierls contour argument [28], applicable in $d \geq 2$, which is the most generally applicable approach in higher dimensions.

In [7], inspired by and extending results of the seminal paper of Fröhlich and Spencer [24], the authors presented a contour argument which works even for long-range models in one dimension, in particular, it worked for one-dimensional Dyson models with $\alpha^* < \alpha \leq 2$. The techniques used in [7] rely on developing a graphical representation of spin configurations in terms of triangular *contours.*

It turns out that for a one-dimensional long-range model in a decaying field, depending on the relation between α and γ, there can be either one or two extremal Gibbs measures. Let us emphasize that we manage to remove technical restrictions both on on α and $J(1)$. We can prove the following theorem.

Theorem 2 ([5]). *Let $\alpha \in (1, 2]$ and $\gamma > \max\{\alpha - 1, \alpha^* - 1\}$ be the exponents of the Dyson model w.r.t. an interaction Φ^D given by (17). Then, there exists $\beta_{c,h}^D > 0$ s.t. for all $\beta > \beta_{c,h}^D$ we have $M_0(\beta, \alpha, \gamma) > 0$, i.e. $\mu^+ \neq \mu^-$.*

Sketch of Proof. The main idea of the proof is to extend the analysis in [7,9], combined with [44]. Consider a finite-volume Gibbs measure on an interval, say $\Lambda = [-N, N]$ and fix +-boundary conditions. Each spin configuration σ can be uniquely mapped into a triangle configuration $\underline{T} = (T_1, ..., T_n)$ where endpoints of the triangles are defined by interface points dividing plus from minus spins. Contours Γ are collections of triangles T_i such that they are in some sense *well separated* from each other and subadditive, so that we obtain a lower bound for the energy of given triangle configuration $\underline{T}$. Phase coexistence will follow from the well-known Peierls argument for $d > 1$, i.e. from the estimate that for β sufficiently large

$$\mu_\Lambda^+[\sigma_o = -1] \leq \mu_\Lambda^+[\{o \in \Gamma\}] \leq \frac{1}{Z_\Lambda^+} \sum_{\Gamma \ni o} \sum_{\Gamma \text{ compatible}} \mathrm{e}^{-\beta H(\underline{T})} < \frac{1}{2}.$$

The main difficulty then is to obtain a *good* energetic lower bound for the Hamiltonian including the effect of the external field. □

Physically, an argument explaining the statement of the theorem goes as follows: There is a competition between the effect of the pair interaction and that of the external field. Having minus boundary conditions means that inserting a large interval $[-L, L]$ of plus spins will cost an energy of order

$$\sum_{|i|<L} \sum_{|j|>L} |i-j|^{-\alpha} = O(L^{2-\alpha}).$$

However, the gain in energy due to the spins following the external magnetic field is of order

$$\sum_{|i|<L} |i|^{-\gamma} = O(L^{1-\gamma}).$$

Thus, (somewhat similar to an Imry–Ma argument), we see that for $\gamma > \alpha - 1$ we should expect that the field is too weak to overcome the boundary conditions and the plus and minus measures are different: $\mu^+ \neq \mu^-$.

When the opposite case pertains, that is $\gamma < \alpha - 1$, there should be a unique Gibbs measure, with a magnetisation in the direction of the field, whatever the boundary conditions employed. We are in the process of rigorising this picture.

In fact, the analogous prediction in the 2-dimensional short-range model has been fully proved by [3,10], also giving that the critical value for γ equals 1, where is possible to prove the phase transition even in the critical case, assuming that h is small enough. Here we have the same situation, we can extend the theorem above for the case when $\gamma = \max\{\alpha - 1, \alpha^* - 1\}$ if we take h small enough.

The restriction on γ involving α^* seems due to technical reasons, since we use arguments developed in [7]. However, from the physical argument sketched above, we expect that these limitations should not be required and the argument should work, just assuming the inequality between γ and α.

Remark Added in Proof: J. Littin has kindly informed us that he has performed a similar analysis in which the signs of the external fields are chosen i.i.d.

random and symmetric. In that case, instead of $\gamma = \alpha - 1$, the threshold value for phase transition stability becomes $\gamma = \alpha - \frac{1}{2}$.

Acknowledgments. We gratefully dedicate this paper to Chuck Newman on the occasion of his 70th birthday. Especially AvE has over many years been inspired by and profited from his scientific contributions, his always friendly stimulating questions and discussions, and his encouraging attitude, whether it was about Dyson models, non-Gibbsian measures, spin glasses or any other topic from statistical mechanics.

We thank J. Littin for making [44] available to us and P. Picco for very helpful discussions.

We were appreciative to be asked and pleased to contribute this article after Vladas Sidoravicius invited us to do so. The more it is a matter of sadness for us all that he will not be able to see this volume come to fruition.

RB is supported by CAPES, FAPESP grant 2011/16265-8 and CNPq grants 453985/2016-5 and 312112/2015-7. EOE is supported by FAPESP Grants 14/10637-9 and 15/14434-8.

We also thank the universities of Delft and Groningen and STAR–NWO for making various research visits possible.

References

1. Aizenman, M., Chayes, J., Chayes, L., Newman, C.M.: Discontinuity of the magnetization in the one-dimensional $1/\mid x-y\mid^2$ percolation, Ising and potts models. J. Stat. Phys. **50**(1–2), 1–40 (1988)
2. Bissacot, R., Cioletti, L.: Phase transition in ferromagnetic Ising models with non-uniform external fields. J. Stat. Phys. **139**(5), 598–617 (2010)
3. Bissacot, R., Cassandro, M., Cioletti, L., Presutti, E.: Phase transition in ferromagnetic Ising models with spatially dependent magnetic fields. Comm. Math. Phys. **337**, 41–53 (2015)
4. Bissacot, R., Endo, E.O., van Enter, A.C.D.: Stability of the phase transition of critical-field Ising model on Cayley trees under inhomogeneous external fields. Stoch. Process. Appl. **127**, 4126–4138 (2017). arXiv:1611.00424
5. Bissacot, R., Endo, E.O., van Enter, A.C.D., Kimura, B., Ruszel, W.M.: Contour methods for long-range Ising models: weakening nearest-neighbor interactions and adding decaying fields. Ann. Henri Poincaré **19**, 2557–2574 (2018)
6. Bricmont, J., Lebowitz, J.L., Pfister, C.-E.: On the equivalence of boundary conditions. J. Stat. Phys. **21**(5), 573–582 (1979)
7. Cassandro, M., Ferrari, P.A., Merola, I., Presutti, E.: Geometry of contours and peierls estimates in $d = 1$ Ising models with long range interactions. J. Math. Phys. **46**(5), 0533305 (2005)
8. Cassandro, M., Olivieri, E.: Renormalization group and analyticity in one dimension: a proof of Dobrushin's theorem. Comm. Math. Phys. **80**, 255–269 (1981)
9. Cassandro, M., Orlandi, E., Picco, P.: Phase transition in the 1D random field Ising model with long range interaction. Comm. Math. Phys. **288**, 731–744 (2009)
10. Cioletti, L., Vila, R.: Graphical representations for Ising and Potts models in general external fields. J. Stat. Phys. **162**, 81–122 (2016)
11. Dobrushin, R.L.: The description of a random field by means of conditional probabilities and conditions of its regularity. Theory Probab. Appl. **13**, 197–224 (1968)

12. Dynkin, E.B.: Sufficient statistics and extreme points. Ann. Probab. **6**(5), 705–730 (1978)
13. Dyson, F.J.: Existence of a phase transition in a one-dimensional Ising ferromagnet. Comm. Math. Phys. **12**, 91–107 (1969)
14. Dyson, F.J.: An Ising ferromagnet with discontinuous long-range order. Comm. Math. Phys. **21**, 269–283 (1971)
15. van Enter, A.C.D., Fernández, R., Sokal, A.D.: Regularity properties and pathologies of position–space R.G. transformations: scope and limitations of Gibbsian theory. J. Stat. Phys. **72**, 879–1167 (1993)
16. van Enter, A.C.D., Külske, C.: Two connections between random systems and non-Gibbsian measures. J. Stat. Phys. **126**, 1007 1024 (2007)
17. van Enter, A.C.D., Le Ny, A.: Decimation of the Dyson-Ising ferromagnet. Stoch. Process. Appl. **127**, 2776–3791 (2017). arXiv:1603.05409
18. Fernández, R.: Gibbsianness and non-Gibbsianness in lattice random fields. In: Bovier, A., van Enter, A.C.D., den Hollander, F., Dunlop, F. (eds.) Mathematical Statistical Physics. Proceedings of the 83rd Les Houches Summer School, July 2005. Elsevier (2006)
19. Fernández, R., Pfister, C.-E.: Global specifications and non-quasilocality of projections of Gibbs measures. Ann. Probab. **25**(3), 1284–315 (1997)
20. Föllmer, H.: Phase transition and martin boundary. In: Séminaires de Probabilités IX, Université de Strasbourg. Lecture Notes in Mathematics, vol. 465, pp. 305–317. Springer, Heidelberg (1975)
21. Föllmer, H.: On the global Markov property. In: Streit, L. (ed.) Quantum fields: Algebras, Processes, pp. 293–302. Springer, New York (1980)
22. Fortuin, C.M., Kasteleyn, P.W., Ginibre, J.: Correlation inequalities on some partially ordered sets. Comm. Math. Phys. **22**, 89–103 (1971)
23. Fröhlich, J., Israel, R.B., Lieb, E.H., Simon, B.: Phase transitions and reflection positivity. I. General theory and long range lattice models. Comm. Math. Phys. **62**, 1–34 (1978)
24. Fröhlich, J., Spencer, T.: The phase transition in the one-dimensional Ising model with $1/r^2$ interaction energy. Comm. Math. Phys. **84**, 87–101 (1982)
25. Friedli, S., Velenik, Y.: Equilibrium Statistical Mechanics: A Concrete Mathematical Introduction. Cambridge University Press, Cambridge (2017). http://www.unige.ch/math/folks/velenik/smbook/index.html
26. Georgii, H.O.: Gibbs Measures and Phase Transitions. De Gruyter Studies in Mathematics, 2nd edn., vol. 9, Berlin, New York, 1988 (2011)
27. Goldstein, S.: Remarks on the global Markov property. Comm. Math. Phys. **74**, 223–234 (1980)
28. Griffiths, R.B.: Peierls proof of spontaneous magnetization in a two-dimensional Ising ferromagnet. Phys. Rev. **136**(2), A437–A439 (1964)
29. Haller, K., Kennedy, T.: Absence of renormalization pathologies near the critical temperature. Two examples. J. Stat. Phys. **85**, 607–637 (1996)
30. Hulse, P.: On the ergodic properties of Gibbs states for attractive specifications. J. Lond. Math. Soc. (2) **43**(1), 119–124 (1991)
31. Imbrie, J.: Decay of correlations in one-dimensional Ising model with $J_{ij} = |i-j|^{-2}$. Comm. Math. Phys. **85**, 491–515 (1982)
32. Imbrie, J., Newman, C.M.: An intermediate phase with slow decay of correlations in one-dimensional $1/\mid x-y\mid^2$ percolation, Ising and Potts models. Comm. Math. Phys. **118**, 303–336 (1988)
33. Johansson, K.: Condensation of a one-dimensional lattice gas. Comm. Math. Phys. **141**, 41–61 (1991)

34. Kac, M., Thompson, C.J.: Critical behaviour of several lattice models with long-range interaction. J. Math. Phys. **10**, 1373–1386 (1969)
35. Kerimov, A.: A block effect of external field in the one-dimensional ferromagnetic Ising model with long-range interaction. J. Phys. A Math. Theor. **40**, 40407–40414 (2007)
36. Kozlov, O.: Gibbs description of a system of random variables. Probl. Peredachi Informatsii **10**, 258–265 (1974)
37. Lanford, O.E., Ruelle, D.: Observables at Infinity and states with short range correlations in statistical mechanics. Comm. Math. Phys. **13**, 194–215 (1969)
38. Lebowitz, J.L.: Coexistence of phases for Ising ferromagnet. J. Stat. Phys. **16**(6), 463–476 (1977)
39. Lebowitz, J.L.: Thermodynamic limit of the free energy and correlation functions of spins systems. Acta Phys. Austr. Suppl. XVI, 201–220 (1976)
40. Lebowitz, J.L., Penrose, O.: Analytic and clustering properties of thermodynamic functions and distribution functions for classical lattice and continuum systems. Comm. Math. Phys. **11**, 99–124 (1968)
41. Lee, T.D., Yang, C.N.: Statistical theory of equations of state and phase transitions. II. Lattice gas and Ising Model. Phys. Rev. **87**, 410–419 (1952)
42. Le Ny, A.: Introduction to generalized Gibbs measures. Ensaios Matemáticos **15** (2008)
43. Le Ny, A.: Almost Gibbsianness and parsimonious description of the decimated 2d-Ising model. J. Stat. Phys. **152**(2), 305–335 (2013)
44. Littin, J., Picco, P.: Quasiadditive estimates on the Hamiltonian for the one-dimensional long-range Ising model. J. Math. Phys. **58**, 073301 (2017)
45. Maes, C., Redig, F., Shlosman, S., Van Moffaert, A.: Percolation, path large deviations and weak Gibbsianity. Comm. Math. Phys **209**(8), 517–545 (1999)
46. Redig, F., Wang, F.: Transformations of one-dimensional Gibbs measures with infinite range interaction. Markov Process. Relat. Fields **16**(4) (2010)
47. Ruelle, D.: Statistical mechanics of a one dimensional lattice gas. Comm. Math. Phys. **9**, 267–278 (1968)
48. Ruelle, D.: On the use of "small external fields" in the problem of symmetry breakdown in statistical mechanics. Ann. Phys. **69**, 364–374 (1972)
49. Rushbrooke, G., Ursell, H.: On one-dimensional regular assemblies. Proc. Camb. Philos. Soc. **44**(2), 263–271 (1948)
50. Simon, B., Sokal, A.D.: Rigorous entropy-energy arguments. J. Stat. Phys. **25**(4), 679–694 (1981)
51. Sokal, A.D.: Existence of compatible families of proper regular conditional probabilities. Z. Wahrsch. verw. Geb. **56**, 537–548 (1981)
52. Sullivan, W.G.: Potentials for almost Markovian random fields. Comm. Math. Phys. **33**, 61–74 (1976)

Complex Burgers Equation: A Probabilistic Perspective

Radu Dascaliuc[1], Nicholas Michalowski[2], Enrique Thomann[1], and Edward C. Waymire[1(✉)]

[1] Department of Mathematics, Oregon State University, Corvallis, OR 97331, USA
{dascalir,waymire}@math.oregonstate.edu
[2] Department of Mathematics, New Mexico State University, Las Cruces, NM 88003, USA

Abstract. In 1997 Yves LeJan and Alain-Sol Sznitman provided a probabilistic gateway in the form of a stochastic cascade model for the treatment of 3d incompressible, Navier–Stokes equations in free space. The equations themselves are noteworthy for the inherent mathematical challenges that they pose to proving existence, uniqueness and regularity of solutions. The main goal of the present article is to illustrate and explore the LeJan–Sznitman cascade in the context of a simpler quasi-linear pde, namely the complex Burgers equation. In addition to providing some unexpected results about these equations, consideration of mean-field models suggests analysis of branching random walks having naturally imposed time delays.

Keywords: Complex Burgers equation · Global well-possedness · Self-similarity · Mean-field equations · Stochastic cascades · Stochastic explosion · Markov evolutionary processes

1 Introduction

Throughout his career Chuck Newman has demonstrated mathematical insights that are remarkable for both their depth and their wide ranging relevance to seemingly diverse areas of mathematics and science. There are many things to admire about the way in which Chuck is able to resolve things mathematically. The following story may be a lesser known example, but it is relevant to the topic of this article, and is mainly being shared in admiring tribute to Chuck. During the early 1980's while still a professor at the University of Arizona, Chuck attended a colloquium talk on fluids featuring a discussion of Burgers equation. As some of us were leaving the talk, Chuck shared a scrap of paper

V. Sidoravicius (Ed.): *Sojourns in Probability Theory and Statistical Physics - I*, PROMS 298, pp. 138–170, 2019.
https://doi.org/10.1007/978-981-15-0294-1_6

in which he had doodled[1] a striking connection between Burgers equation for fluids and one of his first loves, statistical physics. Specifically, in a matter of minutes, he had discovered that the magnetization in a finite volume Curie–Weiss probability model is governed, as a function of inverse temperature and external field, exactly by the same Burgers equation presented by the speaker in a seemingly unrelated context of fluid flow. In addition to providing a rather unique perspective on shocks in quasilinear pde's in relation to spontaneous magnetization, Chuck had quite cleverly provided new insights into a basic model of interest to mathematical physics and probability; see [44]. In particular, early representations of solutions to the Burgers' equation as expected values in an interesting probability model can just as well be attributed to Chuck Newman.

Proposition 1 (Newman, 1986). *Let* $m = m_n(h, \beta) = \mathbb{E}S_1$ *where* $\{S_i : i = 1, \ldots, n\}$ *has the joint distribution* $Z^{-1}\exp[\frac{\beta}{2}\sum J_{ij}s_is_j + \sum h_is_i]\prod\rho(ds_i)$ *with* $J_{ij} = J/2n, (J > 0), h_i = h$ *for all* i, j *and* $\int\exp(Ks^2)\rho(ds) < \infty$ *for all* $K > 0$. *Then*

$$\partial m/\partial\beta = Jm\frac{\partial m}{\partial h} + \frac{J}{2n}\frac{\partial^2 m}{\partial h^2}, \quad m(h, 0) = \frac{d}{dh}\ln\int e^{hs}\rho(ds).$$

As remarked in [44], in the usual spin-1/2 Ising models the measure ρ is given by $\rho(ds) = [\delta(s-1) + \delta(s+1)]\frac{ds}{2}$, and the resulting initial condition is $m(h, 0) = \tanh(h)$. The classic space-time Burgers equation results by defining $t = \beta J$, $x = -h$ and $\nu = 1/n$. Considerations of complex h arise naturally in ([44], Theorem 6), see also [42,45], in connection with Chuck's take on the zeros of the partition function and the Riemann hypothesis. As will be seen, the complexification of Burgers suggested by the LJS-cascade is in the initial data and solutions, rather than their spatial domain.

The (unforced) three-dimensional incompressible Navier–Stokes equations governing fluid velocity $v(x,t) = (v_1(x,t), v_2(x,t), v_3(x,t))$ and (scalar) pressure $p(x,t)$, in free space $x \in \mathbb{R}^3, t \geq 0$, are given for initial data v_0 and viscosity parameter $\nu > 0$, by a system of quasi-linear partial differential equations

$$\frac{\partial v}{\partial t} + v\cdot\nabla v = \nu\Delta v - \nabla p, \quad \nabla\cdot v = 0, \quad v(x, 0^+) = v_0(x), \ x \in \mathbb{R}^3, \ t > 0. \quad (1)$$

The equation $\nabla \cdot v = 0$ defines the incompressibility condition. The nonlinear term $v\cdot\nabla v$ is the result of representing the flow in a Lagrangian coordinate system; i.e., $\frac{\partial v}{\partial t} + v\cdot\nabla v$ is the acceleration in a frame following a moving particle and, as such, mathematically intrinsic to the equations.

Like the Riemann hypothesis, settling the question about the global existence of smooth solutions for smooth initial data ranks among the millennial problems

[1] The reference to "doodle" is deliberate. Once, when asked how he discovered his central limit theorem for associated random variables given in [43], Chuck replied that he had expected for some time that a central limit theorem should be possible for ferromagnetic Ising models at high temperatures as a consequence of correlation decay: "Then, one day I was doodling with the FKG inequalities and out popped just the right correlation inequalities for the characteristic function of the magnetization". This would prove that central limit theorem.

for mathematics. The following is the precise Clay prize formulation of a positive resolution provided by Charles Fefferman [19].

Navier–Stokes Millennial Problem (MP): For divergence-free $v_0 \in \mathcal{S}$ of rapid decay, show that there exists $p, v \in C^\infty(\mathbb{R}^3 \times [0,\infty))$ satisfying (1), and the finite energy condition that, for some $C > 0$,

$$\int_{\mathbb{R}^3} |v(x,t)|^2 dx \le C, \quad \forall t \ge 0.$$

It may be noted that uniqueness is implicit in this formulation of the problem. Also an alternative formulation posed on the torus with periodic boundary conditions can be made in place of the free space formulation. Of course a negative formulation is also possible. Since Fourier transform is a homeomorphism on the Schwarz space $\mathcal{S}$, an equivalent condition follows accordingly for the initial data.

The goal of the present article is to illustrate and explore a probabilistic gateway to three-dimensional incompressible Navier–Stokes equations (1) in Fourier space discovered by Le Jan and Sznitman [37], hereafter referred to as the LJS-cascade, but in the context of stochastic cascades derived from the ostensibly simpler complex Burgers equation, and related mean-field cascades to be defined. This will also give an opportunity to provide some contrasting remarks and/or related results for the case of (1) and the LJS-cascade.

Recent treatments of the complexified Burgers equations are largely focussed on blow-up singularities in finite time using Cole–Hopf transformations, i.e., loss of spatial analyticity; see [PS], [L]. For a comprehensive analytic treatment of complex evolution equations, including Burgers, see [GG]. An approach to singularities in Fourier space is given in [LS]. Our results are consistent, but the methods are probabilistic. Hadamard ill-posedness is shown in Sect. 6.2. In fact, for a mean-field model, compactly supported initial data in Fourier space becomes L^∞ in Fourier space in finite time, but without decay. However, in time the solution actually regains regularity. This seems to be a new phenomenon.

As will be seen, two modifications of the original LJS-cascade are made throughout this paper: (i) to permit the phenomena of *stochastic explosion* in cascades where it may naturally happen in unforced equations, and (ii) to exploit time-space self-similarity representation of solutions to (1) via a modification of the stochastic cascade. The modified cascade will continue to be referred to as the LJS-cascade.

The organization of this paper is as follows. The main features of a mild formulation of (1) based on the LJS-cascade is provided in the next section as motivating background. A few other notable probabilistic approaches to (1) are also cited to provide a slightly broader perspective on (1), but the primary focus here is the nature of the LJS-cascade in a simpler context of a complex Burgers equation; one may also consult [56] for a survey. As in the case of (1), the LJS-cascade dictates a certain probabilistically natural choice for the function spaces in which the Burgers equation can be effectively analyzed. In particular, one is naturally led to Hardy space formulations as defined by functions whose Fourier transform vanish for negative Fourier frequencies and suitably decays for large positive frequencies. For contrast and comparison, the LJS-cascade associated with (1) leads naturally to certain Besov-type solution spaces; e.g., see [4,37]. In any case, once this determination is made, the associated LJS-cascade may be broadly viewed in terms of a time-delayed[2] branching random walk. More specifically, two essential features of the LJS-cascade are (a) the Yule process, viewed as a binary tree-indexed family of i.i.d. mean one exponential random variables, and (b) a corresponding branching random walk in Fourier frequency space.

The LJS-cascade for Burgers leads to the introduction of new related probability models in the forms of a *mean-field Burgers model*, a *β-field Burgers model*, and an *α-delayed Yule process* given in the next two sections. A key feature of the LJS-cascade in general is a certain conservation of *spatial* Fourier frequencies in the corresponding branching random walk. This conservation property persists spatially in the mean-field Burgers model, for which $\beta = \frac{1}{2}$, however is not reflected in the other β-field models. A two-parameter generalization would accommodate the conservation property, however it will not be considered in the present paper beyond a few remarks.

The class of β-field Burgers models conserves temporal frequencies in the case $\beta = \frac{1}{2}$. Due to the difference in scaling of spatial frequencies of a branching random walk and the corresponding scaling of times between movements, one is thus naturally led to a companion purely temporal α-delayed Yule process whose connection to the β-field models can be explicitly expressed via associated Markov semigroups. Temporal frequency conservation is significant among the α-delayed Yule processes and led to the recent discovery that the Poisson process may be realized as an α-delayed Yule process when $\alpha = \frac{1}{2}$, [14]. From a purely probabilistic perspective, this and related results in [14] may be viewed as a variant on an old discovery in [36] identifying the Poisson process as a random time change of a Yule process. It is also noteworthy that $\alpha = \frac{1}{2}$ is a critical value for the α-delayed Yule processes when viewed in terms of boundedness of their infinitesimal generators if and only if $\alpha \leq \frac{1}{2}$. In particular this includes $\alpha = \beta^2 = \frac{1}{4}$ of the mean-field cascade.

Following the brief introduction of these various models, we conclude the paper with a capstone section devoted to the analysis of the four leading questions that prompted their consideration: *existence/uniqueness, well-posedness, regularity, and self-similarity.* Although the probabilistic framework is very close to familiar classic models, the perturbations to existing theory provide interesting new challenges for the resolution of these problems.

2 A Brief Highlight of Probabilistic Approaches to Navier–Stokes Equations and the LJS-Cascade

Let us first agree on the signs and normalizations in the Fourier transform to be used throughout. Namely, for integrable functions and/or tempered distributions, suitably interpreted,

[2] The mean-field models for the Navier–Stokes equation, on the other hand, involve parameters $\beta > 1$ as well; see [15] in this regard.

$$\hat{f}(\xi) = (2\pi)^{-\frac{n}{2}} \int_{\mathbb{R}^n} e^{-i\xi \cdot x} f(x)dx, \quad \xi \in \mathbb{R}^n.$$

Two noteworthy observations about (1) can be made in terms of Fourier transforms, denoted by $\wedge$:

$$\nabla \cdot v = 0 \quad \text{corresponds to} \quad \xi \cdot \hat{v} = 0$$

and

$$\nabla p \quad \text{corresponds to} \quad \hat{p}\xi.$$

As a result of incompressibility, the pressure term $\hat{p}(\xi, t)$ may be removed while retaining the velocity $\hat{v}(\xi, t)$ by a Leray projection of the (Fourier transformed) Eq. (1) in the direction orthogonal to ξ. As a result of this, the Fourier transformation of multiplication to convolution, and the multiplier effect of the Fourier transform of the Laplacian term, the LJS-cascade emerges naturally as a natural stochastic structure associated with the integrated equations, i.e., a mild form of (1). This formulation was generalized in [4], and also developed in [6,52] from a perspective of harmonic analysis.

Remark 1. The modification (i) of the original LJS-cascade noted above essentially involves the representation of the forcing term. Specifically, in the original LJS-cascade, the unforced equations would be viewed as equations *forced by zero*, whereas we elect to view the lack of forcing as simply *ignorable*. This latter view leads to considerations of explosive branching that are not an issue for the original formulation in [37], since a finite explosion time[3] means that there will be infinitely many branchings within a finite time.

To recover p from the projected velocity one notes that, again owing to the incompressibility condition, the divergence of the linear terms in (1) is zero. Thus, taking the divergence followed by the Fourier transform, one arrives at

$$\hat{p}(\xi, \cdot) = \sum_{j,k} \hat{R}_j \hat{v}_j(\xi, \cdot) \hat{R}_k \hat{v}_k(\xi, \cdot),$$

where $\hat{R}_j f(\xi) = -\frac{\xi_j}{|\xi|} \hat{f}(\xi)$ is the Fourier symbol expresses the j-th Riesz transform R_j convolved with f. In particular,

$$p = \sum_{j,k} R_j R_k (v_j v_k).$$

The Gundy–Varapoulos–Silverstein probabilistic representation of Riesz transforms in terms of Brownian motion from infinity is noteworthy in this context. Although originally formulated on a measure space of infinite measure, the construction has been modified in [3] to a measure space with total probability

[3] An unfortunate typo occurs in the Appendix to [13] in which the explosion event should be denoted $[\zeta < \infty]$, not $[\zeta = \infty]$.

one. To appreciate the role of incompressibility from a probabilistic perspective, one may consider the linearized Stokes problem

$$\frac{\partial v}{\partial t} = \Delta v, \quad \nabla \cdot v = 0, \ x \in \mathbb{R}^3, \ t > 0.$$

with arbitrary, not necessarily incompressible, initial data v_0. In [54] the fundamental solution is explicitly computed as

$$v(x,t) = \int_{\mathbb{R}^3} \Gamma(x-z,t)v_0(z)dz,$$

where the semigroup $\Gamma = [I_{3\times 3} + \mathcal{R}](K)$, for the 3×3 identity matrix $I_{3\times 3}$, matrix of Riesz transforms $\mathcal{R} = ((R_j R_k))$, and Gaussian transition kernel K for standard Brownian motion. Of course if v_0 is incompressible then the usual representation of solutions to the heat equation in terms of Brownian motion is recovered. A stochastic calculus that would capture the effect of incompressibility on Brownian motion remains an intriguing challenge.

Recently an alternative probabilistic approach to Navier–Stokes was developed by Constantin and Iyer; [11,33,34]. Their idea is to use a Weber formula to express the velocity of the inviscid equation in terms of the particle paths, being careful to avoid derivatives in time of the particle paths. For given periodic, incompressible, $2+\delta$-Hölder continuous initial data v_0 this leads to an equivalent system for the inviscid equations Navier–Stokes equations, i.e., Euler equations, of the form

$$\begin{aligned} \dot{X} &= v \\ A &= X^{-1} \\ v &= \mathbf{P}\big[(\nabla^{\mathrm{tr}} A)(v_0 \circ A)\big] \\ X(a,0) &= a, \end{aligned} \tag{2}$$

where $\mathbf{P}$ represents the Leray projection onto divergence free vector fields noted earlier, ∇^{tr} is the transpose to the Jacobian, and A_t is the spatial inverse map $A_t(X(t,a)) = a, a \in \mathbb{R}^k (k = 2,3)$. The key idea for the Constantin–Iyer formulation is reflected in their result that, upon replacing the dynamics for the particle trajectories X in (2) by a stochastic differential equation $dX = v\,dt + \sqrt{2\nu}\,dW$, the velocity field $v = \mathbb{E}\mathbf{P}\big[(\nabla^{\mathrm{tr}} A)(u_0 \circ A)\big]$ is a fixed point of this modified system if and only if v solves (1). In particular, if $\nu = 0$ then this is the system (2) for Euler equations. A noteworthy feature of this stochastic framework is that it accommodates domains with boundary conditions beyond periodic [12], while the LJS-cascade theory is restricted to free space and/or periodic boundary conditions by virtue of the Fourier transform.

Earlier probabilistic approaches to (1) were introduced in terms of the corresponding vorticity (curl of velocity) equation in [10]; see [25] for a rigorous treatment. Finally, ergodic theory has also provided a natural framework in which (1) can be viewed as an infinite dimensional dynamical system; [20–22,27,41,50,55].

Given the overall strengths and weaknesses of various probabilistic approaches to (1), we wish to mention that some practical utility has been demonstrated with the LJS-cascade for computing a convergence rate in a related context of the LANS-α (Lagrange Averaged Navier–Stokes) equations for fluids; a problem posed in [40]. These are essentially the Navier–Stokes equations on a torus except that the spatial scales that are in some sense "smaller than α" are strategically filtered for computational purposes at high Reynolds number. Denoting the solution to LANS-α by $v^{(\alpha)}$, the LJS-cascade may be used to show in suitable function spaces that for $T > 0$,

$$\int_0^T ||v^{(\alpha)}(\cdot,t) - v^{(0)}(\cdot,t)||_{L^2(\mathbf{T})} dt \leq A(T)\alpha,$$

for a suitable constant $A(T) > 0$; see [9]. To our knowledge the LJS-cascade has been the only approach to yield a rate in three dimensions, however [7] has subsequently been successful on the two-dimensional problem using more standard pde methods and estimates. In addition to this, the use of LJS-cascades as a numerical Monte Carlo tool has been tested on Burgers equation in [48].

3 Complex Burgers and the LJS-Cascade

The (unforced) viscous Burgers equation is given in free space by

$$\begin{aligned} \frac{\partial v}{\partial t} + v\frac{\partial v}{\partial x} &= \nu\frac{\partial^2 v}{\partial x^2} \\ v(x,0) &= v_0(x) \end{aligned} \tag{3}$$

where $v(x,t)$ represents the 1-dimensional velocity at time $t \geq 0$ at position $x \in \mathbb{R}$, $v_0(x)$ represents the initial velocity, and $\nu > 0$ is the viscosity parameter.

The equation above has the following natural symmetry:

If $v(t,x)$ is a solution to Burgers equation, then $v_\lambda(t,x) := \lambda v(\lambda^2 t, \lambda x)$, for $\lambda > 0$, is also a solution corresponding to the initial velocity $(v_0)_\lambda = \lambda v_0(\lambda x)$.

The quantities that are invariant under the above scaling are called *scaling-critical* or *self-similar*. In particular, a solution to (3) is called *self-similar* if $v(t,x) = v_\lambda(t,x)$ for all t and x. Of course a self-similar solution would arise from a self-similar initial data $v_0 = (v_0)_\lambda$, and therefore self-similar solutions must be viewed in a function space setting that accommodates this scaling. In Fourier terms this may be expressed in integrated (mild) form as

$$\hat{v}(\xi,t) = \hat{v}_0(\xi)e^{-\nu\xi^2 t} + \frac{i}{2\sqrt{2\pi}}\,\xi \int_0^t e^{-\nu\xi^2(t-s)} \int_{-\infty}^{\infty} \hat{v}(\xi - y, s)\hat{v}(y,s)\,dy\,ds. \tag{4}$$

The scaling-symmetry can thusly be expressed in Fourier terms as:

If $\hat{v}(\xi,t)$ is a solution to (4), $\hat{v}_\lambda(\xi,t) := \hat{v}(\xi/\lambda, \lambda^2 t)$, for $\lambda > 0$, is also a solution corresponding to the initial velocity $(\hat{v}_0)_\lambda = \hat{v}_0(\xi/\lambda)$.

Consequently self-similar solutions in Fourier space satisfy $\hat{v}(\xi,t) = \hat{v}_\lambda(\xi,t) = \hat{v}(\xi/\lambda, \lambda^2 t)$. In particular this means that self-similar initial data must be piecewise constant functions of the form:

$$\hat{v}_0(\xi) = \begin{cases} c_1, \ \xi > 0; \\ c_2, \ \xi < 0. \end{cases} \tag{5}$$

This means that in order to accommodate the self-similar case, one must consider settings that include initial data in L^∞ in Fourier space.

For the LJS-cascade we seek to represent the mild formulation (4) in terms of expected values of products of initial data along paths of an evolving random binary tree. The exponential density in t is evident, and the space integral naturally occurs as an expected value under the above-mentioned idea of self-similarity if we set $\hat{v}_0(\xi) = 0$ for $\xi < 0$ ($c_2 = 0$ in (5)). For then $\hat{v}(\xi,t) = 0$ for all $\xi < 0$ and $t > 0$, and the mild formulation (4) becomes

$$\begin{aligned} &\hat{v}(\xi,t) = \hat{v}_0(\xi)e^{-\nu\xi^2 t} \\ &+ \frac{i}{2\sqrt{2\pi\nu}}\int_0^t \nu\xi^2 e^{-\nu\xi^2\tau}\frac{1}{\xi}\int_0^\xi \hat{v}(\eta, t-\tau)\hat{v}(\xi-\eta, t-\tau)d\eta d\tau, \quad \xi > 0, \ t \geq 0. \end{aligned} \tag{6}$$

Note that the convolution integral in the above formulation is an integral with respect to a probability distribution concentrated on $\{(\eta_1,\eta_2) \in (0,\xi)\times(0,\xi) : \eta_1+\eta_2 = \xi\}$ having uniformly distributed marginals on $(0,\xi)$, respectively. These can, in turn, be rescaled in terms of uniform distributions on $(0,1)$. The linear relation between Fourier frequencies (or wave numbers) is referred to as (spatial) *conservation of frequencies*. Of course the implied asymmetry of the Fourier transform necessitates consideration of complex-valued solutions v.

The natural function space settings that is associated with (6) is that of a Hardy-type space

$$\mathcal{H}_\infty = \{v \in \mathcal{D}'(\mathbb{R} : \mathbb{C}) : \ \hat{v}(\xi) = 0 \text{ for } \xi < 0, \hat{v} \in L^\infty([0,\infty), \mathbb{C})\}$$

Also, we set

$$\|v\|_{\mathcal{H}_\infty} = \inf\{M \geq 0 : \ |\hat{v}(\xi)| \leq M \text{ a.e. } \xi > 0\}$$

Remark 2. In fact, one can show that the real self-similar solutions to (3) must be of the form

$$v(x,t) = \begin{cases} \frac{c_1}{x}, \ x > 0; \\ \frac{c_2}{x}, \ x < 0. \end{cases}$$

where $c_1, c_2 \in \{0, -2\}$. Therefore, to obtain a nontrivial LJS-cascade theory that includes the self-similar case, one must consider complex-valued solutions v.

Note that upon rescaling ξ by $\frac{1}{\sqrt{\nu}}\xi$ and multiplying the equation by $\frac{i}{2\sqrt{2\pi\nu}}$; i.e., a dilation-rotation in the complex plane, the factor $\frac{i}{2\sqrt{2\pi\nu}}$ of the integral term is removed and $\nu = 1$ in (6) so transformed. Therefore, to simplify the notation we will adopt the following convention for the rest of the paper.

Convention 1. *From here out, unless otherwise stated, we rescale and assume the convention for a rotation-dilation of the complex plane to both render $\nu = 1$, and to remove the factor $\frac{i}{2\sqrt{2\pi\nu}}$. For notational simplicity, we continue to denote the transformed function by $\hat{v}$.*

Using this Convention, (6) becomes

$$\hat{v}(\xi,t) = \hat{v}_0(\xi)e^{-\xi^2 t} + \int_0^t \xi^2 e^{-\xi^2\tau}\frac{1}{\xi}\int_0^\xi \hat{v}(\eta,t-\tau)\hat{v}(\xi-\eta,t-\tau)d\eta d\tau, \quad \xi > 0,\ t \geq 0. \tag{7}$$

This is the mild formulation of the Fourier transformed complex Burgers equation that will be of main focus in this paper. In addition, related "mean field" equations will also be introduced in due course.

The LJS-cascade corresponding to (7), to be referred to as the Burgers cascade, consists of (i) a Yule process

$$Y = \{T_s : s \in \mathbb{T} = \cup_{n=0}^{\infty}\{1,2\}^n\}$$

defined by a collection of i.i.d. mean one exponentially distributed random variables T_s indexed by vertices s of the full binary tree $\mathbb{T} = \cup_{m=0}^{\infty}\{1,2\}^m, \{1,2\}^0 = \{\theta\}$; (ii) a multiplicative branching random walk starting from $\xi \neq 0$, recursively defined by

$$W_\theta(\xi) = \xi, \quad W_s(\xi) = U_s W_{s|m-1}(\xi), \quad s \in \mathbb{T},\ |s| = m \geq 1,$$

where $\{(U_{s1}, U_{s2}) : s \in \mathbb{T}\}$ is a collection of i.i.d. random vectors, independent of Y, having uniformly distributed marginals on $(0,1)$, and satisfying the conservation of frequency constraint

$$U_{s1} + U_{s2} = 1, \quad s \in \mathbb{T}.$$

Here we use $|\theta| = 0, |s| = |(s_1,\ldots,s_n)| = n$ to denote the generation of $s \in \mathbb{T}$, and for $|s| \geq n$, $s|0 = \theta, s|j = (s_1,\ldots,s_j), j \geq 1$, denotes $s \in \mathbb{T}$ restricted to the first j generations. In the special case $\xi = 1$ we will simply write W in place of $W(1)$.

The following stochastic processes defining the cascade genealogy are convenient for the analysis of the Burgers cascade.

Definition 1. *The genealogy of the* complex Burgers cascade *is the set-valued branching stochastic process*

$$V_{\mathrm{unif}}(\xi,t) \equiv \{s \in \mathbb{T} : \sum_{j=0}^{|s|-1} W_{s|j}^{-2}T_{s|j} \leq \xi^2 t < \sum_{j=0}^{|s|} W_{s|j}^{-2}T_{s|j}\} \in \mathcal{E},$$

where $\mathcal{E}$ *is the space of* evolutionary sets $V \subset \mathbb{T}$ *inductively defined by* $V = \{\theta\}$ *and, for* $\#V \geq 2$, $V = W \backslash s \cup \{s1, s2\}$ *for some* $W \in \mathcal{E}, \#W = \#V - 1, s \in W$, *where* $\#A$ *denotes the cardinality of a set* A. *The* number of cascade leaves at time t *is denoted*

$$N_{\text{unif}}(\xi, t) = \#V_{\text{unif}}(\xi, t), \quad t \geq 0, \xi > 0.$$

The (stochastic) explosion time *for the complex Burgers cascade is the non-negative extended real-valued random variable*

$$\zeta = \lim_{n\to\infty} \min_{|s|=n} \sum_{j=1}^{n} W_{s|j}^{-2} T_{s|j}.$$

The event that explosion occurs *is defined by* $[\zeta < \infty]$.

Note that the state space $\mathcal{E}$ of evolutionary sets is a denumerable set, and $V_{\text{unif}}(\xi, t)$ may be viewed as a random binary subtree of $\mathbb{T}$ rooted at θ.

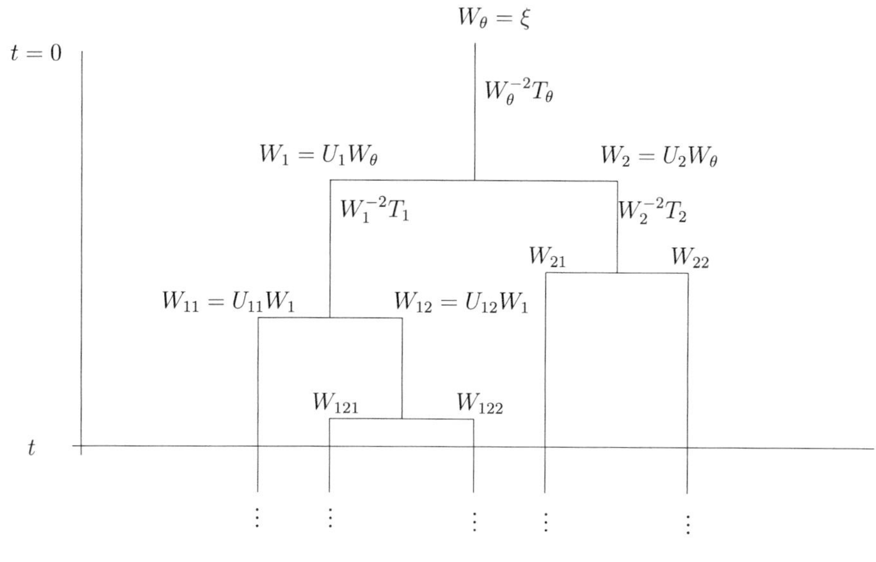

Remark 3. This is essentially the LJS-cascade modified for the absence of forcing and applied to Burgers equation. If the explosion time is finite then the number of branchings within a finite time horizon will be infinite. In such cases the recursive definition below (8) of the cascade must also be modified. In the case of (1), one makes the orthogonal projection noted earlier to remove the pressure term prior to taking Fourier transforms. The convolution term remains as an integral over $\mathbb{R}^3$ with a variety of normalizations (majorizations) available, see [4,37], to obtain an integral with respect to a probability distribution of conserved wave number pairs. This latter flexibility provides alternative function spaces for solutions.

Based on the LJS-cascade described above, a "stochastic solution" may be expressed iteratively as:

$$X(\xi,t)=\prod_{s\in V_{\text{unif}}(\xi,t)}\hat{v}_0(W_s(\xi))$$
$$=\begin{cases}\hat{v}_0(\xi), & T_\theta\geq t\\ X^{(1)}(W_1(\xi),t-T_\theta)X^{(2)}(W_2(\xi)),t-T_\theta), & T_\theta<t,\end{cases}\tag{8}$$

where $X^{(1)}, X^{(2)}$ are independent copies of X.

The LJS-cascade solution to (7) is obtained from the stochastic solution (8) via

$$\hat{v}(\xi,t)=\mathbb{E}X(\xi,t)=\mathbb{E}\prod_{s\in V_{\text{unif}}}\hat{v}_0(W_s).\tag{9}$$

Clearly, if the expected value above is well-defined, then the LJS-cascade solution (9) solves (7).

In Sect. 6 we will consider the existence of the expected value in (9) as well as the following basic questions driving the problems, conjectures and partial results pertaining to (7) and related models.

(Q1) Existence/Uniqueness of Mild Solutions: *A probabilistic rendering of* (7) *naturally suggests analysis in a Hardy-type space* $\mathcal{H}_\infty$. *The questions of existence and uniqueness may be analyzed globally in time for a subset of initial data, or locally in time for all initial data.*

(Q2) Global Well-Posedness (frequency asymptotic): *This question pertains to the identification of linear subspaces of* $\mathcal{H}_\infty$ *for which mild solutions exist for all time and remain in the subspace. Regularity and/or self-similarity considerations, i.e., Q3, Q4, may impose specific conditions on solution spaces for Q1 and Q2.*

(Q3) Regularity: *The positive form of this question involves the* C^∞ *behavior of solutions for all time for smooth (Schwarz space) initial data. This question includes finite time unbounded growth of the Fourier transform.*

(Q4) Self-Similarity: *The positive form of this question involves the uniqueness of solutions under conditions of unique space/time scale-invariant solutions. The absence of such uniqueness defines a notion of symmetry breaking. The solution space is naturally required to contain constants since these provide self-similar solutions.*

We will see that (Q1) can be answered in the affirmative. In particular, for Burgers equation and the related LJS-cascade we will prove the following basic result (see Theorem 2).

Theorem 1. *For* $\xi\neq 0$, *with probability one*

- *No Stochastic Explosion:* $\zeta=\infty$.
- ∞*-Radius of Convergence:* $\sup\{r\geq 1:\mathbb{E}r^{N(\xi,t)}<\infty\}=\infty,\quad t\geq 0$.

Moreover, for $v_0 \in \mathcal{H}_\infty$

$$\hat{v}(\xi, t) = \mathbb{E} \prod_{s \in V_{\text{unif}}(\xi,t)} \hat{v}_0(W_s(\xi)), \quad \xi > 0, t \geq 0,$$

is the unique mild solution to (7).

On the other hand (Q2) and (Q3) are delicate even in this substantially reduced framework of complex Burgers. Their resolution can take interesting twists and turns, suggesting still further questions about the nature of quasilinear equations and their companion delayed branching random walks.

4 Mean-Field Burgers and β-field Burgers

The mean-field Burgers model is defined under Convention 1 by replacing U by the constant $\beta = \mathbb{E}U = \frac{1}{2}$ in the LJS-cascade for complex Burgers (7) defined in the previous section. More generally, the β-field Burgers model is defined by replacing U by a specified constant $\beta \in [0, 1]$ for spatial frequencies, i.e., $W_s(\xi) = \beta^{|s|}\xi$ defines the (multiplicative) branching random walk. As a result $W_s(\xi)^2$ is replaced by $W_s(\xi)^2 = \beta^{2|s|}\xi^2$, $\beta \in [0, 1]$, $s \in \mathbb{T}$, when scaling the temporal frequencies. To keep the notation simple, we let the context indicate the meaning of W, continuing to suppress ξ when $\xi = 1$. Now define

$$V_\beta(\xi, t) = \{s \in \mathbb{T} : \sum_{j=0}^{|s|-1} \beta^{-2j} T_{s|j} \leq \xi^2 t < \sum_{j=0}^{|s|-1} \beta^{-2j} T_{s|j}\}, \quad t \geq 0.$$

Then the explosion time is accordingly replaced by

$$\zeta_\beta = \lim_{n\to\infty} \min_{|s|=n} \sum_{j=0}^{n} \beta^{-2j} T_{s|j}.$$

The mean-field Burgers cascade is the β-field Burgers cascade with $\beta = 1/2$. Observe that the β-field Burgers cascade has the associated mild equation obtained by replacing the uniform distribution by the Dirac distribution $\delta_{\{\beta\}}$:

$$\hat{v}(\xi, t) = \hat{v}_0(\xi)e^{-\xi^2 t} + \int_0^t \xi^2 e^{-\xi^2\tau} \hat{v}^2(\beta\xi, t - \tau)d\tau, \quad \xi \neq 0, \ t \geq 0. \tag{10}$$

The corresponding pde associated with the mild β-field cascade is given by

$$\frac{\partial \hat{v}(\xi, t)}{\partial t} = -\xi^2 \hat{v}(\xi, t) + \xi^2 \hat{v}^2(\beta\xi, t), \ t > 0, \quad \hat{v}(\xi, 0) = \hat{v}_0(\xi). \tag{11}$$

Remark 4. Although it will not be considered beyond the mean-field Burgers cascade in this paper, it is also natural to consider a two-parameter *mixed* (β_1, β_2)-*field cascade* where $0 \leq \beta_1, \beta_2$, corresponding to the equation

$$\hat{v}(\xi, t) = \hat{v}_0(\xi)e^{-\xi^2 t} + \int_0^t \xi^2 e^{-\xi^2\tau} \hat{v}(\beta_1\xi, t - \tau)\hat{v}(\beta_2\xi, t - \tau)d\tau, \quad \xi \neq 0, \ t \geq 0.$$

In cases of the two extremes $\beta = 0$ and $\beta = 1$, one has the following explicit solutions:

$$\beta = 0: \qquad \hat{v}(\xi, t) = e^{-\xi^2 t}\hat{v}_0(\xi) + (1 - e^{-\xi^2 t})\hat{v}_0^2(0), \quad t \geq 0,$$

and, noting for the Yule process that $N_1 = \#V_1(\xi, t)$ has a geometric distribution with parameter $p_t = e^{-\xi^2 t}$, or simply solving (11) in the case $\beta = 1$,

$$\begin{aligned} \beta = 1: \qquad \hat{v}(\xi, t) &\equiv \mathbb{E}\hat{v}_0(\xi)^{\#V_1(\xi,t)-1} \\ &= \frac{\hat{v}_0(\xi)e^{-\xi^2 t}}{1 - \hat{v}_0(\xi) + \hat{v}_0(\xi)e^{-\xi^2 t}}, \quad 0 \leq t < t_\infty(\hat{v}_0(\xi)), \end{aligned}$$

respectively, where

$$t_\infty(\hat{v}_0(\xi)) = \begin{cases} \frac{1}{\xi^2}\ln\left(\frac{\hat{v}_0(\xi)}{\hat{v}_0(\xi)-1}\right), & \text{for} \quad \hat{v}_0(\xi) > 1 \\ \infty, & \text{for} \quad -\infty < \hat{v}_0(\xi) \leq 1. \end{cases}$$

In particular, while the solution in the case $\beta = 0$ preserves the structure of the initial data over all time, in the case $\beta = 1$ there is finite-time blow-up for any initial data with $\hat{v}_0(\xi) > 1$. In fact, if $\hat{v}_0(\xi) = M > 1, \xi > 0$ is constant then $\hat{v}$ instantaneously exits $\mathcal{H}_\infty$, i.e., $\inf_{\xi>0} t_\infty(\hat{v}_0(\xi)) = 0$, making the problem ill-posed in $\mathcal{H}_\infty$. Even if $\hat{v} = M\mathbf{1}_{[a,b]}, M > 1$, is compactly supported on $0 < a < b < \infty$, there is finite time blow-up at

$$t_\infty(\hat{v}_0) = \inf_{\xi>0} t_\infty(\hat{v}_0(\xi)) = b^{-2}\frac{\ln M}{\ln(M-1)}.$$

The mean field value $\beta = \frac{1}{2}$ is of particular interest, but as shown in the next section, the parameter $\beta = \frac{1}{\sqrt{2}}$ also results in distinguished structure.

5 α-Delayed Yule Process

The α-delayed Yule process is defined by the β-field Burgers cascade under Convention 1 with $\alpha = \beta^2, \xi = 1$. With this reduction in parameters, for $\alpha \in (0, 1]$, the α-delayed Yule process is simply denoted

$$V^{(\alpha)}(t) = V_\beta(\xi, t), \; (\beta = \sqrt{\alpha}, \; \xi = 1), \quad t \geq 0. \tag{12}$$

The mean-field model corresponds to $\alpha = \frac{1}{4} < \frac{1}{2}$. The Yule process is then obtained in this context when $\alpha = 1 > \frac{1}{2}$. The parameter value $\alpha = \frac{1}{2}$ may be viewed as a critical value of the β-field evolutions as explained below. Moreover, as also shown in [14] the $\frac{1}{2}$-delayed Yule process is the (shifted) Poisson process with unit intensity; in fact, this extends to a two-parameter delayed Yule process provided $\alpha_1 + \alpha_2 = 1$.

Give $\mathcal{E}$ the discrete topology and let $C_0(\mathcal{E})$ denote the space of (continuous) real-valued functions $f : \mathcal{E} \to \mathbb{R}$ that vanish at infinity; i.e., given $\epsilon > 0$, one has

$|f(V)| < \epsilon$ for all but finitely many $V \in \mathcal{E}$, with the uniform norm $||\cdot||_u$. The subspace $C_{00}(\mathcal{E}) \subset C_0(\mathcal{E})$ of functions with compact (finite) support is clearly dense in $C_0(\mathcal{E})$ for the uniform norm.

Since for each $0 < \alpha \leq 1$, (12) defines a Markov process[4] $V^{(\alpha)}$, one has corresponding semigroups of positive linear contractions $\{T_t^{(\alpha)} : t \geq 0\}$ defined by

$$T_t^{(\alpha)} f(V) = \mathbb{E}_V f(V^{(\alpha)}(t)), \quad t \geq 0, f \in C_0(\mathcal{E}), \tag{13}$$

with the usual branching process convention that given $V^{(\alpha)}(0) = V \in \mathcal{E}$, $V^{(\alpha)}(t)$ is the union of those progeny at time t independently produced by single progenitors at each node $s \in V$.

The connection with the β-field model under Convention 1 and for suitable $\hat{v}_0$ may be expressed in terms of the semigroups as

$$\hat{v}(\xi, t) = T_{\beta^2 t}^{(\beta^2)} \varphi(\hat{v}_0, \xi, \beta; \cdot)(\emptyset),$$

where, for real $\hat{v}_0$ and Convention 1, $\varphi(\hat{v}_0, \xi, \beta; \cdot) : \mathcal{E} \to \mathbb{R}$ is given by

$$\varphi(\hat{v}_0, \xi, \beta; V) = \prod_{s \in V} \hat{v}_0(|s|^{\beta} \xi), \quad V \in \mathcal{E}.$$

The usual considerations imply that the infinitesimal generator $(A^{(\alpha)}, \mathcal{D}_\alpha)$ of $V^{(\alpha)}$ is given on $C_{00}(\mathcal{E})$ via

$$A^{(\alpha)} f(V) = \sum_{s \in V} \alpha^{|s|} \{f(V^s) - f(V)\}, \quad f \in C_{00}(\mathcal{E}),$$

where

$$V^s = V \backslash \{s\} \cup \{< s1, s2 >\}, \quad s \in V.$$

Proposition 2. *The space $C_{00}(\mathcal{E})$ of continuous functions on $\mathcal{E}$ having compact (finite) support is a core for the generator of the semigroup defined by (13). In particular one has*

$$\frac{\partial}{\partial t} T_t^{(\alpha)} f(V) = A^{(\alpha)} T_t^{(\alpha)} f(V) = T_t^{(\alpha)} A^{(\alpha)} f(V), \ t > 0, \quad f \in \mathcal{D}_\alpha \supset C_{00}(\mathcal{E}).$$

Proof. Note that $f \in C_{00}(\mathcal{E})$ if and only if there are $W_1, \ldots, W_m$ in $\mathcal{E}$, and real numbers $f_1, \ldots, f_m$ such that $f = \sum_{j=1}^m f_j \delta_{\{W_j\}}$. Thus, if $\#V > \#W$, then

$$T_t^{(\alpha)} f(V) = \sum_{j=1}^m f_j P(V^{(\alpha)}(t) = W_j | V^{(\alpha)}(0) = V) = 0.$$

In particular, $T_t^{(\alpha)} f \in C_{00}(\mathcal{E})$. Since $C_{00}(\mathcal{E})$ is dense in $C_0(\mathcal{E})$, the assertion follows from standard semigroup theory. □

[4] Another closely related Markov evolution that takes place in the sequence space ℓ_1 is given in [14].

The following result from [14] displays a distinct role of $\alpha = \frac{1}{2}$ as a critical parameter in terms of boundedness of the infinitesimal generators.

Proposition 3. *$(A^{(\alpha)}, \mathcal{D}_\alpha)$, $\mathcal{D}_\alpha \subset C_0(\mathcal{E})$ is a bounded linear operator if and only if $\alpha \le \frac{1}{2}$. In particular, the infinitesimal generator is a bounded operator for the mean-field parameter $\alpha = \frac{1}{4}$.*

Other interesting values of $\alpha \in (0,1)$ arise by consideration of

$$m^{(\alpha)}(t) = \mathbb{E}N^{(\alpha)}(t).$$

Proposition 4. *For $0 < \alpha \le 1$,*

$$\mathbb{E}N^{(\alpha)}(t) = 1 + \sum_{n=1}^{\infty} \prod_{j=0}^{n-1} (2\alpha^j - 1)\frac{t^n}{n!}, \quad t \ge 0.$$

In particular, $t \to \mathbb{E}N^{(\alpha)}(t)$ is a polynomial *in t of degree k for any α that is a k-th root of $\frac{1}{2}$ for some $k = 1, 2, \ldots$.*

Proof. One may readily observe, e.g, by conditioning on T_θ, that

$$\frac{dm^{(\alpha)}}{dt} = -m^{(\alpha)}(t) + 2m^{(\alpha)}(\alpha t), \; m^{(\alpha)}(0) = 1.$$

From here one may either derive the asserted formula by series expansion, or check the assertion directly. The polynomial solutions are made obvious by inspection. □

Remark 5. The positive functions

$$a_\beta(V) = \sum_{s \in V} \beta^{|s|}, \quad V \in \mathcal{E},$$

provide a class of genealogical gauges on evolutionary sets. In particular, under the convention $0^0 = 1$, $a_0(V) = \delta_{\{\theta\}}(V)$, and $a_1(V) = \#V, V \in \mathcal{E}$. Although $a_\beta \notin C_0(\mathcal{E})$ for any $\beta \in (0,1]$, the following formal calculation for $\alpha \in (0,1]$,

$$A^{(\alpha)} a_\beta(V) = (2\beta - 1)a_{\alpha\beta}(V), \quad V \in \mathcal{E},$$

leads to a class of positive martingales associated with the Yule process given by

$$M(t) = e^{(2\beta-1)t} a_\beta(V^{(1)}(t)), \quad t \ge 0,$$

which is shown in [14] to be uniformly integrable if and only if $\beta < \beta_c$, where $\beta_c \approx 0.1867$ is the unique solution in $(0,1]$ to

$$\beta_c \ln \beta_c = \beta_c - 1.$$

The associated semigroup equations become available by the following. Define a sequence $a_\beta^{(n)} \in C_{00}(\mathcal{E})$ by restricting the positive support of a_β to $\mathcal{E}^{(n)} = \{V \in \mathcal{E} : \#V \le n\}$, for $n = 1, 2 \ldots$, respectively. Then, one has $\mathcal{E} = \cup_{n=1}^{\infty} \mathcal{E}^{(n)}, \mathcal{E}^{(1)} \subset \mathcal{E}^{(2)} \subset \cdots$, and

$$A^{(\alpha)} a_\beta^{(n)}(V) = (2\beta - 1)a_{\alpha\beta}(V), \quad V \in \mathcal{E}^{(n-1)}, n = 2, 3, \ldots.$$

6 Basic Problems for Complex Burgers and Mean-Field Burgers: Some Results and Conjectures

We will start with several general results about probabilistic properties of LJS-cascades for Burgers and β-field Burgers equations. In the subsequent subsections we will use these results to analyze well-posedness and regularity issues for the solutions to the corresponding PDE.

As a matter of notation, the evolutionary sets $V_\bullet(\xi, t)$ as well as generic functionals, e.g, $N_\bullet(\xi, t)$, will be denoted without specific subscripts: i.e., V will be used for either V_{unif} or V_β and similarly, $N(\xi, t) = \#V(\xi, t)$ will stand for either N_{unif} or N_β, as dictated by context.

First, we note that the tree structures of the corresponding LJS-cascades are independent on the initial data in (7) or (10), and thus must preserve the scaling-invariance $(\xi, t) \to (\lambda^{-1}\xi, \lambda^2 t)$. As a result, the following self-similarity and monotonicity properties are straightforward to prove and are useful in some of the general analysis:

$$\textbf{Self}-\textbf{Similarity}: \quad \begin{matrix} V(\lambda^{-1}\xi, \lambda^2 t) = V(\xi, t) \equiv V(\tau) \\ N(\lambda^{-1}\xi, \lambda^2 t) = N(\xi, t) \equiv N(\tau) \end{matrix} \quad \lambda > 0, \ \tau = \xi^2 t.$$

Definition 2. *For evolutionary sets $W, V \in \mathcal{E}$ we say V* precedes W, *denoted $V \prec W$, if each $s \in V$ has a (possibly empty) concatenation belonging to W, i.e., for each $s \in V$ either $s \in W$ or there is an $\overline{s} \in \{1, 2\}^m$, for some $m \geq 1$, such that $s * \overline{s} \in W$, where $*$ denotes concatenation.*

It is straightforward to check that

$$V^{(\alpha_1)}(t) \prec V^{(\alpha_2)}(t), \quad 0 < \alpha_2 \leq \alpha_1 \leq 1, \ \xi, t \geq 0.$$

$$V_{(\beta_1)}(\xi, t) \prec V_{(\beta_2)}(\xi, t), \quad 0 < \beta_2 \leq \beta_1 \leq 1, \ \xi, t > 0.$$

$$V(\tau_1) \prec V(\tau_2), \quad 0 < \tau_1 < \tau_2.$$

As a consequence one has the following.

Proposition 5. *For a fixed $\tau > 0$, the following functionals are increasing in $\alpha = \beta^2$.*

$$N_\beta = H^{(\alpha)}(\tau) = \#V_\beta(\tau) \quad \textit{and} \quad H_\beta(\tau) = H^{(\alpha)}(\tau) = \max_{s \in V^{(\alpha)}(\tau)} \sum_{j=0}^{|s|-1} (\frac{1}{\alpha})^j T_{s|j},$$

The next property, specifically the *non-explosion* of the LJS-cascades, is crucial for the analysis in the subsections to follow. The non-explosion becomes obvious if we compare Burgers and β-field cascades to the Yule process.

Proposition 6. *Let* $\tau = \xi^2 t$*,* $\beta \in [0,1]$*. Then*

$$P(N_\beta(\tau) < \infty) = P(N_{\text{unif}}(\tau) < \infty) = 1\,,$$

and consequently, Burgers and β*-Burgers cascades are non-exploding.*

Proof. First, we note that $P(N_1(\tau) < \infty) = 1$ for the Yule process, ($\alpha = \beta = 1$) since, as is well-known and easily checked, $N_1(\tau)$ is distributed geometrically: $P(N_1(\tau) = n) = e^{-\tau}(1 - e^{-\tau})^{n-1}$, and therefore,

$$P(N_1(\tau < \infty) = \sum_{n=1}^{\infty} P(N_1(\tau) = n) = 1.$$

Now observe that since for Burgers and β-field trees, $W_s(\xi) \le \xi$, $N_{\text{unif}}(\tau) \le N_1(\tau)$ and $N_\beta \le N_1(\tau)$ a.s., and so $P(N_{\text{unif}}(\tau) < \infty) = P(N_\beta(\tau) < \infty) = 1$. The non-explosion immediately follows. □

The following estimate on the distribution of N will be used in Subsect. 6.1 to establish finiteness of the expected value (9).

Proposition 7. *Let* $\tau = \xi^2 t$ *and denote* $P_k(\tau) := P(N_{\text{unif}}(\tau) = k)$*, we have*

$$P_k(\tau) \le e^{-\tau/k} \frac{\tau^{k-1}}{(k-1)!}\,, \qquad k \in \mathbb{N}\,. \tag{14}$$

Proof. The estimate (14) is proved by induction. For $n = 1$ we have $P_1(\tau) = P(T_\theta < t) = e^{-\tau}$. Assume (14) holds for $k \le n$. Then, conditioning on the first branching,

$$\begin{aligned} P_{n+1}(\tau) &= \int_0^t \xi^2 e^{-\xi^2 s} \frac{1}{\xi} \int_0^\xi \sum_{k=1}^n P_k(y^2 t - s) P_{n+1-k}((\xi - y)^2 (t - s))\, dy\, ds \\ &= e^{-\tau} \int_0^\tau e^\sigma \int_0^1 \sum_{k=1}^n P_k(\eta^2 \sigma) P_{n+1-k}((1-\eta)^2 \sigma)\, d\eta\, d\sigma \\ &\le e^{-\tau} \int_0^\tau e^\sigma \int_0^1 \sum_{j=0}^{n-1} e^{-\left(\frac{\eta^2}{k} + \frac{1-\eta^2}{n-k}\right)\sigma} \frac{(\eta^2\sigma)^{k-1}}{(k-1)!} \frac{((1-\eta)^2\sigma)^{n-k}}{(n-k)!}\, d\eta\, d\sigma. \end{aligned}$$

We will use the following lemma.

Lemma 1. *For any* $\eta \in [0,1]$*,* $n \in \mathbb{N}$*, and* $k \in \{1, \dots n\}$*, one has:*

$$\frac{1}{n+1} \le \frac{\eta^2}{k} + \frac{(1-\eta)^2}{n+1-k} \quad \left(\le \max\left\{\frac{1}{k}, \frac{1}{n+1-k}\right\}\right) \tag{15}$$

Proof. The lemma follows by considering the extrema of the function $\phi(\eta) = \frac{\eta^2}{k} + \frac{(1-\eta)^2}{n+1-k}$ on $\eta \in [0,1]$. □

Using (15) we obtain

$$P_{n+1}(\tau) \le e^{-\tau} \int_0^\tau e^{\frac{n}{n+1}\sigma} \frac{1}{(n-1)!} \int_0^1 \sum_{j=0}^{n-1} \frac{(n-1)!}{j!(n-1-j)!} \times (\eta^2\sigma)^j ((1-\eta)^2\sigma)^{n-1-j}\, d\eta\, d\sigma$$

$$= \frac{e^{-\tau}}{(n-1)!} \int_0^\tau e^{\frac{n}{n+1}\sigma} \int_0^1 \left((\eta^2 + (1-\eta)^2)\sigma\right)^{n-1} d\eta\, d\sigma$$

$$\le \frac{e^{-\tau}}{(n-1)!} \left(\frac{n+1}{n}\right)^n \int_0^\tau e^{\frac{n}{n+1}\sigma} \left(\frac{n}{n+1}\sigma\right)^{n-1} d\left(\frac{n}{n+1}\sigma\right),$$

where we used $\eta^2 + (1-\eta)^2 \le 1$ on $\eta \in [0,1]$.

To further estimate the integral above we note that by the mean value theorem,

$$\int_0^x e^y y^{n-1}\, dy \le \frac{x^n}{n} e^x \tag{16}$$

for $x \ge 0$ and $n \in \mathbb{N}$.

Thus, using the (16) with $y = \frac{n}{n+1}\sigma$, we obtain:

$$P_{n+1}(\tau) \le \frac{e^{-\tau}}{(n-1)!} \left(\frac{n+1}{n}\right)^n \frac{\left(\frac{n}{n+1}\tau\right)^n}{n} e^{\frac{n}{n+1}\tau} = e^{-\frac{\tau}{n+1}} \frac{\tau^n}{n!},$$

and so (14) holds for $k = n+1$. □

In the case β-field models one has the following bounds.

Proposition 8. *Let $\tau = \xi^2 t$ and $0 < \beta^2 = \alpha \le 1$ and denote $P_k^{(\alpha)}(\tau) := P(N_\beta(\tau) = k)$. Then for $k \ge 2$ we have:*

$$P_k^{(\alpha)}(\tau) \le \frac{e^{-(2\beta^2)^{k-2}\tau}}{2\beta^2} \frac{(2\beta^2\tau)^{k-1}}{(k-1)!}, \qquad \alpha = \beta^2 < \frac{1}{2};$$

$$P_k^{(1/2)}(\tau) = e^{-\tau} \frac{\tau^{k-1}}{(k-1)!}, \qquad \alpha = \beta^2 = \frac{1}{2}. \tag{17}$$

and

$$P_k^{(\alpha)}(\tau) \le e^{-\tau} \frac{\left(1 - e^{-(2\beta^2-1)\tau}\right)^{k-1}}{(2\beta^2-1)^{k-1}}, \qquad \alpha = \beta^2 > \frac{1}{2}; \tag{18}$$

Proof. The statements can be proved by induction analogously to (14). Note that in the β-field case, we have for $\alpha = \beta^2$

$$P_{n+1}^{(\alpha)}(\tau) = e^{-\tau} \int_0^\tau e^\sigma \sum_{k=1}^n P_k^{(\alpha)}(\alpha\sigma) P_{n+1-k}^{(\alpha)}(\alpha\sigma)\, d\sigma,$$

and so, for $\beta = 1/\sqrt{2}$ the exponentials inside the integral disappear, leading to Poisson distribution (17).

In the case $\beta^2 \in (1/2, 1]$ we use bound $(1 - e^{-(2\beta^2-1)\beta^2\sigma}) \le (1 - e^{-(2\beta^2-1)\sigma})$ in the inductive step. □

Remark 6. We note that by Proposition 5 the tails of N_β become thinner as $\beta \to 0$.

The next result will be relevant to establish lack of well-posedness in Hadamard sense.

Proposition 9. *Let $\tau = \xi^2 t$, $\beta \in [0, 1]$. Then, as $\tau \to \infty$*

$$\mathbb{E}\, r^{N_{\mathrm{unif}}(\tau)},\ \mathbb{E}\, r^{N_\beta(\tau)} \to \begin{cases} 0, & 0 \le r < 1 \\ 1, & r = 1 \\ \infty & r > 1. \end{cases}$$

Proof. We note that the numbers of nodes $N_\bullet(\tau) \nearrow \infty$ as $\tau \to \infty$. Then the assertion follows from the monotonicity in τ of the sequence $r^{N_\bullet(\tau)}$ by applying the monotone convergence theorem for $r \ge 1$ and the dominated convergence theorem for $0 \le r < 1$. □

For the analysis of regularity, it is important to obtain estimates on $L(\xi, t)$, and $R(\xi, t)$ – left-most and right-most (delayed) branching random walkers. In particular, the properties below will prove sufficient to capture initial data in $\mathcal{H}_\infty$ with compact support in ξ which will be analyzed in Subsect. 6.3.

Remark 7. A rather complete probabilistic analysis of branching random walks associated with Yule processes ($\beta = 1$) has been evolving in the probability literature over the past several decades. The recent paper [49] is especially appropriate to the present setting in its focus on the *heights, fully saturated trees,* and the *saturation height*; also see [1,5,8,17,26,28,35,46] for related results of various types. However the corresponding problems for the Burgers cascade and/or the β-field cascades involve temporal delays to the Yule structure that make the analysis of the relevant functionals more challenging in cases other than $\beta = 1$.

In view of the results of the previous section on the α-delayed Yule process, the case $\alpha = \frac{1}{2}$, or $\beta = \frac{1}{\sqrt{2}}$, represents relatively tractable cases that will be considered in some detail. Since the α-delayed Yule process, $\alpha = \beta^2$ contains the essential stochastic structure for applications to the β-field equations, the focus is on the former.

Recall that from (17) in the case $\alpha = \frac{1}{2}$, $N^{(\frac{1}{2})}(\tau) = \#V^{(\frac{1}{2})}(\tau)$ is a (shifted) Poisson process

$$P(N^{(\frac{1}{2})}(\tau) = k) = \frac{\tau^{k-1}}{(k-1)!} e^{-\tau},\ k \ge 1, \tau = \xi^2 t.$$

Let

$$p_n^{(\alpha)}(\tau) = P(N^{(\alpha)}(\tau) = 2^n, h^{(\alpha)}(\tau) = n)$$

denote the probability that an $\alpha = \beta^2$-tree originating at ξ is "fully saturated" by time t, i.e., has exactly n full generations (and 2^n branches); the saturated tree height h is defined as the maximal generation of the tree in which all nodes are present. Observe that for $n \geq 1$ the two subtrees resulting from the first branching must also be saturated with $n-1$ full generations, and therefore we have

$$p_0^{(\alpha)}(\xi^2 t) = e^{-\xi^2 t}$$

$$p_n^{(\alpha)}(\xi^2 t) = \xi^2 \int_0^t e^{-\xi^2(t-s)} \left[p_{n-1}^{(\alpha)}(\alpha\xi^2 s)\right]^2 ds, \qquad n \geq 1. \tag{19}$$

The case $\alpha = 1/2$ is amenable to exact calculations that will be useful in the analysis of regularity of solutions of the $\beta-$field equation. The main result is

Proposition 10. *For $n \geq 1$, let*

$$Q_n = \prod_{j=1}^{n} \left(\frac{1}{1-\frac{1}{2^j}}\right)^{\frac{1}{2^j}}. \tag{20}$$

Then

$$p_n^{(\frac{1}{2})}(\xi^2 t) = e^{-\xi^2 t} \left(\frac{\xi^2 t}{2^n}\right)^{2^n - 1} Q_n^{2^n} \tag{21}$$

Proof. The statement is clearly true for $n = 1$. Assuming it holds for $n = k$, we have from (19)

$$\begin{aligned}
p_{k+1}^{(1/2)}(\xi^2 t) &= \xi^2 \int_0^t e^{-\xi^2(t-s)} \left[p_k^{(1/2)}(\xi^2 s/2)\right]^2 ds \\
&= \xi^2 \int_0^t e^{-\xi^2(t-s)} \left(e^{-\xi^2 s/2} \left(\frac{\xi^2 s/2}{2^k}\right)^{2^k - 1} Q_k^{2^k}\right)^2 ds \\
&= e^{-\xi^2 t} Q_k^{2^{k+1}} \left(\frac{1}{2^{k+1}}\right)^{2^{k+1}-2} \int_0^{\xi^2 t} u^{2^{k+1}-2} du \\
&= e^{-\xi^2 t} \left(\frac{1}{2^{k+1}}\right)^{2^{k+1}-2} (\xi^2 t)^{2^{k+1}-1} \frac{1}{2^{k+1}-1} Q_k^{2^{k+1}} \\
&= e^{-\xi^2 t} \left(\frac{\xi^2 t}{2^{k+1}}\right)^{2^{k+1}-1} \frac{1}{1-\frac{1}{2^{k+1}}} Q_k^{2^{k+1}}
\end{aligned}$$

which is the statement for $n = k+1$. □

Remark 8. In the context of binary tree searching, the constant $Q = \prod_{j=1}^{\infty}(1 - 1/2^j)$ is introduced.

Since $1 < Q_n < 1/Q$, and Q_n is increasing in n, $Q^* = \lim_{n\to\infty} Q_n$ is well defined. Furthermore, according to WolframAlpha [57],

$$Q^* = \prod_{k=1}^{\infty} \left(\frac{1}{1 - \frac{1}{2^k}} \right)^{\frac{1}{2^k}} \approx 1.55354. \tag{22}$$

6.1 Existence/Uniqueness of Mild Solutions

In this section we will pose the following question for both complex Burgers and β-field Burgers equations.

Remark 9. Within the rather large literature, e.g. [2,18,23,29–32,51,53], on uniqueness/non-uniqueness to certain parabolic semi-linear equations associated with Markov branching processes, the explosion time distribution (and its complement) are known to play a key role in demonstrating non-uniqueness for initial data 0 (or 1, respectively). While the quasi-linear Burgers equation, and Navier–Stokes equations naturally involve semi-Markov branching processes in defining their genealogy, the consequences of explosion, or its absence, are not obvious for general initial data. Even in the case of the mean-field models, where the genealogy is a Markov branching process, the issues for general initial data are diverse; see [15].

Existence/Uniqueness in $\mathcal{H}_\infty$: Does (3) and (11) have unique global in time mild solution for any $v_0 \in \mathcal{H}_\infty$?

We will give the detailed proofs for the case of Burgers equation, following Convention 1. The corresponding proofs for β-field equations generally proceed similarly, and we will provide indications whenever necessary.

Existence of the solution to (7) and (10) hinges on finiteness of the expected value in (9), while the uniqueness is the consequence of the non-explosion property of the LJS-cascades for (complex) Burgers equation.

Proposition 11. *The expected value in (9) for the stochastic solution X defined either on Burgers or β-field cascades for $\beta \in [0, 1)$ is finite, provided $v_0(\xi) \in \mathcal{H}_\infty$. Thus $\hat{v}(\xi, t)$ provided by (9) is a well-defined solution to (7) and (10) respectively.*

Proof. Suppose $\|v_0\|_{\mathcal{H}_\infty} = M$. Using (14) we can estimate.

$$\mathbb{E}M^{N(\tau)} \le M \sum_{k=0}^{\infty} \frac{(M\tau)^k}{k!} e^{-\frac{\tau}{(k+1)}} \le e^{M\tau} < \infty. \tag{23}$$

Then $|\mathbb{E}X(\xi, t)| \le \mathbb{E}|X(\xi, t)| \le \mathbb{E}M^{N_{\text{unif}}(\xi^2 t)} < \infty$. The case of β-field Burgers equation with $\beta^2 \le 1/2$ is treated similarly.

In the case $\beta^2 > 1/2$ and $M > 2\beta^2 - 1$, estimate (18) only gives local existence:

$$\mathbb{E}M^{N_\beta(\tau)} \leq Me^{-\tau} \sum_{k=0}^{\infty} \left(\frac{M(1 - e^{-(2\beta^2-1)\tau})}{2\beta^2 - 1} \right)^{n-1}$$
$$= \frac{(2\beta^2 - 1)Me^{-\tau}}{(2\beta^2 - 1) - M(1 - e^{-(2\beta^2-1)\tau})}.$$

Clearly, $\mathbb{E}M^{N_\beta(\tau)} < \infty$ for $\tau \in [0, \tau_0]$, provided

$$\tau_0 < \frac{1}{2\beta^2 - 1} \ln \left(\frac{M}{M - (2\beta^2 - 1)} \right).$$

Let

$$C_0 = C_0(\beta, \tau_0) = \frac{(2\beta^2 - 1)}{(2\beta^2 - 1) - M(1 - e^{-(2\beta^2-1)\tau})}.$$

Clearly, since $M > 2\beta^2 - 1$ we have $C_0 > 1$.

Also note that $w(\tau) = \mathbb{E}M^{N_\beta(\tau)}$ solves the self-similar form of the β-field Burgers equation (27) with $w_0 = M$ on any interval containing zero where the expectation is finite.

The key observation is that we can use induction to extend $w(\tau)$ to a solution of (27), defined on the entire $[0, \infty)$. In fact any extension of the solution $w(t)$ extension will satisfy:

$$\begin{gathered} w(t) \leq \gamma_n Me^{-\tau} = \left(\frac{2MC_0}{2\beta^2 - 1} \right)^{2^n} \frac{2\beta^2 - 1}{2} e^{-\tau}, \\ \text{for all } \ \tau \in \left[\frac{\tau_0}{(\beta^2)^{n-1}}, \frac{\tau_0}{(\beta^2)^n} \right], \ n \in \mathbb{N}, \end{gathered} \tag{24}$$

where γ_n is defined by the following recursion:

$$\gamma_0 = C_0, \quad \gamma_{n+1} = \frac{2M}{2\beta^2 - 1}\gamma_n^2, \quad \text{i.e, } \gamma_n = \left(\frac{2M}{2\beta^2 - 1} \right)^{2^n - 1} C_0^{2^n}.$$

Indeed, in the case $n = 1$, we already have $w(t) \leq C_0 Me^{-\tau}$ for $\tau \in [0, \tau_0]$. Now, assuming (24) holds for $k \leq n$, we have for $\sigma \in [0, \tau_0/(\beta^2)^{n+1}]$, $\beta^2\sigma \in [0, \tau_0/(\beta^2)^n]$, and since γ_n is increasing, $w(\alpha\sigma) \leq \gamma_n Me^{-\beta^2\sigma}$. Thus, for $\tau \in [0, \tau_0/(\beta^2)^{n+1}]$:

$$w(\tau) = Me^{-\tau} + e^{-\tau} \int_0^\tau e^\sigma w^2(\beta^2\sigma)\, d\sigma \leq Me^{-\tau} \left(1 + M\gamma_n^2 \int_0^\tau e^{-(2\beta^2-1)\sigma}\, d\sigma \right)$$
$$\leq \left(1 + \frac{M\gamma_n^2}{2\beta^2 - 1} \right) Me^{-\tau} \leq \frac{2M}{2\beta^2 - 1}\gamma_n^2 Me^{-\tau} = \gamma_{n+1} Me^{-\tau}.$$

Thus, (24) holds for all n.

As a consequence $\mathbb{E}M^{N_\beta(\tau)}$ does not blow up in finite "time" τ, and so for $\beta^2 > 1/2$, as in the other cases, $\mathbb{E}X_\beta(\xi, t)$ is a well-defined solution to (10). □

Remark 10. The proof of Proposition 11 yields the following bounds on the growth rates for $\mathbb{E}M^{N_\bullet(\tau)}$, $M>1$ as $\tau\to\infty$ (see Theorem 9), and consequently on the solutions of the corresponding equations with $\|v_0\|_{\mathcal{H}_\infty}=M$:

$$\mathbb{E}M^{N_\bullet(\tau)} \le O\left(Me^{c_\bullet M\tau}\right),$$

for complex Burgers or β-field Burgers, $\beta^2\in(0,1/2)$, and

$$\mathbb{E}M^{N_\bullet(\tau)} \le O\left((c_\bullet M)^{\tau^{\frac{\ln 2}{\ln(1/\beta^2)}}}\right),$$

for β-field Burgers, $\beta^2\in(1/2,1)$, where $c_\bullet$ is a constant that depends on the model, complex or β-field Burgers.

The uniqueness follows from the "martingale method" of Le Jan and Sznitman. Here it also requires the added non-explosion property established in Proposition 6. The proof is presented for sake of completeness.

Proposition 12. *Let $v(x,t)$ be the solution of (3) or (11) with $\beta\in[0,1)$ (equivalently, $\hat{v}(\xi,t)$ is a solution of (7) or (10)) with the initial data $v_0\in\mathcal{H}_\infty$. Then $\hat{v}$ is is given by (9).*

Proof. Without loss of generality, suppose $\hat{v}(\xi,t)$ - a solution to (7), and X be the stochastic solution defined by (8). We aim to show $\hat{v}=\mathbb{E}X$. For this purpose we define recursively the following sequence: $X_0(\xi,t)=\hat{v}(\xi,t)$; Given $X_n(\xi,t)$, define X_{n+1} by

$$X_{n+1}(\xi,t)=\begin{cases}\hat{v}_0(\xi), & T_\theta\ge t\\ X_n(W_1(\xi),t-T_\theta)X_n(W_2(\xi),t-T_\theta), & T_\theta<t.\end{cases}$$

More explicitly,

$$X_n=\left(\prod_{\substack{|s|<n\\ s\in V_{\text{unif}}}} v_0(W_s)\right)\left(\prod_{\substack{|s|=n\\ \exists\tilde{s}\in V_{\text{unif}}\ s=\tilde{s}|n}} v(W_{s|n},t-\sum_{j=0}^{n-1}T_{s|j})\right).$$

By induction, it follows that

$$\mathbb{E}(X_n)=\hat{v},\qquad \forall n\in\mathbb{N}.$$

Fix $\xi,t>0$. Denote

$$M=\max\{\max_{0\le\eta\le\xi}\{|\hat{v}_0(\eta)|\},\max_{\substack{0\le\eta\le\xi,\\ 0\le s\le t}}\{|\hat{v}(\eta,s)|\}\},$$

and let X_M be defined as in (8) but with $\hat{v}_0$ replaced with M. Note that since $M<\infty$, by Proposition 11, $\mathbb{E}(X_M)<\infty$.

Now let

$$A_n=\{s\in V_{\text{unif}}:\ |s|>n\}.$$

Then by the non-explosion,

$$\bigcap_{n\in\mathbb{N}} A_n = \emptyset,$$

and so by the dominated convergence theorem:

$$\mathbb{E}(\mathbf{1}_{A_n}) \to 0, \quad \text{as } n \to \infty.$$

Observe that $X_n|_{A_n^c} = X|_{A_n^c}$ and by the dominated convergence theorem,

$$\mathbb{E}(2X_M \mathbf{1}_{A_n}) \to 0 \text{ as } n \to \infty.$$

Therefore,

$$|\mathbb{E}(X_n) - \mathbb{E}(X)| \leq \mathbb{E}(2X_M \mathbf{1}_{A_n}) \to 0 \quad \text{as } n \to \infty.$$

Thus, $\mathbb{E}(X) = \hat{v}$.

The β-field equation may be treated analogously. □

We collect the results above in the following theorem.

Theorem 2. *Consider either Burgers equation or β-field Burgers equation with $\beta \in [0, 1)$ together with corresponding LJS-cascades. Then, for $\xi > 0$, with probability one*

- *No Stochastic Explosion:* $\zeta = \infty$.
- ∞*-Radius of Convergence:* $\sup\{r \geq 1 : \mathbb{E}\, r^{N_\bullet(\xi,t)} < \infty\} = \infty, \quad t \geq 0.$

Moreover, for any $v_0(\xi) \in \mathcal{H}_\infty$

$$\hat{v}(\xi, t) = \mathbb{E}X(\xi, t), \quad \xi > 0, t \geq 0,$$

is the unique solution to (7) or (10).

Proof. The non-explosion is the consequence of Proposition 6, while existence and uniqueness are established in Propositions 11 and 12. The infinite radius of convergence for $\mathbb{E}\, r^{N_{\text{unif}}(\xi,t)}$ follows from (23) with $M = r$. □

6.2 Global Well-Posedness Issues in $\mathcal{H}_\infty$.

As far as behavior at infinity, we will ask the following well-posedness question.

Well-posedness in $\mathcal{H}_\infty$: Suppose $v_0(x) \in \mathcal{H}_\infty$. If v denotes the solution to (3) or (11), will $v(x, t) \in \mathcal{H}_\infty$ for all $t > 0$?

It turns to that the answer is negative.

Theorem 3. *For either Burgers or β-field Burgers with $\beta \in [0,1)$ equations we have:*

1. *(Lack of well-posedness in $\mathcal{H}_\infty$) For any $M > 1$ there exist initial data v_0 with $\|v_0\|_{\mathcal{H}_\infty} \geq M$, namely,*

$$\hat{v}_0(\xi) \geq M > 1, \qquad \forall \xi \geq 0,$$

such that the corresponding solution $v(\xi, t)$ of (7) satisfies

$$\lim_{\xi \to \infty} \hat{v}(\xi, t) = \infty \qquad \forall t > 0,$$

i.e., Burgers equation (3) and β-field Burgers equations (11) are not well-posed in $\mathcal{H}_\infty$ even locally in time.

2. *(Well-posedness in $\mathcal{H}_\infty$ for small initial data). If $\|v_0\|_{\mathcal{H}_\infty} \leq 1$, then for all $t > 0$, the solution $v(\xi, t) \in \mathcal{H}_\infty$, i.e. the corresponding equations are globally well-posed in the unit ball of $\mathcal{H}_\infty$.*

Proof. The theorem follows immediately from Proposition 9 once we observe that in the case $\hat{v}_0(\xi) \geq M > 1$, $|X(\xi, t)| \geq M^{N_\bullet(\xi^2 t)}$, while in the case $\|v_0\|_{\mathcal{H}_\infty} \leq 1$, $|X(\xi, t)| \leq 1$. □

Remark 11. Note that for the self-similar initial data $\hat{v}_0(\xi) = M > 1$ the solution $\hat{v}(\xi, t) = w(\xi^2 t)$ satisfies

$$\lim_{\xi \to \infty} \hat{v}(\xi, t) = \infty \qquad \lim_{t \to \infty} \hat{v}(\xi, t) = \infty\,.$$

Remark 12. For $\beta = 1$ the global existence and uniqueness holds if and only if $\|v_0\|_{\mathcal{H}_\infty} \leq 1$. When $\|v_0\|_{\mathcal{H}_\infty} > 1$ the solutions in general do not exist even locally in time. Therefore, the corresponding β-field model is automatically ill-posed. Indeed, in this case β-field Burgers becomes

$$\frac{d\hat{v}(\xi, t)}{dt} = -\xi^2 \hat{v}(\xi, t) + \xi^2 \hat{v}^2(\xi, t),$$

and so

$$\hat{v}(\xi, t) = \frac{\hat{v}_0(\xi)}{\hat{v}_0(\xi) - e^{-\xi^2 t}(\hat{v}_0(\xi) - 1)}.$$

Clearly, if, e.g. $\hat{v}_0(\xi) \geq c > 1$, then for any $t > 0$, $\hat{v}(\xi, t)$ becomes infinite at a certain $\xi \in [0, \infty)$.

As will be shown in the next subsection, there is evidence that the lack of well-posedness in Theorem 3 for big initial data cannot be eliminated even if one considers smaller subspaces of $\mathcal{H}_\infty$. In fact, in the case $\beta = 1/\sqrt{2}$ there exist mild solutions of (11) with *compactly-supported* (in Fourier space) initial data that exit $\mathcal{H}_\infty$ in finite time.

6.3 Regularity

The analysis of regularity properties of the solution of the β-field model (10) can also be approached using the probabilistic representation of solutions given by

$$\hat{v}(\xi,t) = \mathbb{E}\left[\prod_{s\in V_\beta(\xi,t)} \hat{v}_0(\beta^{|s|}\xi)\right]$$

As already mentioned, for results pertaining to regularity of solutions, one may wish to consider bounded initial data having compact support on the positive half-line. For this let us consider

$$\hat{v}_0(\xi) = M\mathbf{1}_{[a,b]}, \quad 0 \le a < b < \infty.$$

The particular case of $\beta = 1/\sqrt{2}$ serves to illustrate the lack of regularity in the solution of the corresponding β-field equation. Note that for the particular case that $\hat{v}_0(\xi) = M\mathbf{1}_{[0,\infty)}$ the solution is given by

$$\hat{v}(\xi,t) = \exp((M-1)\,\xi^2 t)$$

This simple example shows that if $M < 1$, the solution gains regularity, indicated by an exponential decay in the Fourier domain but, that for $M > 1$, the solution leaves $\mathcal{H}_\infty$ instantly. This lack of well-posedness in the Hadamard sense is reminiscent of the behavior of solutions of the backward heat equation that is manifested also even in the case of initial data that is of compact support.

The precise statement of this result is as follows.

Proposition 13. *Let $\hat{v}(\xi,t)$ be a solution of (10) with $\beta = 1/\sqrt{2}$ and initial data $\hat{v}_0(\xi) = M\mathbf{1}_{[a,b]}$. Let $T_l = 1/b^2, T_u = 1/a^2$, and Q^* be given in (22). Then, if $M > e/Q^*$,*

$$\limsup_{\xi\to\infty} \hat{v}(\xi,t) = \infty, \text{ for } T_l < t < T_u.$$

Proof. The probabilistic representation of the solution of the $\beta-$field model provides a lower bound that is fundamental to establishing the result. Fix $t \in [T_l, T_u]$ and let $\bar{\xi} = 1/\sqrt{t}$ so that by hypothesis, $\bar{\xi} \in [a,b]$.

Let $\xi_n = 2^{n/2}\bar{\xi}$. Then

$$\hat{v}(\xi_n,t) \ge p_n(2^n\bar{\xi}^2 t)M^{2^n}$$

where p_n is given by (21). Recalling the definition of Q_n given in (20) one has

$$\limsup_{n\to\infty} \hat{v}(\xi_n,t) \ge \lim_{n\to\infty} \left(e^{-1}MQ_n\right)^{2^n} = \infty$$

since Q_n is an increasing sequence, and we are assuming that $MQ^*e^{-1} > 1$. $\square$

We note that the vanishing of the Fourier transform in a neighborhood of the origin plays a distinct role in this problem. Indeed, while the previous result shows that the solution leaves the space $\mathcal{H}_\infty$ in finite time, it does so for a finite time. To be precise,

6.4 Self-Similarity

We note that the existence/uniqueness and well-posedness in $\mathcal{H}_\infty$ analysis of the LJS-cascades, in both complex Burgers and β-field Burgers equations exploited the natural scaling $(\xi, t) \to (\xi/\lambda, \lambda^2 t)$ in the crucial ways, most notably through the scaling-invariance of $V_\bullet$ and $N_\bullet$ in Propositions 6, 11, and 9. Clearly, self-similar solutions in this setting are the unique solutions that arise from the self-similar initial data $\hat{v}_0(\xi) = \hat{v}_0(\xi/\lambda)$, i.e. from *constant* $\hat{v}_0$. Thus, self-similar solutions present the limit-case scenarios for establishing existence (through the finiteness of the expected values (9)), uniqueness (through the non-explosion of the LJS-cascades) as well as lack of well-posedness (the most obvious ill-posed solutions are bounded below by self-similar solutions).

Remark 13. In the case $\hat{v}(\xi, t)$ is a self-similar solution, using the change of variables $\tau = \xi^2 t$ and setting $w(\tau) = v(\xi, t) = v(1, \tau)$ we obtain a *self-similar form* of the complex Burgers Eq. (7):

$$w(\tau) = w_0 e^{-\tau} + \int_0^\tau e^{-\sigma} \int_0^1 w(|\eta|^2(\tau - \sigma)) w(|1-\eta|^2(\tau-\sigma))\, d\eta d\sigma,$$

as well as a *self-similar from* of the β-field Burgers equations (10):

$$w(\tau) = w_0 e^{-\tau} + \int_0^\tau e^{-\sigma} w^2(\beta^2(\tau - \sigma))\, d\sigma. \tag{27}$$

Note that for $\alpha = \beta^2$, the last equation is a mild formulation of the following non-local differential equation:

$$u'(t) = -u(t) + u^2(\alpha t).$$

In [15] we refer to this equation as the *α-Riccati equation* and analyse it's LJS-cascades in the case $\alpha > 1$.

This close connection between well-posedness of self-similar and general (non-symmetric) solutions is more pronounced here than in the Navier-Stokes case treated in [13]. Viewed from the prism of the symmetry breaking question:

Symmetry Breaking: Does the existence and uniqueness, or even well-posedness, of self-similar solutions differ from that of general *non self-similar* solutions in appropriate settings?

For the Navier-Stokes case,[5] lack of symmetry breaking appeared on the level of LJS-cascades, which had the same finiteness and explosion properties for both self-similar and general formulations. For the Burgers equation in $\mathcal{H}_\infty$,

[5] The explosion problem for the self-similar LJS-cascade has been resolved in [16], where it has been shown that indeed explosion occurs.

Theorem 2 establishes that general solutions exhibit exactly the same properties as self-similar ones in this regard, and so there is *no symmetry breaking* in Burgers (nor in β-field Burgers) equations. The following is a stronger and more intriguing formulation of the question:

Symmetry and Regularity: Does well-posedness (or lack of it) of self-similar solutions in a natural scaling-invariant space mirror the persistence of regularity (or loss of it) for general solutions?

As we have seen in the case of β-field Burgers equation for $\beta = 1/\sqrt{2}$, the lack of well-posedness in $\mathcal{H}_\infty$ of self-similar solutions is correlated with a finite-time regularity loss for solutions arising from the smoothest possible initial data, albeit compactly supported in Fourier space. Thus it appears that, at least in this case, existence/uniqueness, well-posedness, *and* regularity properties of mild solutions are mirrored by the existence/uniqueness and well-posedness properties of the self-similar solutions.

Acknowledgments. This work was partially supported by grants DMS-1408947, DMS-1408939, DMS-1211413, and DMS-1516487 from the National Science Foundation.

References

1. Aldous, D.J.: Stochastic models and descriptive statistics for phylogenetic trees, from Yule to today. Stat. Sci. **16**(1), 23–34 (2001)
2. Athreya, K.: Discounted branching random walks. Adv. Appl. Probab. **17**(1), 53–66 (1985)
3. Bass, R.F.: Probabilistic Techniques in Analysis. Springer, New York (1995)
4. Bhattacharya, R., Chen, L., Dobson, S., Guenther, R., Orum, C., Ossiander, M., Thomann, E., Waymire, E.: Majorizing kernels and stochastic cascades with applications to incompressible Navier–Stokes equations. Trans. Am. Math. Soc. **355**(12), 5003–5040 (2003)
5. Biggins, J.D.: Uniform convergence of martingales in the branching random walk. Ann. Probab. **20**(1), 137–151 (1992)
6. Cannone, M.: Harmonic analysis tools for solving the incompressible Navier–Stokes equations. In: Handbook of Mathematical Fluid Dynamics, vol. 3, pp. 161–244 (2004)
7. Cao, Y., Titi, E.S.: On the rate of convergence of the two-dimensional α-models of turbulence to the Navier–Stokes equations. Numer. Funct. Anal. Optim. **30**(11–12), 1231–1271 (2009)
8. Chauvin, B., Klein, T., Marckert, J.F., Rouault, A.: Martingales and profile of binary search trees. Electron. J. Probab. **10**(12), 420–435 (2005)
9. Chen, L., Guenther, R.B., Kim, S.C., Thomann, E.A., Waymire, E.C.: A rate of convergence for the LANS-α regularization of Navier–Stokes equations. J. Math. Anal. Appl. **348**(2), 637–649 (2008)
10. Chorin, A.J.: Numerical study of slightly viscous flow. J. Fluid Mech. **57**(04), 785–796 (1973)

11. Constantin, P., Iyer, G.: A stochastic Lagrangian representation of the three-dimensional incompressible Navier–Stokes equations. Commun. Pure Appl. Math. **61**(3), 330–345 (2008)
12. Constantin, P., Iyer, G.: A stochastic-Lagrangian approach to the Navier–Stokes equations in domains with boundary. Ann. Appl. Probab. **21**(4), 1466–1492 (2011)
13. Dascaliuc, R., Michalowski, N., Thomann, E., Waymire, E.C.: Symmetry breaking and uniqueness for the incompressible Navier–Stokes equations. Chaos Interdiscip. J. Nonlinear Sci. **25**(7), 075402 (2015)
14. Dascaliuc, R., Michalowski, N., Thomann, E., Waymire, E.C.: A delayed Yule process. Proc. Am. Math. Soc. **146**, 1335–1346 (2018)
15. Dascaliuc, R., Thomann, E., Waymire, E.C.: Stochastic explosion and non-uniqueness for an α-Riccati equation. J. Math. Anal. Appl. **476**, 53–85 (2019)
16. Dascaliuc, R., Thomann, E., Waymire, E.C.: Stochastic explosion in the self-similar Le Jan–Sznitman cascade associated to the 3D Navier–Stokes equation. Preprint (2018)
17. Dey, P., Waymire, E.: On normalized multiplicative cascades under strong disorder. Electron. Commun. Probab. **20** (2015)
18. Erickson, K.B.: Uniqueness of the null solution to a nonlinear partial differential equation satisfied by the explosion probability of a branching diffusion. J. Appl. Probab. **53**(3), 938–945 (2016)
19. Fefferman, C.L.: Existence and smoothness of the Navier–Stokes equation. Millenn. Prize. Probl. 57–67 (2006)
20. Foias, C.: Statistical study of Navier–Stokes equations. I. Rend. Sem. Mat. Univ. Padova **48**, 219–348 (1972)
21. Foias, C.: Statistical study of Navier–Stokes equations. II. Rend. Sem. Mat. Univ. Padova **49**, 9–123 (1973)
22. Foias, C., Rosa, R.M.S., Temam, R.: Properties of time-dependent statistical solutions of the three-dimensional Navier–Stokes equations. Annales de l'institut Fourier **63**(6), 2515–2573 (2013)
23. Fujita, H., Watanabe, S.: On the uniqueness and non-uniqueness of solutions of initial value problems for some quasi-linear parabolic equations. Commun. Pure Appl. Math. **21**(6), 631–652 (1968)
24. Gal, C.G., Goldstein, J.A.: Evolution Equations with a Complex Spatial Variable. World Scientific Publishing, Singapore (2014)
25. Goodman, J.: Convergence of the random vortex method. Commun. Pure Appl. Math. **40**(2), 189–220 (1987)
26. Grill, K.: The range of simple branching random walk. Stat. Probab. Lett. **26**(3), 213–218 (1996)
27. Hopf, E.: Statistical hydromechanics and functional calculus. J. Rat. Mech. Anal. **1**(1), 87–123 (1952)
28. Hu, Y., Shi, Z.: Minimal position and critical martingale convergence in branching random walks, and directed polymers on disordered trees. Ann. Probab. **37**(2), 742–789 (2009)
29. Ikeda, N., Nagasawa, M., Watanabe, S.: Branching Markov processes I. J. Math. Kyoto Univ. **8**(2), 233–278 (1968)
30. Ikeda, N., Nagasawa, M., Watanabe, S.: Branching Markov processes II. J. Math. Kyoto Univ. **8**(3), 365–410 (1968)
31. Ikeda, N., Nagasawa, M., Watanabe, S., et al.: Branching Markov processes III. J. Math. Kyoto Univ. **9**(1), 95–160 (1969)
32. Itô, K., Henry Jr., P., et al.: Diffusion Processes and Their Sample Paths. Springer, Heidelberg (2012)

33. Iyer, G.: A stochastic Lagrangian formulation of the incompressible Navier-Stokes and related transport equations. ProQuest LLC, Ann Arbor, MI (2006). Thesis Ph.D.–The University of Chicago
34. Iyer, G.: A stochastic perturbation of inviscid flows. Commun. Math. Phys. **266**(3), 631–645 (2006)
35. Johnson, T.: On the support of the simple branching random walk. Stat. Probab. Lett. **91**, 107–109 (2014)
36. Kendall, D.G.: Branching processes since 1873. J. Lond. Math. Soc. **1**(1), 385–406 (1966)
37. Le Jan, Y., Sznitman, A.: Stochastic cascades and 3-dimensional Navier–Stokes equations. Probab. Theory Relat. Fields **109**(3), 343–366 (1997)
38. Li, D., Sinai, Y.G.: Complex singularities of solutions of some 1D hydrodynamic models. Physica D **237**, 1945–1950 (2008)
39. Li, L.: Isolated singularities of the 1D complex viscous Burgers equation. J. Dyn. Differ. Equ. **21**, 623–630 (2009)
40. Linshiz, J.S., Titi, E.S.: Analytical study of certain magnetohydrodynamic-alpha models. Preprint arXiv:math/0606603 (2006)
41. Monin, A.S., Yaglom, A.M.: Statistical Fluid Mechanic: Mechanics of Turbulence. MIT Press, Cambridge (1975)
42. Newman, C.M.: Fourier transforms with only real zeros. Proc. Am. Math. Soc. **61**(2), 245–251 (1976)
43. Newman, C.M.: Normal fluctuations and the FKG inequalities. Commun. Math. Phys. **74**(2), 119–128 (1980)
44. Newman, C.M.: Percolation theory: a selective survey of rigorous results. Appendix. shock waves and mean field bounds: Concavity and analyticity of the magnetization at low temperature. In: Advances in Multiphase Flow and Related Problems, pp. 147–167 (1986)
45. Odlyzko, A.M.: An improved bound for the de Bruijn–Newman constant. Numer. Algorithms **25**(1–4), 293–303 (2000)
46. Pittel, B.: On growing random binary trees. J. Math. Anal. Appl. **103**(2), 461–480 (1984)
47. Poláčik, P., Šverák, V.: Zeros of complex caloric functions and singularities of complex viscous Burgers equation. J. reine angew. Math. **616**, 205–217 (2008)
48. Ramirez, J.M.: Multiplicative cascades applied to pdes (two numerical examples). J. Comput. Phys. **214**(1), 122–136 (2006)
49. Roberts, M.I.: Almost sure asymptotics for the random binary search tree. Preprint arXiv:1002.3896 (2010)
50. Ruelle, D.: Measures describing a turbulent flow. Ann. N. Y. Acad. Sci. **357**(1), 1–9 (1980)
51. Sawyer, S.A.: A formula for semigroups, with an application to branching diffusion processes. Trans. Am. Math. Soc. **152**(1), 1–38 (1970)
52. Sinai, Y.: Power series for solutions of the 3D-Navier–Stokes system on r^3. J. Stat. Phys. **121**(5), 779–803 (2005)
53. Skorokhod, A.V.: Branching diffusion processes. Theory Probab. Appl. **9**(3), 445–449 (1964)
54. Thomann, E.A., Guenther, R.B.: The fundamental solution of the linearized Navier–Stokes equations for spinning bodies in three spatial dimensions-time dependent case. J. Math. Fluid Mech. **8**(1), 77–98 (2006)
55. Višik, M.I., Fursikov, A.V.: Matematicheskie zadachi statisticheskoi gidromekhaniki. "Nauka", Moscow (1980)

56. Waymire, E.C.: Probability & incompressible Navier–Stokes equations: an overview of some recent developments. Probab. Surv. **2**, 1–32 (2005)
57. Weisstein, E.W.: Tree searching. MathWorld–A Wolfram Web Resource (2016). http://mathworld.wolfram.com/TreeSearching.html

Absence of Chaotic Size Dependence for Spin Glasses on Hierarchical Lattices

Jeffrey Gertler[1] and Jonathan Machta[1,2(✉)]

[1] Department of Physics, University of Massachusetts, Amherst, MA 01003, USA
jeffgertler@gmail.com, machta@physics.umass.edu
[2] Santa Fe Institute, 1399 Hyde Park Road, Santa Fe, NM 87501, USA

To Chuck Newman, on his 70th birthday

Abstract. We investigate the question of whether chaotic size dependence occurs on hierarchical lattices and demonstrate that it is not present in these systems. Our results show that the metastate for spin glasses on hierarchical lattices is simple.

Keywords: Spin glasses · Migdal–Kadanoff · Chaotic size dependence

1 Introduction

Chuck Newman's many contributions to statistical physics include fundamental insights into the proper definitions of "thermodynamic limit" for disordered systems such as spin glasses. He and Dan Stein elucidated the notions of *chaotic size dependence* (CSD) and of the *metastate* [21–24,29]. For systems with CSD, the usual thermodynamic limit fails to exist and one cannot define a unique thermodynamic state of the system. Instead, the infinite volume limit must be described through the metatstate, a probability distribution over thermodynamic states (see also [1]). For a heuristic understanding of CSD, consider correlation functions in a disordered spin system. Specifically, consider spin correlation functions within a system of size L within a large environment of size L' with $L' \gg L$. Now, imagine increasing L' keeping the couplings that have already been determined at smaller sizes and the boundary conditions fixed. How do the spin correlation functions in the system change as L' increases keeping L fixed? If the usual thermodynamic limit exists these correlation functions all converges to a limit. If CSD holds, then some correlation functions in the system fail to settle down as L' increases.

A primary motivation for introducing chaotic size dependence and the metastate was to settle the question of the low-temperature behavior of the Edwards–Anderson model [11], the Ising spin glass on finite-dimensional Euclidean lattices. The mean field Ising spin glass or Sherrington–Kirkpatrick model was solved by Parisi [25,26]. His solution uses the replica trick and requires *replica symmetry*

V. Sidoravicius (Ed.): *Sojourns in Probability Theory and Statistical Physics - I*, PROMS 298, pp. 171–182, 2019.
https://doi.org/10.1007/978-981-15-0294-1_7

breaking (RSB). The RSB solution displays a infinitely many thermodynamic states and a number of other related properties that are quite counter to standard intuition about the nature of the low temperature phase of spin systems. The question that naturally arises is whether the counterintuitive features of the RSB solution also hold for the Edwards–Anderson spin glass. An alternative scenario holds that the mean field solution is misleading and the Edwards–Anderson model behaves more like the ferromagnetic Ising model with a simple thermodynamic limit consisting of a single pair of pure states related by a global spin flip. This simpler picture was by developed by McMillan [19], Bray and Moore [8], and Fisher and Huse [12–14] and goes by the name *droplet-scaling*. After several decades of intense study and controversy, the question of which general scenario is correct remains open. However, Chuck and Dan's work on CSD and the metastate has radically sharpened the question and yielded insights so that at least we now know what the question is and what it would mean for something like RSB to hold for spin systems on finite-dimensional Euclidean lattices. The resulting *non-standard* RSB metastate [22,27] has several strange features, including CSD and support on an uncountable infinity of thermodynamic states. Chuck and Dan also introduced an alternative, perhaps more plausible, scenario also displaying CSD, called the *chaotic pairs* picture [21]. In chaotic pairs, the support of the metastate is thermodynamic states each consisting of a single pair of pure states. In chaotic pairs, for a given large volume, one sees only two pure state related by a spin flip whereas for non-standard RSB, one sees evidence of many pure states. In both cases, there is CSD so that as the system size increases, the observed pure states change.

While we wait for the breakthrough that finally settles the question of the nature of the low temperature phase of the Edwards–Anderson model, it is useful to seek guidance from simpler systems. One such simpler system is the spin glass on a hierarchical lattice. The study of disordered spin systems on hierarchical lattices has a long history [2,4,7,9,10,15–18,20,28]. Analyzing spin systems on a hierarchical lattices often yields better qualitative results than mean field theory, equivalent to spins on the complete graph. For example, hierarchical lattices can be assigned an effective dimensionality and the behavior the system as a function of dimensionality can be studied while the complete graph is effectively infinite dimensional. The behavior of spin systems on hierarchical lattices is usually analytically tractable or at least amenable to simple numerical simulations. A key motivation for studying spin systems on hierarchical lattices is the Migdal–Kadanoff real space renormalization group scheme, which was shown to be equivalent to solving the spin model on a hierarchical lattice [3]. As in the case of mean field theory, the applicability of the results to Euclidean lattices must be treated with skepticism.

There have been a number of studies of the Ising spin glass on hierarchical lattices [7,10,15,18,20,28]. Gardner [15] showed that the overlap distribution for the Ising model on the diamond hierarchical lattice consists of two delta functions suggesting that for each finite size system only a single pair of thermodynamic pure states is present. A second argument against the RSB picture on hierarchical lattices is the fact that the exponent describing the dimension of domain walls is trivially required to be $d-1$ so domains walls cannot be space filling. Later work explains why the RSB scenario can appear to be correct for small sizes despite its absence in the infinite volume limit [20].

Although the above observations would seem to preclude a complex metastate on hierarchical lattices, they do not obviously rule out the chaotic pairs picture [21]. Here we directly confront the question of whether the metastate for the Ising spin glass on hierarchical lattices contains many pairs of pure states by studying chaotic size dependence. In the process of answering this question we must develop new recursion relations that describe the influence of the environment on pair correlations. Our conclusion is that, at least for the two most commonly used examples of hierarchical lattices, there is no chaotic size dependence. We thus rule out both the RSB and the chaotic pairs scenarios for spin glasses on these hierarchical lattices.

2 Spin Glasses on Hierarchical Lattices

A multigraph $\mathcal{G}(V, E, r)$ consists of a set of vertices V, edges E and a function $r : E \to \{\{u, v\} : u, v \in V \text{ and } u \neq v\}$. A multigraph is distinguished from a graph by the possibility of multiple edges connecting the same pair of vertices. The Ising spin glass on a multigraph is defined by the Hamiltonian,

$$-\beta\mathcal{H} = \sum_{\substack{e \in E \\ r(e)=\{u,v\}}} K_e S_u S_v$$

where the summation is over edges e connecting vertices u and v. The set $\{K_e\}$ consists of i.i.d. quenched random couplings on edges e. Here we assume that the distribution of couplings is Gaussian with mean zero. The variance of this Gaussian increases linearly with inverse temperature β. An Ising spin $S_u = \pm 1$ exists on each vertex u.

Hierarchical lattices are multigraphs built recursively by substituting a template for each instance of an edge at the previous level of the construction. The template and the construction process for the "necklace" hierarchical lattice are shown in Fig. 1. The necklace is parameterized by a scale factor b and dimension d. The number of parallel edges connecting adjacent vertices is b^{d-1} while the length of the chain of vertices is b. In each step of the construction, the length scale of the system increases by the factor b. The figure shows the case $b = 2$ and $d = 3$.

The diamond and necklace hierarchical lattices are dual to one another and the results for spin glasses are similar although some exponents differ. Here we focus on the necklace lattice because the analysis is slightly simpler though the qualitative results are the same for diamond lattice.

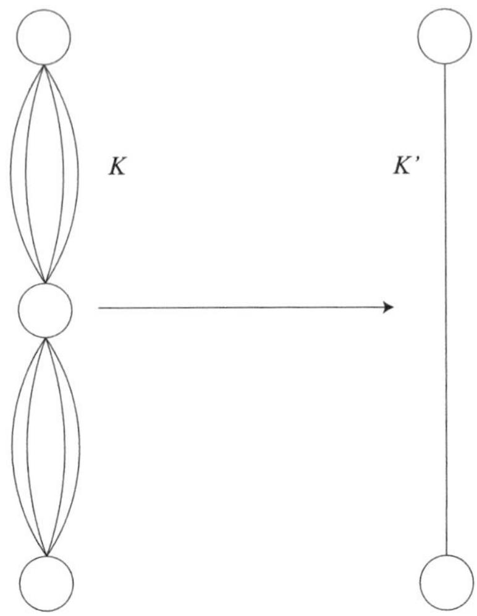

Fig. 1. Construction and decimation of the necklace hierarchical lattice for $b = 2$ and $d = 3$. The arrow points in the direction of decimation where eight bonds K are replaced by a single bond K'. Construction of the hierarchical lattice proceeds in the reverse direction with the template consisting of a single bond connecting two spin replaced by eight bonds connecting three spins.

A key advantage of studying spin systems on hierarchical lattices is that real space renormalization group methods may be implemented exactly. The decimation of spins follows the construction process of the multigraph in reverse, as shown in Fig. 1. Each edge is populated with a random coupling and an effective coupling connecting the two outermost spins in the template is computed by summing over the interior spins. This process consists of two steps. In the first step, all parallel couplings connecting the same pair of spins $\{u, v\}$ are added together,

$$\tilde{K}(u,v) = \sum_{e \in r^{-1}(\{u,v\})} K_e \tag{1}$$

In the second step, the linear chain of spins and couplings $\tilde{K}$ are combined into a single effective coupling K', on the edge e' connecting the outermost vertices in the template, u and v, with intermediate vertex z. For the case studied here, $b = 2$, decimation results in the relation [9],

$$K'_{e'} = \frac{1}{2} \log \left[\frac{\cosh(\tilde{K}(u,z) + \tilde{K}(z,v))}{\cosh(\tilde{K}(u,z) - \tilde{K}(z,v))} \right]. \tag{2}$$

We adopt a shorthand notation for this commutative operation,

$$x \otimes y \equiv \frac{1}{2} \log \left[\frac{\cosh(x+y)}{\cosh(x-y)} \right],$$

so that

$$K'_{e'} = \tilde{K}(u,z) \otimes \tilde{K}(z,v)$$

For spin glasses, where the couplings are chosen from a distribution, the recursion relations act on random variables so (1) and 2 define a functional renormalization group for the distribution of couplings.

For the Ising spin glass on a hierarchical lattice with $d \gtrsim 2.5$ there is a zero temperature or strong disorder fixed point that controls the low temperature spin glass phase. At the strong disorder fixed point, the coupling distribution scales under the recursion relations with a fixed form but increasing variance. In addition to the usual magnetic and thermal exponents, strong disorder fixed points are described by a third independent critical exponent, θ, which characterizes how the variance of the coupling distribution increases under the recursion relations according to,

$$b^{2\theta} = \frac{\mathrm{var}(K')}{\mathrm{var}(K)}. \tag{3}$$

Since the coupling between the two terminal spins of the system characterizes the stiffness of the system, θ can be identified as the stiffness exponent.

Near the strong disorder fixed point the recursion relations simplify since the variance of the coupling distribution is very large. For random variables with very large variance the decimation operator reduces to

$$x \otimes y = \min\big(\,|x|, |y|\,\big)\ \mathrm{sign}(xy), \tag{4}$$

and, specifically, (2) becomes

$$K'_{e'} = \min\big(\,|\tilde{K}(u,z)|, |\tilde{K}(z,v)|\,\big)\ \mathrm{sign}\big(\tilde{K}(u,z)\tilde{K}(z,v)\big). \tag{5}$$

In the limit of large d, the recursion relations simplify further and the stiffness exponent can be evaluated analytically because the distribution of $\tilde{K}$ is Gaussian with variance given by $\sigma^2 = b^{d-1}\mathrm{var}(K)$. For $b = 2$, the variance of K' is obtained from (5) and the Gaussian form for $\tilde{K}$,

$$\begin{aligned}\mathrm{var}(K') &= \frac{1}{\pi\sigma^2}\int_{-\infty}^{\infty} \mathrm{d}x \int_{-|x|}^{|x|} \mathrm{d}y\, y^2 \mathrm{e}^{-(x^2+y^2)/2\sigma^2} \\ &= \left(\frac{\pi-2}{\pi}\right)\sigma^2,\end{aligned}$$

so the stiffness exponent is given by

$$2\theta = d - 1 - \log_2\left(\frac{\pi}{\pi - 2}\right) \approx d - 2.46.$$

Although this results was obtained in the large d limit it remains a good approximation for small d because the fixed distribution is close to Gaussian [6,28]. One conclusion is that the lower critical dimension on the diamond hierarchical lattice is about 2.5 [5].

3 Chaotic Size Dependence on Hierarchical Lattices

In order to investigate CSD we must understand how the environment affects a system embedded in a larger environment. Figure 2 shows such a situation. In this figure, renormalization has been carried out so that the system is represented by its two terminal vertices and a single edge, shown as a dotted line. The environment is shown as extending to five levels in the hierarchy above the level of the system. In shorthand notation used hereafter, K_i is the random variable representing the distribution of renormalized couplings at level i above the level of the system. We suppose the environment is subject to free boundary conditions so the influence of the environment on the system is via coupling through the environment between the terminal spins of the system. If this coupling is at least as strong as the system coupling and changes sign as the environment grows, there is CSD. On the other hand, if this coupling converges to a limit, then CSD does not occur.

Figure 2 shows the system located at the top end of the environment so that it includes the upper terminal spin of the environment. The system may exist anywhere within the environment so this arrangement would seem to be a special case. However, for an environment a given factor larger than the system, the bonds coupling the terminal vertices of the system through the environment have the same connectivity, independent of the location of the system. This "translation invariance" means that we need only carry out the calculation for the system position shown in Fig. 2. This invariance is essentially equivalent to the fact that the decimation relation "$\otimes$" is commutative.

Figure 3 shows the environment in reduced form where parallel edges are combined into a single edge. We are now ready to write down recursion relations for the environmental coupling between the terminal spins of the system. Let B_h be the random variable describing the coupling through the environment between the terminal spins of the system if the environment is h levels larger than the system. It is evident that

$$B_1 = \sum_{j=1}^{n} K_i^{(j)} = S_{w-1}(K_0),$$

where $w = b^{d-1}$ and the notation $S_n(X)$ is a shorthand for the random variable obtained by adding n i.i.d. random variables, $X^{(1)} \ldots X^{(n)}$,

$$S_n(X) \equiv \sum_{j=1}^{n} X^{(j)}. \tag{6}$$

The recursion relations for the environmental couplings are expressed as a pattern replacement rule rather than an equation. The replacement that produces B_{h+1} from B_h is

$$S_{w-1}(K_{h-1}) \rightarrow S_{w-1}(K_{h-1}) + S_w(K_{h-1}) \otimes S_{w-1}(K_h).$$

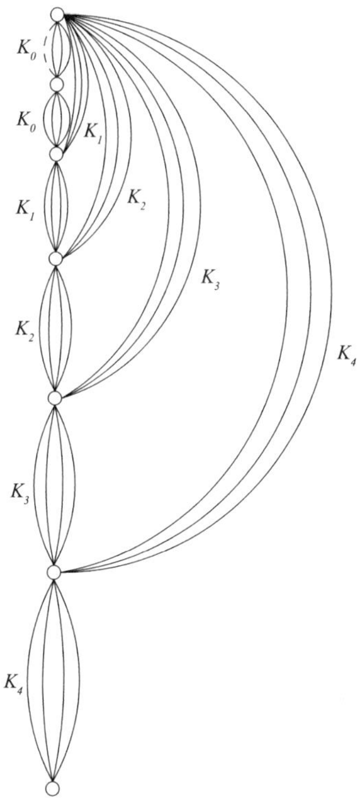

Fig. 2. A system, represented by a single bond K_0 in an environment that is $h = 4$ levels larger than the system.

For example, B_2 is obtained from B_1 using the replacement rule,

$$B_2 = S_{w-1}(K_0) + S_w(K_0) \otimes S_{w-1}(K_1).$$

and, expanding the environment one step further, we obtain

$$B_3 = S_{w-1}(K_0) + S_w(K_0) \otimes \big[S_{w-1}(K_1) + S_w(K_1) \otimes S_{w-1}(K_2)\big].$$

The net coupling K_{tot} between the terminal spins of the system is given by the sum of the internal coupling K_0 and the environmental coupling B_h,

$$K_{\text{tot}} = K_0 + B_h$$

In the low temperature phase of a large system, the distribution of K_{tot} is the sum of two independent random variables each with mean zero and large variance so it also has mean zero and large variance. Thus the correlation between the terminal spins is almost always determined simply by the sign of K_{tot}. The existence of CSD is thus equivalent to K_{tot} changing sign indefinitely as h increases. It is important to note that in studying CSD, we require that the system and

24. Newman, C.M., Stein, D.L.: TOPICAL REVIEW: Ordering and broken symmetry in short-ranged spin glasses. J. Phys. Condens. Matter **15**, 1319–1364 (2003)
25. Parisi, G.: Infinite number of order parameters for spin-glasses. Phys. Rev. Lett. **43**, 1754–1756 (1979)
26. Parisi, G.: The order parameter for spin glasses: a function on the interval 0-1. J. Phys. A **13**, 1101–1112 (1980)
27. Read, N.: Short-range Ising spin glasses: the metastate interpretation of replica symmetry breaking. Phys. Rev. E **90**, 032142 (2014)
28. Southern, B.W., Young, A.P.: Real space rescaling study of spin glass behaviour in three dimensions. J. Phys. C **10**, 2179–2195 (1977)
29. Stein, D.L., Newman, C.M.: Spin Glasses and Complexity. Primers in Complex Systems. Princeton University Press, Princeton (2013)

Mean-Field Bound on the 1-Arm Exponent for Ising Ferromagnets in High Dimensions

Satoshi Handa[1], Markus Heydenreich[2], and Akira Sakai[3(✉)]

[1] Graduate School of Mathematics, Hokkaido University, Sapporo, Japan
handa@math.sci.hokudai.ac.jp
[2] Mathematisches Institut, Ludwig-Maximilians-Universität München, Munich, Germany
m.heydenreich@lmu.de
[3] Faculty of Science, Hokkaido University, Sapporo, Japan
sakai@math.sci.hokudai.ac.jp

Abstract. The 1-arm exponent ρ for the ferromagnetic Ising model on $\mathbb{Z}^d$ is the critical exponent that describes how fast the critical 1-spin expectation at the center of the ball of radius r surrounded by plus spins decays in powers of r. Suppose that the spin-spin coupling J is translation-invariant, $\mathbb{Z}^d$-symmetric and finite-range. Using the random-current representation and assuming the anomalous dimension $\eta = 0$, we show that the optimal mean-field bound $\rho \le 1$ holds for all dimensions $d > 4$. This significantly improves a bound previously obtained by a hyperscaling inequality.

Keywords: Ising model · 1-arm exponent · Random-current representation

We dedicate this work to Chuck Newman on the occasion of his 70th birthday.

A Personal Note. I (AS) met Chuck for the first time when I visited New York for a month in summer 2004. I knew him before, not in person, but for his influential papers on phase transitions and critical behavior of percolation and the Ising model. So, naturally, I felt awful to speak with him (partly because of my poor language skills) and to give a presentation on my work back then. However, he was friendly and positive about my presentation, which he may no longer remember. Since then, I became a big fan of his. There must be many researchers like me who were greatly encouraged by Chuck, and I am sure there will be many more.

The topic of my presentation back in summer 2004 was about the critical exponent ρ for the percolation 1-arm probability in high dimensions [24]. The 1-arm probability is the probability that the center of the ball of radius r is connected to its surface by a path of occupied bonds. Compared with most of the other critical exponents, such as

V. Sidoravicius (Ed.): *Sojourns in Probability Theory and Statistical Physics - I*, PROMS 298, pp. 183–198, 2019.
https://doi.org/10.1007/978-981-15-0294-1_8

β, γ, η and δ, the 1-arm exponent ρ is harder to investigate, not because we still do not know the continuity of the critical percolation probability (= the $r \uparrow \infty$ limit of the critical 1-arm probability), but because ρ is associated with a finite-volume quantity and therefore deals with boundary effects. Even in high dimensions, it is difficult to identify the mean-field value of ρ. However, by the second-moment method [24], it is rather easy to show the one-sided inequality $\rho \leq 2$, if ρ exists and $\eta = 0$. The latter assumption is known to hold in high dimensions, thanks to the lace-expansion results [13,14]. Then, in [22], Kozma and Nachmias finally proved the equality $\rho = 2$ by assuming $\eta = 0$ and using a sophisticated inductive argument.

1 Introduction

We consider the ferromagnetic Ising model at its critical temperature $T = T_c$, and study the 1-spin expectation $\langle\sigma_o\rangle_r^+$ at the center of a ball of radius r surrounded by plus spins. The decreasing limit of $\langle\sigma_o\rangle_r^+$ as $r \uparrow \infty$ is the spontaneous magnetization. Recently, Aizenman, Duminil-Copin and Sidoravicius [4] showed that, if the spin-spin coupling satisfies a strong symmetry condition called reflection-positivity, then the spontaneous magnetization is a continuous function of temperature in all dimensions $d > 2$, in particular $\lim_{r\uparrow\infty}\langle\sigma_o\rangle_r^+ = 0$ at criticality. The present paper gives quantitative bounds on the rate of convergence. The nearest-neighbor model is an example that satisfies reflection-positivity. Also, its spontaneous magnetization on $\mathbb{Z}^2$ is known to be zero at criticality [32]. However, in general, finite-range models do not satisfy reflection-positivity, and therefore we cannot automatically justify continuity of the spontaneous magnetization for, e.g., the next-nearest-neighbor model. Fortunately, by using the lace expansion [7,25], we can avoid assuming reflection-positivity to ensure $\eta = 0$ (as well as $\beta = 1/2$, $\gamma = 1$, $\delta = 3$) and $\lim_{r\uparrow\infty}\langle\sigma_o\rangle_r^+ = 0$ at criticality in dimensions $d > 4$ if the support of J is large enough.

In this paper, we prove that it does not decay very fast whenever $d > 4$; in this case we prove $\langle\sigma_o\rangle_r^+ \geq r^{-1+o(1)}$. The proof relies on the random-current representation, which is a sophisticated version of the high-temperature expansion. It was initiated in [11] to show the GHS inequality. Then, in 1980s, Aizenman revived it to show that the bubble condition (i.e., square-summability of the critical 2-spin expectation) is a sufficient condition for the mean-field behavior [1,3,5]. It is also used in [4,25,26] to obtain many useful results for the Ising and φ^4 models. In combination with the second-moment method, we prove a correlation inequality that involves $\langle\sigma_o\rangle_r^+$ and free-boundary 2-spin expectations. Then, by using this correlation inequality, we derive the desired result.

First, we provide the precise definition of the model.

1.1 The Model

First we define the Ising model on V_R, which is the d-dimensional ball of radius $R > 0$:

$$V_R = \{v \in \mathbb{Z}^d : |v| \leq R\}.$$

It is convenient to use the Euclidean distance $|\cdot|$ here, but our results hold for any norm on the lattice $\mathbb{Z}^d$. We define the Hamiltonian for a spin configuration $\boldsymbol{\sigma} \equiv \{\sigma_v\}_{v \in V_R} \in \{\pm 1\}^{V_R}$ as

$$H^h_{r,R}(\boldsymbol{\sigma}) = - \sum_{\{u,v\} \subset V_R} J_{u,v} \sigma_u \sigma_v - h \sum_{v \in \partial V_r} \sigma_v,$$

where $J_{u,v} \geq 0$ is a translation-invariant, $\mathbb{Z}^d$-symmetric and finite-range coupling, h is the strength of the external magnetic field, and ∂V_r $(r < R)$ is the boundary of V_r:

$$\partial V_r = \{v \in V_R \setminus V_r : \exists u \in V_r \text{ such that } J_{u,v} > 0\}.$$

We note that it is crucial to impose the external magnetic field only on ∂V_r. Due to this slightly unusual setup, we will eventually be able to derive an essential correlation inequality that differs from the one for percolation.

The thermal expectation of a function f on spin configurations at the critical temperature T_c is given by

$$\langle f \rangle^h_{r,R} = \frac{1}{2^{|V_R|}} \sum_{\boldsymbol{\sigma} \in \{\pm 1\}^{V_R}} f(\boldsymbol{\sigma}) \, \frac{e^{-H^h_{r,R}(\boldsymbol{\sigma})/T_c}}{Z^h_{r,R}},$$

$$Z^h_{r,R} = \frac{1}{2^{|V_R|}} \sum_{\boldsymbol{\sigma} \in \{\pm 1\}^{V_R}} e^{-H^h_{r,R}(\boldsymbol{\sigma})/T_c}.$$

The major quantities to be investigated are the 1-spin and 2-spin expectations. Since they are increasing in h by Griffiths' inequality [10], we simply denote their limits by

$$\langle \sigma_x \rangle^+_r = \lim_{h \uparrow \infty} \langle \sigma_x \rangle^h_{r,R} \qquad [x \in V_r \cup \partial V_r],$$

$$\langle \sigma_x \sigma_y \rangle_R = \lim_{h \downarrow 0} \langle \sigma_x \sigma_y \rangle^h_{r,R} \qquad [x, y \in V_R].$$

Since $\langle \sigma_x \sigma_y \rangle_R$ is also increasing in R by Griffiths' inequality, we denote its limit by

$$\langle \sigma_x \sigma_y \rangle = \lim_{R \uparrow \infty} \langle \sigma_x \sigma_y \rangle_R.$$

In the following statement (as well as later in the proofs) we use the notation $f \asymp g$ to mean that the ratio f/g is bounded away from zero and infinity (in the prescribed limit). One assumption that we shall make throughout is the mean-field decay for the critical 2-spin expectation (or often called two-point function)

$$\langle \sigma_o \sigma_x \rangle \asymp |x|^{2-d} \quad \text{as } |x| \uparrow \infty. \tag{1}$$

A sharp asymptotic expression that implies (1) is proven by the lace expansion for a fairly general class of J, whenever the support of J is sufficiently large [25].

We note that reflection positivity has not succeeded in providing the above two-sided x-space bound; only one exception is the nearest-neighbor model, for which a one-sided x-space bound is proven [29]. In dimensions $d < 4$, the exponent on the right-hand side may change. An exact solution for $d = 2$ was identified by Wu et al. [31], which implies $\langle \sigma_o \sigma_x \rangle \asymp |x|^{-1/4}$ as $|x| \uparrow \infty$.

1.2 The Main Result

We are investigating the 1-arm exponent for the Ising model at criticality, informally described as $\langle \sigma_o \rangle_r^+ \approx r^{-\rho}$ as $r \uparrow \infty$. In order to make the symbol $\approx$ precise, we give the formal definition

$$\rho = -\liminf_{r\to\infty} \frac{\log \langle \sigma_o \rangle_r^+}{\log r}. \tag{2}$$

A more conventional way of defining ρ is by letting $\langle \sigma_o \rangle_r^+ \asymp r^{-\rho}$ as $r \uparrow \infty$, which was used to define the percolation 1-arm exponent [22,24]. However, the latter definition does not necessarily guarantee the existence of ρ. To avoid this existence issue, we adopt the former definition (2).

Our main result is the one-sided bound $\rho \le 1$ in the mean-field regime, i.e., when $d > 4$ and (1) holds. Folklore of statistical physics predicts that (2) is actually a limit. The use of limit inferior is somewhat arbitrary (lim sup would be also possible), but this choice gives the strongest result.

Theorem 1. *For the ferromagnetic Ising model on $\mathbb{Z}^d$, $d > 4$, defined by a translation-invariant, $\mathbb{Z}^d$-symmetric and finite-range spin-spin coupling satisfying* (1),

$$\liminf_{r\to\infty} r^{1+\varepsilon} \langle \sigma_o \rangle_r^+ = \infty \tag{3}$$

whenever $\varepsilon > 0$. Consequently, the critical exponent ρ defined in (2) *satisfies $\rho \le 1$.*

Tasaki [30] proved that $\langle \sigma_o \rangle_{|x|/3}^+ \ge \sqrt{\langle \sigma_o \sigma_x \rangle}$ holds at any temperature for sufficiently large $|x|$ (so that $|x|/3$ is larger than the range of the spin-spin coupling). In dimensions $d > 4$, this implies the hyperscaling inequality $\rho \le (d-2)/2$, and our bound in (3) improves on Tasaki's result.

It is a challenge now to prove

$$\limsup_{r\to\infty} r^{1-\varepsilon} \langle \sigma_o \rangle_r^+ = 0$$

for any $\varepsilon > 0$, which implies readily (together with our theorem) that (2) is actually a limit and $\rho = 1$.

Our proof of $\rho \le 1$ uses (1), which requires $d > 4$ (and the support of J to be large), even though the result is believed to be true for all dimensions $d \ge 2$. The aforementioned correlation inequality $\langle \sigma_o \rangle_{|x|/3}^+ \ge \sqrt{\langle \sigma_o \sigma_x \rangle}$ combined with the exact solution for $d = 2$ [31] and numerical predictions for $d = 3, 4$ supports this belief. This is why we call $\rho \le 1$ the optimal mean-field bound.

Another key ingredient for the proof of $\rho \le 1$ is the random-current representation, which provides a translation between spin correlations and percolation-like connectivity events. Then, by applying the second-moment method to the connectivity events as explained below for percolation, we can derive a crucial correlation inequality (cf. (6)) that relates 1-spin and 2-spin expectations. To explain what the second-moment method is and to compare the resulting correlation inequalities for the two models, we spend the next subsection to explain the derivation of the mean-field bound on the percolation 1-arm exponent, i.e., $\rho \le 2$ for $d > 6$.

1.3 Derivation of the Mean-Field Bound for Percolation

We consider the following bond percolation on $\mathbb{Z}^d$. Each bond $\{u, v\} \subset \mathbb{Z}^d$ is either occupied or vacant with probability $pJ_{u,v}$ or $1 - pJ_{u,v}$, independently of the other bonds, where $p \ge 0$ is the percolation parameter. The 2-point function $G_p(x, y)$ is the probability that x is connected to y by a path of occupied bonds ($G_p(x, x) = 1$ by convention). It is well-known that, for any $d \ge 2$, there is a nontrivial critical point p_c such that the susceptibility $\sum_x G_p(o, x)$ is finite if and only if $p < p_c$ [6]. The 1-arm probability θ_r, which is the probability that the center of the ball of radius r is connected to its surface by a path of occupied bonds, also exhibits a phase transition at p_c [2]: $\theta(p) \equiv \lim_{r\uparrow\infty} \theta_r = 0$ if $p < p_c$ and $\theta(p) > 0$ if $p > p_c$. Although the continuity $\theta(p_c) = 0$ has not yet been proven in full generality, it is shown by the lace expansion [9,15] that, if $d > 6$ and the support of J is sufficiently large, then $\theta(p_c) = 0$ and $G_{p_c}(o, x) \asymp |x|^{2-d}$ as $|x| \uparrow \infty$. This Newtonian behavior of G_{p_c} is believed not to hold in lower dimensions.

Fix $p = p_c$ and define the percolation 1-arm exponent ρ by letting $\theta_r \asymp r^{-\rho}$ as $r \uparrow \infty$. Since $\theta_{|x|/3} \ge \sqrt{G_p(o, x)}$ if $|x| \gg 1$ [30], it is known that the same hyperscaling inequality $\rho \le (d-2)/2$ holds for all dimensions $d > 6$. In [24], we were able to improve this to the optimal mean-field bound $\rho \le 2$ for $d > 6$ by using the second-moment method, which we explain now. Let X_r be the random number of vertices on ∂V_r that are connected to the origin o. We note that X_r can be positive only when o is connected to ∂V_r. Then, by the Schwarz inequality,

$$\mathbb{E}_p[X_r]^2 = \mathbb{E}_p\Big[X_r \mathbb{1}_{\{o \text{ is connected to } \partial V_r\}}\Big]^2$$
$$\le \underbrace{\mathbb{E}_p\Big[\mathbb{1}_{\{o \text{ is connected to } \partial V_r\}}\Big]}_{=\theta_r} \mathbb{E}_p[X_r^2],$$

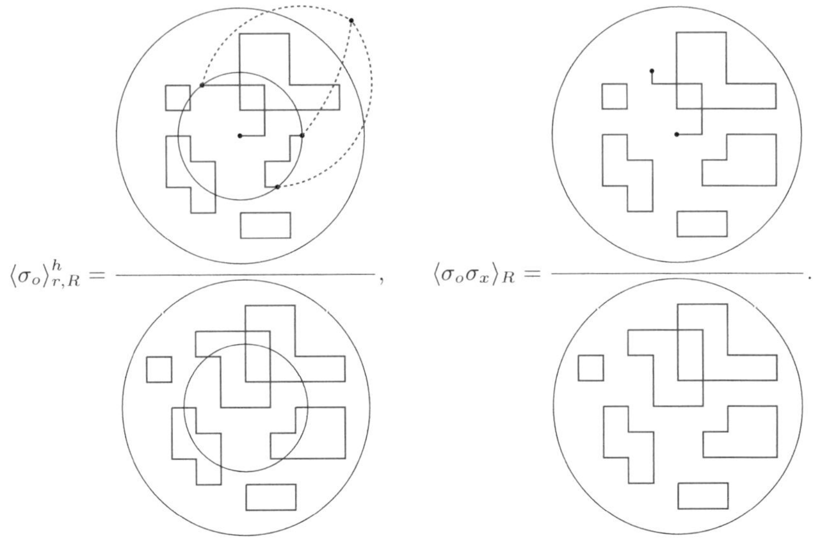

Fig. 1. The random-current representation for $\langle\sigma_o\rangle^h_{r,R}$ and $\langle\sigma_o\sigma_x\rangle_R$. The bonds with even current are all omitted. The vertex connected by dashed line segments is the ghost site g.

an imaginary ghost site. Given a current configuration $\boldsymbol{n}$, we define the source set $\partial\boldsymbol{n}$ as

$$\partial\boldsymbol{n} = \Big\{v \in V_R \cup \{g\} : \sum_{b \ni v} n_b \text{ is odd}\Big\},$$

and the weight functions $w^h_{r,R}(\boldsymbol{n})$ and $w_R(\boldsymbol{n})$ as

$$w^h_{r,R}(\boldsymbol{n}) = \prod_{b \in B_R} \frac{(J_b/T_\mathrm{c})^{n_b}}{n_b!} \prod_{b' \in G_r} \frac{(h/T_\mathrm{c})^{n_{b'}}}{n_{b'}!}, \qquad\qquad w_R(\boldsymbol{n}) = w^0_{r,R}(\boldsymbol{n}).$$

Then, we obtain the following random-current representation (cf. Fig. 1):

$$Z^h_{r,R} = \sum_{\partial\boldsymbol{n}=\varnothing} w^h_{r,R}(\boldsymbol{n}), \qquad\qquad Z_R = \sum_{\partial\boldsymbol{n}=\varnothing} w_R(\boldsymbol{n}),$$

and for $x, y \in V_R$,

$$\langle\sigma_x\rangle^h_{r,R} = \sum_{\partial\boldsymbol{n}=\{x,g\}} \frac{w^h_{r,R}(\boldsymbol{n})}{Z^h_{r,R}}, \qquad\qquad \langle\sigma_x\sigma_y\rangle_R = \sum_{\partial\boldsymbol{n}=\{x\}\triangle\{y\}} \frac{w_R(\boldsymbol{n})}{Z_R},$$

where $\triangle$ represents a symmetric difference.

Given a current configuration $\boldsymbol{n} = \{n_b\}$, we say that x is $\boldsymbol{n}$-connected to y, denoted $x \underset{n}{\longleftrightarrow} y$ if either $x = y \in V_R \cup \{g\}$ or there is a path from x to y

consisting of bonds $b \in B_R \cup G_r$ with $n_b > 0$. For $A \subset V_R \cup \{g\}$, we also say that x is $\boldsymbol{n}$-connected to y in A, denoted $x \underset{n}{\longleftrightarrow} y$ in A, if either $x = y \in A$ or there is a path from x to y consisting of bonds $b \subset A$ with $n_b > 0$.

Given a subset $A \subset V_R$, we define

$$W_A(\boldsymbol{m}) = \prod_{b \subset A} \frac{(J_b/T_c)^{m_b}}{m_b!}, \qquad \mathcal{Z}_A = \sum_{\partial \boldsymbol{m} = \varnothing} W_A(\boldsymbol{m}).$$

The most important feature of the random-current representation is the so-called source-switching lemma (e.g., [25, Lemma 2.3]). We state the version we use the most in this paper as below. This is an immediate consequence from the source-switching lemma.

Lemma 1 (Consequence from the source-switching lemma, [25]). *For any subsets $A \subset V_R$ and $B \subset V_R \cup \{g\}$, any $x, y \in V_R$ and any function f on current configurations,*

$$\sum_{\substack{\partial \boldsymbol{n} = B \\ \partial \boldsymbol{m} = \varnothing}} w_{r,R}^h(\boldsymbol{n}) \, W_A(\boldsymbol{m}) \mathbb{1}\{x \underset{\boldsymbol{n}+\boldsymbol{m}}{\longleftrightarrow} y \text{ in } A\} \, f(\boldsymbol{n} + \boldsymbol{m})$$
$$= \sum_{\substack{\partial \boldsymbol{n} = B \triangle \{x\} \triangle \{y\} \\ \partial \boldsymbol{m} = \{x\} \triangle \{y\}}} w_{r,R}^h(\boldsymbol{n}) \, W_A(\boldsymbol{m}) \, f(\boldsymbol{n} + \boldsymbol{m}).$$

For a proof, we refer to [25, Lemma 2.3].

2.2 A Correlation Inequality

The main technical vehicle in the proof of Theorem 1 is the following correlation inequality that relates $\langle \sigma_o \rangle_r^+$ to the sum of 2-spin expectations.

Proposition 1. *For the ferromagnetic Ising model,*

$$\langle \sigma_o \rangle_r^+ \geq \frac{\left(\sum_{x \in \partial V_r} \langle \sigma_o \sigma_x \rangle \right)^2}{\sum_{x,y \in \partial V_r} \langle \sigma_o \sigma_x \rangle \langle \sigma_x \sigma_y \rangle + \sum_{\substack{u \in \mathbb{Z}^d \\ x,y \in \partial V_r}} \langle \sigma_o \sigma_u \rangle \langle \sigma_u \sigma_x \rangle \langle \sigma_u \sigma_y \rangle \langle \sigma_o \rangle_{\mathrm{dist}(u, \partial V_r)}^+}. \tag{6}$$

Compare this with the correlation inequality (4) for percolation. The extra factor in the denominator of (6), $\langle \sigma_o \rangle_{\mathrm{dist}(u, \partial V_r)}^+$, will eventually be the key to obtain the optimal mean-field bound on the Ising 1-arm exponent.

Proof of Proposition 1. The proof is carried out in four steps.

Step 1: The second-moment method. Let

$$X_r(\boldsymbol{n}) = \sum_{x\in\partial V_r} \mathbb{1}\{o \underset{n}{\longleftrightarrow} x \text{ in } V_R\}.$$

Then, by the Schwarz inequality, we obtain

$$\begin{aligned}
&\sum_{\substack{\partial\boldsymbol{n}=\{o,g\}\\ \partial\boldsymbol{m}=\varnothing}} \frac{w_{r,R}^h(\boldsymbol{n})}{Z_{r,R}^h}\,\frac{w_R(\boldsymbol{m})}{Z_R}\,X_r(\boldsymbol{n}+\boldsymbol{m})\\
&\le \Bigg(\underbrace{\sum_{\partial\boldsymbol{n}=\{o,g\}} \frac{w_{r,R}^h(\boldsymbol{n})}{Z_{r,R}^h}}_{=\langle\sigma_o\rangle_{r,R}^h}\ \underbrace{\sum_{\partial\boldsymbol{m}=\varnothing} \frac{w_R(\boldsymbol{m})}{Z_R}}_{=1}\Bigg)^{1/2}\\
&\quad\times \Bigg(\sum_{\substack{\partial\boldsymbol{n}=\{o,g\}\\ \partial\boldsymbol{m}=\varnothing}} \frac{w_{r,R}^h(\boldsymbol{n})}{Z_{r,R}^h}\,\frac{w_R(\boldsymbol{m})}{Z_R}\,X_r(\boldsymbol{n}+\boldsymbol{m})^2\Bigg)^{1/2}.
\end{aligned}$$

By Lemma 1, we can rewrite the left-hand side as

$$\begin{aligned}
&\sum_{x\in\partial V_r}\sum_{\substack{\partial\boldsymbol{n}=\{o,g\}\\ \partial\boldsymbol{m}=\varnothing}} \frac{w_R(\boldsymbol{n})}{Z_{r,R}}\,\frac{w_R(\boldsymbol{m})}{Z_R}\,\mathbb{1}\{o \underset{n+m}{\longleftrightarrow} x \text{ in } V_R\}\\
&= \sum_{x\in\partial V_r}\sum_{\substack{\partial\boldsymbol{n}=\{x,g\}\\ \partial\boldsymbol{m}=\{o,x\}}} \frac{w_{r,R}^h(\boldsymbol{n})}{Z_{r,R}^h}\,\frac{w_R(\boldsymbol{m})}{Z_R}\\
&= \sum_{x\in\partial V_r} \langle\sigma_x\rangle_{r,R}^h\,\langle\sigma_o\sigma_x\rangle_R. \qquad (7)
\end{aligned}$$

As a result, we obtain

$$\langle\sigma_o\rangle_{r,R}^h \ge \frac{\Big(\sum_{x\in\partial V_r} \langle\sigma_x\rangle_{r,R}^h\,\langle\sigma_o\sigma_x\rangle_R\Big)^2}{\sum_{\substack{\partial\boldsymbol{n}=\{o,g\}\\ \partial\boldsymbol{m}=\varnothing}} \frac{w_{r,R}^h(\boldsymbol{n})}{Z_{r,R}^h}\,\frac{w_R(\boldsymbol{m})}{Z_R}\,X_r(\boldsymbol{n}+\boldsymbol{m})^2}. \qquad (8)$$

Step 2: Switching sources. Next, we investigate the denominator of the right-hand side of (8), which equals

$$\sum_{x,y\in\partial V_r}\sum_{\substack{\partial\boldsymbol{n}=\{o,g\}\\ \partial\boldsymbol{m}=\varnothing}} \frac{w_{r,R}^h(\boldsymbol{n})}{Z_{r,R}^h}\,\frac{w_R(\boldsymbol{m})}{Z_R}\,\mathbb{1}\{o \underset{n+m}{\longleftrightarrow} x, y \text{ in } V_R\}.$$

The contribution from the summands $x = y$ may be rewritten as in (7). Similarly, for the case of $x \neq y$, we use Lemma 1 to obtain

$$\sum_{\substack{x,y\in\partial V_r\\(x\neq y)}}\sum_{\substack{\partial\boldsymbol{n}=\{o,g\}\\\partial\boldsymbol{m}=\varnothing}}\frac{w_{r,R}^h(\boldsymbol{n})}{Z_{r,R}^h}\,\frac{w_R(\boldsymbol{m})}{Z_R}\,\mathbb{1}\{o\underset{\boldsymbol{n}+\boldsymbol{m}}{\longleftrightarrow}x,y\text{ in }V_R\}$$

$$=\sum_{\substack{x,y\in\partial V_r\\(x\neq y)}}\sum_{\substack{\partial\boldsymbol{n}=\{x,g\}\\\partial\boldsymbol{m}=\{o,x\}}}\frac{w_{r,R}^h(\boldsymbol{n})}{Z_{r,R}^h}\,\frac{w_R(\boldsymbol{m})}{Z_R}\,\mathbb{1}\{o\underset{\boldsymbol{n}+\boldsymbol{m}}{\longleftrightarrow}y\text{ in }V_R\}. \tag{9}$$

For the event $o\underset{\boldsymbol{n}+\boldsymbol{m}}{\longleftrightarrow}y$ in V_R to occur under the source constraint $\partial\boldsymbol{n} = \{x, g\}$, $\partial\boldsymbol{m} = \{o, x\}$, either one of the following must be the case:

(i) $o\underset{\boldsymbol{m}}{\longleftrightarrow}y$.

(ii) $o\underset{\boldsymbol{m}}{\not\longleftrightarrow}y$ and $\exists u \in \overline{\mathcal{C}_{\boldsymbol{m}}(o)} \equiv \{v \in V_R : o\underset{\boldsymbol{m}}{\longleftrightarrow}v\}$ that is $(\boldsymbol{n}+\boldsymbol{m}')$-connected to y in $V_R \setminus \overline{\mathcal{C}_{\boldsymbol{m}}(o)}$, where $\boldsymbol{m}'$ is the restriction of $\boldsymbol{m}$ on bonds $b \subset V_R \setminus \overline{\mathcal{C}_{\boldsymbol{m}}(o)}$.

Case (i) is easy; by Lemma 1, the contribution to (9) is bounded as

$$\sum_{\substack{x,y\in\partial V_r\\(x\neq y)}}\underbrace{\sum_{\partial\boldsymbol{n}=\{x,g\}}\frac{w_{r,R}^h(\boldsymbol{n})}{Z_{r,R}^h}}_{=\langle\sigma_x\rangle_{r,R}^h\le 1}\sum_{\partial\boldsymbol{m}=\{o,x\}}\frac{w_R(\boldsymbol{m})}{Z_R}\,\mathbb{1}\{o\underset{\boldsymbol{m}}{\longleftrightarrow}y\}$$

$$\le\sum_{\substack{x,y\in\partial V_r\\(x\neq y)}}\sum_{\substack{\partial\boldsymbol{m}=\{o,x\}\\\partial\boldsymbol{l}=\varnothing}}\frac{w_R(\boldsymbol{m})}{Z_R}\,\frac{w_R(\boldsymbol{l})}{Z_R}\,\mathbb{1}\{o\underset{\boldsymbol{m}+\boldsymbol{l}}{\longleftrightarrow}y\}$$

$$=\sum_{\substack{x,y\in\partial V_r\\(x\neq y)}}\underbrace{\sum_{\partial\boldsymbol{m}=\{x,y\}}\frac{w_R(\boldsymbol{m})}{Z_R}}_{=\langle\sigma_x\sigma_y\rangle_R}\underbrace{\sum_{\partial\boldsymbol{l}=\{o,y\}}\frac{w_R(\boldsymbol{l})}{Z_R}}_{=\langle\sigma_o\sigma_y\rangle_R}. \tag{10}$$

Step 3: Conditioning on clusters. Case (ii) is a bit harder and needs extra care. Here we use the conditioning-on-clusters argument. First, by conditioning on $\overline{\mathcal{C}_{\boldsymbol{m}}(o)}$, we can rewrite the contribution to (9) from case (ii) as

$$\sum_{\substack{u\in V_R\\x,y\in\partial V_r\\(x\neq y)}}\sum_{\substack{A\subset V_R\\(o,u,x\in A)}}\sum_{\substack{\partial\boldsymbol{n}=\{x,g\}\\\partial\boldsymbol{m}=\{o,x\}}}\frac{w_{r,R}^h(\boldsymbol{n})}{Z_{r,R}^h}\,\frac{w_R(\boldsymbol{m})}{Z_R}$$

$$\times\,\mathbb{1}\{\overline{\mathcal{C}_{\boldsymbol{m}}(o)}=A\}\,\mathbb{1}\{u\underset{\boldsymbol{n}+\boldsymbol{m}'}{\longleftrightarrow}y\text{ in }V_R\setminus A\}. \tag{11}$$

Then, the sum over the current configurations in (11) can be rewritten as

$$\sum_{\partial\boldsymbol{m}=\{o,x\}} \frac{W_A(\boldsymbol{m})\,\mathcal{Z}_{V_R\setminus A}}{Z_R}\,\mathbb{1}\{\overline{\mathcal{C}_{\boldsymbol{m}}(o)}=A\}$$

$$\times \sum_{\substack{\partial\boldsymbol{n}=\{x,g\}\\ \partial\boldsymbol{m}'=\varnothing}} \frac{w^h_{r,R}(\boldsymbol{n})}{Z^h_{r,R}}\,\frac{W_{V_R\setminus A}(\boldsymbol{m}')}{\mathcal{Z}_{V_R\setminus A}}\,\mathbb{1}\{u \underset{\boldsymbol{n}+\boldsymbol{m}'}{\longleftrightarrow} y \text{ in } V_R\setminus A\}.$$

Now, by using Lemma 1, the above expression is equal to

$$\sum_{\partial\boldsymbol{m}=\{o,x\}} \frac{W_A(\boldsymbol{m})\,\mathcal{Z}_{V_R\setminus A}}{Z_R}\,\mathbb{1}\{\overline{\mathcal{C}_{\boldsymbol{m}}(o)}=A\}$$

$$\times \underbrace{\sum_{\partial\boldsymbol{n}=\{u,x,y,g\}} \frac{w^h_{r,R}(\boldsymbol{n})}{Z^h_{r,R}}}_{=\langle\sigma_u\sigma_x\sigma_y\rangle^h_{r,R}}\ \underbrace{\sum_{\partial\boldsymbol{m}'=\{u,y\}} \frac{W_{V_R\setminus A}(\boldsymbol{m}')}{\mathcal{Z}_{V_R\setminus A}}}_{=\langle\sigma_u\sigma_y\rangle_{V_R\setminus A}},$$

where $\langle\sigma_u\sigma_y\rangle_{V_R\setminus A}$ is the 2-spin expectation on the vertex set $V_R\setminus A$ under the free-boundary condition, and is bounded by $\langle\sigma_u\sigma_y\rangle_R$ due to monotonicity. As a result, we obtain

$$\begin{aligned}
(11) \le &\sum_{\substack{u\in V_R\\ x,y\in\partial V_r\\ (x\neq y)}} \langle\sigma_u\sigma_x\sigma_y\rangle^h_{r,R}\,\langle\sigma_u\sigma_y\rangle_R \\
&\times \sum_{\substack{A\subset V_R\\ (o,u,x\in A)}} \sum_{\partial\boldsymbol{m}=\{o,x\}} \frac{W_A(\boldsymbol{m})\,\mathcal{Z}_{V_R\setminus A}}{Z_R}\,\mathbb{1}\{\overline{\mathcal{C}_{\boldsymbol{m}}(o)}=A\} \\
= &\sum_{\substack{u\in V_R\\ x,y\in\partial V_r\\ (x\neq y)}} \langle\sigma_u\sigma_x\sigma_y\rangle^h_{r,R}\,\langle\sigma_u\sigma_y\rangle_R \sum_{\partial\boldsymbol{m}=\{o,x\}} \frac{w_R(\boldsymbol{m})}{Z_R}\,\mathbb{1}\{o \underset{\boldsymbol{m}}{\longleftrightarrow} u\} \\
\le &\sum_{\substack{u\in V_R\\ x,y\in\partial V_r\\ (x\neq y)}} \langle\sigma_u\sigma_x\sigma_y\rangle^h_{r,R}\,\langle\sigma_u\sigma_y\rangle_R \sum_{\substack{\partial\boldsymbol{m}=\{o,x\}\\ \partial\boldsymbol{l}=\varnothing}} \frac{w_R(\boldsymbol{m})}{Z_R}\,\frac{w_R(\boldsymbol{l})}{Z_R}\,\mathbb{1}\{o \underset{\boldsymbol{m}+\boldsymbol{l}}{\longleftrightarrow} u\} \\
= &\sum_{\substack{u\in V_R\\ x,y\in\partial V_r\\ (x\neq y)}} \langle\sigma_u\sigma_x\sigma_y\rangle^h_{r,R}\,\langle\sigma_u\sigma_y\rangle_R \underbrace{\sum_{\partial\boldsymbol{m}=\{u,x\}} \frac{w_R(\boldsymbol{m})}{Z_R}}_{=\langle\sigma_u\sigma_x\rangle_R}\ \underbrace{\sum_{\partial\boldsymbol{l}=\{o,u\}} \frac{w_R(\boldsymbol{l})}{Z_R}}_{=\langle\sigma_o\sigma_u\rangle_R},
\end{aligned} \tag{12}$$

where, in the last line, we have used Lemma 1 again.

Step 4: Conclusion. Summarizing (8), (10) and (12), we arrive at

$$\langle\sigma_o\rangle_{r,R}^h \geq \frac{\left(\sum_{x\in\partial V_r} \langle\sigma_x\rangle_{r,R}^h \langle\sigma_o\sigma_x\rangle_R\right)^2}{\sum_{x,y\in\partial V_r} \langle\sigma_o\sigma_x\rangle_R \langle\sigma_x\sigma_y\rangle_R + \sum_{\substack{u\in V_R\\ x,y\in\partial V_r}} \langle\sigma_o\sigma_u\rangle_R \langle\sigma_u\sigma_x\rangle_R \langle\sigma_u\sigma_y\rangle_R \langle\sigma_u\sigma_x\sigma_y\rangle_{r,R}^h}.$$

Now we take $h \uparrow \infty$ in both sides. In this limit, the spins on ∂V_r take on $+1$. Moreover, by Griffiths' inequality, we have $\lim_{h\uparrow\infty}\langle\sigma_u\sigma_x\sigma_y\rangle_{r,R}^h = \langle\sigma_u\rangle_{r,R}^\infty \leq \langle\sigma_o\rangle_{\mathrm{dist}(u,\partial V_r)}^+$. Therefore,

$$\langle\sigma_o\rangle_r^+ \geq \frac{\left(\sum_{x\in\partial V_r} \langle\sigma_o\sigma_x\rangle_R\right)^2}{\sum_{x,y\in\partial V_r} \langle\sigma_o\sigma_x\rangle_R \langle\sigma_x\sigma_y\rangle_R + \sum_{\substack{u\in V_R\\ x,y\in\partial V_r}} \langle\sigma_o\sigma_u\rangle_R \langle\sigma_u\sigma_x\rangle_R \langle\sigma_u\sigma_y\rangle_R \langle\sigma_o\rangle_{\mathrm{dist}(u,\partial V_r)}^+}.$$

Taking $R \uparrow \infty$, we finally obtain (6). □

2.3 Proof of the Main Theorem

Proof of Theorem 1. We proceed indirectly and assume, by contradiction, that (3) is false. Then there exist a constant $K > 0$ and a monotone sequence $(r_k)_{k\in\mathbb{N}}$ diverging to ∞ such that

$$\langle\sigma_o\rangle_{r_k}^+ \leq K r_k^{-(1+\varepsilon)} \tag{13}$$

whenever k is large enough.

We are starting from Proposition 1. We estimate every term in the numerator and the denominator of (6) using (1). Firstly, the numerator of (6) is of the order r^2 since

$$\sum_{x\in\partial V_r} \langle\sigma_o\sigma_x\rangle \asymp \sum_{x\in\partial V_r} |x|^{2-d} \asymp r^{d-1} r^{2-d} = r.$$

Secondly, the first term in the denominator is of order $O(r^2)$ since

$$\sum_{x,y\in\partial V_r} \langle\sigma_o\sigma_x\rangle\langle\sigma_x\sigma_y\rangle \asymp \sum_{x\in\partial V_r} |||x|||^{2-d} \sum_{y\in\partial V_r} |||x-y|||^{2-d} \asymp r^{d-1} r^{2-d} r = r^2,$$

where we have used $|||\cdot||| = |\cdot| \vee 1$ (cf. below (4)). The second term in the denominator is the dominant one. To this end, we fix a sequence r_k satisfying (13). We split the sum over u into three cases: (i) $|u| < r_k/2$, (ii) $r_k/2 \leq |u| < 3r_k/2$, (iii) $3r_k/2 \leq |u|$, and show that it is $O({r_k}^3)$ for any $\varepsilon > 0$.

Case (i):

$$\sum_{\substack{|u|<r_k/2\\ x,y\in\partial V_{r_k}}} |||u|||^{2-d}|||u-x|||^{2-d}|||u-y|||^{2-d}\Big|\Big|\Big|r_k-|u|\Big|\Big|\Big|^{-(1+\varepsilon)}$$

$$\asymp r_k^{2(2-d)-(1+\varepsilon)} \underbrace{\sum_{x,y\in\partial V_{r_k}} \sum_{|u|<r_k/2} |||u|||^{2-d}}_{\asymp r_k^{2(d-1)+2}}$$

$$\asymp r_k^{3-\varepsilon}.$$

Case (ii):

$$\sum_{\substack{r_k/2\le|u|<3r_k/2\\ x,y\in\partial V_{r_k}}} |u|^{2-d}|||u-x|||^{2-d}|||u-y|||^{2-d}\Big|\Big|\Big|r_k-|u|\Big|\Big|\Big|^{-(1+\varepsilon)}$$

$$\asymp r_k^{2-d} \sum_{r_k/2\le|u|<3r_k/2} \Big|\Big|\Big|r_k-|u|\Big|\Big|\Big|^{-(1+\varepsilon)} \underbrace{\sum_{x\in\partial V_{r_k}} |||u-x|||^{2-d} \sum_{y\in\partial V_{r_k}} |||u-y|||^{2-d}}_{\asymp r_k^2}$$

$$\asymp r_k^{4-d}\int_{r_k/2}^{3r_k/2} |||r_k-l|||^{-(1+\varepsilon)} l^{d-1}\mathrm{d}l$$

$$\asymp r_k^3\int_0^{r_k/2} |||l|||^{-(1+\varepsilon)}\mathrm{d}l \asymp {r_k}^3.$$

Case (iii): By the Schwarz inequality,

$$\sum_{\substack{|u|\ge 3r_k/2\\ x,y\in\partial V_{r_k}}} |u|^{2-d}|u-x|^{2-d}|u-y|^{2-d}(|u|-r_k)^{-(1+\varepsilon)}$$

$$\asymp r_k^{2-d-(1+\varepsilon)} \underbrace{\sum_{x,y\in\partial V_{r_k}} \sqrt{\sum_{|u|\ge 3r_k/2} |u-x|^{4-2d}}\sqrt{\sum_{|u|\ge 3r_k/2} |u-y|^{4-2d}}}_{\asymp r_k^{2(d-1)+4-d}}$$

$$\asymp r_k^{3-\varepsilon}.$$

Plugging these estimates into the bound of Proposition 1 obtains

$$\langle\sigma_o\rangle_{r_k}^{+} \ge C\frac{{r_k}^2}{{r_k}^2+{r_k}^3} \asymp r_k^{-1}$$

for any large k and for some C (independent of k), which contradicts (13), and (3) follows.

The claim $\rho \leq 1$ follows straightforwardly: Suppose $\rho > 1+\varepsilon$ for some $\varepsilon > 0$, then there exists a sequence $(r_k)_{k\in\mathbb{N}}$ such that

$$\frac{\log \langle \sigma_o \rangle_{r_k}^{+}}{\log r_k} < -1-\varepsilon,$$

and this contradicts (3). □

Acknowledgements. The work of AS is supported by the JSPS Grant-in-Aid for Challenging Exploratory Research 15K13440. The work of SH is supported by the Ministry of Education, Culture, Sports, Science and Technology through Program for Leading Graduate Schools (Hokkaido University "Ambitious Leader's Program"). We thank Aernout van Enter for providing references about the history of the problem.

References

1. Aizenman, M.: Geometric analysis of ϕ^4 fields and Ising models. Commun. Math. Phys. **86**, 1–48 (1982)
2. Aizenman, M., Barsky, D.J.: Sharpness of the phase transition in percolation models. Commun. Math. Phys. **108**, 489–526 (1987)
3. Aizenman, M., Barsky, D.J., Fernández, R.: The phase transition in a general class of Ising-type models is sharp. J. Stat. Phys. **47**, 343–374 (1987)
4. Aizenman, M., Duminil-Copin, H., Sidoravicius, V.: Random currents and continuity of Ising model's spontaneous magnetization. Commun. Math. Phys. **334**, 719–742 (2015)
5. Aizenman, M., Fernández, R.: On the critical behavior of the magnetization in high-dimensional Ising models. J. Stat. Phys. **44**, 393–454 (1986)
6. Aizenman, M., Newman, C.M.: Tree graph inequalities and critical behavior in percolation models. J. Stat. Phys. **36**, 107–143 (1984)
7. Chen, L.-C., Sakai, A.: Critical two-point functions for long-range statistical-mechanical models in high dimensions. Ann. Probab. **43**, 639–681 (2015)
8. Bissacot, R., Endo, E.O., van Enter, A.C.D.: Stability of the phase transition of critical-field Ising model on Cayley trees under inhomogeneous external fields. Stoch. Proc. Appl. **127**, 4126–4138 (2017)
9. Fitzner, R., van der Hofstad, R.: Mean-field behavior for nearest-neighbor percolation in $d > 10$. Electron. J. Probab. **22**(43), 1–65 (2017). An extended version on arxiv.org/abs/1506.07977
10. Ginibre, J.: General formulation of Griffiths' inequalities. Commun. Math. Phys. **16**, 310–328 (1970)
11. Griffiths, R.B., Hurst, C.A., Sherman, S.: Concavity of magnetization of an Ising ferromagnet in a positive external field. J. Math. Phys. **11**, 790–795 (1970)
12. Grimmett, G.R.: Percolation, 2nd edn. Springer, Berlin (1999)
13. Hara, T.: Decay of correlations in nearest-neighbour self-avoiding walk, percolation, lattice trees and animals. Ann. Probab. **36**, 530–593 (2008)
14. Hara, T., van der Hofstad, R., Slade, G.: Critical two-point functions and the lace expansion for spread-out high-dimensional percolation and related models. Ann. Probab. **31**, 349–408 (2003)
15. Hara, T., Slade, G.: Mean-field critical behaviour for percolation in high dimensions. Commun. Math. Phys. **128**, 333–391 (1990)

16. Heydenreich, M., van der Hofstad, R.: Progress in High-Dimensional Percolation and Random Graphs. Springer International Publishing, Switzerland (2017)
17. Heydenreich, M., van der Hofstad, R., Sakai, A.: Mean-field behavior for long- and finite range Ising model, percolation and self-avoiding walk. J. Stat. Phys. **132**, 1001–1049 (2008)
18. Heydenreich, M., Kolesnikov, L.: The critical 1-arm exponent for the ferromagnetic Ising model on the Bethe lattice. J. Math. Phys. **59**, 043301 (2018)
19. Hulshof, T.: The one-arm exponent for mean-field long-range percolation. Electron. J. Probab. **20**(115), 1–26 (2015)
20. Jonasson, J., Steif, J.E.: Amenability and phase transition in the ising model. J. Theor. Probab. **12**, 549–559 (1999)
21. Katsura, S., Takizawa, M.: Bethe lattice and the Bethe approximation. Progr. Theor. Phys. **51**, 82–98 (1974)
22. Kozma, G., Nachmias, A.: Arm exponents in high dimensional percolation. J. Am. Math. Soc. **24**, 375–409 (2011)
23. Preston, C.J.: Gibbs States on Countable Sets. Cambridge University Press, London (1974)
24. Sakai, A.: Mean-field behavior for the survival probability and the percolation point-to-surface connectivity. J. Stat. Phys. **117**, 111–130 (2004)
25. Sakai, A.: Lace expansion for the Ising model. Commun. Math. Phys. **272**, 283–344 (2007)
26. Sakai, A.: Application of the lace expansion to the φ^4 model. Commun. Math. Phys. **336**, 619–648 (2015)
27. Schonmann, R.H.: Multiplicity of phase transitions and mean-field criticality on highly non-amenable graphs. Commun. Math. Phys. **219**, 271–322 (2001)
28. Simon, B., Griffiths, R.B.: The $(\phi^4)_2$ field theory as a classical Ising model. Commun. Math. Phys. **33**, 145–164 (1973)
29. Sokal, A.D.: An alternate constructive approach to the φ^4_3 quantum field theory, and a possible destructive approach to φ^4_4. Ann. Inst. Henri Poincaré Phys. Théorique **37**, 317–398 (1982)
30. Tasaki, H.: Hyperscaling inequalities for percolation. Commun. Math. Phys. **113**, 49–65 (1987)
31. Wu, T.T., McCoy, B.M., Tracy, C.A., Barouch, E.: Spin-spin correlation functions for the two-dimensional Ising model: exact theory in the scaling region. Phys. Rev. B **13**, 316–374 (1976)
32. Yang, C.N.: The spontaneous magnetization of a two-dimensional Ising model. Phys. Rev. **85**, 808–816 (1952)

Formation of Facets for an Effective Model of Crystal Growth

Dmitry Ioffe[1(✉)] and Senya Shlosman[2,3,4]

[1] Faculty of IE&M, Technion, 32000 Haifa, Israel
ieioffe@ie.technion.ac.il
[2] Skolkovo Institute of Science and Technology, Moscow, Russia
S.Shlosman@skoltech.ru
[3] Aix Marseille Université, Université de Toulon, CNRS, CPT UMR 7332, 13288 Marseille, France
[4] Institute of the Information Transmission Problems, RAS, Moscow, Russia

We dedicate this paper to Chuck Newman on the occasion of his 70th birthday

Abstract. We study an effective model of microscopic facet formation for low temperature three dimensional microscopic Wulff crystals above the droplet condensation threshold. The model we consider is a 2+1 solid on solid surface coupled with high and low density bulk Bernoulli fields. At equilibrium the surface stays flat. Imposing a canonical constraint on excess number of particles forces the surface to "grow" through the sequence of spontaneous creations of macroscopic size monolayers. We prove that at all sufficiently low temperatures, as the excess particle constraint is tuned, the model undergoes an infinite sequence of first order transitions, which traces an infinite sequence of first order transitions in the underlying variational problem. Away from transition values of canonical constraint we prove sharp concentration results for the rescaled level lines around solutions of the limiting variational problem.

Keywords: Equilibrium crystal shapes · Microscopic facets · SOS model with bulk fields · Infinite sequence of first order transitions

D. Ioffe–The research was supported by Israeli Science Foundation grant 1723/14, by The Leverhulme Trust through International Network Grant *Laplacians, Random Walks, Bose Gas, Quantum Spin Systems* and by the Meitner Humboldt Award.
S. Shlosman–The research has been carried out in the framework of the Labex Archimede (ANR-11-LABX-0033) and of the A*MIDEX project (ANR-11-IDEX-0001-02), funded by the "Investissements d'Avenir" French Government programme managed by the French National Research Agency (ANR). Part of this work has been carried out at IITP RAS. The support of Russian Foundation for Sciences (project No. 14-50-00150) is gratefully acknowledged.

V. Sidoravicius (Ed.): *Sojourns in Probability Theory and Statistical Physics - I*, PROMS 298, pp. 199–245, 2019.
https://doi.org/10.1007/978-981-15-0294-1_9

1 Introduction

Low temperature three dimensional equilibrium crystal shapes exhibit flat facets, see e.g. [7,8,31]. It is known that lattice oriented low temperature microscopic interfaces stay flat under Dobrushin boundary conditions [17]. But the $\mathbb{L}_1$-theory, which is the base of the microscopic justification of Wulff construction [2,3,15] in three and higher dimensions, does not directly address fluctuations of microscopic shapes on scales smaller than the linear size of the system. In particular, the existence and the microscopic structure of facets remains an open question even at very low temperatures.

In a sense, this issue is complementary to a large body of works, see for instance [14,16,22,30] and references to later papers, and also a recent review [32], which focus on study of the corners of zero or low temperature microscopic crystals.

For low temperature $2+1$ SOS (solid on solid) interfaces under canonical constraints on the volume below the microscopic surface, existence of flat microscopic facets was established in [4]. Here we consider facet formation for a SOS model coupled to high and low density bulk Bernoulli fields which are supposed to mimic coexisting phases of the three dimensional model.

The phenomenon of droplet condensation in the framework of the Ising model was first described in the papers [19,20]. There it was considered the Ising model at low temperature β^{-1}, occupying a d-dimensional box T_N^d of the linear size $2N$ with periodic boundary conditions. The ensemble was the canonical one: the total magnetization,

$$M_N = \sum \sigma_t,$$

was fixed. In case $M_N = m^*(\beta)\left|T_N^d\right|$, where $m^*(\beta) > 0$ is the spontaneous magnetization, the typical configuration looks as a configuration of the $(+)$-phase: the spins are taking mainly the values $+1$, while the values -1 are seen rarely, and the droplets of minuses in the box T_N^d are at most of the size of $K(d)\ln N$. In order to observe the condensation of small $(-)$-droplets into a big one it is necessary to increase their amount, which can be achieved by considering a different canonical constraint:

$$M_N = m^*(\beta)\left|T_N^d\right| - b_N,$$

$b_N > 0$. It turns out that if $b_N/\left|T_N^d\right|^{\frac{d}{d+1}} \to 0$ as $N \to \infty$, then in the corresponding canonical ensemble all the droplets are still microscopic, not exceeding $K(d)\ln N$ in linear size. On the other hand, once $\liminf_{N\to\infty} b_N/\left|T_N^d\right|^{\frac{d}{d+1}} > 0$, the situation becomes different: among many $(-)$-droplets there is one, $\mathcal{D}$, of the linear size of the order of $(b_N)^{1/d} \geq N^{\frac{d}{d+1}}$, while all the rest of the droplets are still at most logarithmic. Therefore $b_N \sim \left|T_N^d\right|^{\frac{d}{d+1}}$ can be called the *condensation threshold, or dew-point.*

When b_N grows beyond the condensation threshold, the big droplet $\mathcal{D}$ grows as well. To study this growth process, or specifically to try to get an insight of the process of formation of new atomic-scale layers on microscopic facets, we have suggested in our paper "Ising model fog drip: the first two droplets" [24] a simplified growth model, where one puts the observer on the surface $\mathcal{S}$ of $\mathcal{D}$ and studies the evolution of this surface $\mathcal{S}$ as the volume of $\mathcal{D}$ grows.

It was argued in [24] that the evolution of $\mathcal{S}$ proceeds through the spontaneous creations of extra monolayers. Each new monolayer has one-particle thickness, while the breadth of the k-th monolayer is $\sim c_k N$, with $c_k \geq c_{crit} = c_{crit}(\beta) > 0$. It then grows in breadth for some time, until a new monolayer is spontaneously created at its top, of the size of $c_{k+1}N$.

In [24] we were able to analyze this process only for the first two monolayers. Our technique at this time was not sufficient, and we were unable to control the effect of the interaction between the two monolayers when their boundaries come into contact and start to influence each other. This technique was later developed in our paper [25], so we are able now to conclude our studies. The present paper thus contains the material of what we have promised in [24] to publish under the provisional title "Ising model fog drip, II: the puddle".

In the present work we can handle any finite number of monolayers. What we find quite interesting is that the qualitative picture of the process of growth of monolayers changes, as k increases. Namely, for the few initial values of $k = 1, 2, ..., k_c$ the size $c_k N$ of the k-th monolayer at the moment of its creation is strictly smaller than the size of the underlying $(k-1)$-th layer. Thus, the picture is qualitatively the same as that of the lead crystal - there is an extensive physical literature on the latter subject, for instance see Fig. 2 in [5] and discussions in [6,21]. However, for $k > k_c$ this is not the case any more, and the size $c_k N$ is exactly the same as the size of the underlying $(k-1)$-th layer at the creation moment.

Still, creation of each new layer k bears a signature of first order transition - at the moment of creation all the underlying layers shrink. This transition resembles spontaneous creation of mesoscopic size droplets in two-dimensional Ising model [1], and as in the latter work it is related to first order transitions in the underlying variational problem.

Our picture has to be compared with a similar one, describing the layer formation in the SOS model above the wall, studied in a series of works [10–12]. Unlike our model, all the layers of the SOS surface above the wall have different asymptotic shapes. The reason is that the repulsion from the wall results in different area tilts for different layers there, and, accordingly, gives rise to different solutions of the corresponding variational problem. Another important difference is that in the SOS model [10] one never sees the top monolayer detached from the rest of them, as in the model we consider. Nevertheless, we believe that in our model with k monolayers the fluctuations of their boundaries in the vicinity of the (vertical) wall are, as in the case of entropic repulsion [11], of the order of $N^{1/3}$, and their behavior is given, after appropriate scaling, by k non-intersecting Ferrari-Spohn diffusions [23], as in [26,29]. See [28] for a review.

2 The Model and the Results

2.1 The Model

We will study the following simple model of facets formation on interfaces between two coexisting phases which was introduced in [24]: The system is confined to the 3D box

$$\Lambda_N = B_N \times \left\{ -\frac{N+1}{2}, -\frac{N-1}{2}, ..., \frac{N-1}{2}, N + \frac{N+1}{2} \right\},$$

where $N \in 2\mathbb{N}$ is even, B_N is a two-dimensional $N \times N$ box;

$$B_N = \{-N, \dots, N\}^2 = N\mathbb{B}_1 \cap \mathbb{Z}^2,$$

and $\mathbb{B}_1 = [-1,1]^2$. The interface Γ between two phases is supposed to be an SOS-type surface; it is uniquely defined by a function

$$h_\Gamma : \overset{\circ}{B}_N \to \left\{ -\frac{N}{2}, -\frac{N}{2} + 1, ..., \frac{N}{2} \right\}, \tag{1}$$

where $\overset{\circ}{B}_N$ is the interior of B_N. We assume that the interface Γ is pinned at zero height on the boundary ∂B_N, that is $h_\Gamma \equiv 0$ on ∂B_N. Such a surface Γ splits Λ_N into two parts; let us denote by $V_N(\Gamma)$ and $S_N(\Gamma)$ the upper and the lower halves. We suppose that Γ separates the low density phase (vapor) in the upper half of the box from the high density phase (solid) in the lower half. This is modeled in the following fashion: First of all, the marginal distribution of Γ obeys the SOS statistics at an inverse temperature β. That is we associate with Γ a weight,

$$w_\beta(\Gamma) = \exp\left\{ -\beta \sum_{x \sim y} |h_\Gamma(x) - h_\Gamma(y)| \right\}, \tag{2}$$

where we extended $h_\Gamma \equiv 0$ outside B_N, and the sum is over all unordered pairs of nearest neighbors of $\mathbb{Z}^d$.

Next, Γ is coupled to high and low density Bernoulli bulk fields: Let $0 < p_v = p_v(\beta) < p_s = p_s(\beta) < 1$. A relevant choice of p_v, p_s with a simplification of the three dimensional Ising model in mind would be $p_v(\beta) = e^{-6\beta} = 1 - p_s(\beta)$. In the sequel we shall assume[1]

$$\liminf_{\beta \to \infty} \frac{1}{\beta} \log\left(\min\{p_v(\beta), 1 - p_s(\beta)\}\right) > -\infty. \tag{3}$$

At each site $i \in V_N$ we place a particle with probability p_v, while at each site $i \in S_N$ we place a particle with probability p_s. Alternatively, let $\{\xi_i^v\}$ and $\{\xi_i^s\}$ be two independent collection of Bernoulli random variables with parameters p_v and p_s. Then the empirical field of particles *given* interface Γ is

$$\sum_{i \in V_N(\Gamma)} \xi_i^v \delta_i + \sum_{j \in S_N(\Gamma)} \xi_j^s \delta_j.$$

[1] Actually, main results hold even with faster decay than (3).

All together, the joint distribution of the triple (Γ, ξ^v, ξ^s) is given by

$$\mathbb{P}_{N,\beta}(\Gamma, \xi^v, \xi^s) \propto w_\beta(\Gamma) \prod_{i\in V_N} p_v^{\xi_i^v}(1-p_v)^{1-\xi_i^v} \prod_{j\in S_N} p_s^{\xi_j^s}(1-p_s)^{1-\xi_j^s}. \quad (4)$$

We denote the total number of particles in vapor and solid phases, and total number of particles in the system as

$$\Xi_v = \sum_{i\in V_N(\Gamma)} \xi_i^v, \ \Xi_s = \sum_{j\in S_N(\Gamma)} \xi_j^s \quad \text{and} \quad \Xi_N = \Xi_v + \Xi_s$$

respectively. The conditional distributions of Ξ_v and Ξ_s given Γ are binomial $\text{Bin}(|V_N|, p_v)$ and $\text{Bin}(|S_N|, p_s)$.

By the definition of the model, the expected total number of particles

$$\mathbb{E}_{N,\beta}(\Xi_N) = \frac{p^s + p^v}{2} |\Lambda_N| \equiv \rho_\beta N^3.$$

Formation of facets is modeled in the following way: Consider

$$\mathbb{P}^A_{N,\beta}(\,\cdot\,) = \mathbb{P}_{N,\beta}\left(\,\cdot\,\middle|\Xi_N \geq \rho_\beta N^3 + AN^2\right). \quad (5)$$

We claim that the model exhibits a sequence of first order transitions as A in (5) varies. The geometric manifestation of these transitions is the spontaneous creation of macroscopic size monolayers. In [24] we have investigated the creation of the first two monolayers. The task of the current paper is to provide an asymptotic (as $N \to \infty$) description of typical surfaces Γ under $\mathbb{P}_{N,\beta}\left(\cdot \;\middle|\; \Xi_N \geq \rho_\beta N^3 + AN^2\right)$ for *any* A fixed.

To study this conditional distribution we rely on Bayes rule,

$$\mathbb{P}^A_{N,\beta}(\Gamma) = \frac{\mathbb{P}_{N,\beta}\left(\Xi_N \geq \rho_\beta N^3 + AN^2 \mid \Gamma\right)\mathbb{P}_{N,\beta}(\Gamma)}{\sum_{\Gamma'} \mathbb{P}_{N,\beta}\left(\Xi_N \geq \rho_\beta N^3 + AN^2 \mid \Gamma'\right)\mathbb{P}_{N,\beta}(\Gamma')}.$$

The control over the conditional probabilities $\mathbb{P}_{N,\beta}\left(\cdot \mid \Gamma\right)$ comes from volume order local limit theorems for independent Bernoulli variables, whereas a-priori probabilities $\mathbb{P}_{N,\beta}(\Gamma)$ are derived from representation of Γ in terms of a gas of non-interacting contours. Models with bulk fields give an alternative approximation of interfaces in low temperature 3D Ising model, and they enjoy certain technical advantages over the usual SOS model with weights $w_\beta(\Gamma)$ (see (42)). In particular, volume order limit results enable a simple control over the phase of intermediate contours.

2.2 Heuristics and Informal Statement of the Main Result

Let us describe the heuristics behind the claimed sequence of first order transitions: To each surface Γ corresponds a signed volume $\alpha(\Gamma)$. In terms of the height function h_Γ which was defined in (1),

$$\alpha(\Gamma) = \int\!\!\int h_\Gamma(x, y)\mathrm{d}x\mathrm{d}y.$$

Main contribution to $\alpha(\Gamma)$ comes from large microscopic facets, which are encoded by large microscopic level lines $\Gamma_1, \dots \Gamma_\ell$;

$$\alpha(\Gamma) \approx \sum_1^\ell \mathsf{a}\left(\Gamma_i\right) = N^2 \sum_1^\ell \mathsf{a}\left(\frac{1}{N}\Gamma_i\right), \tag{6}$$

where $\mathsf{a}\,(*)$ stands for the area. The notions of level lines are defined and discussed in Sect. 4. Locally large level lines $\Gamma_1, \dots, \Gamma_\ell$ have structure of low temperature Ising polymers, and they give rise to a *two-dimensional* surface tension τ_β. As we shall explain below the a-priori probability of creating surface with a prescribed volume aN^2 is asymptotically given by

$$\log \mathbb{P}_{N,\beta}\left(\alpha(\Gamma) = aN^2\right) \approx -N\tau_\beta(a), \tag{7}$$

where $\tau_\beta(a)$ is the minimal surface tension of a compatible collection of simple curves $\gamma_1, \dots, \gamma_\ell \subset \mathbb{B}_1$ with total area $\sum_i \mathsf{a}(\gamma_i) = a$. The notion of compatibility is explained in the beginning of Sect. 3, which is devoted to a careful analysis of the minimization problem we consider here. In fact the only relevant compatible collections happen to be ordered stacks

$$\mathring{\gamma}_\ell \subseteq \mathring{\gamma}_2 \subseteq \cdots \subseteq \mathring{\gamma}_1 \subseteq \mathbb{B}_1. \tag{8}$$

Informally, the asymptotic relation (7) is achieved when rescaled $\frac{1}{N}\Gamma_i$ microscopic level lines stay close to optimal γ_i-s. On the other hand, the presence of Γ-interface shifts the expected number of particles in the bulk by the quantity $\Delta_\beta \alpha(\Gamma)$, where $\Delta_\beta = (p^s - p^v)$. That is,

$$\mathbb{E}_{N,\beta}\left(\Xi_N \middle| \Gamma\right) = \rho_\beta N^3 + \Delta_\beta \alpha(\Gamma).$$

Therefore, in view of local limit asymptotics for bulk Bernoulli fields,

$$\begin{aligned}\log \mathbb{P}_{N,\beta}\left(\Xi_N \geq \rho_\beta N^3 + AN^2 \middle| \alpha(\Gamma) = aN^2\right) &\approx -\frac{(AN^2 - \Delta_\beta a N^2)^2}{2N^3 R_\beta} \\ &= -N\frac{(\delta_\beta - a)^2}{2D_\beta},\end{aligned} \tag{9}$$

where

$$R_\beta := 2\left(p^s(1-p^s) + p^v(1-p^v)\right), \quad D_\beta := \frac{R_\beta}{\Delta_\beta^2} \quad \text{and} \quad \delta_\beta := A/\Delta_\beta. \tag{10}$$

Consequently, the following asymptotic relation should hold:

$$\frac{1}{N}\log \mathbb{P}_{N,\beta}\left(\Xi \geq \rho_\beta N^3 + AN^2\right) \approx -\min_a \left\{\frac{(\delta_\beta - a)^2}{2D_\beta} + \tau_\beta(a)\right\}, \tag{11}$$

and, moreover, if a^* is the unique minimizer for the right hand side of (11), then the conditional distribution $\log \mathbb{P}_{N,\beta}\left(\cdot \middle| \Xi \geq \rho_\beta N^3 + AN^2\right)$ should concentrate on surfaces Γ which have an optimal volume close to $a^* N^2$ or, in view of (6) and (7) whose rescaled large level lines $\frac{1}{N}\Gamma_1, \dots, \frac{1}{N}\Gamma_\ell$ are macroscopically close to the optimal stack $\mathring{\gamma}_\ell^* \subseteq \cdots \subseteq \mathring{\gamma}_1^*$ which produces $\tau_\beta(a^*)$.

Theorem A. *Assume that bulk occupation probabilities p_v and p_s satisfy* (3). *Then there exists $\beta_0 < \infty$ such that for any $\beta > \beta_0$ there exists a sequence of numbers $0 < A_1(\beta) < A_2(\beta) < A_3(\beta) < \dots$ and a sequence of areas $0 < a_1^-(\beta) < a_1^+(\beta) < a_2^-(\beta) < a_2^+(\beta) < \dots$ which satisfies properties* **A1** *and* **A2** *below:*

A1. *For any $A \in [0, A_1)$ the minimizer in the right hand side* (11) *(with $\delta_\beta = A/\Delta_\beta$ as in* (10)*) is $a^* = 0$ which, in terms of contours in* (8) *corresponds to the empty stack. For any $\ell = 1, 2, \dots$ and for any $A \in (A_\ell, A_{\ell+1})$ the unique minimizer a^* in the right hand side of* (11) *satisfies $a^* \in (a_\ell^-, a_\ell^+)$, and it corresponds to a unique, up to compatible shifts within $\mathbb{B}_1$, stack of exactly ℓ-contours $\mathring{\gamma}_\ell^* \subseteq \dots \subseteq \mathring{\gamma}_1^* \subseteq \mathbb{B}_1$.*

A2. *For any $A \in [0, A_1)$ there are no large level lines (no microscopic facets, that is Γ stays predominately flat on zero-height level) with $\mathbb{P}_{N,\beta}^A$-probability tending to one. On the other hand, for any $\ell = 1, 2, \dots$ and for any $A \in (A_\ell, A_{\ell+1})$, there are exactly ℓ microscopic facets of Γ with $\mathbb{P}_{N,\beta}^A$-probability tending to one and, moreover, the rescaled large level lines $\frac{1}{N}\Gamma_1, \dots, \frac{1}{N}\Gamma_\ell$ concentrate in Hausdorff distance d_H near the optimal stack $\{\gamma_1^*, \dots, \gamma_\ell^*\}$, as described in part* **A1** *of the Theorem, in the following sense: For any $\epsilon > 0$,*

$$\lim_{N\to\infty} \mathbb{P}_{N,\beta}^A \left(\sum_{i=1}^{\ell-1} \mathsf{d}_\mathsf{H} \left(\frac{1}{N}\Gamma_i, \gamma_i^* \right) + \min_{x: x+\gamma_\ell^* \subset \mathbb{B}_1} \mathsf{d}_\mathsf{H} \left(\frac{1}{N}\Gamma_\ell, x + \gamma_\ell^* \right) \geq \epsilon \right) = 0.$$

2.3 Structure of the Paper

Section 3 is devoted to a careful analysis of the multi-layer minimization problem in the right hand side (11). Our results actually go beyond **A1** in Theorem A, namely we give a rather complete description of optimal stacks $\{\gamma_1^*, \dots, \gamma_\ell^*\}$ in terms of Wulff shapes and Wulff plaquettes associated to surface tension τ_β, and in particular we explain geometry behind the claimed infinite sequence of first order transitions.

The notions of microscopic contours and level lines are defined in Sect. 4.

The surface tension τ_β is defined in the very beginning of Sect. 5.

The proof of part **A2** of Theorem A, or more precisely the proof of the corresponding statement for the reduced model of large contours, Theorem C in the end of Sect. 4, is relegated to Sect. 5.

Throughout the paper we rely on techniques and ideas introduced and developed in [18] and [25]. Whenever possible we only sketch proofs which follow closely the relevant parts of these papers.

2.4 Some Notation

Let us introduce the following convenient notation: Given two indexed families of numbers $\{a_\alpha\}$ and $\{b_\alpha\}$ we shall say that $a_\alpha \lesssim b_\alpha$ *uniformly* in α if there exists C such that $a_\alpha \leq C b_\alpha$ for all indices α. Furthermore, we shall say that $a_\alpha \cong b_\alpha$ if both $a_\alpha \lesssim b_\alpha$ and $b_\alpha \lesssim a_\alpha$ hold. Finally, the family $\{a_{\alpha,N}\}$ is $o_N(1)$ if $\sup_\alpha |a_{\alpha,N}|$ tends to zero as N tends to infinity.

3 The Variational Problem

The variational problems we shall deal with are constrained isoperimetric type problems for curves lying inside the box $\mathbb{B}_1 = [-1,1]^2 \subset \mathbb{R}^2$.

Let τ_β be a norm on $\mathbb{R}^2$ which possesses all the lattice symmetries of $\mathbb{Z}^2$. For a closed piecewise smooth Jordan curves (called loops below) $\gamma \subset \mathbb{B}_1$ we define $\mathsf{a}(\gamma)$ to be the area of its interior $\mathring{\gamma}$ and $\tau_\beta(\gamma)$ to be its surface tension,

$$\tau_\beta(\gamma) = \int_\gamma \tau_\beta(\mathfrak{n}_s)\mathrm{d}s.$$

A finite family $\mathcal{L} = \{\gamma_1, \gamma_2, \ldots, \gamma_n\} \subset \mathbb{B}_1$ of loops is said to be compatible if

$$\forall\, i \neq j \text{ either } \mathring{\gamma}_i \cap \mathring{\gamma}_j = \emptyset \text{ or } \mathring{\gamma}_i \subseteq \mathring{\gamma}_j \text{ or } \mathring{\gamma}_j \subseteq \mathring{\gamma}_i.$$

Thus compatible families have a structure of finite number of nested stacks of loops. Given $\mathsf{D}_\beta > 0$ consider the following family, indexed by $\delta > 0$, of minimization problems:

$$\min_{\mathcal{L}} \mathcal{E}_\beta\left(\mathcal{L}\mid\delta\right) := \min_{\mathcal{L}} \left\{ \frac{\left(\delta - \mathsf{a}(\mathcal{L})\right)^2}{2\mathsf{D}_\beta} + \tau_\beta(\mathcal{L}) \right\}, \qquad (\mathrm{VP}_\delta)$$

where

$$\mathsf{a}\left(\mathcal{L}\right) = \sum_{\gamma\in\mathcal{L}} \mathsf{a}\left(\gamma\right) \quad \text{and} \quad \tau_\beta(\mathcal{L}) = \sum_{\gamma\in\mathcal{L}} \tau_\beta(\gamma)$$

3.1 Rescaling of (VP_δ).

Let e be a lattice direction. Set

$$v = \frac{\delta}{\tau_\beta(\mathsf{e})\mathsf{D}_\beta},\quad \sigma_\beta = \mathsf{D}_\beta\tau_\beta(\mathsf{e}) \text{ and } \tau(\cdot) = \frac{\tau_\beta(\cdot)}{\tau_\beta(\mathsf{e})}.$$

Since

$$\mathcal{E}_\beta\left(\mathcal{L}\mid\delta\right) = \frac{\left(\delta - \mathsf{a}(\mathcal{L})\right)^2}{2\mathsf{D}_\beta} + \tau_\beta(\mathcal{L}) = \frac{\delta^2}{2\mathsf{D}_\beta} + \tau_\beta(\mathsf{e})\left\{ -v\mathsf{a}(\mathcal{L}) + \tau(\mathcal{L}) + \frac{\mathsf{a}(\mathcal{L})^2}{2\sigma_\beta} \right\},$$

we can reformulate the family of variational problems (VP_δ) as follows: for $a \geq 0$ define

$$\tau(a) - \min\left\{\tau(\mathcal{L}) \,:\, \mathsf{a}(\mathcal{L}) = a\right\}.$$

Then study

$$\min_{a\geq 0} \left\{ -va + \tau(a) + \frac{a^2}{2\sigma_\beta} \right\}. \qquad (\mathrm{VP}_v)$$

The problem (VP_v) has a clear geometric interpretation: we want to find $a = a(v) \geq 0$ such that the straight line with slope v is the support line at $a(v)$ to the graph $a \mapsto \tau(a) + \frac{a^2}{2\sigma_\beta}$ on $[0, \infty)$.

3.2 Wulff Shapes, Wulff Plaquettes and Pptimal Stacks

By construction τ inherits lattice symmetries of $\mathbb{Z}^d$ and $\tau(\mathsf{e}) = 1$. In this case, the Wulff shape

$$\mathsf{W} \triangleq \partial\left\{x : x \cdot \mathfrak{n} \leq \tau(\mathfrak{n})\ \forall\ \mathfrak{n} \in \mathbb{S}^1\right\}$$

has the following properties: Its radius (half of its projection on the x or y axis) equals to 1. Its area $\mathsf{a}(\mathsf{W})$ satisfies

$$\mathsf{a}(\mathsf{W}) \equiv w = \frac{1}{2}\tau(\mathsf{W}),$$

For any $r > 0$ the radius, the area and the surface tension of the scaled curve $r\mathsf{W}$ equal to r, $r^2 w$ and $2rw$ respectively. The curves $r\mathsf{W}$ are (modulo shifts) unique minimizers of $\tau(\gamma)$ restricted to the loops γ with area $\mathsf{a}(\gamma) = r^2 w$:

$$2rw = \tau(r\mathsf{W}) = \min_{\gamma:\mathsf{a}(\gamma)=r^2 w} \tau(\gamma).$$

Since $\tau(\mathsf{e}) = 1$, the maximal radius and, accordingly, the maximal area of the rescaled Wulff shape which could be inscribed into the box $\mathbb{B}_1$ are precisely 1 and w. For $b \in [0, w]$ let W_b be the Wulff shape of area b. By convention, W_b is centered at the origin. Its radius r_b and its surface tension $\tau(\mathsf{W}_b)$ are given by

$$r_b = \sqrt{\frac{b}{w}} \quad \text{and} \quad \tau(\mathsf{W}_b) = 2r_b w = 2\sqrt{bw}.$$

For $b \in (w, 4]$ the optimal (minimal $\tau(\gamma)$) loop $\gamma \subseteq \mathbb{B}_1$ with $\mathsf{a}(\gamma) = b$ is what we call the Wulff plaquette P_b. It is defined as follows. One takes four Wulff shapes of radius $r \leq 1$ and puts them in four corners of $\mathbb{B}_1$, so that each touches two corresponding sides of $\mathbb{B}_1$; then, one takes the convex envelope of the union of these four Wulff shapes. We will call such Wulff plaquette as having the radius r. It contains four segments of length $2(1 - r)$, and so its surface tension is $8(1 - r) + 2wr$. Its area is $4 - (4 - w)r^2$. In this way, the Wulff plaquette P_b of area $b \in [w, 4]$ has the radius r_b and surface tension $\tau(\mathsf{P}_b)$ given by

$$r_b = \sqrt{\frac{4-b}{4-w}} \quad \text{and} \quad \tau(\mathsf{P}_b) = 8 - 2r_b(4-w) = 8 - 2\sqrt{(4-w)(4-b)}. \qquad (12)$$

Remark 1. By convention $\mathsf{W}_w = \mathsf{P}_w$. Also note that both in the case of Wulff shapes and Wulff plaquettes,

$$\frac{\mathrm{d}}{\mathrm{d}b}\tau\left(\mathsf{W}_b\right) = \frac{1}{r_b} \quad \text{and} \quad \frac{\mathrm{d}}{\mathrm{d}b}\tau\left(\mathsf{P}_b\right) = \frac{1}{r_b}, \qquad (13)$$

for any $b \in (0, w)$ and, respectively, for any $b \in (w, 4)$.

Definition 1. *For $b \in [0, 4]$ define the optimal shape*

$$\mathsf{S}_b = \mathsf{W}_b \mathbb{1}_{b \in [0,w)} + \mathsf{P}_b \mathbb{1}_{b \in [w,4]}.$$

Let now $\mathcal{L}$ be a family of compatible loops. For any $x \in \mathbb{B}_1$ let $n_{\mathcal{L}}(x)$ be the number of loops $\gamma \in \mathcal{L}$ such that $x \in \mathring{\gamma}$. The areas $b_\ell = |x \in \mathbb{B}_1 : n_{\mathcal{L}}(x) \geq \ell|$ form a non-increasing sequence. Therefore $\mathcal{L}^* = \{\mathsf{S}_{b_1}, \mathsf{S}_{b_2}, \ldots\}$ is also a compatible family. Obviously $\mathsf{a}(\mathcal{L}^*) = \mathsf{a}(\mathcal{L})$, but $\tau(\mathcal{L}^*) \leq \tau(\mathcal{L})$. Consequently, we can restrict attention only to compatible families which contain exactly one stack of optimal shapes S_{b_ℓ}.

Furthermore, (13) implies that for each $\ell \in \mathbb{N}$ optimal ℓ-stacks could be only of two types: Let $a \in \mathbb{R}^+$.

Definition 2 (Stacks $\mathcal{L}_\ell^1(a)$ of type 1). *These contain $\ell - 1$ identical Wulff plaquettes and a Wulff shape, all of the same radius $r = r^{1,\ell}(a)$, which should be recovered from*

$$a = (\ell - 1)(4 - (4 - w)r^2) + wr^2 = 4(\ell - 1) - r^2 \left(\ell(4 - w) - 4\right). \tag{14}$$

That is, if $\ell(4 - w) \neq 4$, then

$$r^{1,\ell}(a) = \sqrt{\frac{4(\ell - 1) - a}{4(\ell - 1) - w}}. \tag{15}$$

Remark 2. Of course, if $\frac{4}{4-w} \in \mathbb{N}$, then (14) does not determine r for $\ell^* = \frac{4}{4-w}$. In other words in this case for any $r \in [0, 1]$ the stack of $(\ell^* - 1)$ Wulff plaquettes of radius r and the Wulff shape of the very same radius r on the top of them has area $4(\ell^* - 1)$ and surface tension $8(\ell^* - 1)$. This introduces a certain degeneracy in the problem, but, upon inspection, the inconvenience happens to be of a purely notational nature, and in the sequel we shall ignore this case, and assume that the surface tension τ satisfies

$$\ell^* := \frac{4}{4 - w} \notin \mathbb{N}. \tag{16}$$

We proceed to work under assumption (16). The range $Range(\mathcal{L}_\ell^1)$ of areas a, for which ℓ-stacks of type 1 are defined is:

$$Range(\mathcal{L}_\ell^1) = \begin{cases} [4(\ell - 1), \ell w], & \text{if } \ell < \ell^*, \\ [\ell w, 4(\ell - 1)], & \text{if } \ell > \ell^*. \end{cases}$$

In either of the two cases above the surface tension

$$\tau\left(\mathcal{L}_\ell^1(a)\right) = 8(\ell - 1) + 2\text{sign}(\ell^* - \ell)\sqrt{(4(\ell - 1) - a)(4(\ell - 1) - w)}. \tag{17}$$

In view of (15) and (16)

$$\frac{\mathrm{d}}{\mathrm{d}a}\tau\left(\mathcal{L}_\ell^1(a)\right) = \sqrt{\frac{4(\ell - 1) - w}{4(\ell - 1) - a}} = \frac{1}{r^{1,\ell}(a)}. \tag{18}$$

Definition 3 (Stacks $\mathcal{L}_\ell^2(a)$ of type 2). *These contain ℓ identical Wulff plaquettes of radius*

$$r^{2,\ell}(a) = \sqrt{\frac{4\ell - a}{(4-w)\ell}}, \tag{19}$$

as it follows from $a = \ell(4-(4-w)r^2)$. The range $Range(\mathcal{L}_\ell^2)$ of areas a, for which stacks of type 2 are defined (for a given value of $\ell \in \mathbb{N}$), is:

$$Range(\mathcal{L}_\ell^2) = [\ell w, 4\ell] \tag{20}$$

Substituting the value of the radius (19) into (12) we infer that the surface tension of a stack of type 2 equals to:

$$\tau\left(\mathcal{L}_\ell^2(a)\right) = 8\ell - 2\sqrt{(4\ell - a)(4-w)\ell} \quad \text{and} \quad \frac{\mathrm{d}}{\mathrm{d}a}\tau\left(\mathcal{L}_\ell^2(a)\right) = \frac{1}{r^{2,\ell}(a)}. \tag{21}$$

Note that by definition

$$\mathcal{L}_\ell^1(4(\ell-1)) = \mathcal{L}_{\ell-1}^2(4(\ell-1)) \text{ and } \mathcal{L}_\ell^1(\ell w) = \mathcal{L}_\ell^2(\ell w). \tag{22}$$

Also for notational convenience we set $\mathcal{L}_0^2(0) = \emptyset$.

Set $\tau_\ell(a) = \min\left\{\tau\left(\mathcal{L}_\ell^1(a)\right), \tau\left(\mathcal{L}_\ell^2(a)\right)\right\}$, where we define

$$\tau\left(\mathcal{L}_\ell^i(a)\right) = \infty \quad \text{if } a \notin Range(\mathcal{L}_\ell^i) \tag{23}$$

We can rewrite (VP_v) as

$$\min_{a\geq 0, \ell\in\mathbb{N}}\left\{-va + \tau_\ell(a) + \frac{a^2}{2\sigma_\beta}\right\} = \min_{a\geq 0, \ell\in\mathbb{N}, i=1,2}\left\{-va + \mathcal{G}_\ell^i(a)\right\}, \tag{24}$$

where we put

$$\mathcal{G}_\ell^i(a) = \tau\left(\mathcal{L}_\ell^i(a)\right) + \frac{a^2}{2\sigma_\beta}, \quad i = 1, 2. \tag{25}$$

3.3 Irrelevance of $\mathcal{L}_\ell^1$-stacks for $\ell > \ell^*$

For $\ell > \ell^*$,

$$Range(\mathcal{L}_\ell^1) = [\ell w, 4(\ell-1)] \subset Range(\mathcal{L}_\ell^2) = [\ell w, 4\ell].$$

By the second of (22) the values of $\mathcal{L}_\ell^1$ and $\mathcal{L}_\ell^2$ coincide at the left end point $a = \ell w$. On the other hand, $r_\ell^1(a) \leq r_\ell^2(a)$ for any $a \in [\ell w, 4(\ell-1)]$. Hence, (18) and (21) imply that $\tau\left(\mathcal{L}_\ell^1(a)\right) \geq \tau\left(\mathcal{L}_\ell^2(a)\right)$ whenever $\ell > \ell^*$.

Hence, F_v is strictly convex on $\mathbb{D}_w$ as well. This has the following implication: If $(\ell_1, a_1) \neq (\ell_2, a_2)$ are such that $F_v(\ell_1, a_1) = F_v(\ell_2, a_2)$, then

$$F_v(\ell_1, a_1) > F_v(\lambda \ell_1 + (1-\lambda)\ell_2, \lambda a_1 + (1-\lambda) a_2) \tag{32}$$

for any $\lambda \in (0, 1)$.

Let us go back to (26). Clearly $\min_{a \geq 0, \ell \in \mathbb{N}_0} F_v(\ell, a)$ is attained for all v, and, furthermore $(0,0) = \text{argmin}(F_v)$ for all v sufficiently small. It is also clear that $(0,0) \notin \text{argmin}(F_v)$ whenever v is sufficiently large.

Therefore there exists the unique minimal values $v_1^* > 0$ and $\ell_1^* \geq 1$, and, accordingly the value $a_1^- \in Range\left(\mathcal{L}^2_{\ell_1^*}\right) = [\ell_1^* w, 4\ell_1^*]$ satisfying the condition

$$F_{v_1^*}(0,0) = F_{v_1^*}\left(\ell_1^*, a_1^-\right)$$

Let us show that $\ell_1^* = 1$. Indeed, assume that $\ell_1^* > 1$. But then for the value $\ell = 1$, intermediate between $\ell = 0$ and ℓ_1^*, we have $F_{v_1^*}(1, \frac{a_1^-}{\ell_1^*}) > F_{v_1^*}(0,0)$, which contradicts the convexity property (32). Hence $a_1^- \in [w, 4)$. By the same strict convexity argument,

$$F_{v_1^*}(1, a_1^-) = \min_a F_{v_1^*}(1, a) < \min_{\ell > 1, a} F_{v_1^*}(\ell, a)$$

By continuity the inequality above will persist for $v > v_1^*$ with $v - v_1^*$ sufficiently small. Also, $F_v(1, a_1(v)) < F_v(0,0)$ for every $v > v_1^*$, since the function $a \mapsto \tau\left(\mathcal{L}_1^2(a)\right)$ is strictly convex. This means that there exists the maximal $v_2^* > v_1^*$, $a_1^+ > a_1^-$ and a continuous bijection $a_1 : [v_1^*, v_2^*] \mapsto [a_1^-, a_1^+]$ such that $(1, a(v)) = \text{argmin} F_v$ on (v_1^*, v_2^*).

Now we can proceed by induction. Let us define the segment $[v_\ell^*, v_{\ell+1}^*]$ as the maximal segment of values of the parameter v, for which there exist the corresponding segment $[a_\ell^-, a_\ell^+]$ and a continuous non-decreasing function $a_\ell : [v_\ell^*, v_{\ell+1}^*] \mapsto [a_\ell^-, a_\ell^+]$ such that $(\ell, a_\ell(v)) = \text{argmin} F_v$ for $v \in (v_\ell^*, v_{\ell+1}^*)$. (If there are several such segments for the same value of ℓ, we take for $[v_\ell^*, v_{\ell+1}^*]$, by definition, the leftmost one. Of course, we will show below that it can not be the case, but we do not suppose it now.) Our induction hypothesis is that the open segment $(v_\ell^*, v_{\ell+1}^*)$ is non-empty, and that

$$\min_{m < \ell} \min_a F_v(m, a) > \min_a F_v(\ell, a), \tag{33}$$

for $v > v_\ell^*$. We have already checked it for $\ell = 1$.

Clearly, $(\ell, a) \notin \text{argmin} F_v$ whenever v is sufficiently large. Thus, $v_{\ell+1}^* < \infty$. By induction hypothesis (33),

$$F_{v_{\ell+1}^*}(m, a) = F_{v_{\ell+1}^*}(\ell, a_\ell^+) = \min F_{v_{\ell+1}^*} \tag{34}$$

implies that $m > \ell$. As before, using convexity and continuity arguments we can check that if (34) holds for some $m > \ell$ and a (with $(m, a) \in \mathbb{D}_w$) then necessarily $m = \ell + 1$ and, furthermore,

$$\min_a F_v(\ell+1, a) < \min_{m > \ell+1, a} F_v(m, a)$$

for $v > v^*_{\ell+1}$ with $v - v^*_{\ell+1}$ sufficiently small. The first part of the induction step is justified.

Assume finally that $F_v(\ell, a_\ell) = F_v(\ell+1, a_{\ell+1}) = \min F_v$. Then $a_\ell < a_{\ell+1}$, as it is stated in (31). By the same authority, $b_\ell \geq b_{\ell+1}$, and hence $r_{b_\ell} \leq r_{b_{\ell+1}}$. By convexity of both $\tau\left(\mathcal{L}^2_\ell(a)\right)$ and $\tau\left(\mathcal{L}^2_{\ell+l}(a)\right)$ the inequality $a_\ell < a_{\ell+1}$ implies that

$$\min_a F_u(\ell, a) > \min_a F_u(\ell+1, a) \text{ for any } u > v.$$

Consequently, $\min_a F_u(\ell, a) > \min_a F_u(\ell+1, a)$ for any $u > v^*_{\ell+1}$, and we are home. □

3.5 Analysis of (VP$_v$)

As we already know, ℓ-stacks of type 1 cannot be optimal whenever $\ell > \ell^*$ (see definition (16)). Let us explore what happens if $\ell < \ell^*$. In this case

$$Range(\mathcal{L}^1_\ell) = [4(\ell-1), \ell w] \quad \text{and} \quad Range(\mathcal{L}^2_\ell) = [\ell w, 4\ell].$$

Thus, $Range(\mathcal{L}^1_\ell)$ shares endpoints $4(\ell-1)$ and ℓw with $Range(\mathcal{L}^2_{\ell-1})$ and, respectively, $Range(\mathcal{L}^2_\ell)$, and all these ranges have disjoint interiors. So, in principle, ℓ-stack of type 1 may become optimal. Note that by our convention, (22) and (25),

$$\mathcal{G}^2_{\ell-1}(4(\ell-1)) = \mathcal{G}^1_\ell(4(\ell-1)) \quad \text{and} \quad \mathcal{G}^1_\ell(\ell w) = \mathcal{G}^2_\ell(\ell w),$$

so, for $\ell < \ell^*$ the two families $\mathcal{G}^1_\ell$, $\mathcal{G}^2_\ell$ merge together into a single continuous function. In fact, it is even smooth, except that the tangent to its graph becomes vertical at endpoints 4ℓ. Let v^*_ℓ be the critical slope for variational problem (26), as described in Theorem 3. By construction, there is a line $\mathfrak{l}(v^*_\ell)$ with slope v^*_ℓ which supports both the graphs of $\mathcal{G}^2_{\ell-1}$ and of $\mathcal{G}^2_\ell$ (Fig. 1).

Definition 4. *Let us say that the graph of $\mathcal{G}^1_\ell$ sticks out below $\mathfrak{l}(v^*_\ell)$ if there exists $a \in Range(\mathcal{L}^1_\ell)$ such that*

$$\mathcal{G}^1_\ell(a) < \mathcal{G}^2_\ell(a^-_\ell) - v^*_\ell(a^-_\ell - a).$$

Obviously, there are optimal ℓ-stacks of type 1 iff the graph of $\mathcal{G}^1_\ell$ sticks out below $\mathfrak{l}(v^*_\ell)$.

Proposition 1. *For any $\ell < \ell^*$ the graph of $\mathcal{G}^1_\ell$ sticks out below $\mathfrak{l}(v^*_\ell)$ iff*

$$v^*_\ell < 1 + \frac{\ell w}{\sigma_\beta}. \tag{35}$$

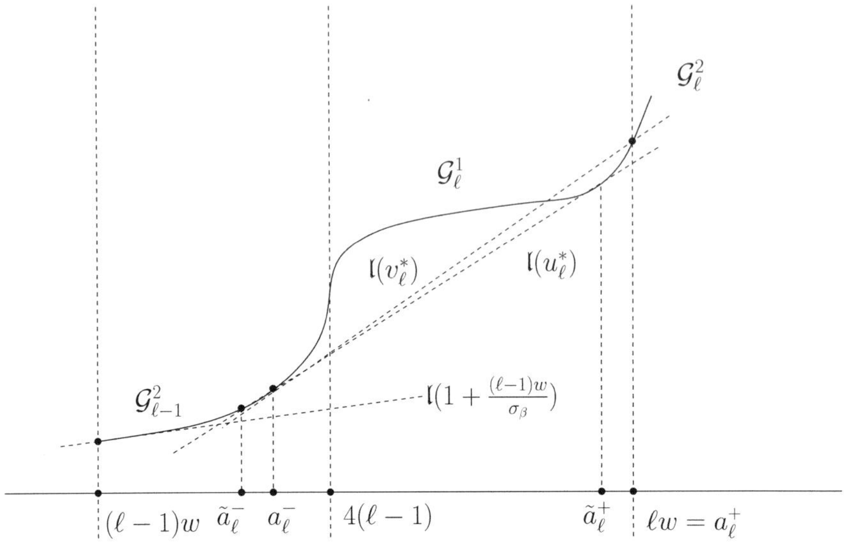

Fig. 1. The graph of $\mathcal{G}_\ell^1$ sticks out below $\mathfrak{l}(v_\ell^*)$. The transition slope $\tilde{v}_\ell^*$ satisfies $1+\frac{(\ell-1)w}{\sigma_\beta} < \tilde{v}_\ell^* < v_\ell^* < 1+\frac{\ell w}{\sigma_\beta}$.

Equivalently, this happens iff

$$\mathcal{G}_{\ell-1}^2(a) > \mathcal{G}_\ell^2(\ell w) - (\ell w - a)\left(1+\frac{\ell w}{\sigma_\beta}\right), \tag{36}$$

for any $a \in Range(\mathcal{L}_{\ell-1}^2) = [(\ell-1)w, 4(\ell-1)]$.

Proof. Note first of all that in view of (27), the condition (35) necessarily implies that $a_\ell^- = \ell w$. Consequently the fact that (35) and (36) are equivalent is straightforward, since by construction $\mathfrak{l}(v_\ell^*)$ supports both graphs.

Note that since $\frac{\mathrm{d}^+}{\mathrm{d}a}\mathcal{G}_{\ell-1}^2(4(\ell-1)) = \infty$ and since $\mathcal{L}_{\ell-1}^2(4(\ell-1)) = \mathcal{L}_\ell^1(4(\ell-1))$ the graph of $\mathcal{G}_\ell^1$ has to stay above $\mathfrak{l}(v_\ell^*)$ for values of $a \in Range(\mathcal{L}_\ell^1)$ which are sufficiently close to $4(\ell-1)$.

Let us compute the second derivative

$$\frac{\mathrm{d}^2}{\mathrm{d}a}\mathcal{G}_\ell^1(a) \overset{(17)}{=} \frac{1}{\sigma_\beta} - \frac{1}{2}\sqrt{\frac{\ell w - (4(\ell-1)}{(a-4(\ell-1))^3}}. \tag{37}$$

The expression on the right hand side above is increasing (from $-\infty$) on $Range(\mathcal{L}_\ell^1)$. If it is non-positive on the whole interval, then the graph of $\mathcal{G}_\ell^1$ is concave and it cannot stick out. Otherwise, the graph of $\mathcal{G}_\ell^1$ is convex near the right end point ℓw and hence its maximal derivative on the convex part is attained at $a = \ell w$ and equals to $1+\frac{\ell w}{\sigma_\beta}$. Therefore, (35) is a necessary condition for the graph of $\mathcal{G}_\ell^1$ to stick out.

To see that it is also a sufficient condition recall once again that $v_\ell^* < 1 + \frac{\ell w}{\sigma_\beta}$ means that $\mathfrak{l}(v_\ell^*)$ supports $\mathcal{G}_\ell^2$ at the left end point $a = \ell w$. But then, $\mathcal{G}_\ell^1$ goes below $\mathfrak{l}(v_\ell^*)$ for all values of $a \in Range(\mathcal{L}_\ell^1)$ which are sufficiently close to ℓw, because "the union of $\mathcal{G}_\ell^1$ and $\mathcal{G}_\ell^2$" is smooth at $a = \ell w$. In particular, it should have a convex part. □

Remark 3. The argument above does not imply that for $\ell < \ell^* = \frac{4}{4-w}$ the graph of $\mathcal{G}_\ell^1$ sticks out if it has a convex part near ℓw. The latter is a necessary condition which gives the following upper bound on the maximal number of layers ℓ such that $\mathcal{G}_\ell^1$ may stick out: Let us substitute $a = \ell w$ into (37):

$$\frac{1}{\sigma_\beta} - \frac{1}{2}\sqrt{\frac{\ell w - 4(\ell-1)}{(\ell w - 4(\ell-1))^3}} > 0 \overset{(16)}{\Leftrightarrow} \ell < \ell^*\left(1 - \frac{\sigma_\beta}{8}\right).$$

Proposition 2. *If $w \le 2\sigma_\beta$, then the graph of $\mathcal{G}_\ell^1$ does not stick out for any value of $\ell < \ell^*\left(1 - \frac{\sigma_\beta}{8}\right)$ (and hence stacks of type 1 are never optimal).*

If, however, $w > 2\sigma_\beta$, then there exists a number k^*, $1 \le k^* < \ell^*\left(1 - \frac{\sigma_\beta}{8}\right)$ such that the graphs of $\mathcal{G}_\ell^1$ stick out below $\mathfrak{l}(v_\ell^*)$-s for any $\ell = 1, \ldots, k^*$, and they do not stick out for $\ell > k^*$.

Proof. The proof comprises two steps.

STEP 1. We claim that the graph of $\mathcal{G}_1^1$ sticks out below $\mathfrak{l}(v_1^*)$ iff $w > 2\sigma_\beta$.

Indeed, recall that $\mathfrak{l}(v_1^*)$ is the line which passes through the origin and which is tangent to the graph of $\mathcal{G}_1^2$. Since the latter is convex and increasing, $v_1^* < 1 + \frac{w}{\sigma_\beta}$ iff

$$\mathcal{G}_1^2(w) < w\left(1 + \frac{w}{\sigma_\beta}\right) \Leftrightarrow 2w + \frac{w^2}{2\sigma_\beta} < w + \frac{w^2}{\sigma_\beta} \Leftrightarrow w > 2\sigma_\beta,$$

so the claim follows from Proposition 1.

STEP 2. We claim that for any $1 \le m < \ell$, if the graph of $\mathcal{G}_\ell^1$ sticks out below $\mathfrak{l}(v_\ell^*)$, then the graph of $\mathcal{G}_m^1$ sticks out below $\mathfrak{l}(v_m^*)$.

Assume that (36) holds. First of all take $a = (\ell-1)w$. Recall that $\mathcal{G}_\ell^2(\ell w) = 2\ell w + (\ell w)^2/2\sigma_\beta$. Therefore,

$$\mathcal{G}_{\ell-1}^2\left((\ell-1)w\right) - \mathcal{G}_\ell^2(\ell w) + w\left(1 + \frac{\ell w}{\sigma_\beta}\right) = \frac{w^2}{2\sigma_\beta} - w > 0. \tag{38}$$

Furthermore, if we record the range $a \in Range(\mathcal{L}_{(\ell-1)}^2)$ as $a = (\ell-1)w + c$; $c \in [0, (4-w)(\ell-1)]$, then, in view of (38), the necessary and sufficient condition (36) for the graph of $\mathcal{G}_\ell^1$ to stick out reads as:

$$\begin{aligned}
&\mathcal{G}_{\ell-1}^2\left((\ell-1)w + c\right) - \mathcal{G}_\ell^2(\ell w) + (w-c)\left(1 + \frac{\ell w}{\sigma_\beta}\right) \\
&= \left(\frac{w^2}{2\sigma_\beta} - w\right) + \int_0^c \left(\frac{\mathrm{d}}{\mathrm{d}\tau}\mathcal{G}_{\ell-1}^2\left((\ell-1)w + \tau\right) - \left(1 + \frac{\ell w}{\sigma_\beta}\right)\right)\mathrm{d}\tau \qquad (39)\\
&\overset{(21)}{=} \left(\frac{w^2}{2\sigma_\beta} - w\right) + \int_0^c \left(\frac{1}{r^{2,\ell-1}((\ell-1)w+\tau)} - \left(1 + \frac{\ell w}{\sigma_\beta}\right)\right)\mathrm{d}\tau > 0.
\end{aligned}$$

for any $c \in [0, (4 - w)(\ell - 1)]$.

Now for any $k \in \mathbb{N}$ (and $\tau \in [0, (4 - w)k]$),

$$r^{2,k}(kw + \tau) \overset{(19)}{=} \sqrt{\frac{4k - (kw + \tau)}{(4 - w)k}} \overset{(16)}{=} \sqrt{1 - \frac{\tau}{(4 - w)k}}.$$

That means that for a fixed τ the function $k \mapsto r^{2,k}(kw + \tau)$ is increasing in k. Therefore, (39) implies that,

$$\left(\frac{w^2}{2\sigma_\beta} - w\right) + \int_0^c \left(\frac{1}{r^{2,m-1}((m-1)w + \tau)} - \left(1 + \frac{mw}{\sigma_\beta}\right)\right) \mathrm{d}\tau > 0,$$

for any $m = 1, \ldots, \ell$ and, accordingly, any $c \in [0, (m-1)(4-w)]$. Consequently, (36) holds for any $m \leq \ell$. □

Assume now that $\ell < \ell^*$ and that the graph of $\mathcal{G}_\ell^1$ sticks out below $\mathfrak{l}(v_\ell^*)$. This means that there exits a range of slopes $(\tilde{v}_\ell^*, v_\ell^*)$ such that for any $v \in (\tilde{v}_\ell^*, v_\ell^*)$, one can find $a = a(v)$, such that $\mathcal{L}_\ell^1(a(v))$ is the unique solution to the variational problem (VP_v). By (18),

$$v = \frac{1}{r^{1,\ell}(a(v))} + \frac{a(v)}{\sigma_\beta} \geq 1 + \frac{4(\ell - 1)}{\sigma_\beta}.$$

Hence, in view of Proposition 1 (and in view of the fact that by Proposition 2 the graph of $\mathcal{G}_{\ell-1}^1$ has to stick out as well and consequently $v_{\ell-1}^* < 1 + \frac{(\ell-1)w}{\sigma_\beta}$),

$$\tilde{v}_\ell^* > 1 + \frac{(\ell - 1)w}{\sigma_\beta} > v_{\ell-1}^*.$$

Which means that in the range of slopes $[1 + \frac{(\ell-1)w}{\sigma_\beta}, \tilde{v}_\ell^*)$ the $(\ell - 1)$-stacks of type 2 continue to be optimal.

The structure of solutions and first order transitions in terms of optimal layers and optimal areas is summarized in Theorem B and depicted on Fig. 2.

Theorem B. *If $w \leq 2\sigma_\beta$, then solutions to the variational problem (VP_v) are as described in Theorem 1.*

If, however, $w > 2\sigma_\beta$, then there exists a number $1 \leq k^* < \ell^* \left(1 - \frac{\sigma_\beta}{8}\right)$ such that the following happens: For every $\ell = 1, \ldots, k^*$ there exists a slope $\tilde{v}_\ell^*$;

$$1 + \frac{(\ell - 1)w}{2\sigma_\beta} < \tilde{v}_\ell^* < v_\ell^* < 1 + \frac{\ell w}{2\sigma_\beta},$$

such that

1. *The empty stack $\mathcal{L}_0^2$ is the unique solution to (VP_v) for $v \in [0, \tilde{v}_1^*)$.*
2. *For every $\ell = 1, \ldots, k^*$ there is an area $\tilde{a}_\ell^- \in (4(\ell - 1), \ell w)$ and a continuous increasing bijection $\tilde{a}_\ell : [\tilde{v}_\ell^*, 1 + \frac{\ell w}{2\sigma_\beta}] \mapsto [\tilde{a}_\ell^-, \ell w]$ such that the ℓ-stack $\mathcal{L}_\ell^1(\tilde{a}_\ell(v))$ of type 1 is the unique solution to (VP_v) for every $v \in (\tilde{v}_\ell^*, 1 + \frac{\ell w}{2\sigma_\beta})$.*

3. *For every* $\ell < k^*$ *the* ℓ*-stack* $\mathcal{L}_\ell^2(a_\ell(v))$ *of type 2 is the unique solution to* (VP_v) *for every* $v \in (1 + \frac{\ell w}{\sigma_\beta}, \tilde{v}_{\ell+1}^*)$*, where* a_ℓ *is the bijection described in Theorem 1.*
4. *The stack* $\mathcal{L}_{k^*}^2(a_{k^*}(v))$ *is the unique solution to* (VP_v) *for every* $v \in [1 + \frac{k^* w}{\sigma_\beta}, v_{k^*+1})$.
5. *Finally, for every* $\ell > k^*$ *the transition slope* $v_\ell^* \geq 1 + \frac{\ell w}{\sigma_\beta}$*, and the stack* $\mathcal{L}_\ell^2(a_\ell(v))$ *of type 2 is the unique solution to* (VP_v) *for every* $v \in (v_\ell^*, v_{\ell+1}^*)$.

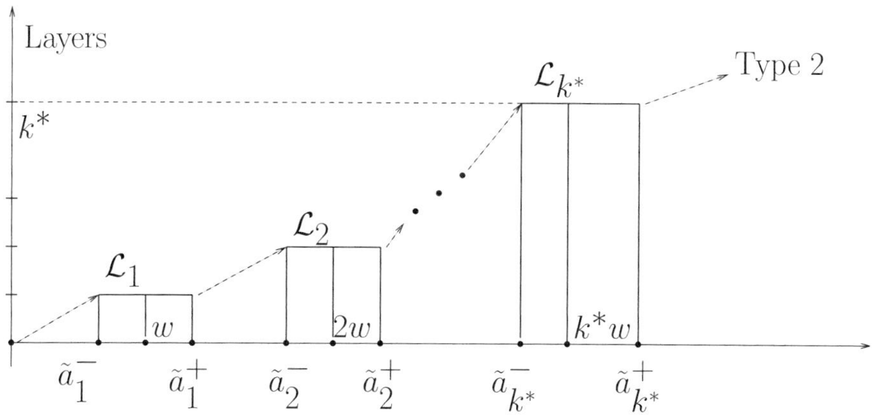

Fig. 2. For $\ell = 1, \ldots, k^* < \ell^*$ families $\mathcal{L}_\ell^1$ of type one are optimal for $a \in (\tilde{a}_\ell^-, \ell w)$, whereas families $\mathcal{L}_\ell^2$ of type 2 are optimal for $a \in (\ell w, \tilde{a}_\ell^+)$. First order transitions - jumps in terms of number of optimal layers from $\ell - 1$ to ℓ, and in terms of sizes of optimal areas from $\tilde{a}_{\ell-1}^+$ to $\tilde{a}_\ell^-$ (where we set $\tilde{a}_0^+ = 0$) - occur at transition slopes $\tilde{v}_\ell^*$. For $\ell > k^*$ only families of type two are optimal, and first order transitions occur as described in Theorem 1.

4 Low Temperature Level Lines

The main contribution to the event $\{\Xi_N \geq \rho_\beta N^3 + AN^2\}$ comes from bulk fluctuations and creations of macroscopic size facets (large contours - see below) of the interface Γ. In order to formulate the eventual reduced model let us first of all collect the corresponding notions and facts from [24].

4.1 Bulk Fluctuations

For each β fixed bulk fluctuations are governed by local limit results for sums of Bernoulli random variables, as the linear size of the system $N \to \infty$. Let us record a more quantitative version of (9): For every K fixed

$$\log \mathbb{P}_{N,\beta}\left(\Xi_N = \rho_\beta N^3 + AN^2 \middle| \Gamma\right) = -N\frac{\left(\delta_\beta - \frac{\alpha(\Gamma)}{N^2}\right)^2}{2D_\beta} + O\left(\log N\right), \qquad (40)$$

uniformly in $A \leq K$ and $|\alpha(\Gamma)| \leq KN^2$.

4.2 Contours and Their Weights

There is a natural contour parametrization of surfaces Γ. Namely, given an interface Γ and, accordingly, the height function h_Γ which, by definition, is identically zero outside $\mathring{\Lambda}_N$, define the following semi-infinite subset $\widehat{\Gamma}$ of $\mathbb{R}^3$,

$$\widehat{\Gamma} = \bigcup_{\substack{(x,y,k)\\ k<h_\Gamma(x,y)}} \left((x,y,k) + \widehat{C}\right),$$

where $\widehat{C} = [-1/2, 1/2]^3$ is the unit cube. The above union is over all $(x, y) \in \mathbb{Z}^2$ and $k \in 1/2 + \mathbb{Z}$.

Consider now the level sets of Γ, i.e. the sets

$$H_k = H_k\left(\widehat{\Gamma}\right) = \left\{(x,y) \in \mathbb{R}^2 : (x,y,k) \in \widehat{\Gamma}\right\}, \; k = -N, -N+1, \ldots, N.$$

We define *contours* as the connected components of sets ∂H_k, subject to south-west splitting rules, see Sect. 2.1 in [25] or [18]. The length $|\gamma|$ of a contour is defined in an obvious way. Since, by construction all contours are closed self-avoiding polygons composed of the nearest neighbor bonds of Λ_N^*, the notions of interior $\mathrm{int}(\gamma)$ and exterior $\mathrm{ext}(\gamma)$ of a contour γ are well defined. A contour γ is called a $\oplus$-contour ($\ominus$-contour), if the values of the function h_Γ at the immediate exterior of γ are smaller (bigger) than those at the immediate interior of γ.

Alternatively, let us orient the bonds of each contours $\gamma \subseteq \partial H_k$ in such a way that when we traverse γ the set H_k remains to the right. Then $\oplus$-contours are those which are clockwise oriented with respect to their interior, whereas $\ominus$-contours are counter-clockwise oriented with respect to their interior.

Let us say that two oriented contours γ and γ' are compatible, $\gamma \sim \gamma'$, if

1. Either $\mathrm{int}(\gamma) \cap \mathrm{int}(\gamma') = \emptyset$ or $\mathrm{int}(\gamma) \subseteq \mathrm{int}(\gamma')$ or $\mathrm{int}(\gamma') \subseteq \mathrm{int}(\gamma)$.
2. Whenever γ and γ' share a bond b, b has the same orientation in both γ and γ'.

A family $\Gamma = \{\gamma_i\}$ of oriented contours is called consistent, if contours of Γ are pair-wise compatible. It is clear that the interfaces Γ are in one-to-one correspondence with consistent families of oriented contours. The height function h_Γ could be reconstructed from a consistent family $\Gamma = \{\gamma\}$ in the following way: For every contour γ the sign of γ, which we denote as $\mathrm{sign}(\gamma)$, could be read from it orientation. Then,

$$h_\gamma(x,y) = \mathrm{sign}(\gamma)\chi_{\mathrm{int}(\gamma)}(x,y) \quad \text{and} \quad h_\Gamma = \sum_{\gamma\in\Gamma} h_\gamma, \qquad (41)$$

where χ_A is the indicator function of A.

In this way the weights $w_\beta(\Gamma)$ which appear in (2) could be recorded as follows: Let $\Gamma = \{\gamma\}$ be a consistent family of oriented (signed) contours, Then,

$$w_\beta(\Gamma) = \exp\left\{-\beta \sum_{\gamma\in\Gamma} |\gamma|\right\}. \tag{42}$$

In the sequel we shall assume that β is sufficiently large. By definition the weight of the flat interface **is** $w_\beta(\Gamma_0) = 1$.

4.3 Absence of Intermediate and Negative Contours

In the sequel a claim that a certain property holds *for all β sufficiently large* means that one can find β_0, such that it holds for all $\beta \geq \beta_0$.

For all β sufficiently large the following rough apriori bound holds (see (Subsection 6.2) in [24]): There exist positive combinatorial (that is independent of β) constant ν such for every $b_0 > 0$ fixed,

$$\log \mathbb{P}_{N,\beta}\left(|\alpha(\Gamma)| > bN^2\right) \lesssim -\nu\beta N\sqrt{b},$$

uniformly in $b \geq b_0$ and in N large enough. A comparison with (40) reveals that we may fix $K_\beta = K_\beta(A)$ and ignore Γ with $\alpha(\Gamma) \geq K_\beta N^2$. Now let the interface Γ with $\alpha(\Gamma) \leq K_\beta N^2$ be given by a consistent collection of contours, and assume that $\gamma \sim \Gamma$. Of course $\alpha(\Gamma \cup \gamma) = \alpha(\Gamma) + \alpha(\gamma)$. Then [24] there exists a constant $c_\beta = c_\beta(A)$ such that

$$-\log \mathbb{P}_{N,\beta}\left(\Gamma \cup \gamma \,\middle|\, \Xi_N \geq \rho_\beta N^3 + AN^2\right) \lesssim c_\beta \frac{|\gamma|^2}{N} \mathbb{1}_{\mathrm{sign}(\gamma)=1} - \beta|\gamma|. \tag{43}$$

Consequently, there exists $\epsilon_\beta = \epsilon_\beta(A) > 0$, such that we can rule out all contours γ with $\epsilon_\beta^{-1} \log N \leq |\gamma| \leq \epsilon_\beta N$, and all negative contours γ with $|\gamma| > \epsilon_\beta N$.

It is not difficult to see that (3) implies that

$$\liminf_{\beta\to\infty} \frac{1}{\beta} \log \epsilon_\beta(A) > -\infty$$

for any A fixed.

Definition 5. *The notion of big and small contours depends on the running value of excess area A, on the linear size of the system N and on the inverse temperature β. Namely, a contour γ is said to be large, respectively small, if*

$$|\gamma| \geq \epsilon_\beta(A)N \quad \textit{and, respectively, if} \quad |\gamma| \leq \frac{1}{\epsilon_\beta(A)} \log N. \tag{44}$$

Since we already know that intermediate contours and large $\ominus$-type contours could be ignored, let us use $\hat{\mathbb{P}}_{N,\beta}$ for the restricted ensemble which contains only $\oplus$-type large or small contours.

4.4 Irrelevance of Small Contours

Let $\Gamma = \Gamma^l \cup \Gamma^s$ is a compatible collection of contours with Γ^l being the corresponding set of large contours of Γ and, accordingly, Γ^s being the collection of small contours of Γ. Clearly,

$$\log \mathbb{P}_{N,\beta}\left(\left|\Gamma^l\right| \geq cN\right) \leq -\beta cN\left(1 - o_N(1)\right),$$

uniformly in c and all β sufficiently large. Hence, again by comparison with (40) we can ignore all collections of large contours with total length $\left|\Gamma^l\right| \geq K_\beta N$.

On the other hand, elementary low density percolation arguments (control of sizes of connected components via Kesten's bound) and the $\pm$-symmetry of height function imply that

$$\log \mathbb{P}_{N,\beta}\left(\left|\alpha(\Gamma^s)\right| \geq b \;\middle|\; \Gamma^l\right) \lesssim -\frac{\mathrm{e}^{\beta} b^2}{N^2} \wedge \frac{\beta \epsilon_\beta b}{\log N},$$

uniformly in Γ^l and in $b \lesssim K_\beta N^2$. Again, a comparison with (40) implies that we may restrict attention to the case of $|\alpha(\Gamma^s)| \lesssim N^{3/2}$. Such corrections to the total value of Ξ are invisible on the scales (5) we work with and, consequently, the area shift induced by small contours may be ignored.

4.5 The Reduced Model of Big Contours

In the sequel we shall employ the following notation: $\mathcal{C}$ for clusters of non-compatible *small* contours and $\Phi_\beta(\mathcal{C})$ for the corresponding cluster weights which shows up in the cluster expansion representation of partition functions. Note that although the family of clusters $\mathcal{C}$ does depend on the running microscopic scale N, the weights $\Phi_\beta(\mathcal{C})$ remains the same. By usual cluster expansion estimates, for all β sufficiently large

$$\sup_{\mathcal{C} \neq \emptyset} \mathrm{e}^{2\beta(\mathrm{diam}_\infty(\mathcal{C})+1)} \left|\Phi_\beta(\mathcal{C})\right| \lesssim 1. \tag{45}$$

We are ready now to describe the reduced model which takes into account only large contours: The probability of a compatible collection $\Gamma = \{\Gamma_1, \dots, \Gamma_k\}$ of *large* contours is given by

$$\mathbb{Q}_{N,\beta}(\Gamma) \cong \exp\left\{-\beta \sum |\Gamma_i| - \sum_{\mathcal{C} \subset \Lambda_N} \mathbb{1}_{\{\mathcal{C} \not\sim \Gamma\}} \Phi_\beta(\mathcal{C})\right\}, \tag{46}$$

The conditional distributions of Ξ_N^v and Ξ_N^s given such Γ^l are still $\mathrm{Bin}\left(|V_N(\Gamma)|, p^v\right)$ and $\mathrm{Bin}\left(|S_N(\Gamma)|, p^s\right)$, and we shall use $\mathbb{Q}_{N,\beta}$ for the corresponding joint distribution.

For future references let us formulate the analog of the bulk fluctuation bound (40) in terms of the reduced measure $\mathbb{Q}_{N,\beta}$: For every K fixed

$$\log \mathbb{Q}_{N,\beta}\left(\Xi_N \geq \rho_\beta N^3 + AN^2 \middle| \Gamma\right) = -N \frac{\left(\delta_\beta - \frac{\alpha(\Gamma)}{N^2}\right)^2}{2D_\beta} + O\left(\log N\right), \tag{47}$$

uniformly in $A \leq K$ and $|\alpha(\Gamma)| \leq KN^2$.

The notation for conditional reduced measures is

$$\mathbb{Q}^A_{N,\beta} = \mathbb{Q}_{N,\beta}\left(\cdot \,\middle|\, \Xi_N \geq \rho_\beta N^3 + AN^2\right).$$

From now on we shall concentrate on proving **A2** of Theorem A for $\mathbb{Q}^A_{N,\beta}$-measures instead of $\mathbb{P}^A_{N,\beta}$-measures. Specifically, we shall prove:

Theorem C. *Assume that bulk occupation probabilities p_v and p_s satisfy (3). Then there exists $\beta_0 < \infty$ such that for any $\beta > \beta_0$ the following holds: Let $0 < A_1(\beta) < A_2(\beta) < A_3(\beta) < \dots$ and, respectively $0 < a_1^-(\beta) < a_1^+(\beta) < a_2^-(\beta) < a_2^+(\beta) < \dots$ be the sequences of as described in part* **A1** *of Theorem A.*

Then, for any $A \in [0, A_1)$ the set Γ of large contours is empty with $\mathbb{Q}^A_{N,\beta}$-probability tending to one. On the other hand, for any $\ell = 1, 2, \dots$ and for any $A \in (A_\ell, A_{\ell+1})$, the set Γ contains, with $\mathbb{Q}^A_{N,\beta}$-probability tending to one, exactly ℓ large contours $\Gamma = \{\Gamma_1, \dots, \Gamma_\ell\}$. Moreover, the rescaled contours from Γ concentrate in Hausdorff distance $\mathsf{d_H}$ near the optimal stack $\{\gamma_1^*, \dots, \gamma_\ell^*\}$, as described in part **A1** *of Theorem A, in the following sense: For any $\epsilon > 0$,*

$$\lim_{N\to\infty} \mathbb{Q}^A_{N,\beta}\left(\sum_{i=1}^{\ell-1} \mathsf{d_H}\left(\frac{1}{N}\Gamma_i, \gamma_i^*\right) + \min_{x\,:\,x+\gamma_\ell`\subset\mathbb{B}_1} \mathsf{d_H}\left(\frac{1}{N}\Gamma_\ell, x+\gamma_\ell^*\right) \geq \epsilon\right) = 0. \tag{48}$$

5 Proofs

5.1 Surface Tension

Let us say that a nearest-neighbor path γ on $\mathbb{Z}^2$ is admissible if it can be suppress realized as a portion of a level line of the height function h_γ in (41). Following (46) we associate with γ its *free* weight

$$w^{\mathsf{f}}_\beta(\gamma) = \mathrm{e}^{-\beta|\gamma| - \sum_{\mathcal{C}\not\sim\Gamma}\Phi_\beta(\mathcal{C})}. \tag{49}$$

We say that an admissible $\gamma = (\gamma_0, \dots, \gamma_n)$ is $\gamma : 0 \to x$ if its end-points satisfy $\gamma_0 = 0$ and $\gamma_n = x$. The corresponding two-point function and the surface tension are

$$G_\beta(x) \triangleq \sum_{\gamma:0\to x} w^{\mathsf{f}}_\beta(\gamma) \quad \text{and} \quad \tau_\beta(x) = -\lim_{n\to\infty}\frac{1}{n}\log G_\beta(\lfloor nx\rfloor). \tag{50}$$

Recall that we are considering only sufficiently large values of β. In particular, (45) applies, and the surface tension τ_β in (50) is well defined.

In the notation of (49) given a large level line $\Gamma \subset B_N$ we define its *free* weight

$$w^{\mathsf{f}}_\beta(\Gamma) = \mathrm{e}^{-\beta|\Gamma| - \sum_{\mathcal{C}\not\sim\Gamma}\Phi_\beta(\mathcal{C})}. \tag{51}$$

In this way the measure $\mathbb{Q}_N$ in (46) describes a gas of large level lines which interact between each other and with the boundary ∂B_N. The statement below is well understood (see e.g. [DKS]), and it holds for all sufficiently low temperatures:

Lemma 1. *Let $\eta \subset \mathbb{B}_1$ be a rectifiable Jordan curve. Given any sequence of positive numbers ϵ_N such that $\lim_{N\to\infty} \epsilon_N = 0$ and $\lim_{N\to\infty} N\epsilon_N = \infty$, the following holds:*

$$\lim_{N\to\infty} \frac{1}{N} \log \left(\sum_{\Gamma \subset B_N} w^{\mathsf{f}}_{\beta}(\Gamma) \mathbb{1}_{\{\mathrm{d}_{\mathsf{H}}\left(\frac{1}{N}\Gamma, \eta\right) \leq \epsilon_N\}} \right) = -\tau_\beta(\eta). \tag{52}$$

5.2 Lower Bounds on $\mathbb{Q}_{N,\beta}\left(\Xi_N \geq \rho_\beta N^3 + AN^2\right)$

Let us apply part **A.1** of Theorem A for the surface tension τ_β defined in (50) and bulk occupation probabilities $p_v(\beta), p_s(\beta)$ which satisfy (3). Assume that $A \in (A_\ell(\beta), A_{\ell+1}(\beta))$ for some $\ell = 0, 1, \ldots$. Let $a^* = 0$ if $\ell = 0$ and $a^* \in (a_\ell^-, a_\ell^+)$ be the optimal rescaled area as described in Theorem A. Then,

Proposition 3. *In the notation of* (10) *and* (VP$_\delta$) *the following lower bound holds:*

$$\begin{aligned} \liminf_{N\to\infty} \frac{1}{N} \log \mathbb{Q}_{N,\beta}\left(\Xi_N \geq \rho_\beta N^3 + AN^2\right) &\geq -\min_{\mathcal{L}} \mathcal{E}_\beta\left(\mathcal{L} \mid \delta_\beta\right) \\ &= -\left(\frac{(\delta_\beta - a^*)^2}{2D_\beta} + \tau_\beta(a^*) \right). \end{aligned} \tag{53}$$

Proof. If $a^* = 0$, then the claim follows from (47).

Assume that $a^* \in (a_\ell^-, a_\ell^+)$ for some $\ell \geq 1$. Let $\gamma_1 \subseteq \gamma_2 \subseteq \cdots \subseteq \gamma_\ell \subset \mathbb{B}_1$ be the optimal stack as described in Theorem B. Pick a sequence ϵ_N which satisfies conditions of Lemma 1 and, for $J = 1, \ldots, \ell$ define tubes

$$A_N^j = (1 - (1 + 3(j-1))\epsilon_N) N\mathring{\gamma}_j \setminus (1 - (2 + 3(j-1))\epsilon_N) N\mathring{\gamma}_j.$$

Lemma 1 implies that for any $j = 1, \ldots, \ell$

$$\lim_{N\to\infty} \frac{1}{N} \log \left(\sum_{\Gamma_j \subset A_N^j} w^{\mathsf{f}}_{\beta}(\Gamma_j) \right) = -\tau_\beta(\gamma_j). \tag{54}$$

By construction A_N^j-s are disjoint and there exists $c_1 = c_1(\beta, a^*) > 0$ such that

$$\min_{1\leq j\ell} \mathrm{d}_{\mathsf{H}}(A_N^{j-1}, A_N^j) \geq c_1 N\epsilon_N$$

where we put $A_N^0 = \partial B_N$. Hence, in view of (45),

$$\exp\left\{ -\beta \sum_{j=1}^{\ell} |\Gamma_j| - \sum_{\substack{\mathcal{C} \not\sim \Gamma \\ \mathcal{C} \subset \Lambda_N}} \Phi_\beta(\mathcal{C}) \right\} \gtrsim \mathrm{e}^{-c_2 \ell N \mathrm{e}^{-2c_1 N\epsilon_N}} \prod_{1}^{\ell} w^{\mathsf{f}}_{\beta}(\Gamma_j) \tag{55}$$

for any collection $\Gamma = (\Gamma_1, \dots, \Gamma_\ell)$ of level lines satisfying $\Gamma_j \subset A_N^j$ for $j = 1, \dots, \ell$.

Note also that, for any $j = 1, \dots \ell$, if a large level line $\Gamma_j \subset A_N^j$, then

$$N^2 \mathsf{a}(\gamma_j)(1 - 3\ell\epsilon_N)^2 \le \mathsf{a}(\Gamma_j) \le N^2 \mathsf{a}(\gamma_j).$$

Hence, by (47),

$$\frac{1}{N} \log \mathbb{Q}_{N,\beta}\left(\Xi_N \ge \rho_\beta N^3 + AN^2 \middle| \Gamma\right) \ge -\frac{(\delta_\beta - a^*)^2}{2D_\beta} - \frac{6\delta_\beta^2 \ell \epsilon_N}{D_\beta}, \tag{56}$$

for any collection $\Gamma = (\Gamma_1, \dots, \Gamma_\ell)$ of such level lines.

Putting (54), (55) and (56) together (and recalling that ϵ_N was chosen to satisfy conditions of Lemma 1) we deduce (3). □

5.3 Strategy for Proving Upper Bounds

Below we explain the strategy which we employ for proving (48). For the rest of the section let us assume that a sufficiently large β and an excess area value $A \in (A_\ell(\beta), A_{\ell+1}(\beta))$ are fixed.

STEP 1 (Hausdorff distance and Isoperimetric rigidity). We shall employ the same notation $\mathsf{d_H}$ for two and three dimensional Hausdorff distances.

Given a family $\mathcal{L}$ of compatible loops define the height function $h[\mathcal{L}] : \mathbb{B}_1 \mapsto \mathbb{N}$,

$$h[\mathcal{L}](y) = \sum_{\gamma \in \mathcal{L}} \mathbb{1}_{\{y \in \mathring{\gamma}\}}.$$

In the case of the optimal stack (recall that we fixed β and A, so the latter notion is well defined) $\mathcal{L}^* = (\gamma_1^*, \dots, \gamma_\ell^*)$ as described in Theorem B, we say that x is admissible if $x + \gamma_\ell^* \subset \mathbb{B}_1$. If x is admissible, then we use $h_x^* = h_x^*(A, \beta)$ for the height function of $(\gamma_1^*, \dots, x + \gamma_\ell^*)$. Of course if $\mathcal{L}^*$ is of type-2, the only admissible x is $x = 0$.

We can think about $h[\mathcal{L}]$ in terms of its epigraph, which is a three dimensional subset of $\mathbb{B}_1 \times \mathbb{R}_+$. In this way the notion of Hausdorff distance $\mathsf{d_H}(h[\mathcal{L}], h_x^*)$ is well defined.

We will need the qualitative stability properties of the minima of the functional

$$\mathcal{E}_\beta(\mathcal{L} \mid \delta_\beta) = \sum_{\gamma \in \mathcal{L}} \tau_\beta(\gamma) + \frac{(\delta_\beta - \mathsf{a}(\mathcal{L}))^2}{2D_\beta}.$$

Namely, we claim that for every $\nu > 0$ there exists $\rho = \rho_\beta(\nu, A) > 0$ such that

$$\mathcal{E}_\beta(\mathcal{L} \mid \delta_\beta) > \tau_\beta(a^*) + \frac{(\delta_\beta - a^*)^2}{2D_\beta} + \rho, \tag{57}$$

whenever $\min_x \mathsf{d_H}(h[\mathcal{L}], h_x^*) > \nu$. Such stability properties are known for the Wulff and constrained Wulff isoperimetric problems. That is, if W is the Wulff loop with area $a(\mathsf{W})$ inside, see Sect. 3.2, and γ is a loop with the same area, $a(\gamma) = a(\mathsf{W})$, which satisfies $\min_x \mathrm{d}_{\mathsf{H}}(\gamma, \mathsf{W} + \mathsf{x}) > \nu$, then for some $\rho = \rho(\nu) > 0$ we have

$$\tau(\gamma) \geq \tau(\mathsf{W}) + \rho.$$

That is the content of the *generalized Bonnesen inequality*, proven in [18], see relation 2.4.1 there. The same stability property holds for the constrained case, that is when we impose the additional constrain that the loop γ has to fit inside a square, and where the role of the Wulff shape W is replaced by the Wulff plaquette P, see [33] for more details. We will show now that the above quantitative stability of the surface tension functional implies the quantitative stability of the functional $\mathcal{E}_\beta(*|\delta_\beta)$.

Let us prove (57). Suppose d_{H} is bigger than ν. There can be several reasons for that. The simplest is that the number of levels in the epigraph of $h[\mathcal{L}]$ is different from that for h^*. (In which case the Hausdorff distance in question is at least 1.) In order to estimate the discrepancy ρ in that case we have to look for the minimizer of the functional $\mathcal{E}_\beta(\mathcal{L}|\delta_\beta)$ under additional constraint on $\mathcal{L}$ to have a different number of levels than the optimal stack h^*. Let $\mathcal{E}_\beta^w(\delta_\beta)$ be the *minimal value* of the functional $\mathcal{E}_\beta(\mathcal{L}|\delta_\beta)$ over these 'wrong' loop configurations. The function $\mathcal{E}_\beta^w(\delta_\beta)$, as a function of δ_β, is piecewise continuous, as is the true optimal value $\mathcal{E}_\beta(\delta_\beta) = \tau_\beta(a^*) + \frac{(\delta_\beta - a^*)^2}{2D_\beta}$ (where $a^* = a^*(\delta_\beta)$ is the optimal area corresponding to the excess value δ_β). The difference $\mathcal{E}_\beta^w(\delta_\beta) - \mathcal{E}_\beta(\delta_\beta)$ is continuous and non-negative, and it vanishes precisely at the values δ_β corresponding to critical values $A_\ell(\beta)$ of the parameter A. Therefore the difference $\mathcal{E}_\beta^w(\delta_\beta) - \mathcal{E}_\beta(\delta_\beta) \equiv \rho_0(A, \beta)$ is positive for our fixed value $A \in (A_\ell(\beta), A_{\ell+1}(\beta))$.

Let now the number of levels l in the epigraph of $h[\mathcal{L}]$ is $l(h^*)$ – i.e. the same as that for h^*. Let $\mathcal{L}_a$ be the minimizer of $\mathcal{E}_\beta(\mathcal{L}|\delta_\beta)$ over all the families with $l(h^*)$ levels and with $a(\mathcal{L}) = a$. If our loop collection $\mathcal{L}$ is far from the optimal stack $\mathcal{L}_{a^*}$ – i.e. if $\min_x \mathrm{d}_{\mathsf{H}}(h[\mathcal{L}], h_x^*) > \nu$ – as well as from all other stacks: $\min_x \mathrm{d}_{\mathsf{H}}(h[\mathcal{L}], h[\mathcal{L}_a]_x) > \frac{\nu}{10}$, then our claim (5.12) follows just from the stability properties of the surface tension functional $\tau(*)$. If, on the other hand, $\mathcal{L}$ is far from the optimal stack $\mathcal{L}_{a^*}$, i.e. $\min_x \mathrm{d}_{\mathsf{H}}(h[\mathcal{L}], h_x^*) > \nu$, but it is close to some other stack $\mathcal{L}_{\bar{a}}$ from 'wrong area stacks': $\min_x \mathrm{d}_{\mathsf{H}}(h[\mathcal{L}], h[\mathcal{L}_{\bar{a}}]_x) < \frac{\nu}{10}$, then we are done, since in that case $\mathrm{d}_{\mathsf{H}}(h[\mathcal{L}_{a^*}], h[\mathcal{L}_{\bar{a}}]) > \frac{9\nu}{10}$, and we already know the minimizer of $\mathcal{E}_\beta(*|\delta_\beta)$ over the set $\left\{\mathcal{L} : \min_x \mathrm{d}_{\mathsf{H}}(h[\mathcal{L}], h[\mathcal{L}_{\bar{a}}]_x) < \frac{\nu}{10}\right\}$ to be far from $\mathcal{L}_{a^*}$.

Step 2 (Upper bound on the number of large contours, compactness considerations and skeleton calculus). In principle there should be a nice way to formulate and prove large deviation results using compactness of the space of closed connected subsets of $\mathbb{B}_1$ endowed with Hausdorff distance, see [13]. However, as in the latter work, we shall eventually resort to skeleton calculus developed in [18].

By (46) and (45),

$$\limsup_{N\to\infty} \frac{1}{N} \log \mathbb{Q}_{N,\beta}\left(\sum |\Gamma_i| \geq KN\right) \leq -K\beta\left(1 - \mathrm{O}\left(\mathrm{e}^{-4\beta}\right)\right).$$

In view of Proposition 3 we can reduce attention to families Γ of large contours which satisfy $\sum |\Gamma_i| \leq K_\beta(A)N$ for some $K_\beta(A)$ large enough. By (44) this means that we can restrict attention to collections Γ of large contours whose cardinality is at most

$$\#\{\Gamma_i \in \Gamma\} \leq n_\beta(A) = K_\beta(A)/\epsilon_\beta(A). \tag{58}$$

Given a compatible collection Γ of large contours we can talk about the rescaled surface $h_N[\Gamma] := h\left[\frac{1}{N}\Gamma\right]$ and, accordingly, consider events

$$\min_x \mathsf{d}_\mathsf{H}\left(h_N[\Gamma], h_x^*(A,\beta)\right) > \nu. \tag{59}$$

Fix a sequence ϵ_N satisfying conditions of Lemma 1 and set $\ell_N = N\epsilon_N$. Consider a large contour $\Gamma_i \in \Gamma$. By (58) there are at most $n_\beta(A)$ such contours, and each of them has a bounded length $|\Gamma_i| \leq KN$. In view of south-west splitting rules we can view Γ_i as a parameterized nearest neighbor loop $\Gamma_i = \{u_0, \ldots, u_n = u_0\}$ with $n \leq KN$.

Definition 6. *The ℓ_N-skeleton γ_i of Γ_i is defined as follows: Set $N\mathsf{u}_0 = u_0$ and $\tau_0 = 0$. Then, given $k = 0, 1, \ldots$ with u_k and τ_k already defined set*

$$\tau_{k+1} = \min\{m > \tau_k \;:\; |u_m - N\mathsf{u}_k|_1 > \ell_N\} \quad \text{and} \quad N\mathsf{u}_{k+1} = u_{\tau_{k+1}},$$

provided $\{m > \tau_k \;:\; |u_m - \mathsf{u}_k| > \ell_N\} \neq \emptyset$. Otherwise stop and set $\gamma_i \subset \mathbb{B}_1$ to be the polygonal approximation through the vertices $\mathsf{u}_0, \ldots, \mathsf{u}_k, \mathsf{u}_0$.

We write $\Gamma_i \overset{\ell_N}{\sim} \gamma_i$ and, accordingly, $\Gamma \overset{\ell_N}{\sim} \mathcal{S}$, if $\mathcal{S} = (\gamma_1, \gamma_2, \ldots)$ is a collection of ℓ_N skeletons compatible with family Γ of large contours. Since $|\Gamma_i| \leq KN$ any compatible ℓ_N skeleton γ_i has at most $\frac{K}{\epsilon_N}$ vertices. Therefore, there are at most

$$\left(\left(N^2\right)^{\frac{K}{\epsilon_N}}\right)^{n_\beta(A)} = \exp\left\{\frac{2KK_\beta(A)\log N}{\epsilon_N \epsilon_\beta(A)}\right\} = \mathrm{e}^{\mathrm{o}_N(1)N}$$

different collections of ℓ_N-skeletons to consider. Thus the entropy of the skeletons does not present an issue, and it would suffice to give uniform upper bounds on

$$\mathbb{Q}_{N,\beta}^A\left(\min_x \mathsf{d}_\mathsf{H}\left(h_N[\Gamma], h_x^*(A,\beta)\right) > \nu; \Gamma \overset{\ell_N}{\sim} \mathcal{S}\right)$$

for fixed ℓ_N-skeleton collections $\mathcal{S}$.

If $\gamma \subset \mathbb{B}_1$ is a closed polygon – for instance an ℓ_N-skeleton of some large contour – then its surface tension $\tau_\beta(\gamma)$ and its signed area $\mathsf{a}(\gamma)$ are well defined. Accordingly one defines $\tau_\beta(\mathcal{S}) = \sum_i \tau_\beta(\gamma_i)$ and $\mathsf{a}(\mathcal{S}) = \sum_i \mathsf{a}(\gamma_i)$ for finite collections $\mathcal{S}$ of polygonal lines. We apply now the isoperimetric rigidity bound (57): For every $\nu > 0$ there exists $\rho = \rho_\beta(\nu, A) > 0$ such that for all N sufficiently large the following holds:

$$\mathcal{E}_\beta\left(\mathcal{S} \mid \delta_\beta\right) = \sum_{\gamma \in \mathcal{S}} \tau_\beta(\gamma) + \frac{(\delta_\beta - \mathsf{a}(\mathcal{S}))^2}{2D_\beta} > \tau_\beta(a^*) + \frac{(\delta_\beta - a^*)^2}{2D_\beta} + \rho, \tag{60}$$

whenever $\mathcal{S}$ is an ℓ_N-skeleton; $\mathcal{S} \overset{\ell_N}{\sim} \Gamma$ of a collection Γ of large contours which satisfies (59).

For $n \leq n_\beta(A)$ (see (58)) and $\rho > 0$ consider the collection $\mathfrak{S}_N(\rho)$ of families of ℓ_N-skeletons $\mathcal{S} = (\gamma_1, \dots, \gamma_n)$ which satisfy (60).

Then (48) would be a consequence of the following statement:

Theorem 2. *There exists a positive function α on $(0, \infty)$ such that the following happens: Fix β sufficiently large and let A be as in the conditions of Theorem C. Then for any $\rho > 0$ fixed,*

$$\max_{\mathcal{S} \in \mathfrak{S}_N(\rho)} \frac{1}{N} \log \mathbb{Q}^A_{\beta,N} \left(\Gamma \overset{\ell_N}{\sim} \mathcal{S} \right) < -\alpha(\rho), \tag{61}$$

as soon as N is sufficiently large.

Similar upper bounds were derived in [18] for collections consisting of one large and several small skeletons. Here we have somewhat more delicate situation, since we need to control the weights of stacks of almost optimal contours, which are interacting. This requires additional tools and efforts.

STEP 3 (Refined estimates in terms of graph structure of Γ).

Let us elaborate on the upper bounds derived in [18]. Consider the ensemble of large microscopic loops Γ with weights $w^\beta_{\mathsf{f}}(\Gamma)$ as in (51). Given a (polygonal) skeleton $\gamma \subset \mathbb{B}_1$ define

$$w^{\mathsf{f}}_\beta \left(\Gamma \overset{\ell_N}{\sim} \gamma \right) := \sum_{\Gamma \overset{\ell_N}{\sim} \gamma} w^{\mathsf{f}}_\beta(\Gamma).$$

More generally, given a function $F(\Gamma_1, \Gamma_2, \dots)$ we put

$$\bigotimes_i w^{\mathsf{f}}_\beta \left(F(\Gamma_1, \Gamma_2, \dots) \right) := \int \bigotimes_i w^{\mathsf{f}}_\beta \left(\mathrm{d}\Gamma_i \right) F(\Gamma_1, \Gamma_2, \dots).$$

Upper bounds derived in [18] imply that here exists a positive non-decreasing function α_0 on $(0, \infty)$ such that the following happens: Fix $a_0 > 0$. Given a closed polygon γ, define its excess surface tension

$$\Omega_\beta(\gamma) = \tau_\beta(\gamma) - \tau_\beta\left(\mathsf{a}(\gamma)\right). \tag{62}$$

Then, for all N and β sufficiently large,

$$\frac{1}{N} \log w^{\mathsf{f}}_\beta \left(\Gamma \overset{\ell_N}{\sim} \gamma \right) < -\left(\tau_\beta\left(\mathsf{a}(\gamma)\right) + \alpha_0\left(\Omega_\beta(\gamma) \right) \right) \left(1 - \mathrm{o}_N(1) \right) \tag{63}$$

uniformly in $\mathsf{a}(\gamma) > a_0$. This estimate, is explained in the beginning of Subsect. 6.1.

Should we be able to bound $\mathbb{Q}_{\beta,N} \left(\Gamma \overset{\ell_N}{\sim} \mathcal{S} \right)$ by product weights

$$\bigotimes w^{\mathsf{f}}_\beta \left(\Gamma \overset{\ell_N}{\sim} \mathcal{S} \right) := \prod_{\gamma_i \in \mathcal{S}} w^{\mathsf{f}}_\beta \left(\Gamma_i \overset{\ell_N}{\sim} \gamma_i \right),$$

then (63) and Proposition 3 would readily imply (61). However, due to cluster sharing in (46) and due to confined geometry of clusters; $\mathcal{C} \subset B_N$, contours in Γ do interact both between each other and with ∂B_N, which a priori may lead to a modification of surface tension. Therefore, one should proceed with care.

A compatible collection $\Gamma = \{\Gamma_v\}_{v\in\mathcal{V}}$ of large level lines has a natural graph structure: Namely let us say that $\Gamma_u \sim \Gamma_v$ if there is a continuous path in $\mathbb{R}^2$ which connects between Γ_u and Γ_v without hitting any other element of Γ. This notion of hitting is ambiguous because by construction different Γ_v-s may share bounds or even coincide. We resolve this ambiguity as follows: If there is a strict inclusion $\mathring{\Gamma}_u \subset \mathring{\Gamma}_v$, then any path from the infinite component of $\mathbb{R}^2 \setminus \Gamma_v$ to Γ_u by definition hits Γ_v. If $\Gamma_{v_1} = \cdots = \Gamma_{v_k}$, then we fix an ordering and declare that any path from the infinite component of $\mathbb{R}^2 \setminus \Gamma_{v_1}$ to Γ_{v_j}; $j > 1$, hits Γ_{v_i} for any $i < j$.

In this way we label collections Γ of large level lines by finite graphs $\mathcal{G} = (\mathcal{V}, \mathcal{E})$. If $\mathcal{S} = \{\gamma_u\}$ is a family of ℓ_N-skeletons of Γ (meaning that $\Gamma_u \overset{\ell_N}{\sim} \gamma_u$ for every $u \in \mathcal{V}$, then by definition $\mathcal{S}$ has the same graph structure.

We write $\Gamma \in \mathcal{G}$ if $\mathcal{G}$ is the above graph of Γ. The chromatic number of this graph plays a role. In Subsect. 5.4 we show how, once the chromatic number is under control, to reduce complex many-body interactions in $\mathbb{Q}_{\beta,N}\left(\Gamma \overset{\ell_N}{\sim} \mathcal{S}\right)$ to upper bounds on pairs of interacting contours.

Step 4 (Entropic repulsion versus interaction). In Subsect. 5.5 we formulate decoupling upper bounds for two interacting contours. In view of these bounds (45) implies that at sufficiently low temperatures entropic repulsion always beats our weak interactions. The proof is relegated to Sect. 6.

Step 5 (Bounds on chromatic numbers). In Subsect. 5.6 we derive exponential upper bounds on chromatic numbers, which enable reduction to the decoupling estimates for pairs of contours.

5.4 A Chromatic Number Upper Bound on a Collection of Large Contours

Let $\mathcal{G} = (\mathcal{V}, \mathcal{E})$ be a finite graph, for instance associated to a collection $\Gamma = \{\Gamma_v\}_{v\in\mathcal{V}}$ of large level lines. Let $\mathcal{S} = \{\gamma_v\}_{v\in\mathcal{V}}$ be a collections of polygonal lines. We wish to derive an upper bound on $\mathbb{Q}_{N,\beta}$-probabilities (see (46))

$$\mathbb{Q}_{N,\beta}\left(\mathbb{1}_{\{\Gamma\in\mathcal{G}\}}\prod_{v\in\mathcal{V}}\mathbb{1}_{\left\{\Gamma_v \overset{\ell_N}{\sim}\gamma_v\right\}}\right)$$
$$\cong \sum_{\Gamma} \mathrm{e}^{-\beta\sum_v|\Gamma_v| - \sum^*_{\mathcal{C}\not\sim\Gamma}\Phi_\beta(\mathcal{C})}\mathbb{1}_{\{\Gamma\in\mathcal{G}\}}\prod_{v\in\mathcal{V}}\mathbb{1}_{\left\{\Gamma_v \overset{\ell_N}{\sim}\gamma_v\right\}}.$$

Above $\sum^*$ restricts summation to connected clusters $\mathcal{C} \subset B_N$. Since we are trying to derive upper bounds in terms of surface tension τ_β which was introduced in (52) in terms of infinite volume weights, it happens to be convenient to augment $\mathcal{V}$ with an additional root vertex 0 which corresponds to $\Gamma_0 = \partial B_N$, and connect it to other vertices of $\mathcal{V}$ using exactly the same rules as specified above (under the convention that if Γ contains other copies of ∂B_N, then Γ_0 is the external one in the ordering of these copies). Let $\mathcal{G}_0 = (\mathcal{V}_0, \mathcal{E}_0)$ to be the augmented graph, and let $\hat{\mathcal{G}}_0$ to be the line graph of $\mathcal{G}_0$. That is vertices of $\hat{\mathcal{G}}_0$ are undirected edges $e = (u, v)$ of $\mathcal{G}_0$, and we say that e and g are neighbors in $\hat{\mathcal{G}}_0$ if they are adjacent in $\mathcal{G}_0$. Let $\kappa_{\mathcal{G}}$ be the chromatic number of $\hat{\mathcal{G}}_0$, and consider a disjoint decomposition $\hat{\mathcal{G}}_0 = \cup_{i=1}^{\kappa_{\mathcal{G}}} \hat{\mathcal{G}}_i$. By definition each class $\hat{\mathcal{G}}_i$ contains pair-wise non-adjacent edges.

Now, if $\Gamma \in \mathcal{G}$, then,

$$\sum_{\substack{\mathcal{C} \not\sim \Gamma \\ \mathcal{C} \subset B_N}} \Phi_\beta(\mathcal{C}) \geq \sum_{v \in \mathcal{V}} \Phi_\beta(\mathcal{C}) \mathbb{1}_{\{\mathcal{C} \not\sim \Gamma_v\}} - \sum_{e \in \mathcal{E}_0} |\Phi_\beta(\mathcal{C})| \, \mathbb{1}_{\{\mathcal{C} \not\sim e\}},$$

where we write $\mathbb{1}_{\{\mathcal{C} \not\sim e\}} = \mathbb{1}_{\{\mathcal{C} \not\sim \Gamma_u\}} \mathbb{1}_{\{\mathcal{C} \not\sim \Gamma_v\}}$ for an undirected edge $e = (u, v) \in \mathcal{E}_0$.

We arrive to the following upper bound in terms of product free weights defined in (51):

$$\mathbb{Q}_{N,\beta}\left(\mathbb{1}_{\{\Gamma \in \mathcal{G}\}} \prod_{v \in \mathcal{V}} \mathbb{1}_{\left\{\Gamma_v \overset{\ell_N}{\sim} \gamma_v\right\}} \right) \lesssim \bigotimes_{v \in \mathcal{V}} w_\beta^{\mathsf{f}} \left(\mathbb{1}_{\{\Gamma \in \mathcal{G}\}} \prod_{v \in \mathcal{V}} \mathbb{1}_{\left\{\Gamma_v \overset{\ell_N}{\sim} \gamma_v\right\}} \exp\left\{ \sum_{i=1}^{\kappa_{\mathcal{G}}} \sum_{\substack{e \in \hat{\mathcal{G}}_i \\ \mathcal{C} \not\sim e}} |\Phi_\beta(\mathcal{C})| \right\} \right).$$

By the (generalized) Hölder inequality,

$$\log \mathbb{E}_{N,\beta}\left(\mathbb{1}_{\{\Gamma \in \mathcal{G}\}} \prod_{v \in \mathcal{V}} \mathbb{1}_{\left\{\Gamma_v \overset{\ell_N}{\sim} \gamma_v\right\}} \right) \lesssim \frac{1}{\kappa_{\mathcal{G}}} \sum_{i=1}^{\kappa_{\mathcal{G}}} \log\left(\bigotimes_{v \in \mathcal{V}} w_\beta^{\mathsf{f}} \left(\mathbb{1}_{\{\Gamma \in \mathcal{G}\}} \prod_{v \in \mathcal{V}} \mathbb{1}_{\left\{\Gamma_v \overset{\ell_N}{\sim} \gamma_v\right\}} \exp\left\{ \kappa_{\mathcal{G}} \sum_{\substack{e \in \hat{\mathcal{G}}_i \\ \mathcal{C} \not\sim e}} |\Phi_\beta(\mathcal{C})| \right\} \right) \right). \quad (64)$$

For each $i = 1, \ldots, \kappa_{\mathcal{G}}$ we can relax constraints and write

$$\mathbb{1}_{\{\Gamma \in \mathcal{G}\}} \leq \prod_{e=(u,v) \in \hat{\mathcal{G}}_i} \mathbb{1}_{\{\Gamma_u \sim \Gamma_v\}}.$$

Above $\Gamma_u \sim \Gamma_v$ just means that Γ_u and Γ_v are two compatible large level lines.

Let us say that $u \notin \hat{\mathcal{G}}_i$ if no edge of $\hat{\mathcal{G}}_i$ contains u as a vertex. Each of the $i = 1, \ldots, \kappa_{\mathcal{G}}$ summands on the right hand side of (64) is bounded above (under notation convention $w_\beta^{\mathsf{f}}(\Gamma_0) = 1$ for the auxiliary vertex $\Gamma_0 = \partial B_N$) by

$$\sum_{u \notin \hat{\mathcal{G}}_i} \log w_\beta^{\mathsf{f}}(\Gamma_u \overset{\ell_N}{\sim} \gamma_u) + \sum_{(u,v) \in \hat{\mathcal{G}}_i} \log w_\beta^{\mathsf{f}} \bigotimes w_\beta^{\mathsf{f}} \left(\mathbb{1}_{\{\Gamma_u \sim \Gamma_v\}} \mathbb{1}_{\left\{\Gamma_v \overset{\ell_N}{\sim} \gamma_v\right\}} \mathbb{1}_{\left\{\Gamma_u \overset{\ell_N}{\sim} \gamma_u\right\}} \mathrm{e}^{\kappa_{\mathcal{G}} \sum_{\mathcal{C} \not\sim e} |\Phi_\beta(\mathcal{C})|} \right). \quad (65)$$

In order to apply (65) we need, first of all, to control the chromatic number $\kappa_{\mathcal{G}}$. After that we shall be left with studying only the case of two compatible contours, and we shall need to show that in the latter case at all sufficiently low temperatures entropic repulsion which is triggered by compatibility constraint $\Gamma_u \sim \Gamma_v$ wins over the attractive potential $\kappa_{\mathcal{G}} \sum_{\mathcal{C} \not\sim e} |\Phi_\beta(\mathcal{C})|$.

5.5 Entropic Repulsion Versus Interaction

In this subsection we formulate upper bounds on probabilities related to two compatible interacting large contours.

Theorem 3. *Assume that a number $\chi > 1/2$ and a sequence $\{\kappa_\beta\}$ are such that*

$$\limsup_{\beta\to\infty} \sup_{\mathcal{C}\neq\emptyset} \kappa_\beta e^{\chi\beta(\mathrm{diam}_\infty(\mathcal{C})+1)} |\Phi_\beta(\mathcal{C})| < 1. \tag{66}$$

Fix $a_0 > 0$. Recall the definition of excess surface tension Ω_β in (62). Then,

$$\begin{aligned}&\frac{1}{N}\log\left(w_\beta^{\mathsf{f}} \bigotimes w_\beta^{\mathsf{f}}\left(\mathbb{1}_{\{\Gamma_u\sim\Gamma_v\}}\mathbb{1}_{\left\{\Gamma_v \overset{\ell_N}{\sim} \gamma_v\right\}}\mathbb{1}_{\left\{\Gamma_u \overset{\ell_N}{\sim} \gamma_u\right\}} \mathrm{e}^{\kappa_\beta \sum_{\mathcal{C}\not\sim e}|\Phi_\beta(\mathcal{C})|}\right)\right)\\ &\quad \leq -\left[\tau_\beta(\mathsf{a}(\gamma_u)) + \tau_\beta(\mathsf{a}(\gamma_v)) + \alpha_0\left(\Omega_\beta(\gamma_u)\right) + \alpha_0\left(\Omega_\beta(\gamma_v)\right)\right]\left(1 - \mathrm{o}_N(1)\right),\end{aligned} \tag{67}$$

uniformly in β and N sufficiently large and uniformly in closed polygonal lines γ_v, γ_u satisfying $\mathsf{a}(\gamma_v), \mathsf{a}(\gamma_u) \geq a_0$. The function α_0 is the function appearing in (63).

We sketch the proof of Theorem 3 in the concluding Sect. 6.

5.6 Upper Bound on $\kappa_{\mathcal{G}}$

In this Subsection we shall show that for all β sufficiently large one can restrict attention to graphs $\mathcal{G}$ satisfying $\kappa_{\mathcal{G}} \leq \kappa_\beta$ where the sequence $\{\kappa_\beta\}$ complies with (66).

We start with a simple general combinatorial observation.

Let G_N be a graph with no loops and double edges, having N vertices. Its edge coloring is called proper, if at every vertex all the bonds entering it have different colors. The minimal number of colors needed for creating a proper edge coloring will be denoted by $\varkappa(G_N)$. It is called the *edge chromatic number.* We need the upper bound on $\varkappa(G_N)$. Evidently, the complete graph with N vertices has the highest edge chromatic number, so it is sufficient to consider only the case of G_N being a complete graph.

Theorem 4. *Let G_N be a complete graph with N vertices. Then*

$$\varkappa(G_N) = \begin{cases} N & \text{if } N \text{ is odd,} \\ N-1 & \text{if } N \text{ is even.} \end{cases}$$

Proof. Let us produce a proper edge coloring of G_N by N colors, for N odd. To do this, let us draw G_N on the plane as a regular N-gon P_N, with all its diagonals. Let us color all the N sides s of P_N using all the N colors. What remains now is to color all diagonals. Note, that every diagonal d of P_N is parallel to precisely one side s_d of P_N, because N is odd. Let us color the diagonal d by the color of the side e_d. Evidently, the resulting coloring $\mathcal{C}_N$ is proper.

Every vertex of G_N has $N-1$ incoming edges, while we have used N colors to color all the edges. Therefore at each vertex v exactly one color c_v is missing – it is the color of the edge opposite v. By construction, all N missing colors c_v are different. Let us use this property to construct a proper edge coloring of G_{N+1} by N colors. First, we color $G_N \subset G_{N+1}$ by the coloring $\mathcal{C}_N$. Let w be the extra vertex, $w = G_{N+1} \setminus G_N$. Let us color the bond (v, w) by the color c_v. Evidently, the resulting coloring $\mathcal{C}_{N+1}$ is again proper.

Finally, for N odd the constructed coloring $\mathcal{C}_N$ is best possible: there is no coloring using $N-1$ colors. Indeed, suppose such a coloring does exist. That means that at each vertex of G_N all $N-1$ colors are present. Therefore, the bonds of the first color, say, define a partition of the set of all vertices into pairs. That, however, is impossible, since N is odd. This nice last argument is due to O. Ogievetsky. □

Next let us record a (straightforward) consequence of (47) as follows: Fix A. Then,

$$\lim_{N\to\infty}\left(\frac{1}{N}\log\left(\mathbb{Q}_{N,\beta}\left(\Xi_N \geq \rho_\beta N^3 + AN^2 \,|\, \Gamma \overset{\ell_N}{\sim} \mathcal{S}\right)\right) + \frac{\left(\delta_\beta - \mathsf{a}\left(\mathcal{S}\right)\right)^2}{2D_\beta}\right) = 0.$$

uniformly in collections $\mathcal{S}$ of closed polygons with $0 \leq \mathsf{a}(\mathcal{S}) \leq \delta_\beta$ (recall (10) for the definition of δ_β).

We proceed with the following two observations concerning the variational problem (VP$_\delta$):

Lemma 2. *Fix β sufficiently large and consider* (VP$_\delta$). *Given $\delta > 0$ let $a^*_\beta(\delta)$ be the optimal area. Then,*

$$\limsup_{\delta\to\infty}\left(\delta - a^*_\beta(\delta)\right) := \xi_\beta < \infty.$$

Furthermore,

Lemma 3. *In the notation of Lemma 2,*

$$\min_\ell\left(\tau_\beta\left(\mathcal{L}^2_\ell(a)\right) + \frac{(\delta - a)^2}{2D_\beta}\right) - \left(\tau_\beta\left(a^*_\beta\right) + \frac{(\delta - a^*_\beta)^2}{2D_\beta}\right) \geq \frac{(a - a^*_\beta)^2}{2D_\beta},$$

for all δ and all $a \in [0, \delta]$

Proof of Lemma 2. We may consider only sufficiently large values of δ, such that solutions to (VP_δ) are given by optimal stacks of type 2. By the second of (21),

$$\frac{\delta - a^*_\beta(\delta)}{D_\beta} = \frac{\tau_\beta(\mathsf{e})}{r^{2,*}(a^*_\beta(\delta))},$$

where $r^{2,*}(a)$ is the radius of the optimal stack of type-2 at given area a. Hence, it would be enough to check that

$$\liminf_{a\to\infty} r^{2,*}(a) > 0. \tag{68}$$

Clearly, if $\ell < m$ and $a \in [mw, 4\ell]$ (that is if area a can be realized by both ℓ and m stacks of type-2), then $r^{2,\ell}(a) < r^{2,m}(a)$. Which means that the map $a \mapsto r^{2,*}(a)$ has the following structure: There is a sequence $w = \hat{a}_1 < \hat{a}_2 < \dots$ of (transition) areas such that:

(a) On each of the intervals $(\hat{a}_\ell, \hat{a}_{\ell+1})$ the optimal radius $r^{2,*}(a) = r^{2,\ell}(a)$ and it is decreasing.
(b) At transition points $r^{2,\ell}(\hat{a}_{\ell+1}) < r^{2,\ell+1}(\hat{a}_{\ell+1})$. Hence we need to show that

$$\liminf_{\ell\to\infty} r^{2,\ell}(\hat{a}_{\ell+1}) > 0.$$

Fix ℓ and define $r_\ell = r^{2,\ell}(\hat{a}_{\ell+1})$ and $\rho_\ell = r^{2,\ell+1}(\hat{a}_{\ell+1})$. Then,

$$\hat{a}_{\ell+1} = \ell\left(4 - (4-w)r_\ell^2\right) = (\ell+1)\left(4 - (4-w)\rho_\ell^2\right). \tag{69}$$

By definition, $\tau\left(\mathcal{L}^2_\ell(\hat{a}_{\ell+1})\right) = \tau\left(\mathcal{L}^2_{\ell+1}(\hat{a}_{\ell+1})\right)$. Which reads (recall (19) and the first of (21)) as

$$\ell\left(8 - 2r_\ell(4-w)\right) = (\ell+1)\left(8 - 2\rho_\ell(4-w)\right) \tag{70}$$

Solving (69) and (70) we recover (68).

□

Remark 4. A slightly more careful analysis implies that under (3) there exists $\nu < \infty$ such that for all sufficiently large values of β,

$$\sup_\delta \left(\delta - a^*_\beta(\delta)\right) \le \mathrm{e}^{\beta\nu}. \tag{71}$$

Proof of Lemma 3. By the second of (21) and then by (19) the function $a \mapsto \tau_\beta\left(\mathcal{L}^2_\ell(a)\right)$ is convex for any $\ell \in \mathbb{N}$. Hence,

$$\frac{\mathrm{d}^2}{\mathrm{d}a^2}\left(\tau_\beta\left(\mathcal{L}^2_\ell(a)\right) + \frac{(\delta-a)^2}{2D_\beta}\right) \ge \frac{1}{D_\beta}$$

uniformly in $a \in (\ell w, \delta \wedge 4\ell)$. The same applies at generic values $a \in (w, \delta)$ for the function

$$\min_\ell \left(\tau_\beta\left(\mathcal{L}^2_\ell(a)\right) + \frac{(\delta-a)^2}{2D_\beta}\right).$$

But the latter attains its minimum at $a^*_\beta(\delta)$. Hence the conclusion. □

Consider now a collection $\mathcal{S} = \{\gamma_v\}_{v\in\mathcal{V}}$ of closed polygonal lines. Given a number $\zeta < 1$, such that $2\zeta w > \mathsf{a}(\mathbb{B}_1) = 4$, let us split $\mathcal{S}$ into a disjoint union,

$$\mathcal{S} = \mathcal{S}_0 \cup \mathcal{S}_1 \cup \dots, \tag{72}$$

where $\mathcal{S}_0$ contains all the polygons γ of $\mathcal{S}$ with area $\mathsf{a}(\gamma) \geq \zeta w$, whereas, for $i = 1, 2, \dots$

$$\mathcal{S}_i = \left\{\gamma \in \mathcal{S} \,:\, \mathsf{a}(\gamma) \in \left[\zeta^{i+1} w, \zeta^i w\right)\right\}.$$

By construction, any compatible collection Γ of large level lines, such that $\Gamma \overset{\ell_N}{\sim} \mathcal{S}_0$ is always an ordered stack. Given numbers $d > 0, m \in \mathbb{N}$ and a value β of the inverse temperature, let us say $\mathcal{S}$ is bad; $\mathcal{S} \in \mathfrak{B}_{d,m}(\beta)$, if either there exists $i > d\beta$ such that $\mathcal{S}_i$ is not empty, or there exists $1 \leq i \leq d\beta$ such that the cardinality

$$|\mathcal{S}_i| = \#\{\gamma \,:\, \gamma \in \mathcal{S}_i\} \geq m. \tag{73}$$

Alternatively, we may think in terms of graphs $\mathcal{G} = (\mathcal{V}, \mathcal{E})$ associated to bad collections $\mathcal{S} = \{\gamma_v\}_{v\in\mathcal{V}}$.

Proposition 4. *There exist $d < \infty$ and $m \in \mathbb{N}$ such that for all sufficiently large values of β the following holds:*

$$\limsup_{N\to\infty} \frac{1}{N} \max_{\mathcal{S}\in\mathfrak{B}_{d,m}(\beta)} \log \mathbb{Q}^A_{N,\beta}\left(\mathbb{1}_{\{\Gamma\in\mathcal{G}\}} \prod_{v\in\mathcal{V}} \mathbb{1}_{\left\{\Gamma_v \overset{\ell_N}{\sim} \mathcal{S}\right\}}\right) < 0, \tag{74}$$

for any excess area $A \geq 0$.

In particular for all sufficiently large β we may restrict attention to graphs $\mathcal{G}$ with chromatic number $\kappa_{\mathcal{G}} \leq \beta dm + 2$, independently of the value of excess area A in $\mathbb{Q}^A_{N,\beta}$.

Proof of Proposition 4. Note that the estimate $\kappa_{\mathcal{G}} \leq \beta dm + 2$ on the chromatic number is obtained from (74) as follows. For good skeleton collections the union $\mathcal{S}_1 \cup \mathcal{S}_2 \cup \dots$ contains at most βdm loops, while the collection of large level lines $\Gamma^{(0)} \overset{\ell_N}{\sim} \mathcal{S}_0$ is an ordered stack. Therefore we have to estimate the edge chromatic number of a graph with at most $\beta dm + 2$ vertices, which we do by using our combinatorial theorem above.

In view of the lower bound (53) it would be enough to prove the following: There exists $c = c_\beta > 0$ such that If $\mathcal{S} \in \mathfrak{B}_{d,m}(\beta)$

$$\limsup_{N\to\infty} \max_{\mathcal{S}\in\mathfrak{B}_{d,m}(\beta)} \left(\frac{1}{N} \log\left(\mathbb{Q}_{N,\beta}\left(\Gamma \overset{\ell_N}{\sim} \mathcal{S}\right)\right) + \tau_\beta\left(\mathsf{a}(\mathcal{S})\right)\right) < -c_\beta \tag{75}$$

Let $\mathcal{S} = \{\gamma_u\} \in \mathfrak{B}_{d,m}(\beta)$ be a bad collection of skeletons. Consider its decomposition (72). For any $k \geq 0$ set $\mathcal{S}^{(k)}_l = \mathcal{S}_0 \cup \dots \cup \mathcal{S}_k$ and $\mathcal{S}^{(k)}_s = \mathcal{S}_{k+1} \cup \dots$.

We shall prove (75) by a gradual reduction procedure using the following identity: Given $k \geq 0$, the decomposition $\mathcal{S} = \mathcal{S}_l^{(k)} \cup \mathcal{S}_s^{(k)}$ induces the decomposition $\Gamma = \Gamma_l^{(k)} \cup \Gamma_s^{(k)}$ of any collection $\Gamma \overset{\ell_N}{\sim} \mathcal{S}$ of large level lines. Then,

$$\mathbb{Q}_{N,\beta}\left(\Xi \geq \rho_\beta N^3 + AN^2\,;\, \Gamma \overset{\ell_N}{\sim} \mathcal{S}\right)$$
$$= \sum_{\Gamma \overset{\ell_N}{\sim} \mathcal{S}} \mathbb{Q}_{N,\beta}\left(\Xi \geq \rho_\beta N^3 + AN^2 \,\middle|\, \Gamma_l^{(k)} \cup \Gamma_s^{(k)}\right) \mathbb{Q}_{\beta,N}\left(\Gamma_s^{(k)} \middle| \Gamma_l^{(k)}\right) \mathbb{Q}_{\beta,N}\left(\Gamma_l^{(k)}\right).$$

The conditional probability $\mathbb{Q}_{N,\beta}\left(\cdot \,\middle|\, \Gamma\right)$ is a straightforward modification of (46): Given a splitting $\Gamma \cup \Gamma'$ of a compatible family of large contours, or, alternatively a splitting $\mathcal{V} \cup \mathcal{V}'$ of the vertices of the associated graph,

$$\mathbb{Q}_{N,\beta}\left(\Gamma' \,\middle|\, \Gamma\right) \overset{\simeq}{=} \exp\left\{ -\beta \sum_{v \in \mathcal{V}'} |\Gamma_v| - \sum_{\substack{\mathcal{C} \subset \Lambda_N \\ \mathcal{C} \sim \Gamma}} \mathbb{1}_{\{\mathcal{C} \not\sim \Gamma'\}} \Phi_\beta(\mathcal{C}) \right\}.$$

We shall rely on the following upper bounds on conditional weight which holds for all β and N sufficiently large:

By (45) and (50),

$$\frac{1}{N} \log \left(\sum_{\Gamma' \overset{\ell_N}{\sim} \mathcal{S}'} \mathbb{Q}_{N,\beta}\left(\Gamma' \middle| \Gamma\right) \right) \leq -\left(1 - \mathrm{O}\left(\mathrm{e}^{-4\beta}\right)\right) \tau_\beta\left(\mathcal{S}'\right). \tag{76}$$

for any Γ and any collection $\mathcal{S}'$ of closed polygonal lines.

Let us now turn to proving (75) (and consequently (74)) proper.

STEP 1. Let us explain how we choose m in $\mathfrak{B}_{d,m}$. We can fix (independently of β two numbers $r > 1$ and $c < \infty$ such that

$$\tau_\beta\left(\mathcal{S}_s^{(k)}\right) \geq r\tau_\beta\left(\mathsf{a}\left(\mathcal{S}_s^{(k)}\right)\right), \tag{77}$$

whenever $\mathcal{S}$ and a number $k \geq 0$ are such that

$$\mathsf{a}\left(\mathcal{S}_s^{(k)}\right) \geq c\zeta^k. \tag{78}$$

Indeed, by construction, the areas of loops from $\mathcal{S}_s^{(k)}$ are bounded above by ζ^{k+1}. Hence,

$$\tau_\beta\left(\mathcal{S}_s^{(k)}\right) \geq 2\tau_\beta(\mathsf{e}) \left\lfloor \frac{a}{\zeta^{k+1}} \right\rfloor \sqrt{w_\beta \zeta^{k+1}}$$

whenever $\mathsf{a}\left(\mathcal{S}_s^{(k)}\right) = a$. This should be compared with $\tau_\beta(a)$ which equals to $2\tau_\beta(\mathsf{e})\sqrt{a w_\beta}$ if $a \leq w_\beta$ and, otherwise, bounded above by $2\tau_\beta(\mathsf{e})\left(\left\lfloor \frac{a}{w_\beta} \right\rfloor + 1\right) w_\beta$.

Note that (78) will hold if $|\mathcal{S}_{k+1}| \geq \frac{c}{\zeta w}$. We set $m = \frac{c}{\zeta w}$.

Step 2. The first consequence of (77) is that we can rule out collections $\mathcal{S}$ with $\mathsf{a}\left(\mathcal{S}_s^{(0)}\right) \geq c$. Indeed, let $\mathcal{S}$ be such a collection of polygonal lines. By construction $\mathcal{S}_0$ is compatible with an ordered stack of large contours, hence its graph is just a line segment. Hence, the decoupling bound (65) and Theorem 3 imply:

$$\frac{1}{N}\log\left(\mathbb{Q}_{N,\beta}\left(\Gamma \overset{\ell_N}{\sim} \mathcal{S}_0\right)\right) \leq -\left(1 - o_N(1)\right)\sum_{\gamma\in\mathcal{S}_0}\left(\tau_\beta\left(\mathsf{a}(\gamma)\right) + \alpha_0\left(\Omega_\beta(\gamma)\right)\right) \quad (79)$$

On the other hand, (76) and (77) imply that

$$\frac{1}{N}\log\left(\mathbb{Q}_{N,\beta}\left(\Gamma_s^{(0)} \overset{\ell_N}{\sim} \mathcal{S}_s^{(0)} \,|\, \Gamma\right)\right) \leq -r\left(1 - \mathrm{O}\left(\mathrm{e}^{-4\beta}\right)\right)\tau_\beta\left(\mathsf{a}\left(\mathcal{S}_s^{(0)}\right)\right). \quad (80)$$

for all compatible collections of large contours $\Gamma \overset{\ell_N}{\sim} \mathcal{S}_0$. The last expression is strictly smaller than $-\tau_\beta\left(\mathsf{a}\left(\mathcal{S}_s^{(0)}\right)\right)$ for all β satisfying

$$r\left(1 - \mathrm{O}\left(\mathrm{e}^{-4\beta}\right)\right) > 1.$$

Hence (75).

There are two implications of the above computation which hold for all β sufficiently large. First of all pick $\nu_1 > \nu$ (see (71)). Then for any A we can restrict attention to skeleton collections $\mathcal{S}$ satisfying

$$\mathsf{a}\left(\mathcal{S}_0\right) \geq \delta_\beta - \mathrm{e}^{\nu_1\beta}. \quad (81)$$

Indeed, recall β-independent constant c which was defined via (77). In view of Proposition 4 we may restrict attention to $\mathsf{a}\left(\mathcal{S}_s^{(0)}\right) \leq c$. Which means that if (81) is violated, then

$$\mathsf{a}\left(\mathcal{S}\right) \leq \mathsf{a}^*_\beta(\delta_\beta) - \mathrm{e}^{\nu\beta} + c$$

for all β sufficiently large. On the other hand, as we have already mentioned, $\mathcal{S}_0$ has graph structure of an ordered stack or, in other words, one-dimensional segment, and (79) holds. Therefore, (81) is secured by Lemma 3 and (71) (and, of course, lower bound (53)).

Next, assume (81). Then, in view of the upper bound on conditional weights (76), and proceeding as in derivation of (43), one can fix $d < \infty$ and rule out skeletons γ with $\mathsf{a}(\gamma) < \xi^{\beta d} w$.

As a result we need to consider only bad collections $\mathcal{S}$ which satisfy $\mathcal{S}_i = \emptyset$ for any $i > \beta d$, but which still violate (73).

Step 3. We proceed by induction. Assume that $\mathcal{S}$ is such that the cardinalities $|\mathcal{S}_1|, \ldots, |\mathcal{S}_k| \leq m$. Then $\mathcal{S}$ may be ignored if $|\mathcal{S}_{k+1}| > m$. Indeed, the latter would imply that $\mathsf{a}\left(\mathcal{S}_s^{(k)}\right) \geq c\zeta^k$, and hence (77) holds. Consequently, as in the case of (80)

$$\limsup_{N\to\infty} \frac{1}{N} \log \left(\mathbb{Q}_{N,\beta} \left(\Gamma_s^{(k)} \overset{\ell_N}{\sim} \mathcal{S}_s^{(k)} \,|\, \Gamma \right) \right) \le -r \left(1 - \mathrm{O}\left(\mathrm{e}^{-4\beta} \right) \right) \tau_\beta \left(\mathsf{a} \left(\mathcal{S}_s^{(k)} \right) \right). \tag{82}$$

for all collection $\Gamma \overset{\ell_N}{\sim} \mathcal{S}_l^{(k)}$.

On the other hand, by induction assumption, the chromatic number of $\mathcal{L}_l^{(k)}$ is bounded above by $km + 2 \le \beta dm + 2 := \kappa_\beta$. Hence, Theorem 3 applies, and in view of the decoupling bound (65), we infer:

$$\frac{1}{N} \log \left(\mathbb{Q}_{N,\beta} \left(\Gamma \overset{\ell_N}{\sim} \mathcal{S}_l^{(k)} \right) \right) \le -\left(1 - \mathrm{o}_N(1) \right) \sum_{\gamma \in \mathcal{S}_l^{(k)}} \left(\tau_\beta \left(\mathsf{a}(\gamma) \right) + \alpha_0 \left(\Omega_\beta(\gamma) \right) \right)$$

Together with (82) this implies (75). □

5.7 Proof of Theorem 2

Let us complete the proof of Theorem 2 and hence of Theorem C. As we have seen in the previous subsection for each β sufficiently large we may ignore skeleton collections with chromatic numbers exceeding $\kappa_\beta = \beta dm + 2$. If, however, the skeleton $\mathcal{S}$ is *good*, that is if the chromatic number of $\mathcal{S}$ is less or equal to κ_β, then (66) is satisfied, and, in view of (64), (65) and (67) we conclude that

$$\frac{1}{N} \log \left(\mathbb{Q}_{N,\beta} \left(\Gamma \overset{\ell_N}{\sim} \mathcal{S} \right) \right) \le -\left(\sum_{\gamma_u \in \mathcal{S}} \left(\tau_\beta \left(\mathsf{a}(\gamma_u) \right) + \alpha_0 \left(\Omega_\beta(\gamma_u) \right) \right) \right) \left(1 - \mathrm{o}_N(1) \right). \tag{83}$$

In (83) there is the same correction term $\mathrm{o}_N(1) \to 0$ for *all* good skeletons. Since,

$$\mathcal{E}_\beta \left(\mathcal{S} \mid \delta_\beta \right) = \sum_{\gamma_u \in \mathcal{S}} \Omega_\beta(\gamma_u) + \left(\sum_{\gamma_u \in \mathcal{L}} \tau_\beta(\mathsf{a}(\gamma_u)) + \frac{(\delta_\beta - \mathsf{a}(\mathcal{L}))^2}{2D_\beta} \right),$$

it remains to apply the quantitative isoperimetric bound (60).

6 Two Interacting Contours

In this concluding section we sketch the proof of Theorem 3. The proof relies on the skeleton calculus developed in [18], Ornstein-Zernike theory and random walk representation of polymer models [27], which, in the particular case of Ising polymers, was refined and adjusted in [25]. We shall repeatedly refer to these papers for missing details.

Throughout this section we shall assume that the constants $\chi > 1/2$ and κ_β are fixed so that (66) holds.

6.1 Low Temperature Skeleton Calculus and Modified Surface Tension

We need to recall some ideas and techniques introduced in [18].

Consider an ℓ_N-skeleton $\gamma = (\mathsf{u}_0, \mathsf{u}_1, \ldots, \mathsf{u}_n)$. It has $n+1$ edges

$$e_0 = (\mathsf{u}_0, \mathsf{u}_1), \ldots, e_n = (\mathsf{u}_n, \mathsf{u}_0).$$

The last edge e_n might have a shorter length than ℓ_N, but for the sake of the exposition we shall ignore the corresponding negligible corrections. The edges of γ are classified into being *good* or *bad* as follows: Fix once and for all some small angle $\theta > 0$ (note that the value of θ and hence the classification of edges we are going to explain does not depend on β). With each edge $e = (\mathsf{u}, \mathsf{v})$ we associate a diamond shape $\mathsf{D}_\theta(e)$,

$$\mathsf{D}(e) = \mathsf{D}_\theta(\mathsf{u}, \mathsf{v}) = (\mathsf{u} + \mathcal{Y}_{\mathsf{v}-\mathsf{u}}) \cap (\mathsf{v} + \mathcal{Y}_{\mathsf{u}-\mathsf{v}}),$$

where for any $\mathsf{x} \in \mathbb{R}^2 \setminus 0$ we use $\mathcal{Y}_\mathsf{x}$ to denote the symmetric cone of opening θ along the ray passing through x.

An edge e_i of γ is called good if

$$\mathsf{D}(e_i) \cap \mathsf{D}(e_j) = \emptyset \quad \text{for any } j \neq i.$$

Otherwise the edge is called *bad*. We use $\mathfrak{g}(\gamma)$ and $\mathfrak{b}(\gamma)$ for, respectively, the sets of good and bad edges of γ.

Let Γ be a large contour, and γ be its skeleton, having the set $\mathfrak{b}(\gamma)$ of bad bonds. We denote by $\Gamma_\mathfrak{b}$ the portion of Γ, corresponding to bonds in $\mathfrak{b}(\gamma)$.

The modified surface tension $\hat{\tau}_\beta(\gamma)$ is defined as follows:

$$\hat{\tau}_\beta(\gamma) = \tau_\beta(\gamma) - \sum_{e \in \mathfrak{b}(\gamma)} \psi_\beta(e).$$

In its turn the function $\psi_\beta \geq 0$ is defined via functions (compare with (50))

$$G^*_\beta(x) := \sum_{\gamma:0\to x} w^\mathsf{f}_\beta(\eta) \mathrm{e}^{\kappa_\beta \sum_{\mathcal{C}\not\sim\eta} |\Phi_\beta(\mathcal{C})|} \quad \text{and} \quad \tau^*_\beta(x) = -\lim_{n\to\infty} \frac{1}{n} \log G^*_\beta(\lfloor nx \rfloor), \tag{84}$$

as

$$\psi_\beta(x) = \tau_\beta(x) - \tau^*_\beta(x).$$

The skeleton calculus developed in [18] implies the following two crucial bounds, which hold uniformly in $\mathsf{a}(\gamma) > a_0$:

$$\hat{\tau}_\beta(\gamma) \geq \tau_\beta(\mathsf{a}(\gamma)) + \alpha_0 (\Omega_\beta(\gamma)). \tag{85}$$

and

$$\left| \ln \frac{w^\mathsf{f}_\beta \left(\mathbb{1}_{\left\{\Gamma \overset{\ell_N}{\sim} \gamma\right\}} \mathrm{e}^{\kappa_\beta \sum_{\mathcal{C}\not\sim\Gamma_\mathfrak{b}} |\Phi_\beta(\mathcal{C})|} \right)}{\prod_{(\mathsf{u},\mathsf{v})\in\mathfrak{g}(\gamma)} G_\beta(\mathsf{v}-\mathsf{u}) \prod_{(\mathsf{u},\mathsf{v})\in\mathfrak{b}(\gamma)} G^*_\beta(\mathsf{v}-\mathsf{u})} \right| \leq o_N(1)\tau_\beta(\gamma)N. \tag{86}$$

The estimate (85) with the function $\alpha_0(x) = \frac{1}{2}x$ is nothing else but the estimate (2.16.1) from the Lemma 2.16 of [18]. (Our Ω is what is called Δ there, and the function n_δ there vanishes in the case of our interest.) The estimate (86) is a very special case of the Theorem 4.16 of [18], which establishes the asymptotic independence of the surface tension on the shape of the volume.

In view of the Ornstein-Zernike asymptotics (for instance (3.4) in [25]) of low temperature two-point functions G_β in (50) and G^*_β in (84) the upper bound (63) readily follows from (85) and (86).

6.2 Decoupling Upper Bound for Two Interacting Skeletons

Consider two ℓ_N-skeletons γ_1 and γ_2 as in the formulation of Theorem 3. Upper bound (86) implies:

$$
\begin{aligned}
& w^{\mathsf{f}}_\beta \otimes w^{\mathsf{f}}_\beta \left(\mathbb{1}_{\{\Gamma_1 \sim \Gamma_2\}} \mathbb{1}_{\left\{\Gamma_1 \overset{\ell_N}{\sim} \gamma_1\right\}} \mathbb{1}_{\left\{\Gamma_2 \overset{\ell_N}{\sim} \gamma_2\right\}} \mathrm{e}^{\kappa_\beta \sum_{\mathcal{C} \not\sim (\Gamma_1, \Gamma_2)} |\Phi_\beta(\mathcal{C})|} \right) \\
& \leq \mathrm{e}^{\mathrm{o}_N(1) N(\tau_\beta(\gamma_1) + \tau_\beta(\gamma_2))} \prod_{(\mathsf{u},\mathsf{v}) \in \mathfrak{b}(\gamma_1) \cup \mathfrak{b}(\gamma_2)} \hat{G}_\beta(\mathsf{v} - \mathsf{u}) \\
& \times \bigotimes_{e \in \mathfrak{g}(\gamma_1), f \in \mathfrak{g}(\gamma_2)} w^{\mathsf{f}}_\beta \left(\prod_{e,f} \mathbb{1}_{\left\{\eta_e \overset{\ell_N}{\sim} e\right\}} \mathbb{1}_{\left\{\eta_f \overset{\ell_N}{\sim} f\right\}} \mathbb{1}_{\{\eta_e \sim \eta_f\}} \mathrm{e}^{\kappa_\beta \sum_{e,f} \sum_{\mathcal{C} \not\sim (\eta_e, \eta_f)} |\Phi_\beta(\mathcal{C})|} \right)
\end{aligned}
$$

Above $\eta_e \overset{\ell_N}{\sim} e = (\mathsf{u}, \mathsf{v})$ means that η_e is an admissible path from u to v which is, in addition, compatible with the ℓ_N-skeleton construction. As before, $\mathcal{C} \not\sim (\eta_e, \eta_f)$ means that $\mathcal{C}$ is not compatible with both η_e and η_f.

As in the derivation of (86) it is possible to check that the contribution coming from $\sum_{\mathcal{C} \not\sim (\eta_e, \eta_f)} |\Phi_\beta(\mathcal{C})|$ could be ignored whenever $\mathsf{D}_\theta(e) \cap \mathsf{D}_\theta(f) = \emptyset$. In the latter case let us say that the edges e and f *are not associated.* Otherwise we say that $e \in \mathfrak{g}(\gamma_1)$ and $f \in \mathfrak{g}(\gamma_2)$ are *associated.* Since by construction $\mathsf{D}_\theta(f) \cap \mathsf{D}_\theta(f') = \emptyset$ for any two good edges $f, f' \in \mathfrak{g}(\gamma_2)$, any good edge $e \in \mathfrak{g}(\gamma_1)$ can be associated to at most m_θ different good edges of $\mathfrak{g}(\gamma_2)$. Proceeding as in the derivation of (65) we conclude that we need to derive the upper bound on

$$
w^{\mathsf{f}}_\beta \otimes w^{\mathsf{f}}_\beta \left(\mathbb{1}_{\left\{\eta_e \overset{\ell_N}{\sim} e\right\}} \mathbb{1}_{\left\{\eta_f \overset{\ell_N}{\sim} f\right\}} \mathbb{1}_{\{\eta_e \sim \eta_f\}} \mathrm{e}^{m_\theta \kappa_\beta \sum_{\mathcal{C} \not\sim (\eta_e, \eta_f)} |\Phi_\beta(\mathcal{C})|}, \right)
$$

uniformly in $\mathsf{D}_\theta(e) \cap \mathsf{D}_\theta(f) \neq \emptyset$.

Since m_θ does not depend on β, $\kappa'_\beta = m_\theta \kappa_\beta$ satisfies (66) with any $\chi' < \chi$. Therefore, Theorem 3 would be a consequence of the following claim:

Proposition 5. *Under the conditions of Theorem 3*

$$
\begin{aligned}
& w^{\mathsf{f}}_\beta \otimes w^{\mathsf{f}}_\beta \left(\mathbb{1}_{\left\{\eta_e \overset{\ell_N}{\sim} e\right\}} \mathbb{1}_{\left\{\eta_f \overset{\ell_N}{\sim} f\right\}} \mathbb{1}_{\{\eta_e \sim \eta_f\}} \mathrm{e}^{\kappa_\beta \sum_{\mathcal{C} \not\sim (\eta_e, \eta_f)} |\Phi_\beta(\mathcal{C})|} \right) \\
& \qquad \leq \mathrm{e}^{-\ell_N \left(\tau_\beta(\mathsf{v} - \mathsf{u}) + \tau_\beta(\mathsf{w} - \mathsf{z})\right)(1 - \mathrm{o}_N(1))}, \qquad (87)
\end{aligned}
$$

uniformly in β large and uniformly in all pairs of ℓ_N-edges $e = (\mathsf{u},\mathsf{v})$ and $f = (\mathsf{v},\mathsf{z})$ such that $\mathsf{D}_\theta(e) \cap \mathsf{D}_\theta(f) \neq \emptyset$.

The estimate (87) is a manifestation of the fact that under (66) entropic repulsion between paths η_e and η_f beats the attractive potential

$$\mathrm{e}^{\kappa_\beta \sum_{\mathcal{C} \not\sim (\eta_e,\eta_f)} |\Phi_\beta(\mathcal{C})|}.$$

As a result, typical paths stay far apart and their contributions to the surface tension just add up.

In the concluding Subsect. 6.3 we shall prove (87) in the most difficult case when e and f stay close to the horizontal axis. This case is the most difficult since it corresponds to the minimal strength of entropic repulsion between η_e and η_f.

6.3 Effective Random Walk Representation

Consider edges $f_1 = (\mathsf{z},\mathsf{w})$ and $f_2 = (\mathsf{u},\mathsf{v})$ with $\mathsf{u} = (0,0) := \mathsf{0}$, $\mathsf{v} = (\ell_N, 0)$, $\mathsf{z} = (0,z)$ and $\mathsf{w} = (\ell_N, z)$. Define the event (collection of paths (γ_1,γ_2))

$$\mathcal{T}_+^2 = \mathcal{T}_+^2(\ell_N \,|\, z) = \left\{ (\gamma_1,\gamma_2) \ : \ \gamma_1 \overset{\ell_N}{\sim} f_1\,;\, \gamma_2 \overset{\ell_N}{\sim} f_2\,;\, \gamma_1 \sim \gamma_2 \right\}.$$

In particular, if $z \geq 0$, $\{\gamma_1,\gamma_2\} \in \mathcal{T}_+^2(f_1,f_2)$ implies that γ_1 "stays above" γ_2. Note, however that they can share edges, and that they might have overhangs.

We claim that the following holds:

Proposition 6. *Assume* (66). *There exist a finite constant c_+ such that for all β sufficiently large,*

$$\sup_{z\geq 0} w_\beta^{\mathsf{f}} \otimes w_\beta^{\mathsf{f}} \left(\mathcal{T}_+^2(\ell_N \,|\, z); \mathrm{e}^{\kappa_\beta \sum_{\mathcal{C}} \mathbb{1}_{\mathcal{C}\not\sim\eta_1} \mathbb{1}_{\mathcal{C}\not\sim\eta_2} |\Phi_\beta(\mathcal{C})|} \right) \leq c_+ \mathrm{e}^{-2\tau_\beta(\mathsf{e}_1)\ell_N} \tag{88}$$

as soon as N is sufficiently large.

Proving that c_+ does not depend on β is the crux of the matter, and it is based on a careful analysis of non-intersection probabilities for effective random walks in a weak attractive potential.

Let $\mathcal{T}^1 = \mathcal{T}^1(\ell_N)$ be the set of paths $\gamma : \mathsf{0} \mapsto \mathsf{v}$ with $\mathsf{v}\cdot\mathsf{e}_1 = \ell_N$. Note that by definition $(\gamma_1,\gamma_2) \in \mathcal{T}_+^2$ implies that $\gamma_2 \in \mathcal{T}^1$ and $\gamma_1 \in (0,z) + \mathcal{T}^1$.

Let $\mathcal{K}$ be a positive (two-dimensional) symmetric cone around e_1 with an opening strictly between $\pi/2$ and π. A high temperature expansions of polymer weights $\mathrm{e}^{\Phi(\mathcal{C})}$ leads (see (4.9) in [25]) to the following irreducible decomposition of decorated (open) contours $[\gamma, \underline{\mathcal{C}}]$, where $\gamma \in \mathcal{T}^1(\ell_N)$ and $\underline{\mathcal{C}}$ is a collection of γ-incompatible clusters:

$$[\gamma, \underline{\mathcal{C}}] = \mathfrak{a}^\ell \circ \mathfrak{a}^1 \circ \ldots \mathfrak{a}^m \circ \mathfrak{a}^r. \tag{89}$$

The irreducible animals $\mathfrak{a}^i = \left[\eta^i, \underline{\mathcal{D}}^i\right]$ belong to the family $\mathsf{A} = \{\mathsf{a} = [\eta, \mathcal{D}]\}$, which could be characterized by the following two properties:

a. If η is a path with endpoints at x, y, then

$$\eta \cup \underline{\mathcal{D}} \subseteq \mathsf{D}(\mathsf{x},\mathsf{y}) := (\mathsf{x}+\mathcal{K}) \cap (\mathsf{y}-\mathcal{K}) . \tag{90}$$

b. $\mathfrak{a}$ could not be split into concatenation of two non-trivial animals satisfying **a** above.

The left and right irreducible animals satisfy one-sided versions of diamond-confinement condition (90). For instance if $\mathfrak{a}^\ell = \left[\eta^\ell, \underline{\mathcal{D}}^\ell\right]$ and y is the right end-points of η^ℓ, then $\eta^\ell \cup \underline{\mathcal{D}}^\ell \subseteq (\mathsf{y}-\mathcal{K})$.

Given an animal $\mathsf{a} = [\eta, \underline{\mathcal{D}}]$ we use $\mathsf{X}(\mathfrak{a}) = \mathsf{X}(\eta)$ to denote the vector which connects the left and right end-points of η. The horizontal and vertical coordinates of X are denoted by $\mathsf{H} = \mathsf{X}\cdot \mathsf{e}_1$ and $\mathsf{V} = \mathsf{X}\cdot\mathsf{e}_2$. By construction, $\mathsf{H}(\mathfrak{a}) \in \mathbb{N}$ for any irreducible animal a.

Returning to the decomposition (89), let us consider for each $k \in \mathbb{N}$ the subset $\mathcal{T}_k^1$ of decorated paths $[\gamma, \underline{\mathcal{C}}]$ from $\mathcal{T}^1$ for which $\mathsf{H}(\mathfrak{a}^\ell) + \mathsf{H}(\mathfrak{a}^r) = k$. Then, by (4.11) in [25] there exists $c_g < \infty$ and $\nu_g > 0$, such that

$$w_\beta^{\mathsf{f}}\left(\mathcal{T}_k^1\left(\ell_N\right)\right) \leq c_g \mathrm{e}^{-\beta\nu_g k} w_\beta^{\mathsf{f}}\left(\mathcal{T}^1\left(\ell_N\right)\right) \tag{91}$$

for all β and N sufficiently large. So in the sequel we shall restrict attention to decorated paths $[\gamma, \underline{\mathcal{C}}] \in \mathcal{T}_0^1$ with empty right and left irreducible animals, that is with $\mathfrak{a}^\ell, \mathfrak{a}^r = \emptyset$ in (89). In particular, we shall restrict attention to $\mathcal{T}_{0,+}^2(\ell_N|z) := \mathcal{T}_+^2 \cap \left((\mathsf{z}+\mathcal{T}_0^1)\times\mathcal{T}_0^1\right)$, and, instead of (88), shall derive an upper bound on

$$\sup_{z\geq 0} w_\beta^{\mathsf{f}} \otimes w_\beta^{\mathsf{f}}\left(\mathcal{T}_{0,+}^2(\ell_N|z); \mathrm{e}^{\kappa_\beta \sum_{\mathcal{C}} \mathbb{1}_{\mathcal{C}\not\sim\gamma_1}\mathbb{1}_{\mathcal{C}\not\sim\gamma_2}|\Phi_\beta(\mathcal{C})|}\right)$$

A general case could be done by a straightforward adaptation based on the mass-gap property (91).

Decorated paths $[\gamma, \underline{\mathcal{C}}] \in \mathcal{T}_0^1$ have an immediate probabilistic interpretation: Set $\tau_\beta = \tau(\mathsf{e}_1)$. Then (see Theorem 5 in [25])

$$\mathbb{P}_\beta(\mathfrak{a}) := \mathrm{e}^{\tau_\beta \mathsf{H}(\mathfrak{a})} w_\beta^{\mathsf{f}}(\mathfrak{a}) \tag{92}$$

is a probability distribution on A with exponentially decaying tail:

$$\sum_{\mathfrak{a}\in\mathsf{A}} \mathbb{1}_{\mathsf{X}(\mathfrak{a})=(h,v)}\mathbb{P}_\beta(\mathfrak{a}) \leq c_g \mathrm{e}^{-\nu_g\beta(h+|v|-1)}, \tag{93}$$

– i.e. no normalization in (92) is needed! Given $\mathsf{x} = (0,x)$ consider random walk $\mathsf{S}_n = \mathsf{x} + \sum_1^n \mathsf{X}^i$, where X^u-s are independent $\mathbb{N}\times\mathbb{Z}$-valued steps distributed according to

$$\mathbb{P}_\beta(\mathsf{X}=\mathsf{y}) = \sum_{\mathfrak{a}\in\mathsf{A}} \mathbb{1}_{\mathsf{X}(\mathfrak{a})=\mathsf{y}}\mathbb{P}_\beta(\mathfrak{a}).$$

Let $\mathbb{P}_{\beta,x}$ be the corresponding measure on random walk paths. In this way $w_\beta^{\mathsf{f}}\left(\mathcal{T}_0^1\right)$ equals to

$$w_\beta^{\mathsf{f}}\left(\mathcal{T}_0^1\right) = \mathrm{e}^{-\ell_N\tau_\beta}\mathbb{P}_{\beta,0}\left(\ell_N \in \mathrm{Range}\left(\mathsf{S}_n\cdot\mathsf{e}_1\right)\right). \tag{94}$$

Let us adapt the above random walk representation of a single decorated path to the case of a pair of decorated paths, from $\left(\mathsf{z}+\mathcal{T}_0^1\right)\times\mathcal{T}_0^1$. These have irreducible decompositions

$$\underline{\mathfrak{a}} = \mathsf{z}+\mathfrak{a}^1\circ\cdots\circ\mathfrak{a}^n \quad \text{and} \quad \underline{\mathfrak{b}} = \mathfrak{b}^1\circ\cdots\circ\mathfrak{b}^m. \tag{95}$$

Following [9] we shall align horizontal projections of underlying random walks. Given (95) consider sets

$$\begin{aligned}\mathcal{H}_a &= \left\{0, \mathsf{H}(\mathfrak{a}_1), \mathsf{H}(\mathfrak{a}_1)+\mathsf{H}(\mathfrak{a}_2), \dots, \ell_N\right\},\\ \mathcal{H}_b &= \left\{0, \mathsf{H}(\mathfrak{b}_1), \mathsf{H}(\mathfrak{b}_1)+\mathsf{H}(\mathfrak{b}_2), \dots, \ell_N\right\}.\end{aligned}$$

Set $\mathcal{H}(\underline{\mathfrak{a}},\underline{\mathfrak{b}}) = \mathcal{H}_a\cap\mathcal{H}_b$. This intersection $\mathcal{H}$ is the set of horizontal projections of end-points of jointly irreducible pairs of strings of animals. The alphabet A^2 of such pairs could be described as follows: $(\underline{\mathfrak{a}},\underline{\mathfrak{b}})\in\mathsf{A}^2$ if $\sum\mathsf{H}(\mathfrak{a}^i) = \sum\mathsf{H}(\mathfrak{b}^j) := \mathsf{H}\left(\underline{\mathfrak{a}},\underline{\mathfrak{b}}\right)$, and

$$\mathcal{H}(\underline{\mathfrak{a}},\underline{\mathfrak{b}}) = \left\{0, \mathsf{H}\left(\underline{\mathfrak{a}},\underline{\mathfrak{b}}\right)\right\}.$$

In the sequel we shall refer to elements $\mathfrak{c}\in\mathsf{A}^2$ as irreducible pairs.

By elementary renewal theory, (92) induces a probability distribution on A^2 which inherits exponential tails from (93). We continue to call this distribution $\mathbb{P}_\beta$. The i.i.d. $\mathbb{N}\times\mathbb{Z}\times\mathbb{Z}$-valued steps of the induced random walk have distribution

$$\mathbb{P}_\beta\left(\mathsf{X} = (\mathsf{H},\mathsf{V}_1,\mathsf{V}_2) = (h,v_1,v_2)\right) = \sum_{\mathfrak{c}\in\mathsf{A}^2}\mathbb{1}_{\mathsf{X}(\mathfrak{c})=(h,v_1,v_2)}\mathbb{P}_\beta(\mathfrak{c}).$$

Decorated paths from $\left(\mathsf{z}+\mathcal{T}_0^1\right)\times\mathcal{T}_0^1$ give rise to random walks

$$(0,z,0)+\sum_1^n\mathsf{X}_i.$$

Note that in this notation

$$\mathcal{T}_{0,+}^2\left(\ell_N\,|\,z\right)\subset\bigcup_n\left\{\sum_1^n\mathsf{H}_i = n; \mathcal{R}_+^n(z)\right\},$$

where

$$\mathcal{R}_+^n(z) = \left\{\mathsf{Z}_k\geq 0 \text{ for } k=0,1,\dots,n\right\} \text{ and } \mathsf{Z}_k = z+\sum_1^k\left(\mathsf{V}_i^1-\mathsf{V}_i^2\right). \tag{96}$$

With a slight abuse of notation we shall use the very same symbol $\mathbb{P}_{\beta,z}$ also for the law of the random walk Z_k in (96).

6.4 Recursion and Random Walk Analysis

Define

$$\mu_\beta(\ell_N|z) = \mathbb{P}_{\beta,z}\left(\mathcal{T}_{0,+}^2(\ell_N\mid z); \mathrm{e}^{\kappa_\beta\sum_{\mathcal{C}}\mathbb{1}_{\mathcal{C}\not\sim\gamma_1}\mathbb{1}_{\mathcal{C}\not\sim\gamma_2}|\Phi_\beta(\mathcal{C})|}\right).$$

Assume (66). Then, following Subsection 6.1 of [25], one can develop the following recursion relation for $\mu_\beta = \sup_N \sup_z \mu_\beta(\ell_N|z)$: There exist β-independent constants c_1 and ν such that

$$\mu_\beta \le 1 + \mu_\beta c_1 \mathrm{e}^{-2\chi\beta} \max_z \sum_n \mathbb{E}_{\beta,z}\left(\mathrm{e}^{-\nu\beta \mathsf{Z}_n}; \mathcal{R}_+^n\right). \tag{97}$$

The importance of (97) is that

$$c_1 \mathrm{e}^{-2\chi\beta} \sup_{z\ge 0} \sum_n \mathbb{E}_{\beta,z}\left(\mathrm{e}^{-\nu\beta \mathsf{Z}_n}; \mathcal{R}_+^n\right) < 1 \tag{98}$$

implies that μ_β is bounded, and, since by (92) and (94),

$$\begin{aligned} &w_\beta^{\mathsf{f}} \otimes w_\beta^{\mathsf{f}}\left(\mathcal{T}_{0,+}^2(\ell_N \mid z); \mathrm{e}^{\kappa_\beta \sum_{\mathcal{C}} \mathbb{1}_{\mathcal{C}\not\sim\gamma_1} \mathbb{1}_{\mathcal{C}\not\sim\gamma_2} |\Phi_\beta(\mathcal{C})|}\right) \\ &= \mathrm{e}^{-2\ell_N \tau_\beta(\mathsf{e}_1)} \mathbb{P}_{\beta,z}\left(\mathcal{T}_{0,+}^2(\ell_N \mid z); \mathrm{e}^{\kappa_\beta \sum_{\mathcal{C}} \mathbb{1}_{\mathcal{C}\not\sim\gamma_1} \mathbb{1}_{\mathcal{C}\not\sim\gamma_2} |\Phi_\beta(\mathcal{C})|}\right) \le \mathrm{e}^{-2\ell_N \tau_\beta(\mathsf{e}_1)} \mu_\beta, \end{aligned} \tag{99}$$

one deduces (88) as an immediate consequence. It remains to prove:

Lemma 4. *If $\chi > 1/2$, then*

$$\lim_{\beta\to\infty} \mathrm{e}^{-2\chi\beta} \sup_{z\ge 0} \sum_n \mathbb{E}_{\beta,z}\left(\mathrm{e}^{-\nu\beta \mathsf{Z}_n}; \mathcal{R}_+^n\right) = 0. \tag{100}$$

In particular, the inequality (98) *holds for all β sufficiently large.*

Proof of Lemma 4. Let us fix $z \ge 0$. Consider decomposition of paths from $\mathcal{R}_+^n(z)$ with respect to the left-most minimum u, $0 \le u \le z$. Define the strict version $\hat{\mathcal{R}}_+^n$ as

$$\hat{\mathcal{R}}_+^n = \{\mathsf{Z}_k > 0 \text{ for } k = 0, 1, \ldots, n\}.$$

Since $\mathsf{W} = \mathsf{V}^1 - \mathsf{V}^2$ has symmetric distribution under $\mathbb{P}_\beta$, we can rewrite

$$\begin{aligned} &\sum_n \mathbb{E}_{\beta,z}\left(\mathrm{e}^{-\nu\beta \mathsf{Z}_n}; \mathcal{R}_+^n\right) \\ &= \sum_{u=0}^{z} \mathrm{e}^{-\nu\beta u} \sum_n \mathbb{P}_{\beta,0}\left(\hat{\mathcal{R}}_+^n; \mathsf{Z}_n = z - u\right) \sum_m \mathbb{E}_{\beta,0}\left(\mathrm{e}^{-\nu\beta \mathsf{Z}_m}; \mathcal{R}_+^m\right). \end{aligned} \tag{101}$$

Below we shall use a shorter notation $\mathbb{P}_\beta$ for $\mathbb{P}_{\beta,0}$.

Following Subsection 7.1 in [25] let us describe in more detail the distribution of steps W under $\mathbb{P}_\beta$. The fact that here we take the cone $\mathcal{K}$ to be symmetric with respect to the e_1-axis simplifies the exposition. In particular, W has a symmetric distribution. The analysis of [25] could be summarized as follows:

$$\begin{aligned} 1 - p_\beta := \mathbb{P}_\beta(\mathsf{W} = 0) = 1 - \mathrm{O}\left(\mathrm{e}^{-\beta}\right), \ \mathbb{P}_\beta(\mathsf{W} = \pm 1) = \mathrm{O}\left(\mathrm{e}^{-\beta}\right) \\ \text{and, for } z \ne 0, \pm 1, \ \mathbb{P}_\beta(\mathsf{W} = z) \le p_\beta \mathrm{e}^{-\nu\beta|z|} \end{aligned} \tag{102}$$

There is a natural Wald-type decomposition of random walk Z_k with i.i.d. steps distributed according to (102): Let $\xi_1, \xi_2, \ldots$ be i.i.d. Bernoulli random variables with probability of success p_β, and let U_k be another independent i.i.d. sequence with

$$\mathbb{P}_\beta\left(\mathsf{U}=z\right)=p_\beta^{-1}\mathbb{P}_\beta\left(\mathsf{W}=z\right)\ \text{ for } z\neq 0. \tag{103}$$

Set $\mathsf{M}_n=\sum_1^n \xi_i$. Then Z_n could be represented as

$$\mathsf{Z}_n=\sum_{k=1}^{\mathsf{M}_n}\mathsf{U}_k:=\mathsf{Y}_{\mathsf{M}_n},$$

where Y_k is a random walk with i.i.d. steps $\mathsf{U}_1,\mathsf{U}_2,\dots$. With a slight abuse of notation we continue to use $\mathcal{R}_+^n$ for the corresponding event for Y_k-walk.

Let $y\in\mathbb{N}$ and consider $\sum_n\mathbb{P}_\beta\left(\hat{\mathcal{R}}_+^n;\mathsf{Z}_n=y\right)$. Let $\hat{\mathcal{L}}_y$ be the event that y is a strict ladder height of Z, or equivalently, of Y. Let $\hat{N}(y)$ be the number of strict ladder heights $v\leq y$. Then (see Subsection 7.3 of [25] for more detail),

$$\mathbb{P}_\beta\left(\hat{\mathcal{R}}_+^n;\mathsf{Z}_n=y\right)=\mathbb{P}_\beta\left(\hat{\mathcal{L}}_y;\mathsf{Z}_n=y\right)=\frac{1}{n}\mathbb{E}_\beta\left(\hat{N}(y);\mathsf{Z}_n=y\right)\leq\frac{y}{n}\mathbb{P}\left(\mathsf{Z}_n=y\right).$$

Hence,

$$\sum_n\mathbb{P}_\beta\left(\hat{\mathcal{R}}_+^n;\mathsf{Z}_n=y\right)\leq y\sum_{\ell=1}^{\infty}\mathbb{P}_\beta(\mathsf{Y}_\ell=y)\sum_{n=1}^{\infty}\frac{1}{n}\mathbb{P}_\beta(\mathsf{M}_n=\ell)$$

Now, since M is a process with Bernoulli steps, there is a combinatorial identity:

$$\sum_{n=1}^{\infty}\frac{1}{n}\mathbb{P}_\beta(\mathsf{M}_n=\ell)=\frac{1}{\ell}\sum_{n=\ell}^{\infty}\binom{n-1}{\ell-1}p_\beta^\ell\left(1-p_\beta\right)^{n-\ell}=\frac{1}{\ell}.$$

We claim that there exists $c_1<\infty$ such that

$$\sum_{\ell=1}^{\infty}\frac{1}{\ell}\mathbb{P}_\beta(\mathsf{Y}_\ell=y)\leq\frac{c_1}{y} \tag{104}$$

holds for every $y\in\mathbb{N}$ and all β sufficiently large. Note that (104) would imply that

$$\sum_n\mathbb{P}_\beta\left(\hat{\mathcal{R}}_+^n;\mathsf{Z}_n=y\right)\leq c_1 \tag{105}$$

for all β sufficiently large.

In its turn (104) follows from routine estimates on characteristic functions. Let ϕ_β be the characteristic function of U in (103). By direct computation,

$$\sum_{\ell=1}^{\infty}\frac{1}{\ell}\mathbb{P}_\beta(\mathsf{Y}_\ell=y)=\frac{1}{y}\cdot\frac{1}{2\pi}\int_0^{2\pi}F_\beta(\theta)\left(\sum_1^{y-1}\mathrm{e}^{-\mathrm{i}k\theta}\right)\mathrm{d}\theta=\frac{1}{y}\sum_1^{y-1}\hat{F}_\beta(k),$$

where,

$$F_\beta(\theta)=F_\beta\left(\mathrm{e}^{i\theta}\right)=\frac{1-\mathrm{e}^{i\theta}}{1-\phi_\beta(\theta)}\phi_\beta'(\theta),$$

and $\hat{F}_\beta$ is its Fourier transform. In view of the exponential decay of probabilities in (102), there exists $r > 0$, such that F_β have uniformly bounded (in large β) analytic extension to the complex annulus $\{\mathsf{z} \in \mathbb{C} \ : \ |\mathsf{z}| \in (1, 1+r)\}$. Hence, $\hat{F}_\beta(k)$ tend to zero uniformly exponentially fast, and in particular $\sum_1^\infty \left|\hat{F}_\beta(k)\right|$ is uniformly bounded in β large. (104) follows.

Let us turn to the second term

$$\sum_m \mathbb{E}_\beta\left(\mathrm{e}^{-\nu\beta \mathsf{Z}_m}; \mathcal{R}_+^m\right) = \sum_{y\geq 0} \mathrm{e}^{-\nu\beta y} \sum_m \mathbb{P}_\beta\left(\mathcal{R}_+^m; \mathsf{Z}_m = y\right)$$

in (101). We claim that there exists a finite constant c_2 such that

$$\sum_m \mathbb{P}_\beta\left(\mathcal{R}_+^m; \mathsf{Z}_m = y\right) \leq \frac{c_2(y+1)}{p_\beta}. \tag{106}$$

for all $y \geq 0$ and β large.

Note that (106) would imply that

$$\sum_m \mathbb{E}_\beta\left(\mathrm{e}^{-\nu\beta \mathsf{Z}_m}; \mathcal{R}_+^m\right) \leq \frac{c_3}{p_\beta}.$$

By (101) and (105) this would mean that

$$\sup_{z\geq 0} \sum_n \mathbb{E}_{\beta,z}\left(\mathrm{e}^{-\nu\beta \mathsf{Z}_n}; \mathcal{R}_+^n\right) \leq \frac{c_1 c_3}{\left(1-\mathrm{e}^{-\nu\beta}\right) p_\beta}$$

Since by (102) the order of $p_\beta = \mathrm{O}\left(\mathrm{e}^{-\beta}\right)$, the limit $\lim_{\beta\to\infty} \frac{\mathrm{e}^{-2\chi\beta}}{p_\beta} = 0$ whenever $\chi > 1/2$. Hence (100).

It, therefore, remains to verify (106). Let N_y be the number of non-strict ladder times for heights between 0 and y. Let M_y be the same variable for the random walk Y. Clearly,

$$N_y = \sum_{i=1}^{M_y} \zeta_i,$$

where ζ_i-s are i.i.d. Geo(p_β)-random variables. Proceeding as in the proof of (105), and in particular relying on Subsection 7.1 in [25], we estimate:

$$\begin{aligned}\sum_m \mathbb{P}_\beta\left(\mathcal{R}_+^m; \mathsf{Z}_m = y\right) &\leq \frac{1}{p_\beta} \sum_n \frac{1}{n} \mathbb{E}_\beta\left(M_y; \mathsf{Y}_n = y\right) \\ &\leq \frac{\left(\mathbb{E}_\beta M_y^k\right)^{1/k}}{p_\beta} \sum_n \frac{1}{n} \left(\mathbb{P}_\beta\left(\mathsf{Y}_n = y\right)\right)^{(k-1)/k}.\end{aligned}$$

Applying Lemma 22 in [25] it is straightforward to see that under (102) for any $k \in \mathbb{N}$ there is a finite constant c_k such that

$$\left(\mathbb{E}_\beta M_y^k\right)^{1/k} \leq c_k(y+1)$$

for all $y \geq 0$ and β sufficiently large. On the other hand, again under (102), it is straightforward to check that

$$\max_{y} \mathbb{P}_{\beta}\left(\mathsf{Y}_n = y\right) \lesssim n^{-1/2}.$$

uniformly in n and in β large. Hence (106), and we are done. □

References

1. Biskup, M., Chayes, L., Kotecký, R.: Critical region for droplet formation in the two-dimensional Ising model. Comm. Math. Phys. **242**(1–2), 137–183 (2003)
2. Bodineau, T.: The Wulff construction in three and more dimensions. Comm. Math. Phys. **207**(1), 197–229 (1999)
3. Bodineau, T., Ioffe, D., Velenik, Y.: Rigorous probabilistic analysis of equilibrium crystal shapes. J. Math. Phys. **41**(3), 1033–1098 (2000). Probabilistic techniques in equilibrium and nonequilibrium statistical physics
4. Bodineau, T., Schonmann, R.H., Shlosman, S.: 3D crystal: how flat its flat facets are? Comm. Math. Phys. **255**(3), 747–766 (2005)
5. Bonzel, H.P.: 3D equilibrium crystal shapes in the new light of STM and AFM. Phys. Rep. **385**(1), 1–67 (2003)
6. Bonzel, H.P., Yu, D.K., Scheffler, M.M.: The three-dimensional equilibrium crystal shape of Pb: recent results of theory and experiment. Appl. Phys. A **87**, 391–397 (2007)
7. Bricmont, J., El Mellouki, A., Fröhlich, J.: Random surfaces in statistical mechanics: roughening, rounding, wetting. J. Statist. Phys. **42**(5–6), 743–798 (1986)
8. Bricmont, J., Fontaine, J.-R., Lebowitz, J.L.: Surface tension, percolation, and roughening. J. Statist. Phys. **29**(2), 193–203 (1982)
9. Campanino, M., Ioffe, D., Louidor, O.: Finite connections for supercritical Bernoulli bond percolation in 2D. Markov Process. Relat. Fields **16**(2), 225–266 (2010)
10. Caputo, P., Lubetzky, E., Martinelli, F., Sly, A., Toninelli, L.F.: Dynamics of (2+1)-dimensional SOS surfaces above a wall: slow mixing induced by entropic repulsion. Ann. Probab. **42**(4), 1516–1589 (2014)
11. Caputo, P., Lubetzky, E., Martinelli, F., Sly, A., Toninelli, L.F.: Scaling limit and cube-root fluctuations in SOS surfaces above a wall. J. Eur. Math. Soc. (JEMS) **18**(5), 931–995 (2016)
12. Caputo, P., Martinelli, F., Toninelli, F.L.: On the probability of staying above a wall for the $(2+1)$-dimensional SOS model at low temperature. Probab. Theor. Relat. Fields **163**, 803–831 (2015)
13. Cerf, R.: Large deviations of the finite cluster shape for two-dimensional percolation in the Hausdorff and L^1 metric. J. Theoret. Probab. **13**(2), 491 517 (2000)
14. Cerf, R., Kenyon, R.: The low-temperature expansion of the Wulff crystal in the 3D Ising model. Comm. Math. Phys. **222**(1), 147–179 (2001)
15. Cerf, R., Pisztora, Á.: On the Wulff crystal in the Ising model. Ann. Probab. **28**(3), 947–1017 (2000)
16. Cohn, H., Kenyon, R., Propp, J.: A variational principle for domino tilings. J. Amer. Math. Soc. **14**(2), 297–346 (2001)
17. Dobrushin, R.L.: Gibbs states describing a coexistence of phases for the three-dimensional ising model. Th. Prob. Appl. **17**(3), 582–600 (1972)

18. Dobrushin, R.L., Kotecký, R., Shlosman, S.: Wulff construction, volume 104 of Translations of Mathematical Monographs. American Mathematical Society, Providence, RI (1992) A global shape from local interaction, Translated from the Russian by the authors
19. Dobrushin, R.L., Shlosman, S.B.: Droplet condensation in the Ising model: moderate deviations point of view. In Probability and phase transition (Cambridge, 1993), volume 420 of NATO Adv. Sci. Inst. Ser. C Math. Phys. Sci., pp. 17–34. Kluwer Acadamic Publisation, Dordrecht (1994)
20. Dobrushin, R. L., Shlosman, S. B.: Large and moderate deviations in the Ising model. In: Probability Contributions to Statistical Mechanics, vol. 20 of Adv. Soviet Math., pp. 91–219. Amer. Math. Soc., Providence, RI (1994)
21. Emundts, A., Bonzel, H.P., Wynblatt, P., Thürmer, K., Reutt-Robey, J., Williams, E.D.: Continuous and discontinuous transitions on 3D equilibrium crystal shapes: a new look at Pb and Au. Surf. Sci. **481**(1–3), 13–24 (2001)
22. Ferrari, P.L., Spohn, H.: Step fluctuations for a faceted crystal. J. Statist. Phys. **113**(1–2), 1–46 (2003)
23. Ferrari, P.L., Spohn, H.: Constrained Brownian motion: fluctuations away from circular and parabolic barriers. Ann. Probab. **33**(4), 1302–1325 (2005)
24. Ioffe, D., Shlosman, S.: Ising model fog drip: the first two droplets. In In and out of equilibrium. 2, vol. 60 of Progr. Probab., pp. 365–381. Birkhäuser, Basel (2008)
25. Ioffe, D., Shlosman, S., Toninelli, F.L.: Interaction versus entropic repulsion for low temperature Ising polymers. J. Stat. Phys. **158**(5), 1007–1050 (2015)
26. Ioffe, D., Shlosman, S., Velenik, Y.: An invariance principle to Ferrari-Spohn diffusions. Comm. Math. Phys. **336**(2), 905–932 (2015)
27. Ioffe, D., Velenik, Y.: Ballistic phase of self-interacting random walks. In: Penrose, M., Schwetlick, H., Mörters, P., Moser, R., Zimmer, J. (eds.) Analysis and Stochastics of Growth Processes and Interface Models, pp. 55–79. Oxford University Press, Oxford (2008)
28. Ioffe, D., Velenik, Y.: Low temperature interfaces: prewetting, layering, faceting and Ferrari-Spohn diffusions. Markov Process. Relat. Fields **24**, 487–537 (2018)
29. Ioffe, D., Velenik, Y., Wachtel, V.: Dyson Ferrari-Spohn diffusions and ordered walks under area tilts. Probab. Theor. Relat. Fields **170**(1), 11–47 (2017)
30. Kenyon, R.: Height fluctuations in the honeycomb dimer model. Comm. Math. Phys. **281**(3), 675–709 (2008)
31. Miracle-Sole, S.: Facet shapes in a Wulff crystal. In: Mathematical Results in Statistical Mechanics (Marseilles, 1998), pp. 83–101. World Scientific Publishing, River Edge, NJ (1999)
32. Okounkov, A.: Limit shapes real and imagined. Bull. Amer. Math. Soc. (N.S.) **53**(2), 187–216 (2016)
33. Schonmann, R.H., Shlosman, S.B.: Constrained variational problem with applications to the Ising model. J. Statist. Phys. **83**(5–6), 867–905 (1996)

Lectures on the Spin and Loop $O(n)$ Models

Ron Peled(✉) and Yinon Spinka

School of Mathematical Sciences, Tel Aviv University, Tel Aviv, Israel
{peledron,yinonspi}@post.tau.ac.il

Dedicated to Chuck Newman on the occasion of his 70th birthday

Abstract. The classical spin $O(n)$ model is a model on a d-dimensional lattice in which a vector on the $(n-1)$-dimensional sphere is assigned to every lattice site and the vectors at adjacent sites interact ferromagnetically via their inner product. Special cases include the Ising model ($n=1$), the XY model ($n=2$) and the Heisenberg model ($n=3$). We discuss questions of long-range order and decay of correlations in the spin $O(n)$ model for different combinations of the lattice dimension d and the number of spin components n.

The loop $O(n)$ model is a model for a random configuration of disjoint loops. We discuss its properties on the hexagonal lattice. The model is parameterized by a loop weight $n \geq 0$ and an edge weight $x \geq 0$. Special cases include self-avoiding walk ($n=0$), the Ising model ($n=1$), critical percolation ($n=x=1$), dimer model ($n=1, x=\infty$), proper 4-coloring ($n=2, x=\infty$), integer-valued ($n=2$) and tree-valued (integer $n >= 3$) Lipschitz functions and the hard hexagon model ($n=\infty$). The object of study in the model is the typical structure of loops. We review the connection of the model with the spin $O(n)$ model and discuss its conjectured phase diagram, emphasizing the many open problems remaining.

Keywords: Spin $O(n)$ model · Loop $O(n)$ model · Ising model · XY model · Heisenberg model · Phase transitions · Spontaneous magnetization · Symmetry breaking · Decay of correlations · Mermin–Wagner · Berezinskii–Kosterlitz–Thouless transition · Reflection positivity · Chessboard estimate · Infra-red bound · Gaussian domination · Graphical representation · Conformal loop ensemble · Schramm–Loewner evolution · Self-avoiding walk · Dilute

R. Peled and Y. Spinka—Research supported by Israeli Science Foundation grant 861/15 and the European Research Council starting grant 678520 (LocalOrder).
Y. Spinka—Research supported by the Adams Fellowship Program of the Israel Academy of Sciences and Humanities.

V. Sidoravicius (Ed.): *Sojourns in Probability Theory and Statistical Physics - I*, PROMS 298, pp. 246–320, 2019.
https://doi.org/10.1007/978-981-15-0294-1_10

potts model · Lipschitz functions · Hard hexagon model · Critical percolation on the triangular lattice · Dimer model · Proper 4-coloring of the triangular lattice · Macroscopic loops · Microscopic loops

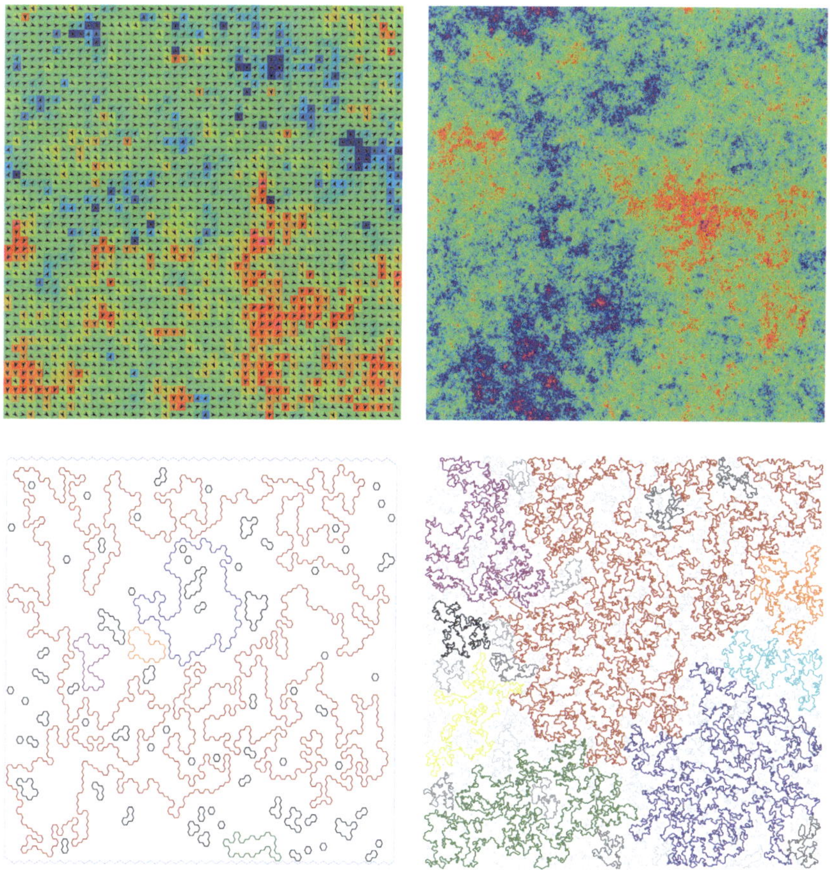

1 Introduction

The classical spin $O(n)$ model is a model on a d-dimensional lattice in which a vector on the $(n-1)$-dimensional sphere is assigned to every lattice site and the vectors at adjacent sites interact ferromagnetically via their inner product. Special cases include the Ising model ($n = 1$), the XY model ($n = 2$) and the Heisenberg model ($n = 3$). We discuss questions of long-range order (spontaneous magnetization) and decay of correlations in the spin $O(n)$ model for different combinations of the lattice dimension d and the number of spin components n. Among the topics presented are the Mermin–Wagner theorem, the Berezinskii–Kosterlitz–Thouless transition, the infra-red bound and Polyakov's conjecture on the two-dimensional Heisenberg model.

The loop $O(n)$ model is a model for a random configuration of disjoint loops. In these notes we discuss its properties on the hexagonal lattice. The model is parameterized by a loop weight $n \geq 0$ and an edge weight $x \geq 0$. Special cases include self-avoiding walk ($n = 0$), the Ising model ($n = 1$), critical percolation ($n = x = 1$), dimer model ($n = 1, x = \infty$), proper 4-coloring ($n = 2, x = \infty$), integer-valued ($n = 2$) and tree-valued (integer $n >= 3$) Lipschitz functions and the hard hexagon model ($n = \infty$). The object of study in the model is the typical structure of loops. We will review the connection of the model with the spin $O(n)$ model and discuss its conjectured phase diagram, emphasizing the many open problems remaining. We then elaborate on recent results for the self-avoiding walk case and for large values of n.

The first version of these notes was written for a series of lectures given at the School and Workshop on Random Interacting Systems at Bath, England in June 2016. The authors are grateful to Vladas Sidoravicius and Alexandre Stauffer for the organization of the school and for the opportunity to present this material there. It is a pleasure to thank also the participants of the meeting for various comments which greatly enhanced the quality of the notes.

Our discussion is aimed at giving a relatively short and accessible introduction to the topics of the spin $O(n)$ and loop $O(n)$ models. The selection of topics naturally reflects the authors' specific research interests and this is perhaps most noticeable in the sections on the Mermin–Wagner theorem (Sect. 2.6), the infrared bound (Sect. 2.7) and the chapter on the loop $O(n)$ model (Sect. 3). The interested reader may find additional information in the recent books of Friedli and Velenik [50] and Duminil-Copin [38] and in the lecture notes of Bauerschmidt [10], Biskup [19] and Ueltschi [119].

2 The Spin $O(n)$ Model

2.1 Definitions

Let $n \geq 1$ be an integer and let $G = (V(G), E(G))$ be a finite graph. A *configuration* of the *spin $O(n)$ model*, sometimes called the *n-vector model*, on G is an assignment $\sigma : V(G) \to \mathbb{S}^{n-1}$ of spins to each vertex of G, where $\mathbb{S}^{n-1} \subseteq \mathbb{R}^n$ is the $(n-1)$-dimensional unit sphere (simply $\{-1, 1\}$ if $n = 1$). We write

$$\Omega := (\mathbb{S}^{n-1})^{V(G)}$$

for the space of configurations. At inverse temperature $\beta \in [0, \infty)$, configurations are randomly chosen from the probability measure $\mu_{G,n,\beta}$ given by

$$d\mu_{G,n,\beta}(\sigma) := \frac{1}{Z^{\text{spin}}_{G,n,\beta}} \exp\left[\beta \sum_{\{u,v\} \in E(G)} \langle \sigma_u, \sigma_v \rangle\right] d\sigma, \tag{1}$$

where $\langle \cdot, \cdot \rangle$ denotes the standard inner product in $\mathbb{R}^n$, the *partition function* $Z^{\text{spin}}_{G,n,\beta}$ is given by

$$Z^{\text{spin}}_{G,n,\beta} := \int_{\Omega} \exp\left[\beta \sum_{\{u,v\}\in E(G)} \langle \sigma_u, \sigma_v \rangle\right] d\sigma \tag{2}$$

and $d\sigma$ is the uniform probability measure on Ω (i.e., the product measure of the uniform distributions on $\mathbb{S}^{n-1}$ for each vertex in G).

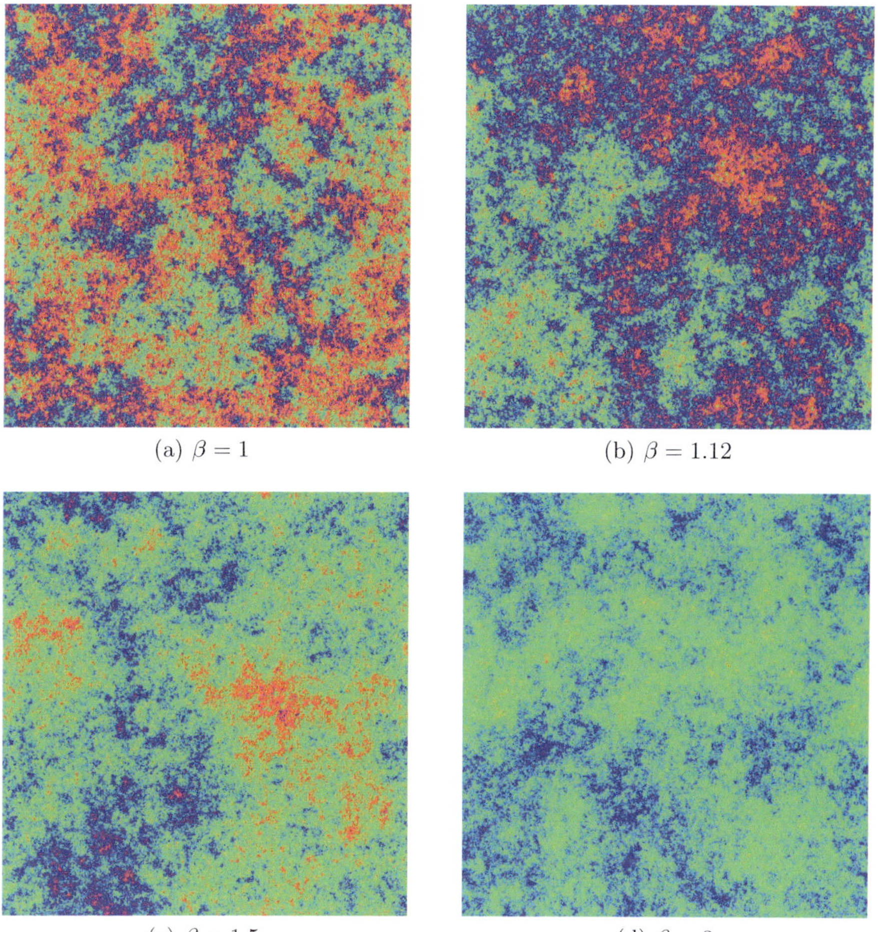

(a) $\beta = 1$

(b) $\beta = 1.12$

(c) $\beta = 1.5$

(d) $\beta = 3$

Fig. 1. Samples of random spin configurations in the two-dimensional XY model ($n = 2$) at and near the conjectured critical inverse temperature $\beta_c \approx 1.1199$ [67,80]. Configurations are on a 500×500 torus. The angles of the spins are encoded by colors, with $0°$, $120°$ and $240°$ having colors green, blue and red, and interpolating in between. The samples are generated using Wolff's cluster algorithm [120].

Special cases of the model have names of their own:

- When $n = 1$, spins take values in $\{-1, 1\}$ and the model becomes the famous *Ising model.* See Fig. 2 for samples from this model.
- When $n = 2$, spins take values in the unit circle and the model is called the *XY model* or the *plane rotator model.* See Fig. 1 for samples from this model. See also the two top figures on the cover page which show samples of the XY model with $\beta = 1.5$.
- When $n = 3$, spins take values in the two-dimensional sphere $\mathbb{S}^2$ and the model is called the *Heisenberg model.* See Fig. 3 for samples from this model.
- In a sense, as n tends to infinity the model approaches the *Berlin–Kac spherical model* (which will not be discussed in these notes), see [18, 76, 113] and [13, Chapter 5].

We will sometimes discuss a more general model, in which we replace the inner product in (1) by a function of that inner product. In other words, when the energy of a configuration is measured using a more general pair interaction term. Precisely, given a measurable function $U : [-1, 1] \to \mathbb{R} \cup \{\infty\}$, termed the *potential function*, we define the *spin $O(n)$ model with potential U* to be the probability measure $\mu_{G,n,U}$ over configurations $\sigma : V(G) \to \mathbb{S}^{n-1}$ given by

$$d\mu_{G,n,U}(\sigma) := \frac{1}{Z^{\text{spin}}_{G,n,U}} \exp\left[-\sum_{\{u,v\}\in E(G)} U(\langle \sigma_u, \sigma_v\rangle)\right] d\sigma, \tag{3}$$

where the partition function $Z^{\text{spin}}_{G,n,U}$ is defined analogously to (2) and where we set $\exp(-\infty) := 0$. Of course, for this to be well defined (i.e., to have finite $Z^{\text{spin}}_{G,n,U}$) some restrictions need to be placed on U but this will always be the case in the models discussed in these notes.

The spin $O(n)$ model defined in (1) with $\beta \in [0, \infty)$ is called *ferromagnetic.* If β is taken negative in (1), equivalently $U(r) = \beta r$ for $\beta > 0$ in (3), the model is called *anti-ferromagnetic.* On bipartite graphs, the ferromagnetic and anti-ferromagnetic versions are isomorphic through the map which sends σ_v to $-\sigma_v$ for all v in one of the partite classes. The two versions are genuinely different on non-bipartite graphs; see Sects. 3.1 and 3.3 for a discussion of the Ising model on the triangular lattice.

The model admits many extensions and generalizations. One may impose *boundary conditions* in which the values of certain spins are pre-specified. An *external magnetic field* can be applied by taking a vector $s \in \mathbb{R}^n$ and adding a term of the form $\sum_{v\in V(G)} \langle \sigma_v, s\rangle$ to the exponent in the definition of the densities (1) and (3). The model can be made *anisotropic* by replacing the standard inner product $\langle \cdot, \cdot \rangle$ in (1) and (3) with a different inner product. A different *single-site distribution* may be imposed, replacing the measure $d\sigma$ in (1) and (3) with another product measure on the vertices of G, thus allowing spins to take values in all of $\mathbb{R}^n$ (e.g., taking the single-site density $\exp(-|\sigma_v|^4)$). We will, however, focus on the versions of the model described above.

The graph G is typically taken to be a portion of a d-dimensional lattice, possibly with periodic boundary conditions. When discussing the spin $O(n)$ model in these notes we mostly take

$$G = \mathbb{T}_L^d,$$

where $\mathbb{T}_L^d$ denotes the *d-dimensional discrete torus of side length* $2L$ defined as follows: The vertex set of $\mathbb{T}_L^d$ is

$$V(\mathbb{T}_L^d) := \{-L+1, -L+2, \ldots, L-1, L\}^d \tag{4}$$

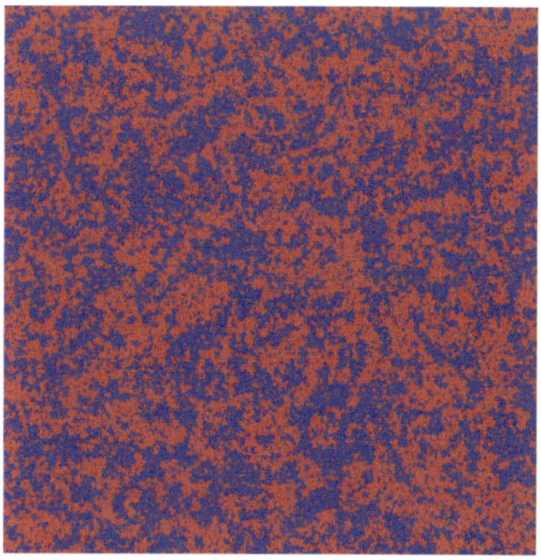

(a) $\beta = 0.4 < \beta_c$

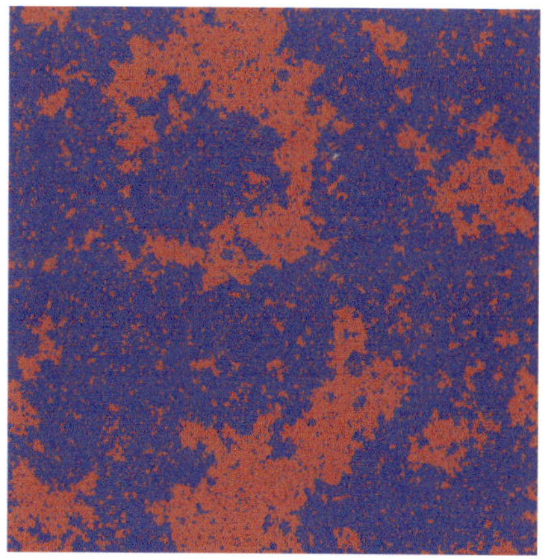

(b) $\beta = \beta_c \approx 0.4407$

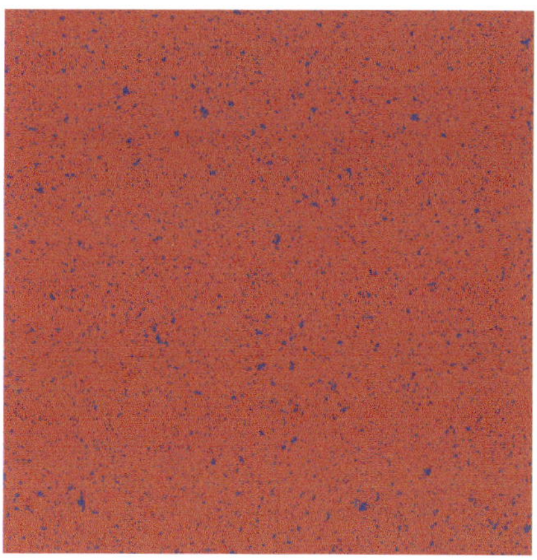

(c) $\beta = 0.5 > \beta_c$

(d) $\beta = 0.5$ with Dobrushin boundary conditions

Fig. 2. Samples of random configurations in the two-dimensional Ising model ($n = 1$) at and near the critical inverse temperature $\beta_c = \frac{1}{2}\log(1+\sqrt{2})$. Configurations are on a 500×500 torus and are generated using Wolff's cluster algorithm [120]. Dobrushin boundary conditions corresponds to fixing the top and bottom halves of the boundary to have different spins.

and a pair $u, v \in V(\mathbb{T}_L^d)$ is adjacent, written $\{u, v\} \in E(\mathbb{T}_L^d)$, if u and v are equal in all but one coordinate and differ by exactly 1 modulo $2L$ in that coordinate. We write $\|x - y\|_1$ for the *graph distance in* $\mathbb{T}_L^d$ of two vertices $x, y \in V(\mathbb{T}_L^d)$ (for brevity, we suppress the dependence on L in this notation).

The results presented below should admit analogues if the graph G is changed to a different d-dimensional lattice graph with appropriate boundary conditions. However, the presented proofs sometimes require the presence of symmetries in the graph G.

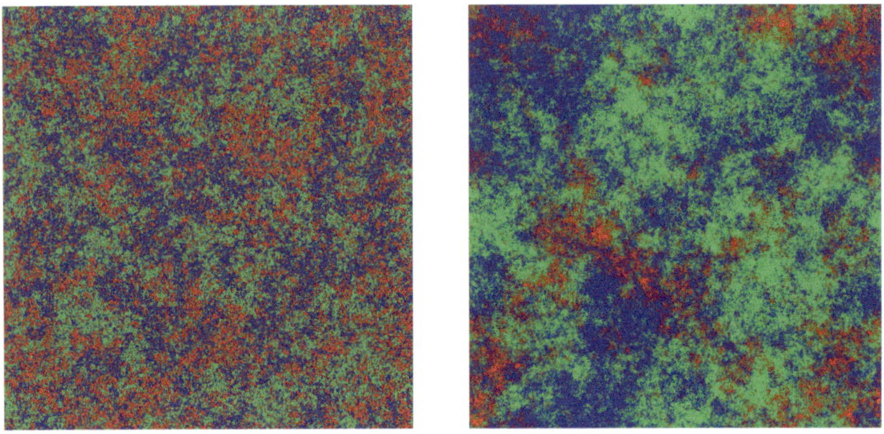

(a) $\beta = 2$ (b) $\beta = 10$

Fig. 3. Samples of random configurations in the two-dimensional Heisenberg model ($n = 3$). Configurations are on a 500×500 torus and are generated using Wolff's cluster algorithm [120]. It is predicted [101] that there is no phase transition for $d = 2$ and $n \geq 3$ so that correlations decay exponentially at any inverse temperature.

2.2 Main Results and Conjectures

We will focus on the questions of existence of long-range order and decay of correlations in the spin $O(n)$ model. To this end we shall study the correlation

$$\rho_{x,y} := \mathbb{E}(\langle \sigma_x, \sigma_y \rangle)$$

for a configuration σ randomly chosen from $\mu_{\mathbb{T}_L^d, n, \beta}$, the (ferromagnetic) spin $O(n)$ model at inverse temperature $\beta \in [0, \infty)$, and two vertices $x, y \in V(\mathbb{T}_L^d)$ with large graph distance $\|x - y\|_1$. The magnitude of this correlation behaves very differently for different combinations of the spatial dimension d, number of spin components n and inverse temperature β. The following list summarizes the main results and conjectures regarding $\rho_{x,y}$. Most of the claims in the list are elaborated upon and proved in the subsequent sections. We use the notation $c_\beta, C_\beta, c_{n,\beta}, \ldots$ to denote positive constants whose value depends only on the parameters given in the subscript (and is always independent of the lattice size L) and may change from line to line.

Non-negativity and Monotonicity. The correlation is always non-negative, that is,

$$d, n \geq 1,\ \beta \in [0, \infty): \quad \rho_{x,y} \geq 0 \quad \text{for all } x, y \in V(\mathbb{T}_L^d).$$

As we shall discuss, this result is a special case of an inequality of Griffiths [63]. It is also natural to expect the correlation to be monotonic non-decreasing in β. A second inequality of Griffiths [63] implies this for the Ising model and was later extended by Ginibre [60] to include the XY model and more general settings. Precisely,

$$d \geq 1,\ n \in \{1, 2\}: \quad \begin{array}{l}\text{for all } x, y \in V(\mathbb{T}_L^d), \rho_{x,y} \text{ is non-decreasing}\\ \text{as } \beta \text{ increases in} [0, \infty).\end{array}$$

It appears to be unknown whether this monotonicity holds also for $n \geq 3$. Counterexamples exist for related inequalities in certain quantum [71] and classical [114] spin systems.

High Temperatures and Spatial Dimension $d = 1$. All the models exhibit *exponential decay of correlations* at high temperature. Precisely, there exists a $\beta_0(d, n) > 0$ such that

$$d, n \geq 1,\ \beta < \beta_0(d, n):$$
$$\rho_{x,y} \leq C_{d,n,\beta} \exp(-c_{d,n,\beta}\|x - y\|_1) \text{ for all } x, y \in V(\mathbb{T}_L^d). \quad (5)$$

This is a relatively simple fact and the main interest is in understanding the behavior at low temperatures. In one spatial dimension ($d = 1$) the exponential decay persists at all positive temperatures. That is,

$$d = 1,\ n \geq 1,\ \beta \in [0, \infty):$$
$$\rho_{x,y} \leq C_{n,\beta} \exp(-c_{n,\beta}\|x - y\|_1) \text{ for all } x, y \in V(\mathbb{T}_L^1). \quad (6)$$

The Ising Model $n = 1$. The Ising model exhibits a *phase transition* in all dimensions $d \geq 2$ at a critical inverse temperature $\beta_c(d)$. The transition is from a regime with exponential decay of correlations [1,2,4,42,45][1],

$$d \geq 2,\ n = 1,\ \beta < \beta_c(d)\colon \rho_{x,y} \leq C_{d,\beta} \exp(-c_{d,\beta}\|x - y\|_1) \text{ for all } x, y \in V(\mathbb{T}_L^d)$$

to a regime with *long-range order*, or *spontaneous magnetization*, which is characterized by

$$d \geq 2,\ n = 1,\ \beta > \beta_c(d): \quad \rho_{x,y} \geq c_{d,\beta} \quad \text{for all } x, y \in V(\mathbb{T}_L^d).$$

[1] Exponential decay is stated in these references in the infinite-volume limit, but is derived as a consequence of a finite-volume criterion and is thus implied, as the infinite-volume measure is unique, also in finite volume.

The behavior of the model at the critical temperature, when $\beta = \beta_c(d)$, is a rich source of study with many mathematical features. For instance, the two-dimensional model is exactly solvable, as discovered by Onsager [95], and has a conformally-invariant scaling limit, features of which were first established by Smirnov [111,112]; see [16,30–32] and references within for recent progress. We mention that it is proved (see Aizenman, Duminil-Copin, Sidoravicius [6] and references within) that the model does not exhibit long-range order at its critical point in all dimensions $d \geq 2$. Moreover, in dimension $d = 2$ it is known [87,100] (see also [31]) that correlations *decay as a power-law* with exponent 1/4 at the critical point, whose exact value is $\beta_c(2) = \frac{1}{2}\log(1+\sqrt{2})$ as first determined by Kramers–Wannier [83] and Onsager [95],

$$d = 2,\ n = 1,\ \beta = \beta_c(2) \colon \mathbb{E}^{\mathbb{Z}^2}(\sigma_x \sigma_y) \sim C\|x-y\|_2^{-\frac{1}{4}}, x, y \in \mathbb{Z}^2, \|x-y\|_2 \to \infty,$$

where we write $\mathbb{E}^{\mathbb{Z}^2}$ for the expectation in the (unique) infinite-volume measure of the two-dimensional critical Ising model, and $\|\cdot\|_2$ denotes the standard Euclidean norm. Lastly, in dimensions higher than some threshold d_0, Sakai [103] proved that

$$d \geq d_0,\ n = 1,\ \beta = \beta_c(d) \colon$$
$$\mathbb{E}^{\mathbb{Z}^d}(\sigma_x \sigma_y) \sim C_d\|x-y\|_2^{-(d-2)},\ x, y \in \mathbb{Z}^d,\ \|x-y\|_2 \to \infty,$$

where, as before, $\mathbb{E}^{\mathbb{Z}^d}$ is the expectation in the (unique) infinite-volume measure of the d-dimensional critical Ising model.

The study of the model at or near its critical temperature is beyond the scope of these notes.

The Mermin–Wagner Theorem: No Continuous Symmetry Breaking in 2*d*. Perhaps surprisingly, the behavior of the two-dimensional model when $n \geq 2$, so that the spin space $\mathbb{S}^{n-1}$ has a continuous symmetry, is quite different from that of the Ising model. The *Mermin–Wagner theorem* [88,89] asserts that in this case there is no phase with long-range order at any inverse temperature β. Quantifying the rate at which correlations decay has been the focus of much research along the years [21,36,54,57,69,72,73,75,90,92,99,101,106,107,109] and is still not completely understood. Improving on earlier bounds, McBryan and Spencer [86] showed in 1977 that the decay occurs at least at a power-law rate,

$$d = 2,\ n \geq 2,\ \beta \in [0, \infty) \colon \quad \rho_{x,y} \leq C_{n,\beta}\|x-y\|_1^{-c_{n,\beta}} \quad \text{for all } x, y \in V(\mathbb{T}_L^2). \tag{7}$$

The sharpness of this bound is discussed in the next paragraphs.

The Berezinskii–Kosterlitz–Thouless Transition for the 2*d* XY Model. It was predicted by Berezinskii [17] and by Kosterlitz and Thouless [81,82] that the XY model ($n = 2$) in two spatial dimensions should indeed exhibit power-law decay of correlations at low temperatures. Thus the model undergoes a phase transition (of a different nature than that of the Ising model) from a phase with exponential decay of correlations to a phase with power-law decay of correlations. This transition is called the Berezinskii–Kosterlitz–Thouless transition. The existence of the transition has been proved mathematically in the celebrated work of Fröhlich and Spencer [54], who show that there exists a β_1 for which

$$d = 2,\ n = 2,\ \beta > \beta_1 : \mathbb{E}^{\mathbb{Z}^2}(\langle \sigma_x, \sigma_y \rangle) \geq c_\beta \|x - y\|_1^{-C_\beta} \text{ for all distinct } x, y \in \mathbb{Z}^2, \tag{8}$$

where $\mathbb{E}^{\mathbb{Z}^2}$ denotes expectation in the unique [22] translation-invariant infinite-volume Gibbs measure of the two-dimensional XY model at inverse temperature β.

A rigorous proof of the bound (8) is beyond the scope of these notes (see [79] for a recent presentation of the proof). In Sect. 2.8 we present a heuristic discussion of the transition highlighting the role of *vortices* - cycles of length 4 in $\mathbb{T}_L^2$ on which the configuration completes a full rotation. We then proceed to present a beautiful result of Aizenman [3], following Patrascioiu and Seiler [96], who showed that correlations decay at most as fast as a power-law in the spin $O(2)$ model with potential U, for certain potentials U for which vortices are deterministically excluded.

Polyakov's Conjecture for the $2d$ Heisenberg Model. Polyakov [101] predicted in 1975 that the spin $O(n)$ model with $n \geq 3$ should exhibit *exponential* decay of correlations in two dimensions at *any* positive temperature. That is, that there is no phase transition of the Berezinskii–Kosterlitz–Thouless type in the Heisenberg model and in the spin $O(n)$ models with larger n. On the torus, this prediction may be stated precisely as

$$d = 2,\ n \geq 3,\ \beta \in [0, \infty) : \rho_{x,y} \leq C_{n,\beta} \exp(-c_{n,\beta} \|x - y\|_1) \text{ for all } x, y \in V(\mathbb{T}_L^2).$$

Giving a mathematical proof of this statement (or its analog in infinite volume) remains one of the major challenges of the subject. The best known results in this direction are by Kupiainen [84] who performed a $1/n$-expansion as n tends to infinity.

The Infra-red Bound: Long-Range Order in Dimensions $d \geq 3$. In three and higher spatial dimensions, the spin $O(n)$ model exhibits long-range order at sufficiently low temperatures for all n. This was established by Fröhlich, Simon and Spencer [53] in 1976 who introduced the powerful method of the *infra-red bound*, and applied it to the analysis of the spin $O(n)$ and other models. They prove that correlations do not decay at temperatures below a threshold $\beta_1(d, n)^{-1}$, at least in the following averaged sense,

$$d \geq 3,\ n \geq 1,\ \beta > \beta_1(d, n) : \quad \frac{1}{|V(\mathbb{T}_L^d)|^2} \sum_{x,y \in V(\mathbb{T}_L^d)} \rho_{x,y} \geq c_{d,n,\beta}.$$

The proof uses the reflection symmetries of the underlying lattice, relying on the tool of *reflection positivity*.

2.3 Non-negativity and Monotonicity of Correlations

In this section we discuss the non-negativity and monotonicity in temperature of the correlations $\rho_{x,y} = \mathbb{E}(\langle \sigma_x, \sigma_y \rangle)$. To remain with a unified presentation, our discussion is restricted to the simplest setup with nearest-neighbor interactions. Many extensions are available in the literature. Recent accounts can be found in the book of Friedli and Velenik [50, Sections 3.6, 3.8 and 3.9] and in the review of Benassi–Lees–Ueltschi [15].

We start our discussion by introducing the spin $O(n)$ model with general non-negative coupling constants. Let $N \geq 1$ be an integer and let $J = (J_{\{i,j\}})_{1 \leq i < j \leq N}$ be non-negative real numbers. The spin $O(n)$ model with coupling constants J is the probability measure on $(\mathbb{S}^{n-1})^N$ defined by

$$d\mu_{n,J}(\sigma) := \frac{1}{Z^{\text{spin}}_{n,J}} \exp\left[\sum_{1 \leq i < j \leq N} J_{\{i,j\}} \langle \sigma_i, \sigma_j \rangle\right] d\sigma, \tag{9}$$

where, as before, $d\sigma$ is the uniform probability measure on $(\mathbb{S}^{n-1})^N$, $Z^{\text{spin}}_{n,J}$ is chosen to normalize $\mu_{n,J}$ to be a probability measure and we refer to the case $n = 1$ as the Ising model. When we speak about the spin $O(n)$ model on a finite graph $G = (V(G), E(G))$ with coupling constants $J = (J_{\{u,v\}})_{\{u,v\} \in E(G)}$, it should be understood that $N = |V(G)|$, that the vertex-set $V(G)$ is identified with $\{1, \ldots, N\}$ and that $J_{\{i,j\}} = 0$ for $\{i, j\} \notin E(G)$. Thus, the standard spin $O(n)$ model (1) on G at inverse temperature β is obtained as the special case in which $J_{\{u,v\}} = \beta$ for $\{u, v\} \in E(G)$.

The following non-negativity result is a special case of Griffiths' first inequality [63].

Theorem 1. *Let $N \geq 1$ be an integer and let $J = (J_{\{i,j\}})_{1 \leq i < j \leq N}$ be non-negative. If σ is sampled from the Ising model with coupling constants J then*

$$\mathbb{E}\left(\prod_{x \in A} \sigma_x\right) \geq 0 \quad \textit{for all } A \subset \{1, \ldots, N\}.$$

Proof. By definition,

$$\mathbb{E}\left(\prod_{x \in A} \sigma_x\right) = \frac{1}{2^N Z^{\text{spin}}_{1,J}} \sum_{\sigma \in \{-1,1\}^N} \left(\prod_{x \in A} \sigma_x\right) \exp\left[\sum_{1 \leq i < j \leq N} J_{\{i,j\}} \sigma_i \sigma_j\right].$$

Using the Taylor expansion $e^t = \sum_{m=0}^{\infty} \frac{t^m}{m!}$, we conclude that $\mathbb{E}(\sigma_x \sigma_y)$ is an absolutely convergent series with *non-negative* coefficients of products of the values of σ on various vertices. That is,

$$\mathbb{E}\left(\prod_{x \in A} \sigma_x\right) = \sum_{\sigma \in \{-1,1\}^N} \sum_{m \in \{0,1,2,\ldots\}^N} C_m \prod_{1 \leq i \leq N} \sigma_i^{m_i},$$

where each $C_m = C_m(A) \geq 0$ and the series is absolutely convergent (in addition, one may, in fact, restrict to $m \in \{0,1\}^N$ as when $\varepsilon \in \{-1,1\}$ we have $\varepsilon^k = \varepsilon$ or $\varepsilon^k = 1$ according to the parity of k). The non-negativity of $\mathbb{E}(\prod_{x\in A}\sigma_x)$ now follows as, for each $m \in \{0,1,2,\ldots\}^N$,

$$\sum_{\sigma\in\{-1,1\}^N}\prod_{1\leq i\leq N}\sigma_i^{m_i} = \prod_{1\leq i\leq N}((-1)^{m_i}+1^{m_i}) = \begin{cases}2^N & m_i \text{is even for all } i\\ 0 & \text{otherwise}\end{cases}. \qquad \square$$

Exercise. Give an alternative proof of Theorem 1 by extending the derivation of the Edwards–Sokal coupling in Sect. 2.4 below to the Ising model with general non-negative coupling constants and arguing similarly to Remark 1.

We now deduce non-negativity of correlations for the spin $O(n)$ models with $n \geq 2$ by showing that conditioning on $n-1$ spin components induces an Ising model with non-negative coupling constants on the sign of the remaining spin component. The argument applies to spin $O(n)$ models with potential $U : [-1,1] \to \mathbb{R}\cup\{\infty\}$ (see (3)) as long as the potential is *non-increasing* in the sense that

$$U(r_1) \geq U(r_2) \text{ when } r_1 \leq r_2. \tag{10}$$

This property implies that configurations in which adjacent spins are more aligned (i.e., have larger inner product) have higher density, a characteristic of ferromagnets.

To state the above precisely, we embed $\mathbb{S}^{n-1}$ into $\mathbb{R}^n$ so as to allow writing the coordinates of a configuration $\sigma : V(G) \to \mathbb{S}^{n-1}$ explicitly as

$$\sigma_v = (\sigma_v^1, \sigma_v^2, \ldots, \sigma_v^n) \quad \text{at each vertex } v \in V(G).$$

For $1 \leq j \leq n$, we write σ^j for the function (σ_v^j), $v \in V(G)$. We also introduce a function $\varepsilon : V(G) \to \{-1,1\}$ defined uniquely by $\sigma_v^1 = |\sigma_v^1|\varepsilon_v$ (when $\sigma_v^1 = 0$, we arbitrarily set $\varepsilon_v := 0$). We note that σ is determined by $(\varepsilon, \sigma^2, \ldots, \sigma^n)$ since $\sigma_v^1 = \varepsilon_v|\sigma_v^1|$ and $|\sigma_v^1|$ is determined from $(\sigma_v^j)_{2\leq j\leq n}$ as $\sigma_v \in \mathbb{S}^{n-1}$.

Theorem 2. *Let $n \geq 2$, $G = (V(G), E(G))$ be a finite graph and let $U : [-1,1] \to \mathbb{R}\cup\{\infty\}$ be* non-increasing. *If σ is sampled from the spin $O(n)$ model on G with potential U, then, conditioned on $(\sigma^2, \sigma^3, \ldots, \sigma^n)$, the random signs ε are distributed as an Ising model on G with coupling constants J given by*

$$J_{\{u,v\}} := -\frac{1}{2}U\left(|\sigma_u^1|\cdot|\sigma_v^1| + \sum_{j=2}^n \sigma_u^j\sigma_v^j\right) + \frac{1}{2}U\left(-|\sigma_u^1|\cdot|\sigma_v^1| + \sum_{j=2}^n \sigma_u^j\sigma_v^j\right).$$

In particular, the coupling constants are non-negative so that for all $x, y \in V(G)$,

$$\mathbb{E}(\langle\sigma_x,\sigma_y\rangle) \geq 0 \quad \textit{and} \quad \mathbb{E}\left(\varepsilon_x\varepsilon_y \mid (\sigma^j)_{2\leq j\leq n}\right) \geq 0 \quad \textit{almost surely.}$$

Proof. Observe that the density of ε conditioned on $(\sigma^j)_{2\le j\le n}$ (with respect to the uniform measure on $\{-1,1\}^{V(G)}$) is proportional to

$$\begin{aligned}
&\exp\left[-\sum_{\{u,v\}\in E(G)} U(\langle\sigma_u,\sigma_v\rangle)\right]\\
&= \exp\left[-\sum_{\{u,v\}\in E(G)} U\left(|\sigma_u^1|\cdot|\sigma_v^1|\varepsilon_u\varepsilon_v + \sum_{j=2}^n \sigma_u^j\sigma_v^j\right)\right]\\
&= \exp\left[\sum_{\{u,v\}\in E(G)} \left(J_{\{u,v\}}\varepsilon_u\varepsilon_v + I_{\{u,v\}}\right)\right]\\
&= I\cdot\exp\left[\sum_{\{u,v\}\in E(G)} J_{\{u,v\}}\varepsilon_u\varepsilon_v\right],
\end{aligned}$$

where $I_{\{u,v\}}$ and I are measurable with respect to $(\sigma^j)_{2\le j\le n}$. We conclude that, conditioned on $(\sigma^j)_{2\le j\le n}$, the signs ε are distributed as an Ising model on G with coupling constants $J=(J_{\{u,v\}})_{\{u,v\}\in E(G)}$.

By the assumption that U is non-increasing, the coupling constants are almost surely non-negative. Thus, Theorem 1 implies that

$$\mathbb{E}\left(\varepsilon_x\varepsilon_y \mid (\sigma^j)_{2\le j\le n}\right)\ge 0 \quad \text{almost surely for every } x,y\in V(G).$$

Finally, to see that $\mathbb{E}(\langle\sigma_x,\sigma_y\rangle)\ge 0$, note that

$$\mathbb{E}(\langle\sigma_x,\sigma_y\rangle) = \mathbb{E}\left(\sum_{j=1}^n \sigma_x^j\sigma_y^j\right) = n\,\mathbb{E}\left(\sigma_x^1\sigma_y^1\right),$$

as the distribution of σ is invariant to global rotations (that is, for any $n\times n$ orthogonal matrix O, σ has the same distribution as $(O\sigma_v)$, $v\in V(G)$, by the choice of density (3)). In particular,

$$\begin{aligned}
\mathbb{E}(\langle\sigma_x,\sigma_y\rangle) &= n\,\mathbb{E}\left(\mathbb{E}\left(\sigma_x^1\sigma_y^1 \mid (\sigma^j)_{2\le j\le n}\right)\right)\\
&= n\,\mathbb{E}\left(|\sigma_x^1|\cdot|\sigma_y^1|\cdot\mathbb{E}\left(\varepsilon_x\varepsilon_y \mid (\sigma^j)_{2\le j\le n}\right)\right)\ge 0.
\end{aligned}$$

□

We remark that Theorem 2 and its proof may be extended in a straightforward manner to the case that different non-increasing potentials are placed on different edges of the graph.

As another remark, we note that the non-negativity of $\mathbb{E}(\langle\sigma_x,\sigma_y\rangle)$ asserted by Theorem 2 may fail for potentials which are not non-increasing. For instance, the discussion of the anti-ferromagnetic spin $O(n)$ model in Sect. 2.1 shows that, on bipartite graphs G and with x and y on different bipartition classes, the sign of $\mathbb{E}(\langle\sigma_x,\sigma_y\rangle)$ in the spin $O(n)$ model is reversed when replacing β by $-\beta$ in (1). A similar remark applies to the assertion of Theorem 1 when some of the coupling constants are negative.

Lastly, we mention that the assumptions of Theorem 2 imply a stronger conclusion than the non-negativity of $\mathbb{E}(\langle\sigma_x, \sigma_y\rangle)$. In [33] it is shown that conditioned on σ_x, there is a version of the density of σ_y (with respect to the uniform measure on $\mathbb{S}^{n-1}$) which is a non-decreasing function of $\langle\sigma_x, \sigma_y\rangle$.

We move now to discuss the monotonicity of correlations with the inverse temperature β in the spin $O(n)$ model. This was first established by Griffiths for the Ising case [63] and is sometimes referred to as Griffiths' second inequality. It was established by Ginibre [60] for the XY case (the case $n = 2$) and in more general settings. Establishing or refuting such monotonicity when $n \geq 3$ is an open problem of significant interest.

We again work in the generality of the spin $O(n)$ model with non-negative coupling constants.

Theorem 3. *Let $n \in \{1, 2\}$, let $N \geq 1$ be an integer and let $J = (J_{\{i,j\}})_{1\leq i<j\leq N}$ be non-negative. If σ is sampled from the spin $O(n)$ model with coupling constants J then*

$$\mathbb{E}\left(\langle\sigma_x, \sigma_y\rangle \cdot \langle\sigma_z, \sigma_w\rangle\right) \geq \mathbb{E}\left(\langle\sigma_x, \sigma_y\rangle\right) \cdot \mathbb{E}\left(\langle\sigma_z, \sigma_w\rangle\right) \text{ for all } 1 \leq x, y, z, w \leq N. \tag{11}$$

In other words, the random variables $\langle\sigma_x, \sigma_y\rangle$ and $\langle\sigma_z, \sigma_w\rangle$ are non-negatively correlated.

The theorem implies that each correlation $\mathbb{E}(\langle\sigma_x, \sigma_y\rangle)$ is a monotone non-decreasing function of each coupling constant $J_{\{z,w\}}$. Indeed, in the setting of the theorem, one checks in a straightforward manner that, for all $1 \leq x, y \leq N$ and $1 \leq z < w \leq N$,

$$\frac{\partial}{\partial J_{\{z,w\}}}\mathbb{E}\left(\langle\sigma_x, \sigma_y\rangle\right) = \mathbb{E}\left(\langle\sigma_x, \sigma_y\rangle \cdot \langle\sigma_z, \sigma_w\rangle\right) - \mathbb{E}\left(\langle\sigma_x, \sigma_y\rangle\right) \cdot \mathbb{E}\left(\langle\sigma_z, \sigma_w\rangle\right) \overset{(11)}{\geq} 0.$$

This monotonicity property is exceedingly useful as it allows to compare the correlations of the spin $O(n)$ model on different graphs by taking limits as various coupling constants tend to zero or infinity (corresponding to deletion or contraction of edges of the graph). For instance, one may use it to establish the existence of the infinite-volume (thermodynamic) limit of correlations in the spin $O(n)$ model ($n \in \{1, 2\}$) on $\mathbb{Z}^d$, or to compare the behavior of the model in different spatial dimensions d.

The following lemma, introduced by Ginibre [60], is a key step in the proof of Theorem 3. Sylvester [114] has found counterexamples to the lemma when $n \geq 3$.

Lemma 1. *Let $n \in \{1, 2\}$ and let $N \geq 1$ be an integer. Then for every choice of non-negative integers $(k_{\{i,j\}}), (\ell_{\{i,j\}})$, $1 \leq i < j \leq N$, we have*

$$\int\int \prod_{1\leq i<j\leq N} \left(\langle\sigma_i, \sigma_j\rangle - \langle\sigma'_i, \sigma'_j\rangle\right)^{k_{\{i,j\}}} \cdot \left(\langle\sigma_i, \sigma_j\rangle + \langle\sigma'_i, \sigma'_j\rangle\right)^{\ell_{\{i,j\}}} d\sigma d\sigma' \geq 0, \tag{12}$$

where, as before, $d\sigma$ and $d\sigma'$ denote the uniform probability measure on $(\mathbb{S}^{n-1})^N$.

Proof. The change of variables $(\sigma,\sigma')\mapsto(\sigma',\sigma)$ preserves the measure $d\sigma d\sigma'$ and reverses the sign of each term of the form $\langle\sigma_i,\sigma_j\rangle-\langle\sigma_i',\sigma_j'\rangle$ while keeping terms of the form $\langle\sigma_i,\sigma_j\rangle+\langle\sigma_i',\sigma_j'\rangle$ fixed. The lemma thus follows in the case that $\sum_{1\le i<j\le N}k_{\{i,j\}}$ is odd as the integral in (12) evaluates to zero. Let us then assume that

$$\sum_{1\le i<j\le N}k_{\{i,j\}}\quad\text{is even.}\tag{13}$$

Identifying $\mathbb{S}^1$ with the unit circle in the complex plane and using that $n\in\{1,2\}$, we may express the spins as $\sigma_j=e^{i\theta_j}$ and $\sigma_j'=e^{i\theta_j'}$. With this notation, we have

$$\begin{aligned}\langle\sigma_i,\sigma_j\rangle-\langle\sigma_i',\sigma_j'\rangle&=\cos(\theta_i-\theta_j)-\cos(\theta_i'-\theta_j')\\&=-2\sin\left(\frac{\theta_i+\theta_i'}{2}-\frac{\theta_j+\theta_j'}{2}\right)\sin\left(\frac{\theta_i-\theta_i'}{2}-\frac{\theta_j-\theta_j'}{2}\right),\end{aligned}\tag{14}$$

and similarly,

$$\langle\sigma_i,\sigma_j\rangle+\langle\sigma_i',\sigma_j'\rangle=2\cos\left(\frac{\theta_i+\theta_i'}{2}-\frac{\theta_j+\theta_j'}{2}\right)\cos\left(\frac{\theta_i-\theta_i'}{2}-\frac{\theta_j-\theta_j'}{2}\right).$$

Thus, using (13) to cancel the minus sign in the right-hand side of (14), we may write

$$\begin{aligned}\int\int\prod_{1\le i<j\le N}(\langle\sigma_i,\sigma_j\rangle-\langle\sigma_i',\sigma_j'\rangle)^{k_{\{i,j\}}}\cdot(\langle\sigma_i,\sigma_j\rangle+\langle\sigma_i',\sigma_j'\rangle)^{\ell_{\{i,j\}}}d\sigma d\sigma'\\=\int\int F(\theta+\theta')F(\theta-\theta')d\sigma d\sigma'=:I\end{aligned}$$

for a real-valued function F, satisfying the condition that $F(\theta+\theta')F(\theta-\theta')$ remains invariant when adding integer multiplies of 2π to any of the coordinates of θ or to any of the coordinates of θ'. We now consider the cases $n=1$ and $n=2$ separately.

Suppose first that $n=2$. Writing $d\theta,d\theta'$ for Lebesgue measure on $\mathbb{R}^N$, and using the above invariance property of F, we have

$$I=\frac{1}{(8\pi^2)^N}\int_{[-2\pi,2\pi]^N}\int_{[-\pi,\pi]^N}F(\theta+\theta')F(\theta-\theta')d\theta d\theta'.$$

One may regard the domain of integration above as $([-2\pi,2\pi]\times[-\pi,\pi])^N$. Consider $E_0:=[-2\pi,2\pi]\times[-\pi,\pi]$, the projection of this domain onto one coordinate of (θ,θ'). We shall split this domain into pieces and then rearrange them so as to obtain a square domain with side-length $2\sqrt{2}\pi$ rotated by 45° and symmetric about the origin, i.e., the domain defined by $E_1:=\{(\theta,\theta')\in\mathbb{R}^2:|\theta\pm\theta'|\le2\pi\}$. Indeed, each of the differences $E_0\setminus E_1$ and $E_1\setminus E_0$ consists of four triangular pieces, each being an isosceles right triangle with side-length π and sides parallel to the axis, so that these pieces can be rearranged to obtain E_1 from E_0. In fact, the only operations involved in this procedure are translations by multiples of 2π in the direction of the axes. Thus, using the above invariance property of F, we conclude that I can be written as

$$I = \frac{1}{(8\pi^2)^N} \int\int_{(E_1)^N} F(\theta+\theta')F(\theta-\theta')d\theta d\theta'.$$

The change of variables $(\theta,\theta') \mapsto (\theta+\theta',\theta-\theta')$ now shows that I is the square of an integral of a real-valued function and hence is non-negative.

The case $n = 1$ is treated similarly, though one must take extra care in handling boundaries between domains of integration, as these no longer need to have measure zero. Writing $d\theta, d\theta'$ for the counting measure on $(\pi\mathbb{Z})^N$, we have

$$I = \frac{1}{8^N} \int_{\{-\pi,0,\pi,2\pi\}^N} \int_{\{0,\pi\}^N} F(\theta+\theta')F(\theta-\theta')d\theta d\theta'.$$

As before, we consider a single coordinate of (θ,θ'). Observe that there is quite some freedom in changing the domain of integration $E_0 := \{-\pi,0,\pi,2\pi\}\times\{0,\pi\}$ without effecting the integral. Consider for instance the domain E_0' obtained from E_0 by removing the points $\{(-\pi,\pi),(2\pi,\pi)\}$ and adding $\{(0,-\pi),(\pi,-\pi)\}$ instead. By the invariance property of F, the integral on E_0' is the same as on E_0. To conclude as before that I is non-negative, it suffices to find a domain of integration E_1, which coincides with E_0' on $(\pi\mathbb{Z})^2$, and which is a $45°$ rotated square (i.e., the product of an interval with itself in the $(\theta+\theta',\theta-\theta')$ coordinates). Indeed, one may easily verify that $E_1 := \{(\theta,\theta') : -3/2 \le \theta \pm \theta' \le 5/2\}$ is such a domain.

Proof of Theorem 3. Let σ and σ' be two independent samples from the spin $O(n)$ model with coupling constants J. Then

$$2\,\mathrm{Cov}\left(\langle\sigma_x,\sigma_y\rangle, \langle\sigma_z,\sigma_w\rangle\right) = \mathbb{E}\left[\left(\langle\sigma_x,\sigma_y\rangle - \langle\sigma_x',\sigma_y'\rangle\right)\cdot\left(\langle\sigma_z,\sigma_w\rangle - \langle\sigma_z',\sigma_w'\rangle\right)\right].$$

Thus, it suffices to show that the expectation on the right-hand side is non-negative. Indeed, denoting $S^{\pm}_{\{i,j\}} := \langle\sigma_i,\sigma_j\rangle \pm \langle\sigma_i',\sigma_j'\rangle$, this expectation is equal to

$$\frac{1}{\left(Z^{\mathrm{spin}}_{n,J}\right)^2} \int\int S^-_{\{x,y\}}\cdot S^-_{\{z,w\}}\cdot\exp\left[\sum_{1\le i<j\le N} J_{\{i,j\}}S^+_{\{i,j\}}\right] d\sigma d\sigma',$$

which, by expanding the exponent into a Taylor's series, is equal to

$$\frac{1}{\left(Z^{\mathrm{spin}}_{n,J}\right)^2} \sum_{m\in\{0,1,2,\dots\}^{\{\{i,j\}:1\le i<j\le N\}}} C_m \int\int S^-_{\{x,y\}}\cdot S^-_{\{z,w\}} \times \prod_{1\le i<j\le N}\left(S^+_{\{i,j\}}\right)^{m_{\{i,j\}}} d\sigma d\sigma',$$

where each C_m is non-negative and the series is absolutely convergent. The desired non-negativity now follows from Lemma 1.

We are not aware of other proofs for Griffiths' second inequality, Theorem 3, for the XY model ($n = 2$). The above proof may also be adapted to treat clock models, models of the XY type in which the spin is restricted to roots of unity of a given order (the ticks of the clock), see [60]. Alternative approaches are available in the Ising case ($n = 1$): One proof relies on positive association (FKG) for the corresponding random-cluster model (see also Remark 1). A different argument of Ginibre [59] deduces Theorem 3 directly from Theorem 1.

2.4 High-Temperature Expansion

At infinite temperature ($\beta = 0$) the models are completely disordered, having all spins independent and uniformly distributed on $\mathbb{S}^{n-1}$. In this section we show that the disordered phase extends to high, but finite, temperatures (small positive β). Specifically, we show that the models exhibit exponential decay of correlations in this regime, as stated in (5) and (6).

We begin by expanding the partition function of the model on an arbitrary finite graph $G = (V(G), E(G))$ in the following manner. Denoting $f_\beta(s,t) := \exp\big[\beta\big(\langle s,t\rangle + 1\big)\big] - 1$ for $s,t \in \mathbb{S}^{n-1}$, we have

$$\begin{aligned} Z^{\text{spin}}_{G,n,\beta} &= \int_\Omega \prod_{\{u,v\}\in E(G)} \exp\left[\beta\left\langle \sigma_u, \sigma_v\right\rangle\right] d\sigma \\ &= e^{-\beta|E(G)|} \int_\Omega \prod_{\{u,v\}\in E(G)} \exp\left[\beta\left(\langle \sigma_u, \sigma_v\rangle + 1\right)\right] d\sigma \\ &= e^{-\beta|E(G)|} \int_\Omega \prod_{\{u,v\}\in E(G)} \big(1 + f_\beta(\sigma_u, \sigma_v)\big) d\sigma \\ &= e^{-\beta|E(G)|} \sum_{E\subset E(G)} \int_\Omega \prod_{\{u,v\}\in E} f_\beta(\sigma_u, \sigma_v) d\sigma. \end{aligned} \tag{15}$$

Exercise. Verify the last equality in the above expansion by showing that for any $(x_e)_{e\in\mathcal{E}}$,

$$\prod_{e\in\mathcal{E}} (1 + x_e) = \sum_{E\subset\mathcal{E}} \prod_{e\in E} x_e.$$

Thus, we have

$$Z^{\text{spin}}_{G,n,\beta} = e^{-\beta|E(G)|} \sum_{E\subset E(G)} Z(E), \tag{16}$$

where

$$Z(E) := \int_\Omega \prod_{\{u,v\}\in E} f_\beta(\sigma_u, \sigma_v) d\sigma. \tag{17}$$

Since f_β is non-negative, we may interpret (16) as prescribing a probability measure on (spanning) subgraphs of G, where the subgraph $(V(G), E)$ has probability proportional to $Z(E)$. Furthermore, given such a subgraph, we may interpret (17) as prescribing a probability measure on spin configurations σ, whose density with respect to $d\sigma$ is proportional to

$$Z(E,\sigma) := \prod_{\{u,v\}\in E} f_\beta(\sigma_u,\sigma_v).$$

Remark 1. For the Ising model ($n = 1$), the above joint distribution on the graph $(V(G), E)$ and spin configuration σ is called the *Edwards–Sokal coupling* [47]. Here, the marginal probability of E is proportional to

$$q^{N(E)} p^{|E|} (1-p)^{|E(\mathbb{T}_L^d)|\setminus|E|} \quad \text{with } q = 2 \text{ and } p = 1 - \exp(-2\beta), \tag{18}$$

where $N(E)$ stands for the number of connected components in $(V(G), E)$. Moreover, given E, the spin configuration σ is sampled by independently assigning to the vertices in each connected component of $(V(G), E)$ the same spin value, picked uniformly from $\{-1, 1\}$. The marginal distribution (18) of E is the famous *Fortuin–Kasteleyn (FK) random-cluster model*, which makes sense also for other values of p and q [65]. Both the Edwards–Sokal coupling and the FK model are available also for the more general *Potts models*.

The Edwards–Sokal coupling immediately implies that, for the Ising model, the correlation $\rho_{x,y} = \mathbb{E}(\sigma_x\sigma_y)$ equals the probability that x is connected to y in the graph $(V(G), E)$. In particular, $\rho_{x,y}$ is non-negative (as in Theorem 1) and, as connectivity probabilities in the FK model (with $q \geq 1$) are non-decreasing with p [65, Theorem 3.21], it follows also that $\rho_{x,y}$ is non-decreasing with the inverse temperature β (as in Theorem 3).

Remark 2. Conditioned on E, the spin configuration σ may be seen as a sample from the spin $O(n)$ model on the graph $(V(G), E)$ with potential $U(x) := -\log(\exp(\beta(1+x)) - 1)$. That is, conditioned on E, the distribution of σ is given by $\mu_{(V(G),E),n,U}$.

It follows from the last remark that, conditioned on E,

If $x \in V(G)$ then σ_x is distributed uniformly on $\mathbb{S}^{n-1}$.
If $x, y \in V(G)$ are not connected in $(V(G), E)$ then σ_x and σ_y are independent.

Hence, we deduce that $\mathbb{E}(\langle\sigma_x, \sigma_y\rangle \mid E) = 0$ when x and y are not connected in $(V(G), E)$. Since $|\langle\sigma_x, \sigma_y\rangle| \leq 1$, we obtain

$$|\rho_{x,y}| \leq \mathbb{P}(x \text{ and } y \text{ are connected in } (V(G), E)),$$

where E is a random subset of $E(G)$ chosen according to the above probability measure. Thus, to establish the decay of correlations, it suffices to show that long connections in E are very unlikely. We first show that

$$\text{for any } e \in E(G) \text{ and } E_0 \subset E(G) \setminus \{e\},\ \mathbb{P}(e \in E \mid E \setminus \{e\} = E_0) \leq 1 - e^{-2\beta}. \tag{19}$$

Indeed,

$$\mathbb{P}(e \in E \mid E \setminus \{e\} = E_0) = \frac{Z(E_0 \cup \{e\})}{Z(E_0 \cup \{e\}) + Z(E_0)} = \frac{1}{1 + \frac{Z(E_0)}{Z(E_0 \cup \{e\})}},$$

and denoting $e = \{u, v\}$ and noting that $f_\beta(s,t) \le \exp(2\beta) - 1$,

$$\frac{Z(E_0 \cup \{e\})}{Z(E_0)} = \frac{\int_\Omega Z(E_0 \cup \{e\}, \sigma) d\sigma}{\int_\Omega Z(E_0, \sigma) d\sigma} = \frac{\int_\Omega Z(E_0, \sigma) f_\beta(\sigma_u, \sigma_v) d\sigma}{\int_\Omega Z(E_0, \sigma) d\sigma} \le e^{2\beta} - 1.$$

Repeated application of (19) now yields that the probability that E contains any fixed k edges is exponentially small in k. Namely,

$$\text{for any } e_1, \ldots, e_k \in E(G), \quad \mathbb{P}(e_1, \ldots, e_k \in E) \le \left(1 - e^{-2\beta}\right)^k.$$

We now specialize to the case $G = \mathbb{T}_L^d$ (in fact, the only property of $\mathbb{T}_L^d$ we use is that its maximum degree is $2d$). Since the event that x and y are connected in $(V(G), E)$ implies the existence of a simple path in E of some length $k \ge \|x - y\|_1$ starting at x, and since the number of such paths is at most $2d(2d-1)^{k-1} \le 2(2d-1)^k$, we obtain

$$\begin{aligned}\mathbb{P}(x \text{ and } y \text{ are connected in } (V(G), E)) &\le \sum_{k=\|x-y\|_1}^{\infty} 2(2d-1)^k (1 - e^{-2\beta})^k \\ &\le C_{d,\beta} \Big((2d-1)(1 - e^{-2\beta}) \Big)^{\|x-y\|_1},\end{aligned}$$

when $(2d-1)(1 - e^{-2\beta}) < 1$. Thus, we have established that

$$|\rho_{x,y}| \le C_{d,\beta} \exp\left(-c_{d,\beta} \|x - y\|_1\right) \quad \text{when } \beta < \frac{1}{2} \log\left(\frac{2d-1}{2d-2}\right).$$

Remark 3. This gives exponential decay in dimension $d \ge 2$ whenever $\beta \le 1/4d$ and in one dimension for all finite β.

Fisher [49] established an improved lower bound for the critical inverse temperature $\beta_c(d)$ for long-range order in the d-dimensional Ising model, showing that $\tanh(\beta_c(d)) \ge \frac{1}{\mu(d)}$, where $\mu(d)$ is the connective constant of $\mathbb{Z}^d$ (the exponential growth rate of the number of self-avoiding walks of length n on $\mathbb{Z}^d$ as $n \to \infty$). Since there are fewer self-avoiding walks than non-backtracking walks, we have the simple bound $\mu(d) \le 2d - 1$, which implies that $\beta_c(d) \ge \frac{1+o(1)}{2d}$ as $d \to \infty$. A similar bound was proved by Griffiths [64]. Simon [108] establishes a bound of the same type for spin $O(n)$ models with $n \ge 2$, proving the absence of spontaneous magnetization when $\beta < \frac{n}{2d}$. An upper bound with matching asymptotics as $d \to \infty$ is proved via the so-called infra-red bound in Sect. 2.7 below.

Fisher's technique is based on a Kramers–Wannier [83] expansion of the Ising model partition function. This expansion, different from (15), relates the model to a probability distribution over even subgraphs (subgraphs in which the degrees of all vertices are even). A special case of the expansion is described in Sect. 3.2 (see remark there).

2.5 Low-Temperature Ising Model - The Peierls Argument

One can approach the low-temperature Ising model using the Kramers–Wannier expansion mentioned in Remark 3 and Sect. 3.2. Here, however, we follow a slightly different route, presenting the classical Peierls argument [97] which is useful in many similar contexts.

Let G be a finite connected graph and let $x, y \in V(G)$ be two vertices. We begin by noting that in the Ising model, since spins take values in $\{-1, 1\}$, we may write the correlations in the following form:

$$\rho_{x,y} = \mathbb{E}(\sigma_x \sigma_y) = \mathbb{P}(\sigma_x = \sigma_y) - \mathbb{P}(\sigma_x \neq \sigma_y) = 1 - 2\mathbb{P}(\sigma_x \neq \sigma_y).$$

Thus, to establish a lower bound on the correlation, we must provide an upper bound on the probability that the spins at x and y are different. To this end, we require some definitions. Given a set of vertices $A \subset V(G)$, we denote the *edge-boundary* of A, the set of edges in $E(G)$ with precisely one endpoint in A, by ∂A. A *contour* is a set of edges $\gamma \subset E(G)$ such that $\gamma = \partial A$ for some $A \subset V(G)$ satisfying that both A and $A^c := V(G) \setminus A$ are induced connected (non-empty) subgraphs of G. Thus, a contour can be identified with a partition of the set of vertices of G into two connected sets. We say that γ *separates* two vertices x and y if they belong to different sets of the corresponding partition. The *length of a contour* is the number of edges it contains.

Exercise. A set of edges γ is a contour if and only if γ is cutset (i.e., the removal of γ disconnects the graph) which is minimal with respect to inclusion (i.e., no proper subset of γ is also a cutset).

Let σ be a spin configuration. We say that γ is an *interface* (with respect to σ) if γ is a contour separating x and y such that

$$\sigma_u \neq \sigma_v \qquad \text{for all } \{u, v\} \in \gamma.$$

The first step in the proof is the following observation:

$$\text{if } \sigma_x \neq \sigma_y \text{ then there exists an interface.} \tag{20}$$

Indeed, if $\sigma_x \neq \sigma_y$ then the connected component of $\{u \in V(G) : \sigma_u = \sigma_x\}$ containing x, which we denote by B, does not contain y. Hence, if we denote the connected component of B^c containing y by A, then $\gamma := \partial A$ is a contour separating x and y. Moreover, it is easy to check that $\sigma_u = \sigma_x$ and $\sigma_v = \sigma_y$ for all $\{u, v\} \in \gamma$ such that $u \in A^c$ and $v \in A$, so that γ is an interface.

Next, we show that for any fixed contour γ of length k,

$$\mathbb{P}(\gamma \text{ is an interface}) \leq e^{-2\beta k}. \tag{21}$$

To see this, let $\{A, A^c\}$ be the partition corresponding to γ and, given a spin configuration σ, consider the modified spin configuration σ' in which the spins in A are flipped, i.e.,

$$\sigma'_u := \begin{cases} -\sigma_u & \text{if } u \in A \\ \sigma_u & \text{if } u \in A^c \end{cases}.$$

Observe that if γ is an interface with respect to σ then

$$\sum_{\{u,v\}\in E(G)} \sigma'_u\sigma'_v - \sum_{\{u,v\}\in E(G)} \sigma_u\sigma_v = \sum_{\{u,v\}\in\gamma} (\sigma'_u\sigma'_v - \sigma_u\sigma_v) = 2|\gamma|.$$

Thus, denoting $F := \{\sigma \in \Omega : \gamma \text{ is an interface with respect to } \sigma\}$ and noting that $\sigma \mapsto \sigma'$ is injective on Ω (in fact, an involution of Ω), we have

$$\begin{aligned}\mathbb{P}(\gamma \text{ is an interface}) &= \frac{\sum_{\sigma\in F} \exp\left[\beta\sum_{\{u,v\}\in E(G)} \sigma_u\sigma_v\right]}{\sum_{\sigma\in\Omega} \exp\left[\beta\sum_{\{u,v\}\in E(G)} \sigma_u\sigma_v\right]}\\ &\le \frac{\sum_{\sigma\in F} \exp\left[\beta\sum_{\{u,v\}\in E(G)} \sigma_u\sigma_v\right]}{\sum_{\sigma\in F} \exp\left[\beta\sum_{\{u,v\}\in E(G)} \sigma'_u\sigma'_v\right]} = e^{-2\beta|\gamma|}.\end{aligned}$$

The final ingredient in the proof is an upper bound on the number of contours of a given length. For this, we henceforth restrict ourselves to the case $G = \mathbb{T}_L^d$, for which we use the following fact:

$$\begin{aligned}&\text{The number of contours of length } k\\ &\qquad\qquad \text{separating two given vertices is at most } e^{C_d k}.\end{aligned} \tag{22}$$

The proof of this fact consists of the following two lemmas.

Lemma 2. *Let γ be a set of edges and consider the graph $\mathcal{G}_\gamma$ on γ in which two edges $e, f \in \gamma$ are adjacent if the $(d-1)$-dimensional faces corresponding to e and f share a common $(d-2)$-dimensional face. If γ is a contour then either $\mathcal{G}_\gamma$ is connected or every connected component of $\mathcal{G}_\gamma$ has size at least L^{d-1}.*

Although intuitively clear, the proof of the above lemma is not completely straightforward. Timár gave a proof [118] of the analogous statement in $\mathbb{Z}^d$ (in which case the graph $\mathcal{G}_\gamma$ is always connected) via elementary graph-theoretical methods. In our case, there is an additional complication due to the topology of the torus (indeed, the graph $\mathcal{G}_\gamma$ need not be connected - although it can have only two connected components - a fact for which we do not have a simple proof). We refer the reader to [98] for a proof.

Lemma 3. *Let $\mathcal{G}$ be a graph with maximum degree Δ. The number of connected subsets of $V(\mathcal{G})$ which have size k and contain a given vertex is at most $a(\Delta)^k$, where $a(\Delta)$ is a positive constant depending only on Δ.*

This lemma has several simple proofs. One may for instance use a depth-first-search algorithm to provide a proof with the constant $a(\Delta) = \Delta^2$. We refer the reader to [20, Chapter 45] for a proof yielding the constant $a(\Delta) = e(\Delta - 1)$ (which is optimal when $\Delta \ge 3$ as can be seen by considering the case when $\mathcal{G}$ is a regular tree).

Exercise. Deduce fact (22) from the two lemmas.

Finally, putting together (20), (21) and (22), when $\beta \geq C_d$, we obtain

$$\mathbb{P}(\sigma_x \neq \sigma_y) \leq \mathbb{P}(\text{there exists an interface}) \leq \sum_{\substack{\gamma \text{ contour} \\ \text{separating } x \text{ and } y}} \mathbb{P}(\gamma \text{ is an interface})$$
$$\leq \sum_{k=1}^{\infty} e^{C_d k} e^{-2\beta k} \leq C_d e^{-2\beta}.$$

Thus, in terms of correlations, we have established that

$$\rho_{x,y} \geq 1 - C_d e^{-2\beta} \geq c_{d,\beta} \quad \text{when } \beta \geq C_d.$$

Remark 4. Specializing Lemma 3 to the relevant graph in our situation, one may obtain an improved and explicit bound of $\exp(Ck\log(d+1)/d)$ on the number of contours of length k separating two given vertices [9,85]. This gives that $\beta_c(d) \leq C\log(d+1)/d$. In fact, the correct asymptotic value is $\beta_c(d) \sim 1/2d$ as $d \to \infty$, as follows by combining Fisher's bound in Remark 3 with Theorem 5 below.

Aizenman, Bricmont and Lebowitz [5] point out that a gap between the true value of $\beta_c(d)$ and the bound on $\beta_c(d)$ obtained from the Peierls argument is unavoidable in high dimensions. They point out that the Peierls argument, when it applies, excludes the possibility of *minority percolation*. That is, the possibility to have an infinite connected component of -1 spins in the infinite-volume limit obtained with $+1$ boundary conditions. However, as they show, such minority percolation does occur in high dimensions when $\beta \leq c\frac{\log d}{d}$, yielding a lower bound on the minimal inverse temperature at which the Peierls argument applies.

2.6 No Long-Range Order in Two Dimensional Models with Continuous Symmetry - The Mermin–Wagner Theorem

In this section, we establish power-law decay of correlations for the two-dimensional spin $O(n)$ model with $n \geq 2$ at any positive temperature. The proof applies in the generality of the spin $O(n)$ model with potential U, where U satisfies certain assumptions, and it is convenient to present it in this context, to highlight the core parts of the argument. The fact that there is no long-range order was first established by Mermin and Wagner [88,89][2], with later works providing upper bounds on the rate of decay of the correlations. Power-law decay of correlations for the standard XY model was first established by McBryan and Spencer [86] who used analytic function techniques. The following theorem which generalizes the result to C^2 potentials was subsequently proved by Shlosman [107] using methods developed by Dobrushin and Shlosman [36].

Theorem 4. *Let $n \geq 2$. Let $U : [-1,1] \to \mathbb{R}$ be twice continuously differentiable. Suppose that $\sigma : V(\mathbb{T}_L^2) \to \mathbb{S}^{n-1}$ is randomly sampled from the two-dimensional*

[2] A related intuition was mentioned earlier by Herring and Kittel [68, Footnote 8a].

spin $O(n)$ model with potential U (see (3)). Then there exist $C_{n,U}, c_{n,U} > 0$ such that

$$|\rho_{x,y}| = |\mathbb{E}(\langle \sigma_x, \sigma_y \rangle)| \le C_{n,U} \|x-y\|_1^{-c_{n,U}} \quad \textit{for all } x,y \in V(\mathbb{T}_L^2). \tag{23}$$

The proof presented below combines elements of the Dobrushin–Shlosman [36] and Pfister [99] approaches to the Mermin–Wagner theorem. The idea to combine the approaches is introduced in a forthcoming paper of Gagnebin, Miłoś and Peled [56], where it is pushed further to prove power-law decay of correlations for *any* measurable potential U satisfying only very mild integrability assumptions. The work [56] relies further on ideas used by Ioffe–Shlosman–Velenik [72], Richthammer [102] and Miłoś–Peled [91].

For simplicity, we will prove Theorem 4 in the special case that $n = 2$, $x = (1,0)$ and $y = (2^m, 0)$ for some positive integer m (assuming, implicitly, that $L \ge 2^m$). We briefly explain the necessary modifications for the general case after the proof.

Fix a C^2 function $U : [-1,1] \to \mathbb{R}$. Suppose that $\sigma : V(\mathbb{T}_L^2) \to \mathbb{S}^1$ is randomly sampled from the two-dimensional spin $O(2)$ model with potential U. It is convenient to parametrize configurations differently: Identifying $\mathbb{S}^1$ with the unit circle in the complex plane, we consider the angle θ_v that each vector σ_v forms with respect to the x-axis. Precisely, for the rest of the argument, we let $\theta : V(\mathbb{T}_L^2) \to [-\pi, \pi)$ be randomly sampled from the probability density

$$t(\phi) := \frac{1}{Z} \exp\left[- \sum_{\{u,v\} \in E(\mathbb{T}_L^2)} U(\cos(\phi_u - \phi_v)) \right] \prod_{v \in V(\mathbb{T}_L^2)} \mathbf{1}_{(\phi_v \in [-\pi,\pi))}, \tag{24}$$

with respect to the product uniform measure, where Z is a normalization constant. One checks simply that then (σ_v) is equal in distribution to $(\exp(i\theta_v))$. Thus, with our choice of the vertices x and y, the estimate (23) that we would like to prove becomes

$$|\rho_{(1,0),(2^m,0)}| = |\mathbb{E}(\cos(\theta_{(2^m,0)} - \theta_{(1,0)}))| \le C_{n,U} \cdot 2^{-c_{n,U} \cdot m}. \tag{25}$$

Step 1: Product of Conditional Correlations. We start by pointing out a conditional independence property inherent in the distribution of θ, which is a consequence of the domain Markov property and the fact that the interaction term in (24) depends only on the difference of angles in ϕ (the gradient of ϕ). This part of the argument is inspired by the technique of Dobrushin and Shlosman [36].

We divide the domain into "layers", where the ℓ-th layer, $0 \le \ell \le m$, corresponds to distance 2^ℓ from the origin. Denote the values and the gradients of θ on the ℓ-th layer by

$$\theta_{=\ell} := \left(\theta_v \colon \|v\|_1 = 2^\ell\right) \quad \text{and} \quad \nabla\theta_{=\ell} := \left(\theta_u - \theta_v \colon \|u\|_1, \|v\|_1 = 2^\ell\right).$$

Similarly, we write $\theta_{\le \ell}$ and $\nabla\theta_{\le\ell}$ for the values/gradients of θ inside the ℓ-th layer (i.e., for u,v with $\|u\|_1, \|v\|_1 \le 2^\ell$) and $\theta_{\ge\ell}$ and $\nabla\theta_{\ge\ell}$ for the values/gradients outside (i.e., for u, v with $\|u\|_1, \|v\|_1 \ge 2^\ell$).

Proposition 1. *Conditioned on* $\nabla\theta_{=\ell}$, *we have that* $\nabla\theta_{\leq\ell}$ *and* $\theta_{\geq\ell}$ *are independent.*

Proof. Consider the random variables $\theta_{<\ell}$ and $\nabla\theta_{<\ell}$, defined in the obvious way. It is straightforward from the definition of the density (24) of θ that, conditioned on $\theta_{\geq\ell}$, $\theta_{<\ell}$ almost surely has a density and that this density depends only on $\theta_{=\ell}$. In particular, conditioned on $\theta_{\geq\ell}$, $\nabla\theta_{<\ell}$ has a density which depends only on $\theta_{=\ell}$. Finally, since the interaction term in (24) depends only on the gradients $\nabla\theta$, we conclude that the density of $\nabla\theta_{<\ell}$ given $\theta_{\geq\ell}$ depends only on the gradients $\nabla\theta_{=\ell}$. Therefore, $\nabla\theta_{<\ell}$, and hence $\nabla\theta_{\leq\ell}$, is conditionally independent of $\theta_{\geq\ell}$ given $\nabla\theta_{=\ell}$.

Proposition 1 implies in particular that, conditioned on $\nabla\theta_{=\ell}$, the gradients $\nabla\theta_{\leq\ell}$ and $\nabla\theta_{\geq\ell}$ are independent. It follows from abstract arguments that, conditioned on $(\nabla\theta_{=k})_{1\leq k\leq m}$, the gradients $\nabla\theta_{\leq\ell}$ and $\nabla\theta_{\geq\ell}$ are independent. For convenience, we state this claim in a general form in the following exercise.

Exercise. Suppose X, Y, Z are random variables satisfying that X is conditionally independent of Y given Z. Then for every two measurable functions f and g, X is conditionally independent of Y given $(f(X), g(Y), Z)$.

In particular, conditioned on $(\nabla\theta_{=k})_{1\leq k\leq m}$, the random variables $(\theta_{(2^k,0)} - \theta_{(2^{k-1},0)})_{1\leq k\leq\ell}$ and $(\theta_{(2^k,0)} - \theta_{(2^{k-1},0)})_{\ell<k\leq m}$ are independent. Since this holds for all $1\leq\ell\leq m$, it follows again by abstract arguments that, conditioned on $(\nabla\theta_{=k})_{1\leq k\leq m}$, the random variables $(\theta_{(2^\ell,0)} - \theta_{(2^{\ell-1},0)})_{1\leq\ell\leq m}$ are mutually independent[3]. Once again, we state this general claim in an exercise.

Exercise. Suppose $X_1, \ldots, X_m, Z$ are random variables satisfying that, for any $1\leq\ell\leq m$, $(X_1, \ldots, X_\ell)$ is conditionally independent of $(X_{\ell+1}, \ldots, X_m)$ given Z. Then $(X_1, \ldots, X_m)$ are mutually conditionally independent given Z.

The above conditional independence therefore allows us to reexpress the quantity of interest to us as a product of expectations in the following way:

$$\begin{aligned}
&\mathbb{E}(\cos(\theta_{(2^m,0)} - \theta_{(1,0)}))\\
&= \Re\mathbb{E}\left(e^{i\left(\theta_{(2^m,0)}-\theta_{(1,0)}\right)}\right)\\
&= \Re\mathbb{E}\left(\prod_{\ell=1}^{m} e^{i\left(\theta_{(2^\ell,0)}-\theta_{(2^{\ell-1},0)}\right)}\right)\\
&= \Re\mathbb{E}\left(\prod_{\ell=1}^{m} \mathbb{E}\left(e^{i\left(\theta_{(2^\ell,0)}-\theta_{(2^{\ell-1},0)}\right)} \mid (\nabla\theta_{=k})_{1\leq k\leq m}\right)\right). \qquad (26)
\end{aligned}$$

This will be the starting point for our next step.

Step 2: Upper Bound on the Conditional Correlations. In this step, we estimate the individual conditional expectations in (26), proving that there exists an absolute constant $\varepsilon > 0$ for which, almost surely,

[3] In fact, more is true, conditioned on $(\nabla\theta_{=k})_{1\leq k\leq m}$, the σ-algebras of $\nabla\theta_{\ell-1\leq\cdot\leq\ell}$ are independent for $1\leq\ell\leq m$, where $\nabla\theta_{\ell-1\leq\cdot\leq\ell}$ is the collection of gradients $\theta_u - \theta_v$ with $2^{\ell-1}\leq\|u\|_1, \|v\|_1\leq 2^\ell$.

$$\left|\mathbb{E}\left(e^{i\left(\theta_{(2^\ell,0)}-\theta_{(2^{\ell-1},0)}\right)} \mid (\nabla\theta_{=k})_{1\le k\le m}\right)\right| \le 1-\varepsilon \quad \text{for all } 1\le \ell\le m, \tag{27}$$

immediately implying the required bound (25) as, from (26),

$$\begin{aligned}\left|\mathbb{E}(\cos(\theta_{(2^m,0)}-\theta_{(1,0)}))\right| &\le \mathbb{E}\left(\prod_{\ell=1}^m\left|\mathbb{E}\left(e^{i\left(\theta_{(2^\ell,0)}-\theta_{(2^{\ell-1},0)}\right)} \mid (\nabla\theta_{=k})_{1\le k\le m}\right)\right|\right)\\ &\le (1-\varepsilon)^m.\end{aligned}$$

This part of the argument is inspired by the technique of Pfister [99], and the variants used in [91, 102]. The idea of introducing a *spin wave* which rotates slowly (our function τ below and its property (33)) is at the heart of the Mermin–Wagner theorem.

Define a vector-valued function g on $\mathbb{R}^{V(\mathbb{T}_L^2)}$ by

$$g(\phi) := \left(\phi_u-\phi_v\colon u,v\in V(\mathbb{T}_L^2),\ \exists\, 1\le k\le m,\ \|u\|_1=\|v\|_1=2^k\right),$$

so that $g(\theta)$ and $(\nabla\theta_{=k})_{1\le k\le m}$ represent the same random variable. Write dm_{g_0} for the lower-dimensional Lebesgue measure supported on the affine subspace of $\mathbb{R}^{V(\mathbb{T}_L^2)}$ where $g(\phi)=g_0$. Standard facts (following from Fubini's theorem) imply that conditioned on $g(\theta)=g_0$, for almost every value of g_0 (with respect to the distribution of $g(\theta)$), the density of θ exists with respect to dm_{g_0} and is of the form (as in (24))

$$\begin{aligned}t_{g_0}(\phi) &= \frac{1}{Z_{g_0}}\exp\left[-\sum_{\{u,v\}\in E(\mathbb{T}_L^2)} U(\cos(\phi_u-\phi_v))\right]\prod_{v\in V(\mathbb{T}_L^2)}\mathbf{1}_{(\phi_v\in[-\pi,\pi))}\\ &= \frac{1}{Z_{g_0}}\exp\left[-\sum_{\{u,v\}\in E(\mathbb{T}_L^2)} \tilde U(\phi_u-\phi_v)\right]\prod_{v\in V(\mathbb{T}_L^2)}\mathbf{1}_{(\phi_v\in[-\pi,\pi))},\end{aligned}$$

where $\tilde U:\mathbb{R}\to\mathbb{R}$ is the 2π-periodic C^2 function defined by

$$\tilde U(\alpha) := U(\cos(\alpha)). \tag{28}$$

In particular,

$$\tilde U(x+\delta)\le \tilde U(x)+\tilde U'(x)\delta+\frac{\sup_y \tilde U''(y)}{2}\delta^2 \quad \text{for all } x,\delta\in\mathbb{R}. \tag{29}$$

Fix $1\le\ell\le m$. Define a function $\tau: V(\mathbb{T}_L^2)\to\mathbb{R}$ by

$$\tau_v := \begin{cases} 1/2 & \|v\|_1\le 2^{\ell-1}\\ 1-\frac{\|v\|_1}{2^\ell} & 2^{\ell-1}\le\|v\|_1\le 2^\ell\\ 0 & \|v\|_1\ge 2^\ell\end{cases} \tag{30}$$

and define for each $\phi: V(\mathbb{T}_L^2)\to[-\pi,\pi)$ its perturbations $\phi^+,\phi^-: V(\mathbb{T}_L^2)\to[-\pi,\pi)$ by

$$\phi_v^+ := \phi_v+\tau_v \pmod{2\pi}, \quad \phi_v^- := \phi_v-\tau_v \pmod{2\pi}. \tag{31}$$

We shall need the following two simple properties of τ (which the reader may easily verify):

$$g(\phi^+) = g(\phi^-) = g(\phi) \quad \text{for every } \phi : V(\mathbb{T}_L^2) \to \mathbb{R}, \tag{32}$$

$$\sum_{\{u,v\}\in E(\mathbb{T}_L^2)} (\tau_u - \tau_v)^2 \leq C \tag{33}$$

for some absolute constant C.

The following is the key calculation of the proof. For every $\phi : V(\mathbb{T}_L^2) \to [-\pi, \pi)$, setting $g_0 := g(\phi)$,

$$\begin{aligned}
&\sqrt{t_{g_0}(\phi^+)t_{g_0}(\phi^-)} \\
&= \frac{1}{Z_{g_0}} \exp\Big[-\frac{1}{2}\sum_{\{u,v\}\in E(\mathbb{T}_L^2)} \big(\tilde{U}(\phi_u - \phi_v + \tau_u - \tau_v) + \tilde{U}(\phi_u - \phi_v - \tau_u + \tau_v)\big)\Big] \\
&\overset{(29)}{\geq} \frac{1}{Z_{g_0}} \exp\Big[-\sum_{\{u,v\}\in E(\mathbb{T}_L^2)} \tilde{U}(\phi_u - \phi_v) - \frac{\sup_y \tilde{U}''(y)}{2} \sum_{\{u,v\}\in E(\mathbb{T}_L^2)} (\tau_u - \tau_v)^2\Big] \\
&\overset{(33)}{\geq} c \cdot t_{g_0}(\phi)
\end{aligned} \tag{34}$$

for an absolute constant $c > 0$.

We wish to convert the inequality (34) into an inequality of probabilities rather than densities. To this end, define for $a \in \mathbb{R}$,

$$E_a := \left\{\phi : V(\mathbb{T}_n^2) \to [-\pi, \pi) \colon \left|\Re e^{i(\phi_{(2\ell,0)} - \phi_{(2\ell-1,0)} - a)}\right| \geq \tfrac{9}{10}\right\}, \tag{35}$$

and, for almost every g_0 with respect to the distribution of $g(\theta)$,

$$I_{a,g_0} := \int_{E_a} \sqrt{t_{g_0}(\phi^+)t_{g_0}(\phi^-)}dm_{g_0}(\phi).$$

On the one hand, by (34),

$$I_{a,g_0} \geq c\int_{E_a} t_{g_0}(\phi)dm_{g_0}(\phi) = c \cdot \mathbb{P}\left(\theta \in E_a \mid g(\theta) = g_0\right). \tag{36}$$

On the other hand, the Cauchy–Schwartz inequality and a change of variables using (31) and (32) yields

$$\begin{aligned}
I_{a,g_0}^2 &\leq \int_{E_a} t_{g_0}(\phi^+)dm_{g_0}(\phi) \cdot \int_{E_a} t_{g_0}(\phi^-)dm_{g_0}(\phi) \\
&= \mathbb{P}\big(\theta - \tau \in E_a \mid g(\theta) = g_0\big) \cdot \mathbb{P}\big(\theta + \tau \in E_a \mid g(\theta) = g_0\big).
\end{aligned} \tag{37}$$

Putting together (36) and (37) and recalling (30) and (35) we obtain that, almost surely,

$$\mathbb{P}\left(\left|\Re e^{i(\theta_{(2\ell,0)}-\theta_{(2\ell-1,0)}+1/2-a)}\right| \ge \tfrac{9}{10} \mid g(\theta)\right)$$
$$\times\, \mathbb{P}\left(\left|\Re e^{i(\phi_{(2\ell,0)}-\phi_{(2\ell-1,0)}-1/2-a)}\right| \ge \tfrac{9}{10} \mid g(\theta)\right)$$
$$\ge c^2 \cdot \mathbb{P}\left(\left|\Re e^{i(\phi_{(2\ell,0)}-\phi_{(2\ell-1,0)}-a)}\right| \ge \tfrac{9}{10} \mid g(\theta)\right)^2.$$

As this inequality holds for any $a \in \mathbb{R}$, it implies that, conditioned on $g(\theta)$, $e^{i\left(\theta_{(2\ell,0)}-\theta_{(2\ell-1,0)}\right)}$ cannot be concentrated around any single value, proving the inequality (27) that we wanted to show.

General Vertices x and y and Larger Values of n. The inequality (23) for arbitrary vertices x and y follows easily from what we have already shown. Indeed, by symmetry, there is no loss of generality in assuming as before that $x = (1,0)$ and $y \neq (0,0)$. Set m to be the integer satisfying that $2^m \le \|y\|_1 < 2^{m+1}$, so that it suffices to show that $|\rho_{x,y}| \le C_{n,U} \cdot 2^{-c_{n,U}\cdot m}$. Indeed, by Proposition 1,

$$\mathbb{E}(\cos(\theta_y - \theta_x)) = \Re\mathbb{E}\left(\mathbb{E}\left(e^{i\left(\theta_y - \theta_{(2^m,0)}\right)} \mid \nabla\theta_{=m}\right) \cdot \mathbb{E}\left(e^{i\left(\theta_{(2^m,0)}-\theta_x\right)} \mid \nabla\theta_{=m}\right)\right).$$

Thus, the required estimate follows from (27) following the decomposition (26) (done conditionally on $\nabla\theta_{=m}$).

Let us briefly explain how to adapt the proof to the case that $n \ge 3$. Write $(\sigma^1, \ldots, \sigma^n)$ for the components of σ. The idea is to condition on $(\sigma^3, \ldots, \sigma^n)$ and apply the previous argument to the conditional distribution of the remaining two coordinates (σ^1, σ^2). In more detail, conditioned on $(\sigma^3, \ldots, \sigma^n) = h$, the random variable (σ^1, σ^2) almost surely has a density with respect to the product over $v \in V(\mathbb{T}_L^2)$ of uniform distributions on $r_v\mathbb{S}^1$, where $r_v := \sqrt{1 - \|h(v)\|_2^2} \in [0,1]$. Moreover, after passing to the angle representation θ_v for each (σ_v^1, σ_v^2), this density has the form

$$t_h(\theta) := \frac{1}{Z_h} \exp\left[-\sum_{\{u,v\}\in E(\mathbb{T}_L^2)} U\Big(r_u r_v \cos(\theta_u - \theta_v) + \langle h(u), h(v)\rangle\Big)\right] \times \prod_{v \in V(\mathbb{T}_L^2)} \mathbf{1}_{(\theta_v \in [-\pi,\pi))}.$$

In particular, we see from this expression that, conditioned on $(\sigma^3, \ldots, \sigma^n)$, the distribution of (σ^1, σ^2) is invariant to global rotations and has the domain Markov property (just as in the $n = 2$ case). This allows the first step of the proof to go through essentially without change. In the second step, the function $\tilde{U}$ defined in (28) should be replaced by a collection of functions $\tilde{U}_{\{u,v\}}$, one for each edge $\{u,v\} \in E(\mathbb{T}_L^2)$, defined by $\tilde{U}_{\{u,v\}}(\alpha) := U(r_u r_v \cos(\alpha) + \langle h(u), h(v)\rangle)$. It is straightforward to check that, since U is a C^2 function and $r_u, r_v \in [0,1]$, the second derivative of $\tilde{U}_{\{u,v\}}$ is bounded above (and below) uniformly in $\{u,v\}$ and h. The argument in the second step of the proof now goes through as well, replacing each appearance of $\tilde{U}$ with the suitable $\tilde{U}_{\{u,v\}}$.

2.7 Long-Range Order in Dimensions $d \geq 3$ - The Infra-red Bound

In this section we prove that the spin $O(n)$ model in spatial dimensions $d \geq 3$ exhibits long-range order at sufficiently low temperatures. This was first proved by Fröhlich, Simon and Spencer [53] who introduced the method of the *infra-red bound* to this end. Our exposition benefitted from the excellent 'Marseille notes' of Daniel Ueltschi [119, Lecture 2, part 2], the recent book of Friedli and Velenik [50, Chapter 10] and discussions with Michael Aizenman. We prove the following result.

Theorem 5. *For any $d \geq 3$ and any $n \geq 1$ there exists a constant $\beta_1(d,n)$ such that the following holds. Suppose $\sigma : V(\mathbb{T}_L^d) \to \mathbb{S}^{n-1}$ is randomly sampled from the d-dimensional spin $O(n)$ model at inverse temperature $\beta \geq \beta_1(d,n)$. Then*

$$\frac{1}{|V(\mathbb{T}_L^d)|^2} \sum_{x,y \in V(\mathbb{T}_L^d)} \mathbb{E}(\langle \sigma_x, \sigma_y \rangle) \geq c_{d,n,\beta}.$$

Moreover, for any $d \geq 1$, $n \geq 1$ and $\beta > 0$, we have the limiting inequality

$$\liminf_{L \to \infty} \frac{1}{|V(\mathbb{T}_L^d)|^2} \sum_{x,y \in \Lambda} \mathbb{E}\left(\langle \sigma_x \sigma_y \rangle\right) \geq 1 - \frac{n}{2\beta} \int_{[0,1]^d} \frac{1}{\sum_{j=1}^d (1 - \cos(\pi t_j))} dt.$$

Lastly, the above integral is finite when $d \geq 3$ and is asymptotic to $1/d$ as $d \to \infty$.

Of course, long-range order for the Ising model ($n = 1$) has already been established in Sect. 2.5 so that our main interest is in the case of continuous spins, when $n \geq 2$. The last part of the theorem establishes long-range order in the spin $O(n)$ model for $\beta \geq \frac{n}{2d}(1 + o(1))$ as $d \to \infty$. Comparing with Remark 3, we see that this bound has the correct asymptotic dependence on both n and d.

The proof presented below, relying on the original paper of Fröhlich, Simon and Spencer [53] and making use of reflection positivity, remains the main method to establish Theorem 5. Fröhlich and Spencer [55] developed an alternative approach for the XY model ($n = 2$) which relies on the duality transformation explained in Sect. 2.9 below; see also [10, Section 5.5]. Kennedy and King [78] provided a second alternative approach for the XY model. However, these alternatives do not apply to the model with larger values of n ($n \geq 3$) where the symmetry group acting on the spins is non-Abelian. For these larger values the only alternative to reflection positivity is due to Balaban who has made rigorous elements of the renormalization group approach to the problem in a formidable series of papers, starting with [8]; see also Dimock's review starting with [35]. Nevertheless, it would be highly desirable to have additional approaches to prove continuous-symmetry breaking as many questions in this direction are still open, most prominently to establish a phase transition for the *quantum* Heisenberg ferromagnet in dimensions $d \geq 3$ (current techniques allow to prove this only for the *antiferromagnet*; See Dyson, Lieb and Simon [46]).

In our treatment we provide additional background on reflection positivity than strictly necessary for the proof of Theorem 5 in order to place the arguments in a wider context and to highlight the use of reflection positivity as a general-purpose tool applicable in many settings. The reader is referred to [119, Lecture 2, part 2] for a direct route to the proof.

Introduction to Reflection Positivity. In this section we provide an introduction to reflection positivity for rather general nearest-neighbor models. Extensions of the theory to certain next-nearest-neighbor and certain long-range interactions are possible and the reader is referred to [19,51,52] and [50, Chapter 10] for alternative treatments.

We allow spins to take values in an arbitrary measure space $(S, \mathcal{S}, \lambda)$. We also consider a general interaction between different values of spins, prescribed by a symmetric measurable function $h\colon S \times S \to [0, \infty)$ which is not essentially zero. Here symmetric means that

$$h(a, b) = h(b, a) \quad \text{for all } a, b \in S$$

and not essentially zero means that $\iint h(a, b) d\lambda(a) d\lambda(b) > 0$. For simplicity, we assume $(S, \mathcal{S}, \lambda)$ to be a finite measure space and h to be bounded.

The spin space $(S, \mathcal{S}, \lambda)$ and the interaction h define a spin model on a finite graph G as follows. The set of configurations is $S^{V(G)}$ and the density of a configuration $\sigma\colon V(G) \to S$ with respect to $d\lambda(\sigma) := \prod_{v \in V(G)} d\lambda(\sigma_v)$ is

$$\frac{1}{Z_{G,S,\lambda,h}} \prod_{\{u,v\} \in E(G)} h(\sigma_u, \sigma_v), \tag{38}$$

where the normalizing constant is given by

$$Z_{G,S,\lambda,h} := \int \prod_{\{u,v\} \in E(G)} h(\sigma_u, \sigma_v) d\lambda(\sigma).$$

For this definition to make sense it is required that $0 < Z_{G,S,\lambda,h} < \infty$. The upper bound follows from our assumptions that λ is finite and h is bounded. For bipartite G, the case of interest to us here, the lower bound follows from the assumption that h is not essentially zero[4] (the assumption does not suffice for general graphs).

The spin $O(n)$ model with potential U can be recovered in this setting by taking the spin space $(S, \mathcal{S}, \lambda)$ to be the uniform probability space on $\mathbb{S}^{n-1}$ and defining the interaction h by $h(a, b) := \exp(-U(\langle a, b \rangle))$.

[4] It suffices to show that $\iint \prod_{i,j=1}^n h(s_i, t_j) d\lambda(s_i) d\lambda(t_j) > 0$ for $n \geq 1$. Fubini's theorem reduces this to $\iint \prod_{i=1}^n h(s_i, t) d\lambda(s_i) d\lambda(t) > 0$, which then follows from Fubini's theorem and the assumption on h.

In order to discuss reflections, the graph G should have suitable symmetries. From here on, we consider only the torus graph $G = \mathbb{T}_L^d$. Denote the vertices of the torus by

$$\Lambda := V(\mathbb{T}_L^d) = \{-L+1, \ldots, L\}^d.$$

The torus graph admits hyperplanes of reflection which pass through vertices and hyperplanes of reflection which pass through edges. We discuss these two cases separately.

Reflections Through Vertices. We split the vertices of the torus into two partially overlapping subsets V_0 and V_1 of vertices, the 'left' and 'right' halves, by defining

$$V_0 := \big\{v \in \Lambda : v_1 \notin \{1, \ldots, L-1\}\big\} \quad \text{and} \quad V_1 := \big\{v \in \Lambda : v_1 \in \{0, 1, \ldots, L\}\big\},$$

where we write each $v \in \Lambda$ as $(v_1, v_2, \ldots, v_d)$. Note that $V_0 \cup V_1 = \Lambda$ and

$$V_0 \cap V_1 = \big\{v \in \Lambda : v_1 \in \{0, L\}\big\}.$$

Define a function $R\colon \Lambda \to \Lambda$ by

$$Rv := \begin{cases} (-v_1, v_2, \ldots, v_n) & \text{if } v_1 \neq L \\ (L, v_2, \ldots, v_n) & \text{if } v_1 = L \end{cases}.$$

Thus, R is the reflection through the vertices $V_0 \cap V_1$. Note in particular that R is an involution which fixes $V_0 \cap V_1$. Geometrically, the reflection is done across the hyperplane orthogonal to the x-axis which passes through the vertices having x-coordinate 0 (or equivalently, the hyperplane passing through the vertices having x-coordinate L). One may similarly consider reflections through other planes orthogonal to one of the coordinate axes, however, for concreteness, we focus on the reflection above. We denote by R also the naturally induced mapping on configurations $\sigma \in S^\Lambda$ which is defined by $(R\sigma)_v := \sigma_{Rv}$.

Let $\mathcal{F}$ denote the set of bounded measurable functions $f\colon S^\Lambda \to \mathbb{C}$. Let $\mathcal{F}_0 \subset \mathcal{F}$ be the subset of functions f which depend only on the values of the spins in V_0, i.e., $f(\sigma)$ is determined by $\sigma|_{V_0}$. We define a bilinear form on $\mathcal{F}_0$ by

$$(f, g) := \mathbb{E}\left(f(\sigma)\overline{g(R\sigma)}\right) \quad \text{for } f, g \in \mathcal{F}_0.$$

Proposition 2 (Reflection positivity through vertices). *The bilinear form defined above is positive semidefinite, i.e.,*

$$(f, f) \geq 0 \quad \textit{for all } f \in \mathcal{F}_0. \tag{39}$$

Proof. The domain Markov property implies that after conditioning on $\sigma|_{V_0 \cap V_1}$ the random variables $\sigma|_{V_0}$ and $(R\sigma)|_{V_0}$ become independent and identically distributed. Thus,

$$\begin{aligned}(f, f) = \mathbb{E}\left(f(\sigma)\overline{f(R\sigma)}\right) &= \mathbb{E}\left(\mathbb{E}\left(f(\sigma)\overline{f(R\sigma)} \mid \sigma|_{V_0 \cap V_1}\right)\right) \\ &= \mathbb{E}\left(\mathbb{E}\left(f(\sigma) \mid \sigma|_{V_0 \cap V_1}\right) \cdot \overline{\mathbb{E}\left(f(R\sigma) \mid \sigma|_{V_0 \cap V_1}\right)}\right) \\ &= \mathbb{E}\left(\left|\mathbb{E}\left(f(\sigma) \mid \sigma|_{V_0 \cap V_1}\right)\right|^2\right) \geq 0.\end{aligned}$$

□

The reflection positivity property (39) (used for all hyperplanes of reflection passing through vertices) implies a version of the important "chessboard estimate". We do not state this estimate here, as a version of it for reflections through edges is given in Proposition 4 below, and refer instead to [19] and [50, Chapter 10] for more details.

Reflections Through Edges. We split the vertices of the torus into two non-overlapping subsets V_0 and V_1 of vertices, the 'left' and 'right' halves, by

$$V_0 := \{v \in \Lambda : v_1 \leq 0\} \quad \text{and} \quad V_1 := \{v \in \Lambda : v_1 \geq 1\}.$$

Note that $V_0 \cup V_1 = \Lambda$ and that $V_0 \cap V_1 = \emptyset$. Define a function $R\colon \Lambda \to \Lambda$ by

$$Rv := (1 - v_1, v_2, \ldots, v_n).$$

Thus, R is the reflection through the edges between V_0 and V_1. Note in particular that R is an involution with no fixed points. Geometrically, the reflection is done across the hyperplane orthogonal to the x-axis which passes through the edges between x-coordinate 0 and x-coordinate 1 (or equivalently, the hyperplane passing through the edges between x-coordinate L and x-coordinate $-L+1$). One may similarly consider reflections through other planes orthogonal to one of the coordinate axes, however, for concreteness, we focus on the reflection above. We again denote by R also the naturally induced mapping on configurations $\sigma \in S^\Lambda$ which is defined by $(R\sigma)_v := \sigma_{Rv}$.

Let $\mathcal{F}$ denote the set of bounded measurable functions $f\colon S^\Lambda \to \mathbb{C}$. Let $\mathcal{F}_0 \subset \mathcal{F}$ be the subset of functions f which depend only on the values of the spins in V_0, i.e., $f(\sigma)$ is determined by $\sigma|_{V_0}$. We define a bilinear form on $\mathcal{F}_0$ by

$$(f, g) := \mathbb{E}\left(f(\sigma)\overline{g(R\sigma)}\right) \quad \text{for } f, g \in \mathcal{F}_0. \tag{40}$$

Proposition 3 (Reflection positivity through edges). *Suppose that the interaction h may be written as follows: there exists a measure space $(T, \mathcal{T}, \nu)$, where ν is a finite (non-negative) measure, and a bounded measurable function $\alpha\colon T \times S \to \mathbb{C}$ such that*

$$h(a, b) = \int \alpha(t, a)\overline{\alpha(t, b)}d\nu(t) \quad \textit{for } (\lambda \times \lambda)\textit{-almost every } a, b \in S. \tag{41}$$

Then the bilinear form defined above is positive semidefinite, i.e., $(f, f) \geq 0$ for all $f \in \mathcal{F}_0$.

We remark that for finite spin spaces S the assumption in the proposition holds if and only if the interaction h, regarded as a real symmetric $S \times S$ matrix, is positive semidefinite. Indeed, if h has eigenvalues (λ_t) and associated (real) eigenvectors (α_t) then $h(a, b) = \sum_t \alpha_t(a)\alpha_t(b)\lambda_t$ so that being positive semidefinite yields a representation of the form (41). Conversely, having such a representation implies that $\sum_{a,b \in S} v(a)h(a, b)v(b) \geq 0$ for all $v : S \to \mathbb{R}$ whence h

is positive semidefinite. This argument may be viewed as saying that, for finite spin spaces, having a representation of the form (41) is a *necessary* condition for the conclusion that $(\cdot,\cdot)$ is positive semidefinite when the graph G is the single-edge graph $\mathbb{T}_1^1$. Further details on necessary and sufficient conditions for reflection positivity may be found in [19, 51, 52] and [50, Chapter 10].

Proof of Proposition 3. By the definition (38) of the density of σ, we have

$$\mathbb{E}\left(f(\sigma)\overline{f(R\sigma)}\right) = \int f(\sigma)\overline{f(R\sigma)}h_0(\sigma)h_0(R\sigma) \prod_{\{u,v\}\in E(V_0,V_1)} h(\sigma_u,\sigma_v)d\lambda(\sigma),$$

where $h_0 \in \mathcal{F}_0$ accounts for the part of the interaction coming from edges within V_0, and $E(V_0, V_1)$ denotes the set of edges between V_0 and V_1. Using the assumption (41) and writing $\alpha_t := \alpha(t,\cdot)$, we see that the above is equal to

$$\begin{aligned}
&\iint f(\sigma)\overline{f(R\sigma)}h_0(\sigma)h_0(R\sigma) \\
&\quad\times \prod_{\{u,v\}\in E(V_0,V_1)} \alpha_{t_{\{u,v\}}}(\sigma_u)\overline{\alpha_{t_{\{u,v\}}}(\sigma_v)}d\nu(t_{\{u,v\}})d\lambda(\sigma) \\
&= \int \prod_{\{u,v\}\in E(V_0,V_1)} d\nu(t_{\{u,v\}}) \int f(\sigma)h_0(\sigma) \\
&\quad\times \prod_{\substack{\{u,v\}\in E(V_0,V_1)\\ u\in V_0}} \alpha_{t_{\{u,v\}}}(\sigma_u) \prod_{u\in V_0} d\lambda(\sigma_u) \\
&\quad\times \int \overline{f(R\sigma)}h_0(R\sigma) \prod_{\substack{\{u,v\}\in E(V_0,V_1)\\ v\in V_1}} \overline{\alpha_{t_{\{u,v\}}}(\sigma_v)} \prod_{v\in V_1} d\lambda(\sigma_v) \\
&= \int \prod_{\{u,v\}\in E(V_0,V_1)} d\nu(t_{\{u,v\}}) \\
&\quad\times \left| \int f(\sigma)h_0(\sigma) \prod_{\substack{\{u,v\}\in E(V_0,V_1)\\ u\in V_0}} \alpha_{t_{\{u,v\}}}(\sigma_u) \prod_{u\in V_0} d\lambda(\sigma_u)\right|^2 \ge 0,
\end{aligned}$$

where in the second equality we used the fact that $Rv = u$ when $\{u,v\} \in E(V_0, V_1)$ to write

$$\begin{aligned}
&\int \overline{f(R\sigma)}h_0(R\sigma) \prod_{\substack{\{u,v\}\in E(V_0,V_1)\\ v\in V_1}} \overline{\alpha_{t_{\{u,v\}}}(\sigma_v)} \prod_{v\in V_1} d\lambda(\sigma_v) \\
&\qquad = \overline{\int f(\sigma)h_0(\sigma) \prod_{\substack{\{u,v\}\in E(V_0,V_1)\\ u\in V_0}} \alpha_{t_{\{u,v\}}}(\sigma_u) \prod_{u\in V_0} d\lambda(\sigma_u)}
\end{aligned}$$

and in the last inequality we used that ν is a non-negative measure. □

Let us consider an important example of a representation of the form (41).

Example. Let $S = \mathbb{R}^n$. Let $\tilde{h} : \mathbb{R}^n \to [0, \infty)$ be a continuous positive-definite function (in particular, $\tilde{h}(-x) = \tilde{h}(x)$). This is equivalent, by Bochner's theorem, to $\tilde{h}$ being the Fourier transform of a finite (non-negative) measure ν on $\mathbb{R}^n$. Suppose that $h \colon \mathbb{R}^n \times \mathbb{R}^n \to [0, \infty)$ is given by $h(a, b) := \tilde{h}(a - b)$. Then we may write

$$h(a, b) = \tilde{h}(a - b) = \int e^{ita}\overline{e^{itb}} d\nu(t), \quad a, b \in \mathbb{R}^n,$$

yielding a representation of the form (41). We remark that the example generalizes to the case that S is a locally compact Abelian group.

A particular function $\tilde{h}$ which we will be interested in later on is the one arising from the Gaussian interaction, namely, $\tilde{h}(x) := e^{-\frac{\beta}{2}\|x\|_2^2}$. In this case, the Fourier transform of $\tilde{h}$ is itself a scaled Gaussian density which is, in particular, non-negative. Thus $h(a, b) := \tilde{h}(a - b)$ admits a representation of the form (41).

The reflection positivity property allows to prove the following "chessboard estimate".

Proposition 4 (Chessboard estimate). *Let σ be sampled from the density* (38) *and suppose that the bilinear form defined in* (40) *is positive semidefinite. Then for any collection of real-valued bounded measurable functions $(f_v)_{v\in\Lambda}$ on $(S, \mathcal{S})$, we have*

$$\left|\mathbb{E}\left(\prod_{v\in\Lambda} f_v(\sigma_v)\right)\right|^{|\Lambda|} \le \prod_{w\in\Lambda} \mathbb{E}\left(\prod_{v\in\Lambda} f_w(\sigma_v)\right).$$

More general versions of the chessboard estimate are available and we refer the reader once again to [19] and [50, Chapter 10] for more details.

Proof of Proposition 4. Let $(f_v)_{v\in\Lambda}$ be real-valued bounded measurable functions on $(S, \mathcal{S})$. We first prove the following weaker inequality:

$$\left|\mathbb{E}\left(\prod_{v\in\Lambda} f_v(\sigma_v)\right)\right| \le \max_{w\in\Lambda} \mathbb{E}\left(\prod_{v\in\Lambda} f_w(\sigma_v)\right). \tag{42}$$

For every $\tau \colon \Lambda \to \Lambda$, denote

$$P(\tau) := \mathbb{E}\left(\prod_{v\in\Lambda} f_{\tau(v)}(\sigma_v)\right) \quad \text{and} \quad M(\tau) := \left|\{\{u, v\} \in E(\mathbb{T}_L^d) : \tau(u) \ne \tau(v)\}\right|.$$

Note that $\prod_{v\in\Lambda} f_v(\sigma_v)$ changes sign under the substitution $f_v \mapsto -f_v$ for any single v, while $\prod_{v\in\Lambda} f_w(\sigma_v)$ remains unchanged (since $|\Lambda|$ is even). Thus, (42) amounts to showing that some minimizer of M is a maximizer of P, i.e., that there exists τ_* such that $M(\tau_*) = 0$ and $P(\tau) \le P(\tau_*)$ for all τ. Let τ_* be a maximizer of P having $M(\tau_*)$ as small as possible among such maximizers. Assume towards a contradiction that $M(\tau_*) \ge 1$. Then there exist $u, w \in \Lambda$

such that $\tau(u) \neq \tau(w)$ and $\{u,w\} \in E(\mathbb{T}_L^d)$. Since rotations and translations of $\mathbb{T}_L^d$ preserve the distribution of σ we may assume without loss of generality that $u \in V_0$ and $w \in V_1$. Define two functions $F_\tau^0(\sigma) := \prod_{v\in V_0} f_{\tau(v)}(\sigma_v)$ and $F_\tau^1(\sigma) := \prod_{v\in V_1} f_{\tau(v)}(\sigma_{Rv})$, and observe that both functions belong to $\mathcal{F}_0$. Thus,

$$P(\tau) = \mathbb{E}\Big(F_\tau^0(\sigma)\cdot F_\tau^1(R\sigma)\Big) = (F_\tau^0, F_\tau^1).$$

Since the above bilinear form is positive semidefinite by assumption, the Cauchy–Schwartz inequality and the fact that τ_* is a maximizer of P imply that

$$P(\tau_*) \le \sqrt{(F_{\tau_*}^0, F_{\tau_*}^0)\cdot(F_{\tau_*}^1, F_{\tau_*}^1)} = \sqrt{P(\tau_0)\cdot P(\tau_1)} \le P(\tau_*),$$

where τ_0 and τ_1 are defined by $\tau_i|_{V_i} = \tau_*|_{V_i}$ and $\tau_i = \tau_i \circ R$. Thus, both τ_0 and τ_1 are maximizers of P. Since $\tau_0(u') = \tau_0(w')$ and $\tau_1(u') = \tau_1(w')$ for $\{u',w'\} \in E(\mathbb{T}_L^d)$, $u' \in V_0$ and $w' \in V_1$, we see that $M(\tau_0) + M(\tau_1) < 2M(\tau_*)$, which is a contradiction to the choice of τ_*.

We now show how to obtain the proposition from (42). For $w \in \Lambda$, define

$$a_w := \mathbb{E}\left(\prod_{v\in\Lambda} f_w(\sigma_v)\right),$$

and note that (42) implies that $a_w \ge 0$ (as can be seen by taking all functions to be equal). Let $\varepsilon > 0$ and define functions $(g_v)_{v\in\Lambda}$ on $(S,\mathcal{S})$ by

$$g_v(s) := \frac{f_v(s)}{(a_v+\varepsilon)^{1/|\Lambda|}}.$$

As these functions are bounded and measurable, (42) implies that

$$\left|\mathbb{E}\left(\prod_{v\in\Lambda} f_v(\sigma_v)\right)\right| \le \prod_{v\in\Lambda}(a_v+\varepsilon)^{1/|\Lambda|}\cdot\max_{w\in\Lambda}\frac{a_w}{a_w+\varepsilon} \le \prod_{v\in\Lambda}(a_v+\varepsilon)^{1/|\Lambda|}.$$

Letting ε tend to zero now yields the proposition.

In the special case when each f_v is taken to be the indicator of some $E_v \in \mathcal{S}$, the chessboard estimate implies that the probability that "E_v occurs at v for all v" is maximized when all the sets $\{E_v\}_v$ are equal. For convenience, and as this is the only use we make of the chessboard estimate in the next section, we state this explicitly as a corollary.

Corollary 1. *Let σ be sampled from the density* (38) *and suppose that the bilinear form defined in* (40) *is positive semidefinite. Then for any collection of measurable sets $(E_v)_{v\in\Lambda}$ in $(S,\mathcal{S})$, we have*

$$\mathbb{P}\Big(\sigma_v \in E_v \text{ for all } v\in\Lambda\Big) \le \max_{w\in\Lambda}\mathbb{P}\Big(\sigma_v \in E_w \text{ for all } v\in\Lambda\Big).$$

Gaussian Domination. Recall that Λ denotes the set of vertices of $\mathbb{T}_L^d$. For $\tau\colon \Lambda \to \mathbb{R}^n$, denote

$$W(\tau) := \exp\left[-\frac{\beta}{2}\sum_{\{u,v\}\in E(\mathbb{T}_L^d)} \|\tau_u - \tau_v\|_2^2\right], \tag{43}$$

where $\|\cdot\|_2$ denotes the Euclidean norm of a vector. Recall that $\Omega = \left(\mathbb{S}^{n-1}\right)^\Lambda$ denotes the space of configurations of the spin $O(n)$ model on $\mathbb{T}_L^d$, and note that since $\|\sigma_v\|_2^2 = 1$ at each vertex v for $\sigma \in \Omega$, the function W is closely related to the density of the spin $O(n)$ model (see (1)), namely,

$$\exp\left[\beta \sum_{\{u,v\}\in E(G)} \langle \sigma_u, \sigma_v\rangle\right] = e^{-\beta d|\Lambda|}\cdot W(\sigma) \quad \text{for all } \sigma \in \Omega. \tag{44}$$

A key part of the argument is the study of the function $Z : (\mathbb{R}^n)^\Lambda \to \mathbb{R}$ defined by

$$Z(\tau) := \int_\Omega W(\sigma+\tau)d\sigma.$$

Using (44), we see that the function $Z(\tau)$ at the zero function $\tau = 0$ is closely related to the partition function of the spin $O(n)$ model (see (2)), namely,

$$Z(0) = e^{\beta d|\Lambda|}\cdot Z^{\text{spin}}_{\mathbb{T}_L^d,n,\beta}. \tag{45}$$

The main step in the proof of Theorem 5 is the verification of the following *Gaussian domination* inequality,

$$Z(\tau) \le Z(0) \quad \text{for all } \tau : \Lambda \to \mathbb{R}^n, \tag{46}$$

which may be reinterpreted as an inequality of expectations in the spin $O(n)$ model. Indeed, if σ is sampled from the spin $O(n)$ model on $\mathbb{T}_L^d$ at inverse temperature β, then, by (44), (45) and (46),

$$\mathbb{E}\left(\frac{W(\sigma+\tau)}{W(\sigma)}\right) = \frac{Z(\tau)}{Z(0)} \le 1. \tag{47}$$

We establish (46) using the method of reflection positivity as described in the previous section, or, more precisely, using the chessboard estimate given in Proposition 4 and Corollary 1. To this end we first define a suitable spin system specified by a finite measure space $(S,\mathcal{S},\lambda)$ and bounded symmetric interaction $h\colon S\times S \to [0,\infty)$ to which we can apply the results of the previous section.

Consider the spin system on $\mathbb{T}_L^d$ whose configurations are pairs $\bar\sigma = (\sigma,\tau)$, where for each $v \in \Lambda$, the spins σ_v and τ_v take values in $\mathbb{R}^n$. Let η_0 be the Lebesgue measure on a bounded open set in $\mathbb{R}^n$ containing the origin. Denote $S := \mathbb{R}^n \times \mathbb{R}^n$ (with Borel σ-algebra) and let λ be the product of the uniform probability measure on $\mathbb{S}^{n-1}$ and η_0. Let the interaction $h\colon S\times S \to [0,\infty)$ be

$h((a,a'),(b,b')) := \tilde{h}(a+a'-b-b')$, where $\tilde{h}(x) := e^{-\frac{\beta}{2}\|x\|_2^2}$. Suppose $\bar{\sigma} = (\sigma,\tau)$ is sampled according to the density (38) with respect to $d\lambda(\bar{\sigma})$, and observe that this density is exactly given by $W(\sigma+\tau)/Z_{\mathbb{T}_L^d,S,\lambda,h}$. In particular, the marginal distribution of τ has density $Z(\tau)/Z_{\mathbb{T}_L^d,S,\lambda,h}$ with respect to $\eta := \prod_{v\in V(\mathbb{T}_L^d)} \eta_0$. For a function $t\colon \Lambda \to \mathbb{R}^n$ and $\varepsilon > 0$, define the event

$$E_{t,\varepsilon} := \{|\tau_v - t_v| < \varepsilon \text{ for all } v \in \Lambda\}.$$

It follows that, for η-almost every t, we have

$$\frac{Z(t)}{Z_{\mathbb{T}_L^d,S,\lambda,h}} = \lim_{\varepsilon\downarrow 0} \frac{\mathbb{P}(E_{t,\varepsilon})}{\eta(E_{t,\varepsilon})} = \lim_{\varepsilon\downarrow 0} (C_n\varepsilon^n)^{-|\Lambda|}\cdot \mathbb{P}(E_{t,\varepsilon}), \tag{48}$$

where C_n is a positive constant depending only on n and the second equality uses that η_0 is supported on an open set.

Proposition 3 and the example following its proof imply that the bilinear form defined by (40), with $\bar{\sigma}$ substituted for σ, is positive semidefinite. Thus Corollary 1 may be used for the distribution of $\bar{\sigma}$. It implies that for each $t\colon \Lambda \to \mathbb{R}^n$, $\mathbb{P}(E_{t,\varepsilon}) \le \mathbb{P}(E_{c(t),\varepsilon})$, where $c(t) : \Lambda \to \mathbb{R}^n$ is a constant function. Combining this with (48) and using that $Z(0) = Z(c)$ for any constant c, we obtain $Z(t) \le Z(0)$ for η-almost every t. The Gaussian domination inequality (46) now follows from the continuity of Z and the fact that η_0 had an arbitrarily large support.

Where the Name "Gaussian domination" Comes From. Let us give a short explanation as to the why (46) is referred to as Gaussian domination. In the previous section, we considered a general spin model with density (38) with respect to the product of some a priori finite measure space. In fact, even when the a priori measure is not finite, in certain cases one can still make sense of the same density. For instance, if this a priori measure space is the Lebesgue measure on $\mathbb{R}^n$ and the interaction h is of the same form as considered above, i.e., $h(a,b) := e^{-\frac{\beta}{2}\|a-b\|_2^2}$, then the distribution of σ is well-defined when considered up to a global addition of a constant (i.e., σ takes values in the quotient space $(\mathbb{R}^n)^\Lambda/\mathbb{R}^n$ in which two configurations are equivalent if they differ by a constant; alternatively, one could introduce a boundary condition by normalizing σ to be 0 at some vertex). This model is called the *discrete Gaussian free field* (see also Sect. 2.8 below). Since the Lebesgue measure is invariant to translations, it follows that the function Z corresponding to this model satisfies $Z(\tau) = Z(0)$ for all τ. For this reason, (46) may be viewed as a comparison to the Gaussian case. Indeed, (46) implies that certain quantities in the spin $O(n)$ model are dominated by the corresponding quantities in the discrete Gaussian free field. For instance, the infra-red bound given by (52) below becomes equality in the Gaussian case.

The Infra-red Bound. In this section, we prove an upper bound on the Fourier transform of the correlation function.

Recall that $\Lambda = \{-L+1, \dots, L\}^d$ is the set of vertices of $\mathbb{T}_L^d$. We begin by introducing the *discrete Laplacian operator* Δ on $\mathbb{C}^\Lambda$ defined by

$$(\Delta f)_u := \sum_{v\colon \{u,v\}\in E(\mathbb{T}_L^d)} (f_v - f_u), \quad \text{for } f \in \mathbb{C}^\Lambda.$$

Thus, one may regard Δ as a $\Lambda \times \Lambda$ matrix given by

$$\Delta_{xy} := \begin{cases} -2d & \text{if } x = y \\ 1 & \text{if } \{x,y\} \in E(\mathbb{T}_L^d) \\ 0 & \text{otherwise} \end{cases}.$$

Denote the inner-product on $\mathbb{C}^\Lambda$ by $(\cdot,\cdot)$, i.e.,

$$(f,g) := \sum_{u\in\Lambda} f_u \overline{g_u}, \quad \text{for } f,g \in \mathbb{C}^\Lambda.$$

Recall now the *discrete Green identity*:

$$\sum_{\{u,v\}\in E(\mathbb{T}_L^d)} (f_u - f_v)\overline{(g_u - g_v)} = (f, -\Delta g), \text{ for } f,g \in \mathbb{C}^\Lambda.$$

With a slight abuse of notation, we also write Δ and $(\cdot,\cdot)$ for the Laplacian and inner-product on $(\mathbb{C}^n)^\Lambda$, so that $\Delta f = (\Delta f^1, \dots, \Delta f^n)$ and $(f,g) = \sum_{j=1}^n (f^j, g^j)$ for $f,g \in (\mathbb{C}^n)^\Lambda$. Using this notation, we can rewrite (43) as

$$W(\tau) = \exp\left[\tfrac{1}{2}\beta(\tau, -\Delta\tau)\right], \quad \text{for } \tau \in (\mathbb{R}^n)^\Lambda.$$

Thus, if σ is sampled from the d-dimensional spin $O(n)$ model on $\mathbb{T}_L^d$ at inverse temperature β, then the Gaussian domination inequality (47) becomes

$$\mathbb{E}\left(\exp\left[-\frac{\beta}{2}\Big((\sigma+\tau, -\Delta\sigma - \Delta\tau) - (\sigma, -\Delta\sigma)\Big)\right]\right) \le 1, \quad \text{for } \tau \in (\mathbb{R}^n)^\Lambda,$$

or, equivalently, using that σ and τ are real-valued and that Δ is symmetric,

$$\mathbb{E}\Big(\exp\left[\beta(\sigma, \Delta\tau)\right]\Big) \le \exp\Big[-\tfrac{1}{2}\beta(\tau, \Delta\tau)\Big], \quad \text{for } \tau \in (\mathbb{R}^n)^\Lambda. \tag{49}$$

Substituting $\alpha\tau$ in (49) for $\alpha > 0$, and expanding both sides of the inequality using the Taylor's series for e^t, yields

$$1 + \alpha\beta \cdot \mathbb{E}\left((\sigma, \Delta\tau)\right) + \tfrac{1}{2}\alpha^2\beta^2 \cdot \mathbb{E}\left((\sigma, \Delta\tau)^2\right) + O(\alpha^3) \le 1 - \tfrac{1}{2}\alpha^2\beta(\tau, \Delta\tau) + O(\alpha^4).$$

Letting α tend to zero, and using that $\mathbb{E}((\sigma, \Delta\tau)) = 0$ by the invariance of the measure to the transformation $\sigma \mapsto -\sigma$, we obtain

$$\mathbb{E}\left((\sigma, \Delta\tau)^2\right) \le \frac{(\tau, -\Delta\tau)}{\beta}, \quad \text{for } \tau \in (\mathbb{R}^n)^\Lambda. \tag{50}$$

At this point, it seems reasonable that diagonalizing the Laplacian may prove useful, and indeed we proceed to do so. As the Laplacian matrix Δ is cyclic, it is diagonalized in the *Fourier basis*, which we now define. Let $\Lambda^* := \frac{\pi}{L}\Lambda$ denote the vertices of the dual torus. The Fourier basis elements are $\{F^k\}_{k\in\Lambda^*}$, where

$$F^k_v := e^{i\langle k,v\rangle}, \quad k \in \Lambda^*, \ v \in \Lambda,$$

and where we use the notation $\langle\cdot,\cdot\rangle$ also for the inner-product on $\mathbb{R}^d$. A straightforward computation now yields that each F^k is an eigenvector of $(-\Delta)$ with eigenvalue λ_k given by

$$\lambda_k := 2\sum_{j=1}^{d}(1-\cos(k_j)). \tag{51}$$

It is also straightforward to check that $(F^k, F^k) = |\Lambda|$ and that $(F^k, F^{k'}) = 0$ for $k \neq k'$, so that the Fourier basis is an orthogonal basis. Thus, we may write any $f \in \mathbb{C}^\Lambda$ in this basis as

$$f_v = \frac{1}{|\Lambda|}\sum_{k\in\Lambda^*}\hat{f}_k e^{i\langle k,v\rangle}, \quad v \in \Lambda,$$

where $\{\hat{f}_k\}_{k\in\Lambda^*}$ are the Fourier coefficients of f given by

$$\hat{f}_k := (f, F^k) = \sum_{v\in\Lambda} f_v e^{-i\langle k,v\rangle}, \quad k \in \Lambda^*.$$

Returning to the inequality (50), we now substitute a particular choice for τ. Let $k \in \Lambda^*$ and let $j \in \{1,\ldots,n\}$. Define $\tau := e_j F^k$, i.e., $\tau_u = e^{i\langle k,u\rangle}\cdot e_j$ for all $u \in \Lambda$, where e_j is the j-th standard basis vector in $\mathbb{R}^n$. Then, since $\Delta\tau = -\lambda_k\tau$, $(\tau,\tau) = |\Lambda|$ and $(\sigma,\tau) = \hat{\sigma}^j_k$, applying (50) to both the real and imaginary parts of τ, we obtain

$$\mathbb{E}\left(|(\sigma,\Delta\tau)|^2\right) = \lambda_k^2\cdot\mathbb{E}\left(|\hat{\sigma}^j_k|^2\right) \le \frac{(\tau,-\Delta\tau)}{\beta} = \frac{\lambda_k|\Lambda|}{\beta}.$$

Hence,

$$\mathbb{E}\left(|\hat{\sigma}^j_k|^2\right) \le \frac{|\Lambda|}{\beta\lambda_k}, \quad \text{for any } k \in \Lambda^*\setminus\{0\},\ 1 \le j \le n. \tag{52}$$

This inequality is called the infra-red bound.

The inequality (52) can be expressed as an upper bound on the Fourier transform of the two-point correlation function $\rho_{x-y} := \mathbb{E}(\langle\sigma_x,\sigma_y\rangle)$. Indeed, for any $k \in \Lambda^*$,

$$\begin{aligned}\hat{\rho}_k = \sum_{v\in\Lambda}\rho_v e^{-i\langle k,v\rangle} &= \frac{1}{|\Lambda|}\sum_{x,y\in\Lambda}\mathbb{E}(\langle\sigma_x,\sigma_y\rangle)e^{-i\langle k,(x-y)\rangle}\\ &= \sum_{j=1}^{n}\mathbb{E}\Big|\sum_{x\in\Lambda}\sigma^j_x e^{-i\langle k,x\rangle}\Big|^2 = \sum_{j=1}^{n}\mathbb{E}\left(|\hat{\sigma}^j_k|^2\right).\end{aligned}$$

As we will see in the next section, it implies that at low temperature, the Fourier transform of the two-point function in the infinite-volume limit must have a non-trivial atom at $k = 0$, implying long-range order.

Long-Range Order. By Parseval's identity,

$$\|f\|_2^2 = (f, f) = \frac{1}{|\Lambda|} \sum_{k \in \Lambda^*} |\hat{f}_k|^2, \quad \text{for } f \in \mathbb{C}^\Lambda.$$

Thus,

$$1 = \frac{1}{|\Lambda|} \sum_{v \in \Lambda} \|\sigma_v\|_2^2 = \frac{1}{|\Lambda|} \sum_{j=1}^{n} \|\sigma^j\|_2^2 = \frac{1}{|\Lambda|^2} \sum_{j=1}^{n} \sum_{k \in \Lambda^*} (\hat{\sigma}_k^j)^2, \text{ for } \sigma \in \left(\mathbb{S}^{n-1}\right)^\Lambda.$$

Therefore, the infra-red bound (52) implies that

$$\frac{1}{|\Lambda|^2} \sum_{j=1}^{n} \mathbb{E}\left((\hat{\sigma}_0^j)^2\right) \geq 1 - \frac{1}{|\Lambda|} \sum_{k \in \Lambda^* \setminus \{0\}} \frac{n}{\beta \lambda_k}. \tag{53}$$

Note that the left-hand side of (53) is precisely the quantity we want to estimate, i.e., the quantity appearing in the statement of Theorem 5, as can be seen from

$$\sum_{j=1}^{n} \mathbb{E}\left((\hat{\sigma}_0^j)^2\right) = \sum_{j=1}^{n} \mathbb{E}\left(\left(\sum_{v \in \Lambda} \sigma_v^j\right)^2\right) = \sum_{j=1}^{n} \sum_{x,y \in \Lambda} \mathbb{E}\left(\sigma_x^j \sigma_y^j\right) = \sum_{x,y \in \Lambda} \mathbb{E}\left(\langle \sigma_x \sigma_y \rangle\right).$$

Plugging in the value of λ_k from (51) into the right-hand side of (53), we identify a Riemann sum, and thus obtain

$$\begin{aligned}
\liminf_{L \to \infty} \frac{1}{|\Lambda|^2} \sum_{x,y \in \Lambda} \mathbb{E}\left(\langle \sigma_x \sigma_y \rangle\right) &\geq 1 - \frac{n}{2\beta(2\pi)^d} \int_{[-\pi,\pi]^d} \frac{1}{\sum_{j=1}^{d}(1 - \cos(t_j))} dt \\
&= 1 - \frac{n}{2\beta} \int_{[0,1]^d} \frac{1}{\sum_{j=1}^{d}(1 - \cos(\pi t_j))} dt.
\end{aligned}$$

This completes the proof of the moreover part of Theorem 5. To deduce the first part of the theorem, note that the integral is finite in dimensions $d \geq 3$, since $1 - \cos(t)$ is of order t^2 when $|t|$ is small. Thus, in dimensions $d \geq 3$, when β is sufficiently large, the quantity of interest, $|\Lambda|^{-2} \sum_{x,y \in \Lambda} (\langle \sigma_x \sigma_y \rangle)$, is bounded from below uniformly in L (for bounded values of L, we appeal directly to (53) without taking a limit). Finally, we note that the latter integral is asymptotic to $1/d$ as $d \to \infty$, as one can deduce using the law of large numbers.

As a final remark we note that the proof of Theorem 5 adapts verbatim to other a priori single-site measures (other than the uniform measure on $\mathbb{S}^{n-1}$), with the only change being the bound on $\sum_j \mathbb{E}((\hat{\sigma}_0^j)^2)$ in (53), due to the fact that we can no longer use Parseval's identity to obtain a simple deterministic bound on the sum of squares of the Fourier coefficients of σ. See, e.g., [10, Section 3.2] for details.

2.8 Slow Decay of Correlations in Spin $O(2)$ Models - Heuristic for the Berezinskii–Kosterlitz–Thouless Transition and a Theorem of Aizenman

In this section we consider the question of proving a *power-law lower bound* on the decay of correlations in the two-dimensional spin $O(2)$ model. As described in Sect. 2.2, this was achieved for the XY model at sufficiently low temperatures in the celebrated work of Fröhlich and Spencer on the Berezinskii–Kosterlitz–Thouless transition [54]. The proof is too difficult to present within the scope of our notes (see [79] for a recent presentation) and instead we start by giving a heuristic reason for the existence of the transition. The heuristic suggests that a power-law lower bound on correlations will *always* hold in the spin $O(2)$ model with a potential U of bounded support (as explained below). We then proceed by presenting a theorem of Aizenman [3], following earlier predictions by Patrascioiu and Seiler [96], who made rigorous a version of the last statement.

Heuristic for the Berezinskii–Kosterlitz–Thouless Transition and Vortices in the XY Model. To motivate the result, let us first give a heuristic argument for the Berezinskii–Kosterlitz–Thouless phase transition. Let $h : V(\mathbb{T}_L^2) \to \mathbb{R}$ be a randomly sampled *discrete Gaussian free field*. By this, we mean that $h((0,0)) := 0$ and h is sampled from the probability measure

$$\frac{1}{Z^{\mathrm{DGFF}}_{\mathbb{T}_L^2,\beta}} \exp\left[-\beta \sum_{\{u,v\}\in E(G)} (h_u - h_v)^2\right] \prod_{\substack{v\in V(\mathbb{T}_L^2)\\ v\neq(0,0)}} dm(h_v), \tag{54}$$

with $Z^{\mathrm{DGFF}}_{\mathbb{T}_L^2,\beta}$ a suitable normalization constant and dm standing for the Lebesgue measure on $\mathbb{R}$. As the expression in the exponential is a quadratic form in h, it follows that h has a multi-dimensional Gaussian distribution with zero mean. Moreover, the matrix of this quadratic form is proportional to the graph Laplacian of $\mathbb{T}_L^2$, whence the covariance structure of h is proportional to the Green's function of $\mathbb{T}_L^2$. In particular,

$$\mathrm{Var}(h_x) = \mathrm{Var}(h_x - h_0) \approx \frac{a}{\beta} \log \|x-y\|_1 \tag{55}$$

for large $\|x-y\|_1$, with a specific constant $a > 0$. Now consider the random configuration $\sigma : V(\mathbb{T}_L^2) \to \mathbb{S}^1$, with $\mathbb{S}^1$ identified with the unit circle in the complex plane, obtained by setting

$$\sigma_v := \exp(ih_v). \tag{56}$$

This configuration has some features in common with a sample of the XY model (normalized to have $\sigma_{(0,0)} = 1$). Although its density is not a product of nearest-neighbor terms, one may imagine that the main contribution to it does come from nearest-neighbor interactions, at least for large β when the differences $h_u - h_v$ of nearest neighbors tend to be small. The interaction term $-\beta(h_u - h_v)^2$ in (54) is then rather akin to an interaction term of the form $\frac{\beta}{2}\langle\sigma_u, \sigma_v\rangle$ as in the XY model (as $\langle s, t\rangle$ is the cosine of the difference of arguments between s and t and one may consider its Taylor expansion around $s = t$). The main advantage in this definition of σ is that it allows a precise calculation of correlation. Indeed, as h_x has a centered Gaussian distribution with variance given by (55), it follows that

$$\rho_{x,(0,0)} := \mathbb{E}(\langle\sigma_x, \sigma_{(0,0)}\rangle) = \mathbb{E}(\cos(h_x)) = e^{-\frac{\mathrm{Var}(h_x)}{2}} \approx \|x - y\|_1^{-\frac{a}{\beta}}, \tag{57}$$

and thus σ exhibits power-law decay of correlations.

There are many reasons why the analogy between the definition (56) and samples of the XY model should not hold. Of these, the notion of vortices has been highlighted in the literature. Suppose now that $\sigma : V(\mathbb{T}_L^2) \to \mathbb{S}^1$ is an *arbitrary* configuration. Associate to each directed edge (u, v), where $\{u, v\} \in E(\mathbb{T}_L^2)$, the difference $\theta_{(u,v)}$ in the arguments of σ_u and σ_v, with the convention that $\theta_{(u,v)} \in [-\pi, \pi)$. Call a 2×2 'square' in the graph $\mathbb{T}_L^2$ a *plaquette* (these are exactly the simple cycles of length 4 in $\mathbb{T}_L^2$). For a plaquette P, set s_P to be the sum of $\theta_{(u,v)}$ on the edges around the plaquette going in 'clockwise' order, say. We necessarily have that $s_P \in \{-2\pi, 0, 2\pi\}$ and one says that there is a *vortex* at P if $s_P \neq 0$, with charge plus or minus according to the sign of s_P. Vortices form an obstruction to defining a *height function* h for which (56) holds, as one would naturally like the differences of this h to be the $\theta_{(u,v)}$, but then one must have $s_P = 0$ for all plaquettes. Existence of vortices means that h needs to be a multi-valued function, with a non-trivial *monodromy* around plaquettes with $s_P \neq 0$.

Now take σ to be a sample of the XY model on $\mathbb{T}_L^2$ at inverse temperature β. When β is small, the model is disordered as one may deduce from the high-temperature expansion (Sect. 2.4) and there are vortices of both charges in a somewhat chaotic fashion (a 'plasma' of vortices), making the analogy with the definition (56) rather weak. Indeed, in this case there is exponential decay of correlations violating (57). However, when β is large, it can be shown (e.g., by a version of the chessboard estimate, see Sect. 2.7) that large differences $\theta_{(u,v)}$ in the angles are rare, whence vortices are rare too. Thus, one may hope vortices to *bind* together, coming in structures of small diameter of overall neutral charge (the smallest structure is a *dipole*, having one plus and one minus vortex). When this occurs, the height function h can be defined as a single-valued function at most vertices and one may hope that the analogy (56) is of relevance so that, in particular, power-law decay of correlations holds. This gives a heuristic reason for the Berezinskii–Kosterlitz–Thouless transition.

Slow Decay of Correlations for Lipschitz Spin $O(2)$ Models. The above heuristic suggests the consideration of the spin $O(2)$ model with a potential U of *bounded support*. By this we mean a measurable $U : [-1, 1] \to (-\infty, \infty]$ (allowing here $U(r) = \infty$) which satisfies

$$U(r) = \infty \quad \text{when } r < r_0 \in (-1, 1).$$

This property constrains the corresponding $O(2)$ model so that adjacent spins have difference of arguments at most $\arccos(r_0)$. Such a spin configuration may naturally be called *Lipschitz* (as in a Lipschitz function). If $r_0 \geq 0$, the maximal difference allowed is at most $\frac{\pi}{2}$ which implies that the spin configuration is *free of vortices* with probability one. If indeed vortices are the reason behind the Berezinskii–Kosterlitz–Thouless transition, then one may expect such models to always exhibit power-law decay of correlations. Patrascioiu and Seiler [96] predicted, based on rigorous mathematical statements and certain yet unproven conjectures, that a phenomenon of this kind should hold. Aizenman [3] then gave a beautiful proof of a version of the above statement, which we now proceed to present.

Theorem 6. *Let $U : [-1, 1] \to (-\infty, \infty]$ be non-increasing and satisfy*

$$U(r) = \infty \quad \text{when } r < \frac{1}{\sqrt{2}}. \tag{58}$$

Suppose that $\sigma : V(\mathbb{T}_L^2) \to \mathbb{S}^1$ is randomly sampled from the two-dimensional spin $O(2)$ model with potential U. Then, for any integer $1 \leq \ell \leq L$,

$$\max_{\substack{x,y \in V(\mathbb{T}_L^2) \\ \|x-y\|_1 \geq \ell}} \rho_{x,y} = \max_{\substack{x,y \in V(\mathbb{T}_L^2) \\ \|x-y\|_1 \geq \ell}} \mathbb{E}(\langle \sigma_x, \sigma_y \rangle) \geq \frac{1}{2\ell^2}. \tag{59}$$

We make a few remarks regarding the statement. First, one would expect that $\rho_{x,y}$ is at least a power of $\|x - y\|_1$ for all $x, y \in V(\mathbb{T}_L^2)$. The bound (59) is a little weaker in that it only shows existence of a pair x, y with this property (the proof actually gives a slightly stronger statement, see (63) below), but is still enough to rule out exponential decay of correlations in the sense we saw occurs at high temperatures (see Sect. 2.4). Second, the bound (59) can be said to hold at all temperatures in that it will continue to hold if we multiply the potential U by any constant. Third, the constraint (58) is stronger than the constraint discussed above which would prohibit vortices ($U(r) = \infty$ if $r < 0$). This stronger assumption is used in the proof and it remains open to understand the behavior with other versions of the constraint. Lastly, the fact that correlations decay *at least* as fast as a power-law under the assumptions of the theorem is a special case of the results of [56].

We proceed to the proof of Theorem 6. Let U be a potential as in the theorem and $\sigma : V(\mathbb{T}_L^2) \to \mathbb{S}^1$ be randomly sampled from the two-dimensional spin $O(2)$ model with potential U.

Step 1: Passing to $\{-1, 1\}$-Valued Random Variables. A main idea in the proof, suggested in the work of Patrascioiu and Seiler [96], is to consider the configuration σ conditioned on the y coordinate of each spin and identify an Ising-type model which is embedded in the configuration. In fact, we have already used this same idea in Sect. 2.3 when proving the non-negativity of correlations for the spin $O(n)$ model with $n \geq 2$. Recall the definitions of the signs $\varepsilon = (\varepsilon_v)_{v \in V(\mathbb{T}_L^2)}$ and the coordinate spin values $(\sigma^1, \sigma^2) = (\sigma_v^1, \sigma_v^2)_{v \in V(\mathbb{T}_L^2)}$ given just prior to Theorem 2. Recall also that $|\sigma^1|$ is determined by σ^2 and that σ is determined by (ε, σ^2). By Theorem 2, we have

$$\mathbb{E}(\varepsilon_x \varepsilon_y \mid \sigma^2) \geq 0 \quad \text{for every } x, y \in V(\mathbb{T}_L^2), \text{ almost surely.} \tag{60}$$

Moreover, as in the proof of Theorem 2, we have

$$\rho_{x,y} = 2\mathbb{E}(|\sigma_x^1| \cdot |\sigma_y^1| \cdot \mathbb{E}(\varepsilon_x \varepsilon_y \mid \sigma^2)), \tag{61}$$

Step 2: A Lower Bound on Correlations in Terms of Connectivity. A key idea in the analysis of Aizenman [3] is the consideration of the following random set of vertices

$$V_0 := \left\{ v \in V(\mathbb{T}_L^2) \colon |\sigma_v^1| \geq \frac{1}{\sqrt{2}} \right\}.$$

Note that this set is measurable with respect to σ^2. Let us consider the relevance of this set to the conditional correlations $\mathbb{E}(\varepsilon_x \varepsilon_y \mid \sigma^2)$ discussed above.

For reasons that will become clear in the next step, we introduce a second adjacency relation on the vertices $V(\mathbb{T}_L^2)$. We say that $u, v \in V(\mathbb{T}_L^2)$ are $\boxtimes$-adjacent if $\{u, v\} \in E(\mathbb{T}_L^2)$ or u, v are next-nearest-neighbors in $\mathbb{T}_L^2$ which differ in both coordinates (they are diagonal neighbors). Now observe that, almost surely,

$$\text{if } u, v \text{ are } \boxtimes\text{-adjacent and both } u, v \in V_0 \text{ then } \varepsilon_u = \varepsilon_v.$$

This is a consequence of the bounded support constraint (58) and it is here that the number $\frac{1}{\sqrt{2}}$ in that constraint is important (as we are allowing next-nearest-neighbors). Together with the non-negativity property (60), it follows that

$$\mathbb{E}(\varepsilon_x \varepsilon_y \mid \sigma^2) \geq \mathbf{1}(E_{x,y}) \quad \text{for every } x, y \in V(\mathbb{T}_L^2), \text{ almost surely,}$$

where we write $\mathbf{1}(E_{x,y})$ for the indicator function of the event

$$E_{x,y} := \{x \text{ and } y \text{ are connected in the graph on } V_0 \subseteq V(\mathbb{T}_L^2) \text{ with the } \boxtimes\text{-adjacency}\}.$$

Plugging this relation back into the identity (61) for the correlation $\rho_{x,y}$ shows that

$$\rho_{x,y} \geq 2\mathbb{E}(|\sigma_x^1| \cdot |\sigma_y^1| \cdot \mathbf{1}(E_{x,y})) \geq \mathbb{P}(E_{x,y}), \tag{62}$$

where we used that $|\sigma_x^1| \cdot |\sigma_y^1| \geq \frac{1}{2}$ when $x, y \in V_0$. We now proceed to deduce Theorem 6 from this lower bound.

Step 3: Duality for Vertex Crossings. Fix an integer $1 \leq \ell \leq L$ and define the discrete square $R := \{1, \ldots, \ell\}^2 \subseteq V(\mathbb{T}_L^2)$.

Geometric Fact: For any subset $R_0 \subseteq R$, either there is a top-bottom crossing of R with vertices of R_0 and the $\boxtimes$-adjacency or there is a left-right crossing of R with vertices of $R \setminus R_0$ and the standard nearest-neighbor adjacency (that of $\mathbb{T}_L^2$).

The fact is intuitive, though finding a simple proof requires some ingenuity. We refer the reader to Timár [118] for this and related statements.

Now consider the two events

$$E := \{\text{there is a top-bottom crossing of } R \\ \text{with vertices of } V_0 \text{ and the } \boxtimes\text{-adjacency}\},$$

$$F := \{\text{there is a left-right crossing of } R \\ \text{with vertices of } V(\mathbb{T}_L^2) \setminus V_0 \text{ and the standard adjacency}\}.$$

By rotational-symmetry of the configuration σ (its distribution is invariant under applying a global rotation of the spins), we have $\mathbb{P}(F) = \mathbb{P}(\tilde{F})$, where

$$\tilde{F} := \{\text{there is a left-right crossing of } R \\ \text{with vertices of } V_0 \text{ and the standard adjacency}\}.$$

In particular, as R is a square and since it easier to be connected in the $\boxtimes$-adjacency than in the nearest-neighbor adjacency, we conclude that

$$\mathbb{P}(E) \geq \mathbb{P}(F).$$

Lastly, the geometric fact implies that $\mathbb{P}(E \cup F) = 1$, whence

$$\begin{aligned} 1 = \mathbb{P}(E \cup F) \leq \mathbb{P}(E) + \mathbb{P}(F) \leq 2\mathbb{P}(E) &\leq 2 \sum_{\substack{x=(a,1),\, 1\leq a\leq \ell \\ y=(b,\ell),\, 1\leq b\leq \ell}} \mathbb{P}(E_{x,y}) \\ &\overset{(62)}{\leq} 2 \sum_{\substack{x=(a,1),\, 1\leq a\leq \ell \\ y=(b,\ell),\, 1\leq b\leq \ell}} \rho_{x,y}, \end{aligned} \tag{63}$$

from which Theorem 6 follows.

2.9 Exact Representations

In this section, we show that the XY model in two dimensions admits an exact representation as an integer-valued height function. Such representations are sometimes called *dual models*. We mention as another example that the dual model of the *Villain model* is the integer-valued (discrete) Gaussian free field. The reader may also consult [54, Appendix A] or [79, Section 6.1] for additional details. We mention also in this regard that the loop $O(n)$ model, discussed in Sect. 3 below, may be regarded as an *approximate* (graphical) representation for the spin $O(n)$ model (an exact representation for $n = 1$); See Sect. 3.2 for details. Another exact representation for the spin $O(n)$ model, which is not discussed here, is the Brydges–Fröhlich–Spencer random walk representation, inspired by pioneering work of Symanzik [115]; see [23, 48] for details.

We begin the treatment here in the general context of the spin $O(n)$ model with $n = 2$ and potential U on an arbitrary finite graph G as defined in (3). As the spins take values in the unit circle, we may reparameterize the spin variables according to their angle, to obtain

$$Z^{\text{spin}}_{G,n,U} = \int_{\Omega} \prod_{\{u,v\}\in E(G)} \exp\Big[-U(\langle \sigma_u, \sigma_v\rangle)\Big] d\sigma = \int_{\Omega'} \prod_{\{u,v\}\in E(G)} g(\theta_u - \theta_v) d\theta, \tag{64}$$

where $d\theta$ is the Lebesgue measure on $\Omega' := [0,1)^{V(G)}$ and $g\colon \mathbb{R} \to \mathbb{R}$ is the 1-periodic function defined by $g(t) := \exp[-U(\cos(2\pi t))]$. When the potential U is sufficiently nice, g has a Fourier expansion:

$$g(t) = \sum_{k=-\infty}^{\infty} \hat{g}(k) e^{2\pi i k t}, \quad \text{where} \quad \hat{g}(k) := \int_0^1 g(t) e^{-2\pi i k t} dt.$$

Note that, since g is real and even, we have that $\hat{g}$ is real and symmetric. Having in mind that we want to plug the Fourier series of g into (64), we note that $\theta_u - \theta_v$ is defined for $\{u, v\} \in E(G)$ up to its sign. For this reason, it is convenient to work with the directed edges of G, which we denote by $\vec{E} := \{(u, v) : \{u, v\} \in E(G)\}$. We say a function $k\colon \vec{E} \to \mathbb{Z}$ is *anti-symmetric* if $k_{(u,v)} = -k_{(v,u)}$ for all $(u, v) \in \vec{E}$. Note that for such a function, $k_{(u,v)}(\theta_u - \theta_v)$ is well-defined for any undirected edge $\{u, v\} \in E(G)$. Now, plugging in the Fourier series of g into (64) yields

$$\begin{aligned} Z^{\text{spin}}_{G,n,U} &= \sum_{\substack{k\colon \vec{E}\to\mathbb{Z} \\ k \text{ anti-symmetric}}} \int_{\Omega'} \prod_{\{u,v\}\in E(G)} \hat{g}(k_{(u,v)}) e^{2\pi i k_{(u,v)}(\theta_u - \theta_v)} d\theta \\ &= \sum_{\substack{k\colon \vec{E}\to\mathbb{Z} \\ k \text{ anti-symmetric}}} \omega_k I_k, \end{aligned}$$

where

$$\omega_k := \prod_{\{u,v\}\in E(G)} \hat{g}(k_{(u,v)}) \quad \text{and} \quad I_k := \int_{\Omega'} \prod_{\{u,v\}\in E(G)} e^{2\pi i k_{(u,v)}(\theta_u - \theta_v)} d\theta.$$

Denoting $k_u := \sum_{\{u,v\}\in E(G)} k_{(u,v)}$ for $u \in E(G)$, we may rewrite I_k as

$$I_k = \prod_{u\in V(G)} \int_{\Omega'} e^{2\pi i k_u \theta_u} d\theta.$$

From this we see that I_k is either 1 or 0 according to whether k is a *flow*, i.e., it satisfies $k_u = 0$ for all $u \in V(G)$. Therefore, we have shown that

$$Z^{\text{spin}}_{G,n,U} = \sum_{\substack{k\colon \vec{E}\to\mathbb{Z} \\ k \text{ flow}}} \omega_k.$$

When the weights ω_k are non-negative, we interpret this relation as prescribing a probability measure on flows, where the probability of a flow k is proportional to ω_k.

In order to obtain a model of height functions, we henceforth assume that G is a finite planar graph (embedded in the plane). In this case, the set of flows on G are in a 'natural' bijection with (suitably normalized) integer-valued height functions on the dual graph of G. The dual graph of G, denoted by G^*, is the planar graph obtained by placing a vertex at the center of every face of G, so that each (directed) edge e in G corresponds to the unique (directed) edge e^* in G^* which intersects e (and is rotated by 90° in the clockwise direction). Note that G^* has a distinguished vertex x_0 corresponding to the unique infinite face of G. Let $\mathcal{F}$ be the set of functions $f\colon V(G^*) \to \mathbb{Z}$ having $f(x_0) = 0$, which we call *height-functions*. Given a function $f \in \mathcal{F}$, define $k^f : \vec{E} \to \mathbb{Z}$ by $k^f_{(x,y)^*} := f(x) - f(y)$. It is straightforward to check that k^f is a flow and that $f \mapsto k^f$ is injective. It remains to show that any flow is obtained in this manner. Let k be a flow and define $f\colon V(G^*) \to \mathbb{Z}$ as follows. For any path $p = (x_0, x_1, \ldots, x_m)$ in G^* starting at x_0, we define $f(x_m) := \phi(p)$, where $\phi(p) := \sum_{i=1}^m k_{(x_{i-1},x_i)^*}$. To show that f is well-defined, we must check that $\phi(p) = \phi(p')$ for any two paths p and p' starting at x_0 and ending at the same vertex. This in turn, is the same as checking that $\phi(q) = 0$ for any path q starting and ending at x_0. It is easy to see that it suffices to check this only for any cycle q in a set of cycles Q which generates the *cycle space* of G^*. To this end, we use the fact that the cycle space of a planar graph is generated by the basic cycles which correspond to the faces. Thus, we may take Q to be the basic cycles in G^* corresponding to the *vertices* of G. That is, for every vertex $v \in V(G)$, we have a cycle $q_v \in Q$ which consists of the dual edges e^* of the edges e incident to v. Finally, the property $\phi(q_v) = 0$ is precisely the defining property $k_u = 0$ of a flow. It is now straightforward to verify that $k^f = k$. Thus, when $\hat{g}$ is non-negative, we obtain a probability measure on height-functions, where the probability of $f \in \mathcal{F}$ is proportional to

$$\omega_f := \prod_{\{x,y\}\in E(G^*)} \hat{g}(f(x) - f(y)).$$

We now specialize to the XY model, i.e., the ordinary spin $O(2)$ model as defined in (1), in which case the relevant potential is $U(t) = -\beta t$ so that $g(t) = \exp(\beta\cos(2\pi t))$. In this case, the Fourier coefficients are given by the modified Bessel functions:

$$\hat{g}(k) = I_k(\beta) := \sum_{m=0}^{\infty} \frac{1}{m!\cdot(m+k+1)!}(\beta/2)^{k+2m}.$$

Since these are positive, we have indeed found a random height-function representation for the XY model in two dimensions.

As mentioned above, the Villain model also admits a similar representation. The model is defined through (64) by taking the function g to be the "periodized Gaussian" given by

$$g(t) := \sum_{k=-\infty}^{\infty} e^{-\beta(t+k)^2/2}.$$

In this case, the Fourier coefficients are themselves Guassian,

$$\hat{g}(k) = \sqrt{2\pi/\beta} \cdot e^{-2\pi^2 k^2/\beta},$$

thus yielding a height-function representation for the Villain model in two dimensions.

3 The Loop $O(n)$ Model

3.1 Definitions

Let $\mathbb{H}$ denote the hexagonal lattice. A *loop* is a finite subgraph of $\mathbb{H}$ which is isomorphic to a simple cycle. A *loop configuration* is a spanning subgraph of $\mathbb{H}$ in which every vertex has even degree; see Fig. 4. The *non-trivial finite* connected components of a loop configuration are necessarily loops, however, a loop configuration may also contain isolated vertices and infinite simple paths. We shall often identify a loop configuration with its set of edges, disregarding isolated vertices. A *domain* H is a non-empty finite connected induced subgraph of $\mathbb{H}$ whose complement $V(\mathbb{H})\setminus V(H)$ induces a connected subgraph of $\mathbb{H}$ (in other words, it does not have "holes"). Given a domain H, we denote by $\mathsf{LoopConf}(H)$ the collection of all loop configurations ω that are contained in H. Finally, for a loop configuration ω, we denote by $L(\omega)$ the number of loops in ω and by $o(\omega)$ the number of edges of ω.

Let H be a domain and let n and x be positive real numbers. The loop $O(n)$ measure on H with edge weight x is the probability measure $\mathbb{P}_{H,n,x}$ on $\mathsf{LoopConf}(H)$ defined by

$$\mathbb{P}_{H,n,x}(\omega) := \frac{x^{o(\omega)} n^{L(\omega)}}{Z^{\text{loop}}_{H,n,x}}, \tag{65}$$

where $Z^{\text{loop}}_{H,n,x}$, the partition function, is given by

$$Z^{\text{loop}}_{H,n,x} := \sum_{\omega \in \mathsf{LoopConf}(H)} x^{o(\omega)} n^{L(\omega)}.$$

The $x = \infty$ Model. We also consider the limit of the loop $O(n)$ model as the edge weight x tends to infinity. This means restricting the model to 'optimally packed loop configurations', i.e., loop configurations having the maximum possible number of edges.

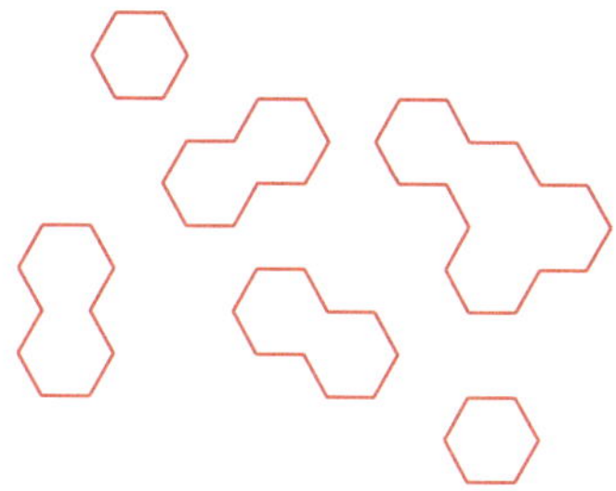

Fig. 4. A loop configuration is a subgraph of the hexagonal lattice in which every vertex has even degree.

Let H be a domain and let $n > 0$. The loop $O(n)$ measure on H with edge weight $x = \infty$ is the probability measure on $\mathsf{LoopConf}(H)$ defined by

$$\mathbb{P}_{H,n,\infty}(\omega) := \lim_{x\to\infty} \mathbb{P}_{H,n,x}(\omega) = \begin{cases} \frac{n^{L(\omega)}}{Z_{H,n,\infty}} & \text{if } o(\omega) = o_H \\ 0 & \text{otherwise} \end{cases},$$

where $o_H := \max\{o(\omega) : \omega \in \mathsf{LoopConf}(H)\}$ and $Z_{H,n,\infty}$ is the unique constant making $\mathbb{P}_{H,n,\infty}$ a probability measure. We note that if a loop configuration $\omega \in \mathsf{LoopConf}(H)$ is *fully-packed*, i.e., every vertex in $V(H)$ has degree 2, then ω is optimally packed, i.e., $o(\omega) = o_H$. In particular, if such a configuration exists for the domain H, then the measure $\mathbb{P}_{H,n,\infty}$ is supported on fully-packed loop configurations.

Like in the spin model, special cases of the loop $O(n)$ model have names of their own:

- When $n = 0$, one formally obtains the self-avoiding walk (SAW); see Sect. 3.5.
- When $n = 1$, the model is equivalent to the Ising model on the triangular lattice under the correspondence $x = e^{-2\beta}$ (the loops represent the interfaces between spins of different value), which in turn is equivalent via the Kramers–Wannier duality [83] to an Ising model on the dual hexagonal lattice.
  ∘ The special case $x = 1$, corresponding to the Ising model at infinite temperature, is critical site percolation on the triangular lattice.
  ∘ The special case $x = \infty$, corresponding to the anti-ferromagnetic Ising model at zero temperature, is a uniformly picked fully-packed loop configuration, whence its complement is a uniformly picked perfect matching of the vertices in the domain. The model is thus equivalent to the dimer model.
- When $n \geq 2$ is an integer, the model is a marginal of a discrete random Lipschitz function on the triangular lattice. When $n = 2$ this function takes integer values and when $n \geq 3$ it takes values in the n-regular tree. See Sect. 3.4 for more details. The special case $n = 2$ and $x = \infty$ is equivalent to uniform proper 4-colorings of the triangular lattice [11] (the loops are

obtained from a proper coloring with colors $\{0, 1, 2, 3\}$ as the edges bordering hexagons whose colors differ by ± 1 modulo 4).

- When $n = \infty$ and $nx^6 = \text{const}$, the model becomes the hard-hexagon model. See Sect. 3.4 for more details.
- When n is the square root of a positive integer, the model is a marginal of the dilute Potts model on the triangular lattice. See Sect. 3.4 for more details.

3.2 Relation to the Spin $O(n)$ Model

We reiterate that the loop $O(n)$ model is defined for any positive *real* n, whereas the spin $O(n)$ model is only defined for positive *integer* n. For integer n, there is a connection between the loop and the spin $O(n)$ models on a domain $H \subset \mathbb{H}$. Rewriting the partition function $Z^{\text{spin}}_{H,n,\beta}$ given by (2) using the approximation $e^t \approx 1 + t$ gives

$$\begin{aligned} Z^{\text{spin}}_{H,n,\beta} &= \int_\Omega \prod_{\{u,v\} \in E(H)} \exp\left[\beta \langle \sigma_u, \sigma_v \rangle\right] d\sigma \\ &\approx \int_\Omega \prod_{\{u,v\} \in E(H)} (1 + \beta \langle \sigma_u, \sigma_v \rangle)\, d\sigma \\ &= \sum_{\omega \subset E(H)} (\beta/n)^{o(\omega)} \int_\Omega \prod_{\{u,v\} \in E(\omega)} \langle \sqrt{n} \cdot \sigma_u, \sqrt{n} \cdot \sigma_v \rangle\, d\sigma, \\ &= \sum_{\omega \in \mathsf{LoopConf}(H)} (\beta/n)^{o(\omega)} n^{L(\omega)}, \end{aligned}$$

where the last equality follows by splitting the integral into a product of integrals on each connected component of ω and then using the following calculation.

Exercise. Let $E \subset E(\mathbb{H})$ be finite and connected. Show that

$$\int_\Omega \prod_{\{u,v\} \in E} \langle \sqrt{n} \cdot \sigma_u, \sqrt{n} \cdot \sigma_v \rangle\, d\sigma = \begin{cases} n & \text{if } E \text{ is a loop} \\ 0 & \text{otherwise} \end{cases}.$$

(see [41, Appendix A] for the calculation)

Therefore, substituting x for β/n, we obtain

$$Z^{\text{spin}}_{H,n,nx} \approx Z^{\text{loop}}_{H,n,x}.$$

In the same manner, the correlation $\rho_{u,v}$ for $u, v \in V(H)$ in the spin $O(n)$ model at inverse temperature $\beta = nx$ may be approximated as follows.

$$\rho_{u,v} = \frac{\displaystyle\int_\Omega \langle \sigma_u, \sigma_v \rangle \prod_{\{w,z\} \in E(H)} \exp\left[\beta \langle \sigma_w, \sigma_z \rangle\right]}{Z^{\mathrm{spin}}_{H,n,\beta}} \approx n \cdot \frac{\displaystyle\sum_{\lambda \in \mathsf{LoopConf}(H,u,v)} x^{o(\lambda)} n^{L'(\lambda)} J(\lambda)}{Z^{\mathrm{loop}}_{H,n,x}}, \tag{66}$$

where $\mathsf{LoopConf}(H, u, v)$ is the set of spanning subgraphs of H in which the degrees of u and v are odd and the degrees of all other vertices are even. Here, for $\lambda \in \mathsf{LoopConf}(H, u, v)$, $o(\lambda)$ is the number of edges of λ, $L'(\lambda)$ is the number of loops in λ after removing an arbitrary simple path in λ between u and v, and $J(\lambda) := \frac{3n}{n+2}$ if there are three disjoint paths in λ between u and v and $J(\lambda) := 1$ otherwise (in which case, there is a unique simple path in λ between u and v).

Exercise. Use the approximation $e^t \approx 1+t$ to obtain the asserted representation in (66) (see [41, Appendix A] for the calculation).

We remark that for $n = 1$, since $e^{\beta s} = \cosh(\beta)(1+s\cdot\tanh(\beta))$ for $s \in \{-1, 1\}$, the above expansion can be made exact by choosing $x = \tanh(\beta)$. This yields an exact duality between the ferromagnetic Ising model on the hexagonal lattice and the ferromagnetic Ising model on the triangular lattice; a special case of the so-called Kramers–Wannier duality [83]. Such duality maps the high-temperature region to the low-temperature region providing a self-dual point on self-dual lattices such as $\mathbb{Z}^2$ (and also for the hexagonal-triangular lattice pair, using an auxiliary star-triangle transformation), which turns out to be the critical point [95].

Unfortunately the above approximation is not justified for any $x > 0$ when $n > 1$. Nevertheless, (66) provides a heuristic connection between the spin and the loop $O(n)$ models and suggests that both these models reside in the same universality class. For this reason, it is natural to ask whether the prediction about the absence of phase transition is valid for the loop $O(n)$ model.

Question: Does the quantity on the right-hand side of (66) decay exponentially fast in the distance between u and v, uniformly in the domain H, whenever $n > 2$ and $x > 0$?

This question is partially answered in [41], where it is shown that for all sufficiently large n and any $x > 0$, the quantity on the right-hand side of (66) decays exponentially fast for a large class of domains H. The result is a consequence of a more detailed understanding of the loop $O(n)$ model with large n, which we elaborate on in Sect. 3.6.

3.3 Conjectured Phase Diagram and Rigorous Results

It is predicted [77, 93, 111] that the loop $O(n)$ model exhibits critical behavior when $n \in [0, 2]$; see Fig. 5. In this regime, the model should have a critical value $x_c(n)$ with the formula

$$x_c(n) := \frac{1}{\sqrt{2+\sqrt{2-n}}}. \tag{67}$$

The prediction is that for $x < x_c$ the model is *sub-critical* in the sense that the probability that a loop passing through a given point has length longer than t decays *exponentially* in t. When $x \geq x_c$, the model should be *critical*, with the same probability decaying only as a *power-law* in t and with the model exhibiting a conformally-invariant scaling limit. Furthermore, there should be two critical regimes: when $x = x_c$ and $x > x_c$, each characterized by its own conformally-invariant scaling limit (the same one for all $x > x_c$ and a different one for $x = x_c$). Kager and Nienhuis [77, Section 5.6] predict that in both cases, the loops should scale in a suitable limit to random *Schramm Löwner evolution* (SLE) curves, introduced by Schramm [104], with parameter κ satisfying

$$n = -2\cos\left(\frac{4\pi}{\kappa}\right), \tag{68}$$

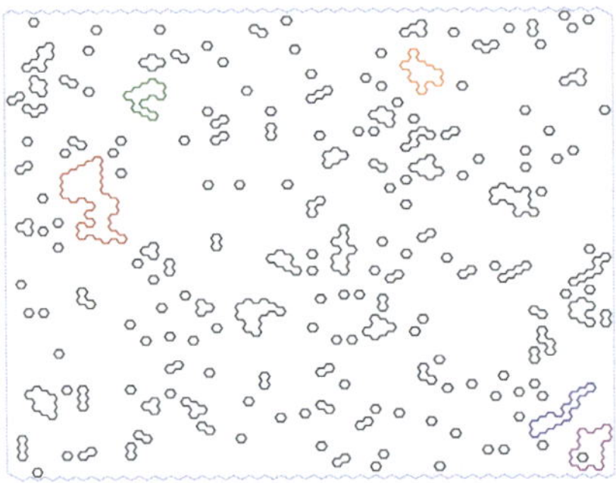

(a) $n = 1.4$ and $x = 0.57 < x_c(n)$.

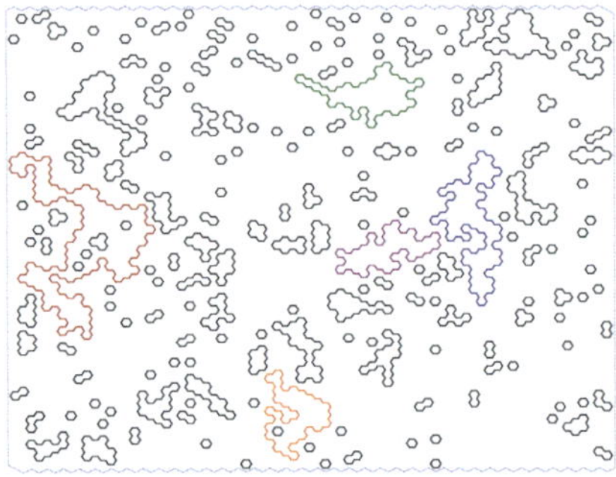

(b) $n = 1.4$ and $x = x_c(n) \approx 0.6$.

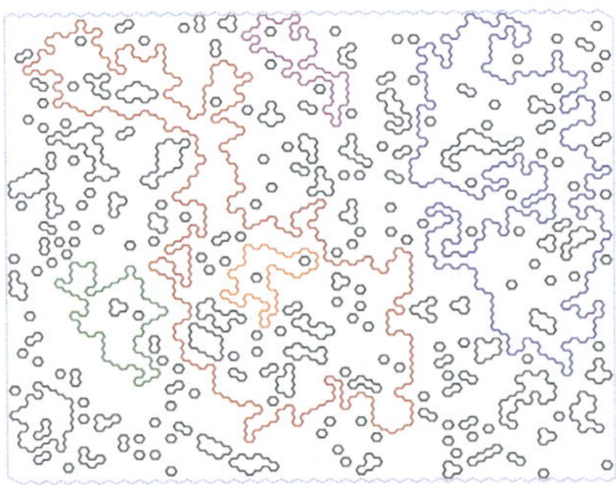

(c) $n = 1.4$ and $x = 0.63 > x_c(n)$.

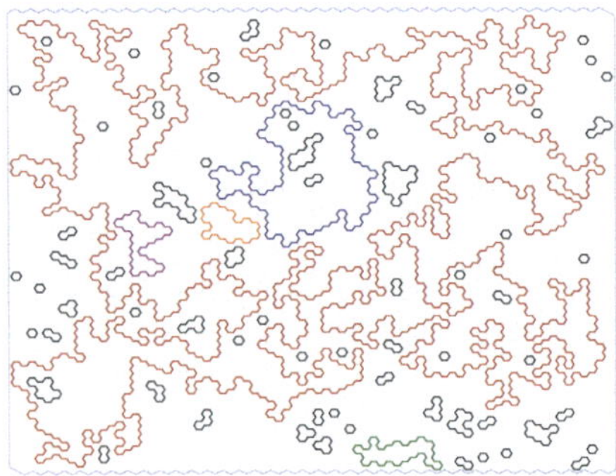

(d) $n = 0.5$ and $x = 0.6 > x_c(n)$.

Fig. 5. Samples of random loop configurations on and around the critical line. Configurations are on a 80×60 rectangular-shaped domain and are sampled via Glauber dynamics for 100 million iterations started from the empty configuration. The longest loops are highlighted (from longest to shortest: red, blue, green, purple, orange).

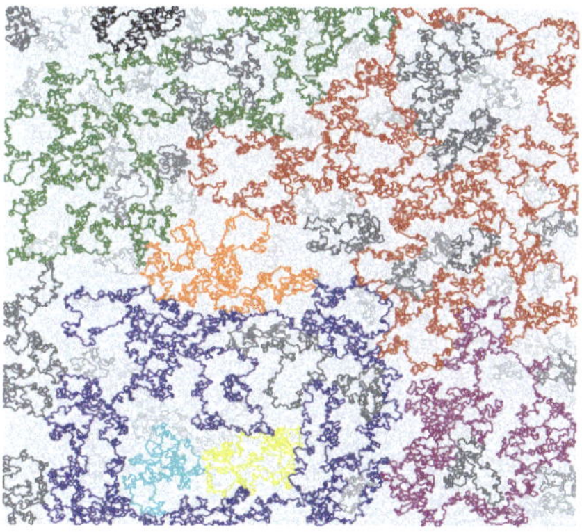

(a) $n = 1.5$ and $x = 1$.

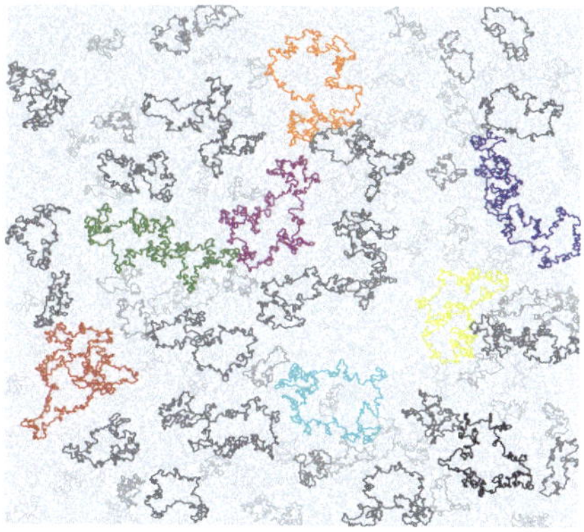

(b) $n = 2.5$ and $x = 1$.

Fig. 6. Samples of random loop configurations on a 340×300 rectangular-shaped domain.

where, however, we take the solution of the above equation to satisfy $\kappa \in [\frac{8}{3}, 4]$ when $x = x_c$ and $\kappa \in [4, 8]$ when $x > x_c$. When the parameter n satisfies $n > 2$, it is predicted that the model is always sub-critical in the sense of exponential decay of loop lengths described above. These predictions have been mathematically validated only in very special cases. See Figs. 5 and 6 for samples from the loop $O(n)$ model. See also the two bottom figures on the cover page which show samples of the model with $n = 0.5$ and $x = 0.6$.

The physics literature considers the loop $O(n)$ model also with negative n, where the model is still defined by (65) but is now a signed measure. Critical behavior is then predicted for $n \in [-2, 2]$, with the same critical value (67) for x; see [93]. Presumably formula (68) continues to describe the parameter κ of the scaling limit of the model throughout this range. However, the precise meaning of these predictions for negative n is less clear.

We list the main rigorous results on the loop $O(n)$ model (Fig. 7).

In the critical percolation case, $n = x = 1$, Smirnov [110] proved that crossing probabilities have a conformally-invariant scaling limit (given by Cardy's formula) and sketched a proof [110, 111] for convergence of the exploration path to SLE(6), following an argument of Schramm [104]. Camia and Newman [24] proved this latter convergence and further showed [25–27] that the full scaling limit is CLE(6), a member of the family of *conformal loop ensembles* introduced by Sheffield [105].

In the Ising model case, $n = 1$, it is known that $x = x_c(1) = \frac{1}{\sqrt{3}}$ is critical [95] with its interface scaling to SLE(3) [30, 32, 70, 74, 112] and its loops scaling to CLE(3) [16].

In the self-avoiding walk case, $n = 0$, it was proved by Duminil-Copin and Smirnov [44] that $x = x_c(0) = 1/\sqrt{2 + \sqrt{2}}$ is critical (it is the inverse of the connective constant of the hexagonal lattice; see Sect. 3.5), though conformal invariance and convergence to SLE have not been established. Furthermore, it was shown that for $x > x_c(0)$ the self-avoiding walk is space-filling [40].

For large values of n, it has been shown by Duminil-Copin–Peled–Samotij–Spinka [41] that there is exponential decay of loop lengths for all values of x (see Sect. 3.6 for details).

For $n \in [1,2]$, it has been shown by Duminil-Copin–Glazman–Peled–Spinka [39] that the model exhibits macroscopic loops at the critical point $x = x_c(n)$. A main tool in the proof is the observation that the Ising-type *spin representation* of the loop $O(n)$ model, in which the loops form the interfaces between ± 1 spins on the triangular lattice, satisfies the FKG lattice condition (i.e., has strong positive association) when $n \geq 1$ and $nx^2 \leq 1$. Based on this and ideas from [43] the authors deduce a *dichotomy* theorem when n and x are in this range: Either the length of the loop passing through a given vertex has exponential tail decay, or the model satisfies Russo–Seymour–Welsh (RSW) type estimates, i.e., for some $c \in (0,1)$ and any given annulus whose outer radius is twice its inner radius, the probability to see a loop which winds around the annulus is between c and $1-c$. In this range of parameters, using a technique of [58], it is further shown that the loop $O(n)$ model has a unique Gibbs measure.

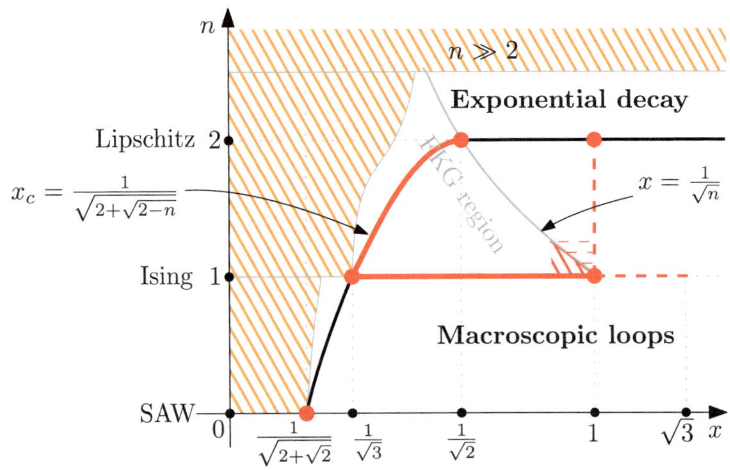

Fig. 7. The predicted phase diagram for the loop $O(n)$ model. The critical line x_c separating the regime of exponential decay from the regime of macroscopic loops is plotted. The region where a dichotomy between the two behaviors is proved is denoted FKG regime. Orange lines illustrate regions where exponential decay is proved. Red dots or lines mark regions where macroscopic loops are proven to occur. Dotted red lines denote regions where exponential decay is ruled out. Picture adapted from Glazman–Manolescu [61].

The high-temperature (ferromagnetic) Ising case, when $n = 1$ and $\frac{1}{\sqrt{3}} < x < 1$, can be shown to exhibit RSW type estimates using the techniques of [117].

Another possibility is to rely on the aforementioned dichotomy: Exponential tail decay can be ruled out by noting that the Ising model has a unique Gibbs measure in this range [50, Theorem 3.25], which then cannot have an infinite connected component of +'s, nor an infinite connected component of −'s (by an argument of Zhang which rules out coexistence of the infinite components; see, e.g., [66, Theorem 14.3]), so that every vertex is surrounded by infinitely many loops (domain walls). Remarkably, the question of convergence of the loops to CLE(6) for the high-temperature Ising model remains open.

We briefly mention some very recent developments: Beffara–Gayet [14] prove that RSW type estimates hold for the (very-)high-temperature antiferromagnetic Ising model, when $n = 1$ and $1 < x < 1 + \varepsilon$ for some small $\varepsilon > 0$. Glazman–Manolescu [62] prove RSW type estimates for $n = 2$ and $x = 1$, where the loops can be viewed as the level lines of a uniform Lipschitz function (see Sect. 3.4 below). Crawford–Glazman–Harel–Peled [34] rule out the possibility that the length of loops has exponential tail decay for $1 \leq n \leq 1 + \varepsilon$ and $1 - \varepsilon \leq x \leq 1$ for some small $\varepsilon > 0$. This implies RSW type estimates for $1 \leq n \leq 1 + \varepsilon$ and $1 - \varepsilon \leq x \leq \frac{1}{\sqrt{n}}$ by the aforementioned dichotomy. They further show that when $1 \leq n \leq 2$ and $x = 1$ the model on a toroidal domain has a *non-contractible* loop with non-negligible probability. Lastly, it is proved there that when $n = 1$ and $1 < x \leq \sqrt{3}$ (antiferromagnetic Ising) the model has loops of large diameter (comparable to that of the domain) with non-negligible probability. On the other side, Taggi [116] established exponential decay of loop lengths when $n > 0$ and $x \leq (\sqrt{2 + \sqrt{2}})^{-1} + \varepsilon(n)$, with $\varepsilon(n) > 0$ some function of n. Glazman–Manolescu [61] further showed exponential decay for any $n > 1$ and $x < \frac{1}{\sqrt{3}} + \varepsilon(n)$, with $\varepsilon(n) > 0$ another function of n.

Many interesting questions remain open for the loop $O(n)$ model, with some of the more notable ones being: proving conformal invariance at any point except $n = 1, x = \frac{1}{\sqrt{3}}$ and $n = x = 1$, and showing the existence of large loops in the remaining parts of the phase diagram: for $0 < n < 1$ and any x, or for $n = 1$ and any $x \in (\sqrt{3}, \infty]$ (it is unknown even for the dimer model case, $x = \infty$), or for $1 < n \leq 2$ and $x > x_c(n)$ (apart from the case $n = 2, x = 1$ and from the neighborhood of $n = x = 1$ mentioned above).

3.4 Equivalent Models

The equivalent models discussed below do not reside on the hexagonal lattice, but rather on its dual, the triangular lattice $\mathbb{T}$, which is obtained by placing a vertex at the center of every face (hexagon) of $\mathbb{H}$, so that each edge e of $\mathbb{H}$ corresponds to the unique edge e^* of $\mathbb{T}$ which intersects e. Since vertices of $\mathbb{T}$ are identified with faces of $\mathbb{H}$, they will be called *hexagons* instead of vertices. We also say that a vertex or an edge of $\mathbb{H}$ *borders* a hexagon if it borders the corresponding face of $\mathbb{H}$.

The Hard-Hexagon Model. As noted already in the paper [37] where the loop $O(n)$ model was introduced, taking the limit $n \to \infty$ and $nx^6 \to \lambda$ leads formally to the hard-hexagon model. As non-trivial loops (loops having length longer than 6) become less and less likely in this limit, hard-hexagon configurations consist solely of trivial loops, with each such loop contributing a factor of λ to the weight. Thus, the hard-hexagon model is the hard-core lattice gas model on the triangular lattice $\mathbb{T}$ with fugacity λ. For this model, Baxter [12] (see also [13, Chapter 14]) computed the critical fugacity

$$\lambda_c = \left(2\cos\left(\frac{\pi}{5}\right)\right)^5 = \frac{1}{2}\left(11+5\sqrt{5}\right) \approx 11.09017,$$

and showed that as λ increases beyond the threshold λ_c, the model undergoes a fluid-solid phase transition, from a homogeneous phase in which the sublattice occupation frequencies are equal, to a phase in which one of the three sublattices is favored. Additional information is obtained on the critical behavior, including the fact that the mean density of hexagons is equal for each of the three sublattices [12, Equation (13)] and the fact that the transition is of second order [12, Equation (9)]. Baxter's arguments use certain assumptions on the model which appear not to have been mathematically justified. Still, this exact solution may suggest that the loop $O(n)$ model with large n will also have a unique transition point $x_c(n)$, that $nx_c(n)^6$ will converge to λ_c as n tends to infinity and that the transition in x is of second order, with the model having a unique Gibbs state when $x = x_c(n)$.

Exact Representations as Spin Models with Local Interactions. As explained in the previous section, the loop $O(n)$ is an *approximation* of the spin $O(n)$ model, a spin model on $\mathbb{H}$ with local interactions. Here we develop *exact* representations of the loop $O(n)$ model as spin models on $\mathbb{T}$ with local interactions (see also [28]).

The spin space here will always be a discrete set S (finite or countably infinite) and we shall restrict ourselves to the set Φ of spin configurations $\varphi \in S^{\mathbb{T}}$ satisfying the condition that $|\{\varphi(y), \varphi(z), \varphi(w)\}| \le 2$ for any three mutually adjacent hexagons $y, z, w \in \mathbb{T}$. Define the 'domain walls' of a configuration $\varphi \in \Phi$ by

$$\omega_\varphi := \big\{e \in E(\mathbb{H}) : \text{the edge } e \text{ borders hexagons } y, z \in \mathbb{T} \text{ satisfying } \varphi(y) \neq \varphi(z)\big\},$$

and observe that ω_φ is a loop configuration. For a domain $H \subset \mathbb{H}$ and a fixed $s_0 \in S$, let $\Phi(H)$ be the set of $\varphi \in \Phi$ satisfying the boundary condition $\varphi(z) = s_0$ for any hexagon $z \in \mathbb{T}$ which is not entirely contained in H. Note that $\omega_\varphi \in \mathsf{LoopConf}(H)$ for $\varphi \in \Phi(H)$.

We now define a spin model in a similar manner as in Sect. 2, with one important difference: as we are now working on the triangular lattice, rather than the square lattice, it is natural to consider triangular interactions, rather than pairwise interactions. Precisely, given a (non-zero) symmetric interaction $h\colon S^3 \to [0,\infty)$, i.e., $h = h \circ \tau$ for any permutation $\tau \in S_3$, we consider the probability distribution on $\Phi(H)$ in which the probability of a configuration $\varphi \in \Phi(H)$ is proportional to

$$\prod_{\{y,z,w\}} h(\varphi(y), \varphi(z), \varphi(w)), \tag{69}$$

where the product is over triples $\{y, z, w\}$ of mutually adjacent hexagons $y, z, w \in \mathbb{T}$, at least one of which has an edge in H. We note that in order for this distribution to be well-defined when S is infinite, one must impose an implicit condition on h to ensure that the sum of the above weights is finite. Note also that this distribution is entirely defined by the collection of numbers $(h_a)_{a \in S}$ and $(h_{a,b})_{a,b \in S, a \neq b}$, where $h_a := h(a, a, a)$ and $h_{a,b} := h(a, b, b)$ ($h_{a,b}$ need not equal $h_{b,a}$ in general).

Any such choice of spin space S and interaction h, gives rise via the map $\varphi \mapsto \omega_\varphi$ to a probability distribution on $\mathsf{LoopConf}(H)$. The goal is then to choose S and h in such a manner that this distribution coincides with the loop $O(n)$ measure $\mathbb{P}_{H,n,x}$. As we now show, there is in fact a general recipe for constructing such examples.

Let G be a simple graph on vertex set S. We focus on the case that h imposes the hard-core constraint that $h_{a,b} = 0$ unless $\{a, b\}$ is an edge of G. In order words, the corresponding distribution on $\Phi(H)$ is supported on $\mathrm{Lip}(G)$, the set of configurations $\varphi \in S^{\mathbb{T}}$ satisfying the Lipschitz condition: if $y, z \in \mathbb{T}$ are adjacent hexagons then either $\varphi(y) = \varphi(z)$ or $\varphi(y)$ is adjacent to $\varphi(z)$ in G. We note that, in general, neither Φ nor $\mathrm{Lip}(G)$ is contained in the other. However, in the case that G contains no triangles, we have $\mathrm{Lip}(G) \subset \Phi$.

For simplicity, we now restrict ourselves to the case that S is finite. Let A be the adjacency matrix of the graph G, i.e., A is a real symmetric matrix, indexed by the set S and defined by $A_{a,b} := \mathbf{1}_{\{\{a,b\} \in E(G)\}}$ for $a, b \in S$. Let ψ be the Perron–Frobenius eigenvector corresponding to the largest eigenvalue λ of A, i.e., the components of ψ are non-negative and $A\psi = \lambda\psi$. We now choose the interaction to be $h_a := 1$ for all $a \in S$ and $h_{a,b} := x(\psi_a/\psi_b)^{1/6}$ for all adjacent $a, b \in S$.

Let us now show that if φ is a random spin configuration sampled according to the distribution corresponding to the above choice of h, then ω_φ is distributed according to $\mathbb{P}_{H,\lambda,x}$. To this end, we must show that, for any fixed $\omega \in \mathsf{LoopConf}(H)$, the sum of weights in (69) over configurations in $\varphi \in \Phi(H)$ having $\omega_\varphi = \omega$ is proportional to $x^{o(\omega)} n^{L(\omega)}$. To see this, observe that ω may have nested loops and that by considering these loops one-by-one, from the innermost to the outermost, it suffices to show that the contribution of any single loop ℓ is $x^{|\ell|}\lambda$. More precisely, for any $a \in S$, the sum of weights in (69) over configurations φ which equal a on the exterior side of ℓ and satisfy $\omega_\varphi = \ell$ is $x^{|\ell|}\lambda$. Indeed, this sum is precisely $\sum_{b \in S} h_{b,a}^m h_{a,b}^{m'}$, where m and m' are the number of vertices of ℓ which are incident to an edge in the exterior and interior sides of ℓ, respectively. Geometrically, if one traverses ℓ in counterclockwise direction, then m and m' are the number of left-hand and right-hand turns, respectively.

In particular, it always holds that $m = m' + 6$. Thus,

$$\sum_{b \in S} h_{b,a}^{m} h_{a,b}^{m'} = x^{m+m'} \sum_{b:\{a,b\} \in E(G)} (\psi_b/\psi_a)^{(m-m')/6} = x^{|\ell|} \frac{(A\psi)_a}{\psi_a} = x^{|\ell|}\lambda.$$

We have thus shown that if there exists a finite graph whose adjacency matrix has maximum eigenvalue n, then one may find an exact representation of the loop $O(n)$ model with any value of the edge-weight x as a spin model with local interactions (and finite spin space). Not all values of $n > 0$ are obtainable as such. The set of possible n in $(0, 2)$ is known; They are the eigenvalues of the ADE diagrams and form an infinite set in $[1, 2)$ having 2 as its sole accumulation point.

We remark that the above construction can sometimes be extended to the case when G is an infinite, locally finite graph (i.e., all vertices have finite degrees). In this case, the Perron–Frobenius eigenvector ψ is replaced by a non-zero element $\psi \in \mathbb{R}^S$ such that $\psi \geq 0$ and $\lambda\psi_a = \sum_{b:\{a,b\}\in E(G)} \psi_b$ for some $\lambda > 0$ and all $a \in S$. If such a ψ exists, then the arguments above continue to hold without change.

Lipschitz Functions. When n is a positive integer, the loop $O(n)$ model admits a height function representation [37]. Let $G = T_n$ be the n-regular tree (so that $T_1 = \{+, -\}$ and $T_2 = \mathbb{Z}$) rooted at an arbitrary vertex ρ. Here, $\text{Lip}(T_n)$ is the set of 1-Lipschitz functions from $\mathbb{T}$ to T_n (where the metrics are the graph distances), and moreover, $\text{Lip}(T_n) \subset \Phi$ as T_n does not contain triangles. In this case, one may regard ω_φ as the 'level lines' of the height function $\varphi \in \text{Lip}(T_n)$. Since $\lambda = n$ is an eigenvalue of T_n (in the sense discussed above; the eigenvector ψ is the constant function), we see that if one samples a random function $\varphi \in \text{Lip}(T_n) \cap \Phi(H)$ with probability proportional to $x^{|\omega_\varphi|}$, then ω_φ is distributed according to $\mathbb{P}_{H,n,x}$. In particular, the height function representation of the loop $O(1)$ model is an Ising model (which may be either ferromagnetic or antiferromagnetic according to whether $x < 1$ or $x > 1$) and the height function representation of the loop $O(2)$ model is a restricted Solid-On-Solid model (an integer-valued Lipschitz function). Andrews–Baxter–Forrester [7] studied a related type of restricted Solid-On-Solid models.

The Dilute Potts Model. Let $q \geq 1$ be an integer and set $S := \{0, 1, \ldots, q\}$. Let G be star graph on S in which 0 is the center, i.e., the edges of G are $\{0, i\}$ for $1 \leq i \leq q$. Here, $\text{Lip}(G) \subset \Psi$ and the elements of this set can be thought of as configurations in a *dilute Potts model*: the value 0 represents a vacancy and a positive value i represent a particle/spin of type i. In this case, the Perron–Frobenius eigenvector is given by $\psi(0) := \sqrt{q}$ and $\psi(i) := 1$ for $1 \leq i \leq q$, and its corresponding eigenvalue is $\lambda = \sqrt{q}$. Thus, this model gives a representation of the loop $O(n)$ model for any n which is the square root of an integer.

Nienhuis [94] proposed a slightly different version of the dilute Potts model, similar to the above representation. A configuration of this model in a domain of the triangular lattice is an assignment of a pair (s_z, t_z) to each vertex z of the domain, where $s_z \in \{1, \dots, q\}$ represents a spin and $t_z \in \{0, 1\}$ denotes an occupancy variable. The probability of configurations involves a hard-core constraint that nearest-neighbor occupied sites must have equal spins (reminiscent of the Edwards–Sokal coupling of the Potts and random-cluster models) and single-site, nearest-neighbor and triangle interaction terms involving the occupancy variables as in (69). With a certain choice of coupling constants, the marginal of the model on the product variables $(s_z t_z)_z$ has the same distribution as φ (with the above choice of G), and thus, the marginal on the occupancy variables is equivalent to the loop $O(n)$ model (with $n = \sqrt{q}$). Nienhuis predicts this choice of parameters to be part of the critical surface of the dilute Potts model. This prediction is partially confirmed in [39] for the loop $O(n)$ model with parameters $n \geq 1$ and $nx^2 \leq 1$.

3.5 Self-avoiding Walk and the Connective Constant

The loop $O(n)$ model as defined in Sect. 3.1 is said to have *vacant boundary conditions*. In this case, the probability of any non-empty loop configuration tends to zero as n tends to zero. Thus, under vacant boundary conditions, the $n = 0$ model is trivial. However, as can be done for the spin $O(n)$ model, here too one may impose different *boundary conditions* on the model, where the states of certain edges are pre-specified. Taking boundary conditions for which precisely two edges e_1 and e_2 on the boundary of the domain H are present, one forces a self-avoiding path between these two edges within the domain (in addition to possible loops). Under such boundary conditions, in the limit as $n \to 0$, one obtains a random self-avoiding walk. The probability of such a given self-avoiding walk γ is proportional to $x^{\text{length}(\gamma)}$. The partition function, $Z^{\text{saw}}_{H,x,e_1,e_2}$, is given by

$$Z^{\text{saw}}_{H,x,e_1,e_2} := \sum_{\substack{\gamma: e_1 \to e_2 \\ \gamma \subset H}} x^{\text{length}(\gamma)} = \sum_{k=0}^{\infty} s_{H,e_1,e_2,k} x^k,$$

where s_{H,k,e_1,e_2} is the number of self-avoiding walks of length k from e_1 to e_2 in H.

We consider the related partition function of all self-avoiding walks starting at a fixed vertex v, given by

$$Z^{\text{saw}}_x := \sum_{\gamma:\, \gamma_0 = v} x^{\text{length}(\gamma)} = \sum_{k=0}^{\infty} s_k x^k,$$

where s_k is the number of self-avoiding walks of length k starting at v. The series defining Z^{saw}_x has a radius of convergence $x_c \in [0, \infty]$ so that $Z^{\text{saw}}_x < \infty$ when $0 < x < x_c$ and $Z^{\text{saw}}_x = \infty$ when $x > x_c$. This is the critical point of the model. The critical value x_c is directly related to the exponential rate of growth of s_k.

An important and simple observation is that s_k is sub-multiplicative. That is,

$$s_{k+m} \leq s_k s_m.$$

It follows that the limit

$$\mu := \lim_{k\to\infty} s_k^{1/k}$$

exists and is finite. The number μ, called the *connective constant* of the hexagonal lattice, clearly relates to the critical value via $\mu = 1/x_c$.

Exercise. Show that μ is well-defined and that $\mu = \inf_k s_k^{1/k}$.

Exercise. Show that $2^{k/2} \le s_k \le 3 \cdot 2^{k-1}$ and deduce that $\sqrt{2} \le \mu \le 2$.

Recently, Duminil-Copin and Smirnov [44] showed the following remarkable result.

Theorem 7. *The connective constant of the hexagonal lattice is*

$$\mu = \sqrt{2 + \sqrt{2}}.$$

We do not give the proof in these notes and refer the interested reader to [44].

3.6 Large n

It is believed that the loop $O(n)$ model, although only an approximation of the spin $O(n)$ model, resides in the same universality class as the spin $O(n)$ model. Thus, as in the case of the spin $O(n)$ model, it has been conjectured that the loop $O(n)$ model exhibits exponential decay of correlations when $n > 2$. Duminil-Copin, Peled, Samotij and Spinka [41] established this for large n, showing that long loops are exponentially unlikely to occur, uniformly in the edge weight x. This result is the content of the first theorem below.

We begin with some definitions (see Fig. 8 for their illustration). Recall that the triangular lattice $\mathbb{T}$ is the dual of the hexagonal lattice. Fix a proper 3-coloring of $\mathbb{T}$ (there is a unique such coloring up to permutations of the colors), and let $\mathbb{T}^0$, $\mathbb{T}^1$ and $\mathbb{T}^2$ denote the color classes of this coloring. The 0-phase ground state ω^0_{gnd} is defined to be the (fully-packed) loop configuration consisting of trivial loops (loops of length 6) around each hexagon in $\mathbb{T}^0$. A domain $H \subset \mathbb{H}$ is said to be *of type 0* if no edge on its boundary belongs to ω^0_{gnd}, or equivalently, if every edge bordering a hexagon in $\mathbb{T}^0$ has either both or neither of its endpoints in $V(H)$. Finally, we say that a loop surrounds a vertex u of $\mathbb{H}$ if any infinite simple path in $\mathbb{H}$ starting at u intersects a vertex of this loop. In particular, if a loop passes through a vertex then it surrounds it as well.

Theorem 8. *There exist $n_0, c > 0$ such that for any $n \ge n_0$, any $x \in (0, \infty]$ and any domain H of type 0 the following holds. Suppose ω is sampled from the loop $O(n)$ model in domain H with edge weight x. Then, for any vertex $u \in V(H)$ and any integer $k > 6$,*

$$\mathbb{P}(\textit{there exists a loop of length } k \textit{ surrounding } u) \le n^{-ck}.$$

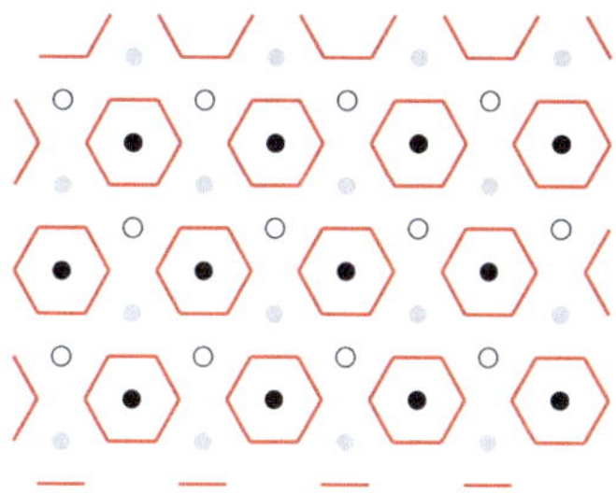

Fig. 8. A proper 3-coloring of the triangular lattice $\mathbb{T}$ (the dual of the hexagonal lattice $\mathbb{H}$), inducing a partition of $\mathbb{T}$ into three color classes $\mathbb{T}^0$, $\mathbb{T}^1$, and $\mathbb{T}^2$. The 0-phase ground state ω^0_{gnd} is the (fully-packed) loop configuration consisting of trivial loops around each hexagon in $\mathbb{T}^0$.

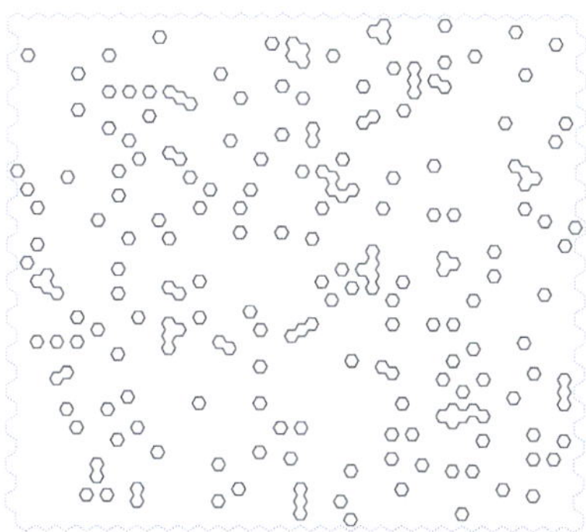

(a) $n = 8$ and $x = 0.5$. When x is small, the limiting measure is unique for domains with vacant boundary conditions, and the model is in a dilute, disordered phase.

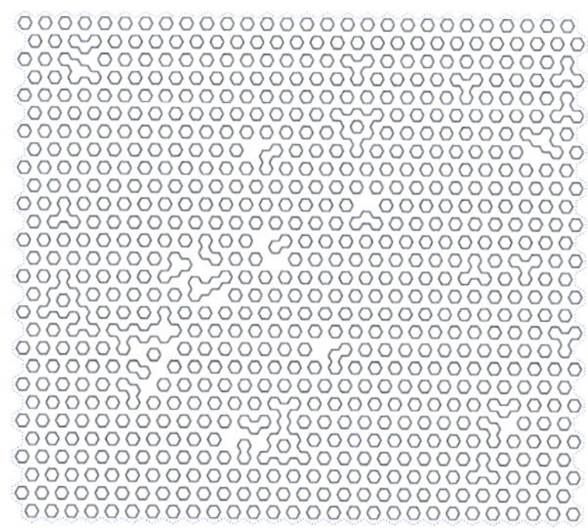

(b) $n = 8$ and $x = 2$. When n is large and x is not small, the model is in an ordered phase where typical configurations are small perturbations of the ground state.

Fig. 9. Two samples of random loop configurations with large n. Configurations are on a 60×45 domain of type 0 and are sampled via Glauber dynamics for 100 million iterations started from the empty configuration.

The reasons behind this exponential decay are quite different when x is small or large. While there is no transition to slow decay of loop lengths as x increases, there is a different kind of transition in terms of the structure of the random loop configuration and, in particular, in how the loops pack in the domain. When x is small, the model is dilute and disordered, whereas, when x is large, the model is dense and ordered (a small perturbation of the 0-phase ground state ω^0_{gnd}); these behaviors are depicted in Fig. 9. We remark that it is this latter behavior that makes the assumption that $k > 6$ necessary in the above theorem. The next theorem makes these statements precise.

Given a loop configuration ω and two vertices u and v in $\mathbb{H}$, we say that u and v are *loop-connected* if there exists a path between u and v consisting only of vertices which belong to loops in ω, and we say that u and v are *ground-connected* if there exists a path between u and v consisting only of vertices which belong to loops in $\omega \cap \omega^0_{\text{gnd}}$.

Theorem 9. *There exist $C, c > 0$ such that for any $n > 0$, any $x \in (0, \infty]$ and any domain H of type 0 the following holds. Suppose ω is sampled from the loop $O(n)$ model in domain H with edge weight x. Then, for any vertex $u \in V(H)$, on the one hand,*

$$\mathbb{P}(u \text{ is loop-connected to a vertex at distance } k \text{ from } u) \leq (C(n+1)x^6)^{ck}, k \geq 1,$$

and, on the other hand,

$$\mathbb{P}(u \text{ is ground-connected to a vertex on the boundary of } H) \geq 1 - C(n \min\{x^6, 1\})^{-c}.$$

Note that the first bound is non-trivial when both x and nx^6 are sufficiently small, while the second bound is non-trivial when both n and nx^6 are sufficiently large. Thus, when n is large, the theorem establishes a change in behavior as nx^6 transitions from small to large values. In particular, when nx^6 is small, any fixed vertex is unlikely to be surrounded by a loop (of any size). On the other hand, when nx^6 is large, any fixed hexagon in $\mathbb{T}^0$ is very likely to be surrounded by a trivial loop. The proof for small x is very similar in nature to the high-temperature case of the spin $O(n)$ model, as described in Sect. 2.4, while the proof for large x is more intricate.

We remark that several rigorous results on the behavior of general loop models on the $\mathbb{Z}^d$ lattice were obtained by Chayes, Pryadko and Shtengel [29]. These include theorems of a similar nature to our Theorems 8 and 9. The proofs there rely on reflection positivity and are thus tied to the $\mathbb{Z}^d$ lattice structure and require as well that n be integer (which is a built-in feature of the loop models studied in [29]). As we have not found a representation for the loop $O(n)$ model (even with integer n) which is reflection positive for large values of n and x, our proofs proceed by different means.

In these notes, we give an extended overview of the proofs of Theorems 8 and 9, omitting most of the technical details. The techniques of the proofs are combinatorial in nature and rely on a general principle captured by the following simple lemma.

Lemma 4. *Let $p, q > 0$ and let E and F be two events in a discrete probability space. If there exists a map $\mathsf{T}\colon E \to F$ such that $\mathbb{P}(\mathsf{T}(e)) \geq p \cdot \mathbb{P}(e)$ for every $e \in E$, and $|\mathsf{T}^{-1}(f)| \leq q$ for every $f \in F$, then*

$$\mathbb{P}(E) \leq \frac{q}{p} \cdot \mathbb{P}(F).$$

Proof. We have

$$p \cdot \mathbb{P}(E) \le \sum_{e \in E} \mathbb{P}(\mathsf{T}(e)) = \sum_{e \in E} \sum_{f \in F} \mathbb{P}(f) \mathbf{1}_{\{\mathsf{T}(e)=f\}}$$
$$= \sum_{f \in F} |\mathsf{T}^{-1}(f)| \cdot \mathbb{P}(f) \le q \cdot \mathbb{P}(F).$$

□

The results for small x are obtained via a fairly standard, and short, Peierls argument, by applying the above lemma to a map which removes loops (see Lemma 6 below). Thus, the primary focus here lies in the study of the loop $O(n)$ model for large x. In this regime, the main idea is to identify the region having an atypical structure (which is called the breakup) and apply the above lemma to a suitably defined 'repair map'. This map takes a configuration ω sampled in a domain of type 0 and having a large breakup, and returns a 'repaired' configuration in which the breakup is significantly reduced (see Fig. 11). In order to use Lemma 4, it is important that the number of preimages of a given loop configuration is exponentially smaller than the probability gain. This yields the main lemma, Lemma 5, from which the results for large x are later deduced.

Basic Definitions. A *circuit* is a simple closed path in $\mathbb{T}$ of length at least 3. We may view a circuit γ as a sequence of hexagons $(\gamma_0, \ldots, \gamma_m)$ with $\gamma_0 = \gamma_m$. Define γ^* to be the set of edges $\{\gamma_i, \gamma_{i+1}\}^* \in E(\mathbb{H})$ for $0 \le i < m$. We now state two standard geometric facts regarding circuits and domains, which may be seen as a discrete version of the Jordan curve theorem. Proofs of these facts can be found in [41, Appendix B].

Fact 1. *If γ is a circuit then the removal of γ^* splits $\mathbb{H}$ into exactly two connected components, one of which is infinite, denoted by* $\mathrm{Ext}(\gamma)$*, and one of which is finite, denoted by* $\mathrm{Int}(\gamma)$*. Moreover, each of these are induced subgraphs of $\mathbb{H}$.*

Fact 2. *Circuits are in one-to-one correspondence with domains via $\gamma \leftrightarrow \mathrm{Int}(\gamma)$.*

Hence, every domain H may be written as $H = \mathrm{Int}(\gamma)$ for some circuit γ. Note also that H is of type 0 if and only if $\gamma \subset \mathbb{T} \setminus \mathbb{T}^0$. We denote the vertex sets and edge sets of $\mathrm{Int}(\gamma), \mathrm{Ext}(\gamma)$ by $\mathrm{Int}^{\mathrm{V}}(\gamma), \mathrm{Ext}^{\mathrm{V}}(\gamma)$ and $\mathrm{Int}^{\mathrm{E}}(\gamma), \mathrm{Ext}^{\mathrm{E}}(\gamma)$, respectively. Note that

$$\{\mathrm{Int}^{\mathrm{V}}(\gamma), \mathrm{Ext}^{\mathrm{V}}(\gamma)\} \text{ is a partition of } V(\mathbb{H})$$

and that

$$\{\mathrm{Int}^{\mathrm{E}}(\gamma), \mathrm{Ext}^{\mathrm{E}}(\gamma), \gamma^*\} \text{ is a partition of } E(\mathbb{H}).$$

We also define $\mathrm{Int}^{\mathrm{hex}}(\gamma)$ to be the set of faces of $\mathrm{Int}(\gamma)$, i.e., the set of hexagons $z \in \mathbb{T}$ having all their six bordering vertices in $\mathrm{Int}^{\mathrm{V}}(\gamma)$. Since $\mathrm{Int}(\gamma)$ is induced, this is equivalent to having all six bordering edges in $\mathrm{Int}^{\mathrm{E}}(\gamma)$.

Definition 1. (c-flower, c-garden, c-cluster, vacant circuit; see Fig. 10**).** *Let* $\mathsf{c} \in \{0,1,2\}$ *and let* ω *be a loop configuration. A hexagon* $z \in \mathbb{T}^{\mathsf{c}}$ *is a* c-flower of ω *if it is surrounded by a trivial loop in* ω*. A subset* $E \subset E(\mathbb{H})$ *is a* c-garden of ω *if there exists a circuit* $\sigma \subset \mathbb{T} \setminus \mathbb{T}^{\mathsf{c}}$ *such that* $E = \mathrm{Int}^{\mathrm{E}}(\sigma) \cup \sigma^*$ *and every* $z \in \mathbb{T}^{\mathsf{c}} \cap \partial \mathrm{Int}^{\mathrm{hex}}(\sigma)$ *is a* c*-flower of* ω*. In this case, we denote* $\sigma(E) := \sigma$*. A garden of* ω *is a* c*-garden of* ω *for some* $\mathsf{c} \in \{0,1,2\}$*. A subset* $E \subset E(\mathbb{H})$ *is a* c-cluster *of* ω *if it is a* c*-garden of* ω *and it is not contained in any other garden of* ω*. A cluster of* ω *is a* c*-cluster of* ω *for some* $\mathsf{c} \in \{0,1,2\}$*. A circuit* σ *is* vacant *in* ω *if* $\omega \cap \sigma^* = \emptyset$*.*

We stress the fact that a garden/cluster is a *subset of the edges of* $\mathbb{H}$. We remark that distinct clusters of ω are edge disjoint and that, moreover, distinct c-clusters (for some c) are slightly separated from one another. Here and below, when A is a subset of vertices of a graph G, we use ∂A to denote the *(vertex) boundary* of A, i.e.,

$$\partial A := \big\{ u \in A \ : \ \{u,v\} \in E(G) \text{ for some } v \notin A \big\}.$$

Statement of the Main Lemma. For a loop configuration ω and a vacant circuit γ in ω, denote by $V(\omega,\gamma)$ the set of vertices $v \in \mathrm{Int}^{\mathrm{V}}(\gamma)$ such that the three edges of $\mathbb{H}$ incident to v are not all contained in the same cluster of $\omega \cap \mathrm{Int}^{\mathrm{E}}(\gamma)$. One may check that a vertex $v \in \mathrm{Int}^{\mathrm{V}}(\gamma)$ satisfies $v \in V(\omega,\gamma)$ if and only if v is incident to an edge which is not in any such cluster or each of its incident edges lies in a different such cluster. The set $V(\omega,\gamma)$ specifies the deviation in ω from the 0-phase ground state along the interior boundary of γ. The main lemma shows that having a large deviation is exponentially unlikely.

Lemma 5. *There exists* $c > 0$ *such that for any* $n > 0$*, any* $x \in (0,\infty]$ *and any circuit* $\gamma \subset \mathbb{T} \setminus \mathbb{T}^0$ *the following holds. Suppose* ω *is sampled from the loop* $O(n)$ *model in domain* $\mathrm{Int}(\gamma)$ *with edge weight* x*. Then, for any positive integer* k*,*

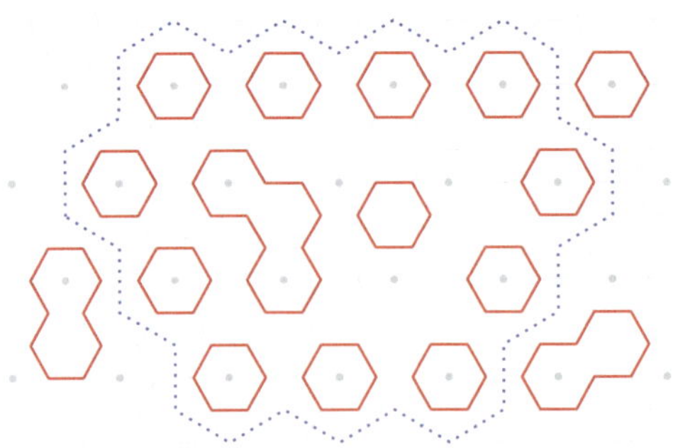

Fig. 10. A garden. The dashed line denotes a vacant circuit $\sigma \subset \mathbb{T} \setminus \mathbb{T}^{\mathsf{c}}$, where $\mathsf{c} \in \{0,1,2\}$. The edges inside σ, along with the edges crossing σ, then comprise a c-garden of ω, since every hexagon in $\mathbb{T}^{\mathsf{c}} \cap \partial \mathrm{Int}^{\mathrm{hex}}(\sigma)$ is surrounded by a trivial loop.

$$\mathbb{P}\big(\partial \mathrm{Int}^{\mathrm{V}}(\gamma) \subset V(\omega,\gamma) \text{ and } |V(\omega,\gamma)| \geq k\big) \leq (cn \cdot \min\{x^6, 1\})^{-k/15}.$$

Definition of the Repair Map. Fix a circuit $\gamma \subset \mathbb{T} \setminus \mathbb{T}^0$ and set $H := \mathrm{Int}(\gamma)$. Consider a loop configuration ω such that γ is vacant in ω. The idea of the repair map is to modify ω in the interior of γ, keeping the configuration unchanged in the exterior of γ, as follows (see Fig. 11 for an illustration):

- Edges in 1-clusters are shifted down "into the 0-phase".
- Edges in 2-clusters are shifted up "into the 0-phase".
- Edges in 0-clusters are left untouched.
- The remaining edges which are not inside (the shifted) clusters, but are in the interior of γ (these edges will be called *bad*), are overwritten to "match" the 0-phase ground state, ω^0_{gnd}.

In order to formalize this idea, we need a few definitions. A *shift* is a graph automorphism of $\mathbb{T}$ which maps every hexagon to one of its neighbors. We henceforth fix a shift $\uparrow$ which maps $\mathbb{T}^0$ to $\mathbb{T}^1$ (and hence, maps $\mathbb{T}^1$ to $\mathbb{T}^2$ and $\mathbb{T}^2$ to $\mathbb{T}^0$), and denote its inverse by $\downarrow$. A shift naturally induces mappings on the vertices and edges of $\mathbb{H}$. We shall use the same symbols, $\uparrow$ and $\downarrow$, to denote these mappings. Endow $\mathbb{T}$ with the coordinate system given by $(0,2)\mathbb{Z} + (\sqrt{3},1)\mathbb{Z}$ and recall that $(\mathbb{T}^0, \mathbb{T}^1, \mathbb{T}^2)$ are the color classes of an arbitrary proper 3-coloring of $\mathbb{T}$. In the figures, we make the choice that $(0,0) \in \mathbb{T}^0$ and $(0,2) \in \mathbb{T}^1$ so that $\uparrow$ is the map $(a,b) \mapsto (a, b+2)$.

For a loop configuration $\omega \in \mathsf{LoopConf}(H)$ and $\mathsf{c} \in \{0,1,2\}$, let $E^{\mathsf{c}}(\omega) \subset E(\mathbb{H})$ be the union of all c-clusters of ω, and define

$$E^{\mathrm{bad}}(\omega) := \big(\mathrm{Int}^{\mathrm{E}}(\gamma) \cup \gamma^*\big) \setminus \big(E^0(\omega) \cup E^1(\omega)^{\downarrow} \cup E^2(\omega)^{\uparrow}\big), \tag{70}$$

$$\overline{E}(\omega) := \big(\mathrm{Int}^{\mathrm{E}}(\gamma) \cup \gamma^*\big) \setminus \big(E^0(\omega) \cup E^1(\omega) \cup E^2(\omega)\big). \tag{71}$$

One may check that $\{E^0(\omega), E^1(\omega), E^2(\omega), \overline{E}(\omega)\}$ is a partition of $\mathrm{Int}^{\mathrm{E}}(\gamma) \cup \gamma^*$ so that $\omega \cap E^0(\omega)$, $\omega \cap E^1(\omega)$, $\omega \cap E^2(\omega)$ and $\omega \cap \overline{E}(\omega)$ are pairwise disjoint loop configurations. Finally, we define the *repair map*

$$R \colon \mathsf{LoopConf}(H) \to \mathsf{LoopConf}(H)$$

by

$$R(\omega) := \big(\omega \cap E^0(\omega)\big) \cup \big(\omega \cap E^1(\omega)\big)^{\downarrow} \cup \big(\omega \cap E^2(\omega)\big)^{\uparrow} \cup \big(\omega^0_{\mathrm{gnd}} \cap E^{\mathrm{bad}}(\omega)\big).$$

The fact that the mapping is well-defined, i.e., that $R(\omega)$ is indeed in $\mathsf{LoopConf}(H)$, is not completely straightforward. However, it is indeed well-defined and, moreover,

$$\omega \cap E^0(\omega), \quad (\omega \cap E^1(\omega))^{\downarrow} \cup (\omega \cap E^2(\omega))^{\uparrow} \quad \text{and} \quad \omega^0_{\mathrm{gnd}} \cap E^{\mathrm{bad}}(\omega)$$

are pairwise disjoint loop configurations in $\mathsf{LoopConf}(H)$.

Proof of Lemma 5. Let V be such that $\partial \mathrm{Int}^{\mathrm{V}}(\gamma) \subset V \subset \mathrm{Int}^{\mathrm{V}}(\gamma)$. We first bound the probability of the event

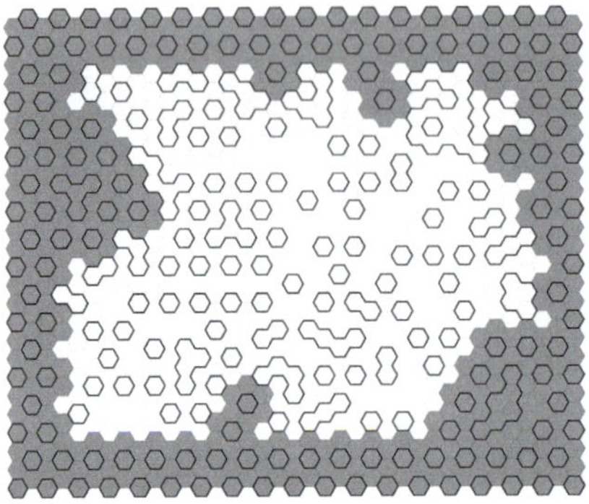

(a) The breakup is found by exploring 0-flowers from the boundary.

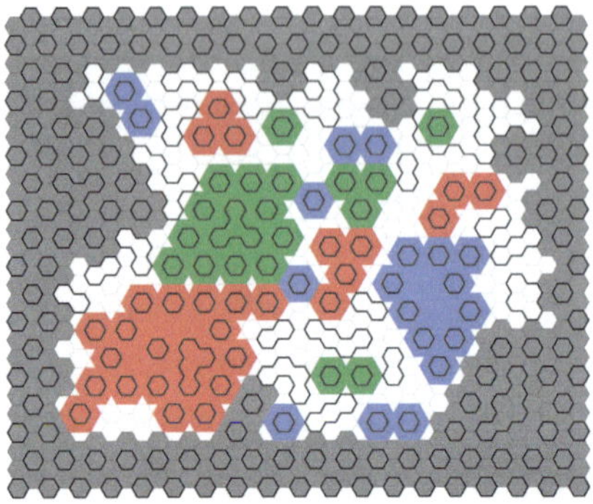

(b) The clusters are found within the breakup (with 0/1/2-clusters shown in green/red/blue).

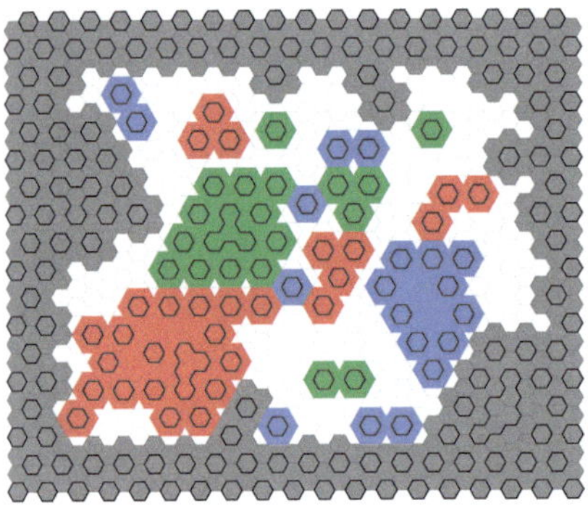

(c) Bad edges are discarded.

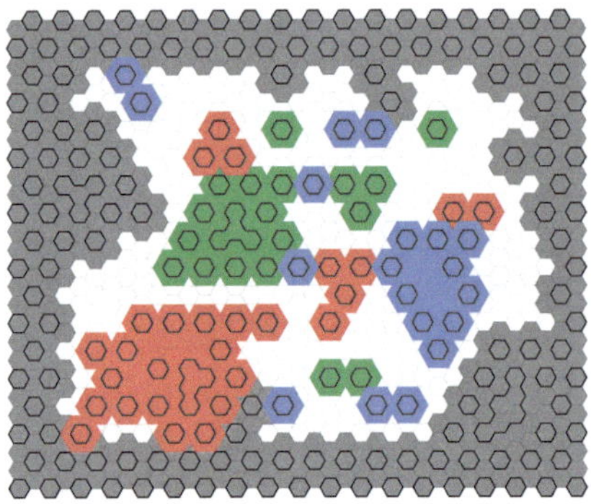

(d) The clusters are shifted into the 0-phase.

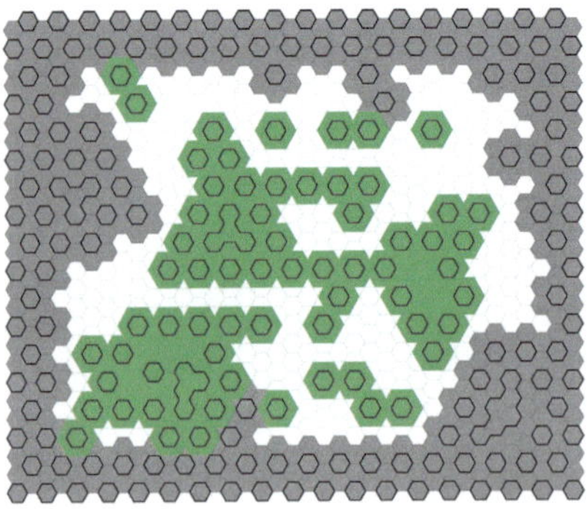

(e) The empty area outside the shifted clusters is now compatible with the 0-phase ground state.

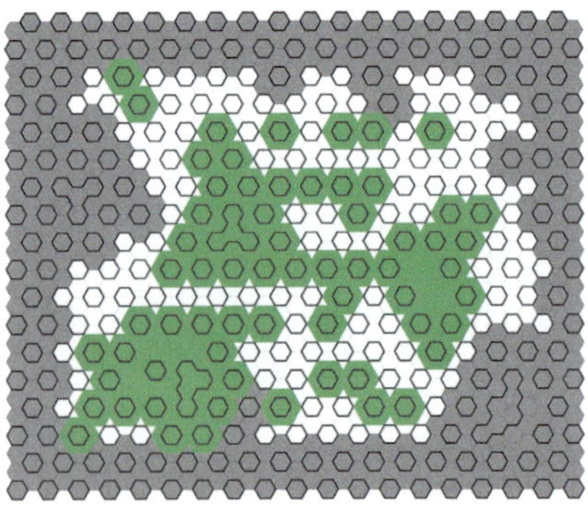

(f) Trivial loops are packed in the empty area outside the shifted clusters.

Fig. 11. An illustration of finding the breakup and applying the repair map in it. The initial loop configuration is modified step-by-step, resulting in a loop configuration with many more loops and at least as many edges.

$$E_V := \{\omega \in \mathsf{LoopConf}(H) \ : \ V(\omega,\gamma) = V\}.$$

To do so, we wish to apply Lemma 4 to the repair map. To this end, we must estimate the gain in probability (parameter p in Lemma 4) and the number of preimages of a given configuration (parameter q in Lemma 4). Let $n > 0$ and $x > 0$. We may assume that $n \geq 1$ and $nx^6 \geq 1$ as otherwise the lemma is trivial. Then

$$\mathbb{P}(R(\omega)) \geq (n \cdot \min\{x^6, 1\})^{|V|/15} \cdot \mathbb{P}(\omega) \qquad \text{for } \omega \in E_V, \tag{72}$$

$$|E_V \cap R^{-1}(\omega')| \leq (2\sqrt{2})^{|V|} \qquad \text{for } \omega' \in \mathsf{LoopConf}(H). \tag{73}$$

The proof of (72) is based on a precise understanding of the change in the number of edges $\Delta o := o(R(\omega)) - o(\omega)$ and in the number of loops $\Delta L := L(R(\omega)) - L(\omega)$. Namely, one may show (see Fig. 11(c)) that

$$\Delta o = |V| - |\omega \cap \overline{E}(\omega)| \quad \text{and} \quad \Delta L = |V|/6 - L(\omega \cap \overline{E}(\omega)).$$

Using this, one deduces that

$$0 \leq \Delta o \leq |V| \quad \text{and} \quad \Delta L \geq \tfrac{|V|}{15} + \tfrac{|\Delta o|}{10},$$

from which (72) easily follows.

The proof of (73) relies on the fact that the only loss of information incurred by the repair map is in the bad edges (see Fig. 11(c)). More precisely, the mapping $\omega \mapsto (R(\omega), \omega \cap E(V))$ is injective on E_V. Thus, the size of $E_V \cap R^{-1}(\omega')$ is at most the number of subsets of $E(V)$. Since $|E(V)| \leq 3|V|/2$, we obtain (73).

Now, using (72) and (73), Lemma 4 implies that

$$\mathbb{P}(E_V) \leq (2\sqrt{2})^{|V|} \cdot (n \cdot \min\{x^6, 1\})^{-|V|/15}.$$

To complete the proof, we must sum over the possible choices for V. For this, we use a connectivity property of $V(\omega,\gamma)$. Let $\mathbb{H}^\times$ be the graph obtained from $\mathbb{H}$ by adding an edge between each pair of opposite vertices of every hexagon, so that $\mathbb{H}^\times$ is a 6-regular non-planar graph. One may show that $V(\omega,\gamma)$ is connected in $\mathbb{H}^\times$ whenever $\partial\mathrm{Int}^{\mathrm{V}}(\gamma) \subset V(\omega,\gamma)$. Thus, recalling Lemma 3, when $n \cdot \min\{x^6, 1\}$ is sufficiently large, we have

$$\begin{aligned}
&\mathbb{P}\big(\partial\mathrm{Int}^{\mathrm{V}}(\gamma) \subset V(\omega,\gamma) \text{ and } |V(\omega,\gamma)| \geq k\big) \\
&\leq \sum_{\substack{V:\ |V|\geq k \\ V \text{ connected in } \mathbb{H}^\times \\ \partial\mathrm{Int}^{\mathrm{V}}(\gamma) \subset V \subset \mathrm{Int}^{\mathrm{V}}(\gamma)}} \mathbb{P}(E_V) \\
&\leq \sum_{\ell=k}^{\infty} C^\ell \cdot (2\sqrt{2})^\ell \cdot (n \cdot \min\{x^6, 1\})^{-\ell/15} \\
&\leq (cn \cdot \min\{x^6, 1\})^{-k/15}.
\end{aligned}$$

□

Proofs of Main Theorems. The proofs of the theorems for large x mostly rely on the main lemma, Lemma 5. The results for small x follow via a Peierls argument, the basis of which is given by the following lemma that gives an upper bound on the probability that a given collection of loops appears in a random loop configuration.

Lemma 6. *Let H be a domain, let $n, x > 0$ and let ω be sampled from the loop $O(n)$ model in domain H with edge weight x. Then, for any $A \in \mathsf{LoopConf}(H)$, we have*

$$\mathbb{P}(A \subset \omega) \le n^{L(A)} x^{o(A)}.$$

Proof. Consider the map

$$\mathsf{T}\colon \{\omega \in \mathsf{LoopConf}(H) : A \subset \omega\} \to \mathsf{LoopConf}(H)$$

defined by

$$\mathsf{T}(\omega) := \omega \setminus A.$$

Clearly, T is well-defined and injective. Moreover, since $L(\mathsf{T}(\omega)) = L(\omega) - L(A)$ and $o(\mathsf{T}(\omega)) = o(\omega) - o(A)$, we have

$$\mathbb{P}(\mathsf{T}(\omega)) = \mathbb{P}(\omega) \cdot n^{-L(A)} x^{-o(A)}.$$

Hence, the statement follows from Lemma 4.

Recall the notion of a loop surrounding a vertex given prior to Theorem 8.

Corollary 2. *Let H be a domain, let $n, x > 0$ and let ω be sampled from the loop $O(n)$ model in domain H with edge weight x. Then, for any vertex $u \in V(H)$ and any positive integer k, we have*

$$\mathbb{P}(\textit{there exists a loop of length } k \textit{ surrounding } u) \le kn(2x)^k.$$

Moreover, for any $u_1, \dots, u_m \in V(H)$ and $k_1, \dots, k_m \ge 1$ with $k = k_1 + \dots + k_m$, we have

$$\mathbb{P}(\forall i \textit{ there exists a distinct loop of length } k_i \textit{ passing through } u_i) \le (2n)^m (2x)^k,$$

Proof. Denote by a_k the number of simple paths of length k in $\mathbb{H}$ starting at a given vertex. Clearly, $a_k \le 3 \cdot 2^{k-1}$. It is then easy to see that the number of loops of length k surrounding u is at most $ka_{k-1} \le k2^k$. Thus, the result follows by the union bound and Lemma 6.

The moreover part follows similarly from Lemma 6 by noting that there are at most $a_{k_1} \cdots a_{k_m} \le 2^{m+k}$ loop configurations A consisting of exactly k loops with the i-th loop having length k_i and passing through u_i.

The main lemma, Lemma 5, shows that for a given circuit γ (which is contained in $\mathbb{T} \setminus \mathbb{T}^{\mathsf{c}}$ for some c), it is unlikely that the set $V(\omega, \gamma)$ is large. The set $V(\omega, \gamma)$ specifies deviations from the ground states which are 'visible' from γ, i.e., deviations which are not 'hidden' inside clusters. In Theorem 8, we claim that it is unlikely to see long loops surrounding a given vertex. Any such long loop constitutes a deviation from all ground states. Thus, the theorem would follow from the main lemma (in the main case, when x is large) if the long loop was captured in $V(\omega,\gamma)$. The next lemma (whose proof we omit) bridges the gap between the main lemma and the theorem, by showing that even when a deviation is not captured by $V(\omega,\gamma)$, there is necessarily a smaller circuit σ which captures it in $V(\omega,\sigma)$.

Lemma 7. *Let ω be a loop configuration, let $\gamma \subset \mathbb{T} \setminus \mathbb{T}^0$ be a vacant circuit in ω and let L be a non-trivial loop of ω in* $\mathrm{Int}(\gamma)$. *Then there exists* $\mathsf{c} \in \{0,1,2\}$ *and a circuit $\sigma \subset \mathbb{T} \setminus \mathbb{T}^{\mathsf{c}}$ such that* $\mathrm{Int}(\sigma) \subset \mathrm{Int}(\gamma)$, *$\sigma$ is vacant in ω and* $V(L) \cup \partial\mathrm{Int}^{\mathrm{V}}(\sigma) \subset V(\omega,\sigma)$.

Proof of Theorem 8. Suppose that n_0 is a sufficiently large constant, let $n \geq n_0$, let $x \in (0,\infty]$, let H be a domain of type 0 and let $u \in V(H)$. Let ω be sampled from the loop $O(n)$ model in domain H with edge weight x. We shall estimate the probability that u is surrounded by a non-trivial loop of length k. We consider two cases, depending on the relative values of n and x.

Suppose first that $nx^6 < n^{1/50}$. Since $n \geq n_0$, we may assume that $2x \leq n^{-4/25}$ and that $kn^{-k/120} \leq 1$ for all $k > 0$. By Corollary 2, for every $k \geq 7$,

$$\mathbb{P}(\text{there exists a loop of length } k \text{ surrounding } u) \leq kn(2x)^k \leq kn^{1-4k/25} \leq kn^{-k/60} \leq n^{-k/120}.$$

Suppose now that $nx^6 \geq n^{1/50}$. Since $n \geq n_0$, we may assume that $n \cdot \min\{x^6,1\}$ is sufficiently large for our arguments to hold. Let $L \subset H$ be a non-trivial loop of length k surrounding u. Note that if $L \subset \omega$ then by Lemma 7, for some $\mathsf{c} \in \{0,1,2\}$, there exists a circuit $\sigma \subset \mathbb{T} \setminus \mathbb{T}^{\mathsf{c}}$ such that $\mathrm{Int}(\sigma) \subset H$, σ is vacant in ω and $V(L) \cup \partial\mathrm{Int}^{\mathrm{V}}(\sigma) \subset V(\omega,\sigma)$. Using the fact that H is of type 0, the domain Markov property and Lemma 5 imply that for every fixed circuit $\sigma \subset \mathbb{T} \setminus \mathbb{T}^{\mathsf{c}}$ with $\mathrm{Int}(\sigma) \subset H$,

$$\mathbb{P}\big(\sigma \text{ vacant and } V(L) \cup \partial\mathrm{Int}^{\mathrm{V}}(\sigma) \subset V(\omega,\sigma)\big) \leq (cn \cdot \min\{x^6,1\})^{-|V(L)\cup\partial\mathrm{Int}^{\mathrm{V}}(\sigma)|/15}.$$

Thus, denoting by $\mathcal{G}(u)$ the set of circuits σ contained in $\mathbb{T} \setminus \mathbb{T}^{\mathsf{c}}$ for some $\mathsf{c} \in \{0,1,2\}$ and having $u \in \mathrm{Int}^{\mathrm{V}}(\sigma)$, we obtain

$$\begin{aligned}\mathbb{P}(L \subset \omega) &\leq \sum_{\sigma \in \mathcal{G}(u)} (cn \cdot \min\{x^6,1\})^{-|V(L)\cup\partial\mathrm{Int}^{\mathrm{V}}(\sigma)|/15} \\ &\leq \sum_{\ell=1}^{\infty} D^{\ell} (cn \cdot \min\{x^6,1\})^{-\max\{k,\ell\}/15} \leq (cn \cdot \min\{x^6,1\})^{-k/15},\end{aligned}$$

where we used the facts that the length of a circuit σ such that $|\partial\mathrm{Int}^{\mathrm{V}}(\sigma)| = \ell$ is at most 3ℓ, that the number of circuits σ of length at most 3ℓ with $u \in$

$\mathrm{Int}^{\mathrm{V}}(\sigma)$ is bounded by D^ℓ for some sufficiently large constant D, and in the last inequality we used the assumption that $n \cdot \min\{x^6, 1\}$ is sufficiently large. Since the number of loops of length k surrounding a given vertex is smaller than $k2^k$, our assumptions that $nx^6 \geq n^{1/50}$ and $n \geq n_0$ yield

$$\mathbb{P}(\text{there exists a loop of length } k \text{ surrounding } u) \leq k2^k (cn^{1/50})^{-k/15} \leq n^{-k/800}.$$

□

Proof of Theorem 9. Let $n > 0$, let $x \in (0, \infty]$, let H be a domain of type 0 and let $u \in V(H)$. Let ω be sampled from the loop $O(n)$ model in domain H with edge weight x.

We first prove the upper bound on the probability that u is loop-connected to distance k. For this, we may assume that x and nx^6 are sufficiently small as the bound is trivial otherwise. Denote $u_0 := u$ and observe that if u is loop-connected to some vertex at distance k from u, then there exist integers $m \geq 1$, $\ell_1, \ldots, \ell_m \geq 6$ and vertices $u_1, \ldots, u_m \in V(H)$ such that $k \leq \ell := \ell_1 + \cdots + \ell_m$ and, for all $1 \leq i \leq m$, $\mathrm{dist}(u_i, u_{i-1}) \leq \ell_i$ and u_i belongs to a distinct loop of ω of length ℓ_i. Thus, summing over the possible choices (for brevity, we omit the conditions on ℓ_i and u_i in the sum below) and applying Corollary 2, we obtain

$$\begin{aligned}
&\mathbb{P}\begin{pmatrix} u \text{ is loop-connected} \\ \text{to distance } k \end{pmatrix} \\
&\leq \sum_{\substack{\ell \geq k \\ \ell/6 \geq m \geq 1}} \sum_{\substack{\ell_1, \ldots, \ell_m \\ u_1, \ldots, u_m}} \mathbb{P}(\forall i \ u_i \text{ belongs to a distinct loop of length } \ell_i) \\
&\leq \sum_{\ell \geq k} \ell \cdot 2^\ell \cdot 3^\ell \cdot (2n+1)^{\ell/6} (2x)^\ell \\
&\leq \sum_{\ell \geq k} (C(n+1)x^6)^{\ell/6} \leq C(C(n+1)x^6)^{k/6}.
\end{aligned}$$

We now prove the lower bound on the probability that u is ground-connected to the boundary of H, i.e., that u and v are ground-connected for some $v \in \partial V(H)$. For this, we may assume that both n and nx^6 are sufficiently large as the bound is trivial otherwise. Assume that u is not ground-connected to the boundary of H. Let $A(\omega)$ be the set of vertices of $\mathbb{H}$ belonging to loops in $\omega \cap \omega^0_{\mathrm{gnd}}$ and let $B(\omega)$ be the unique infinite connected component of $A(\omega) \cup (V(\mathbb{H}) \setminus V(H))$. Note that $u \notin B(\omega)$ by assumption and define the *breakup* $\mathcal{C}$ to be the connected component of $\mathbb{H} \setminus B(\omega)$ containing u. One may check that the subgraph induced by $\mathcal{C}$ is a domain of type 0, and that the enclosing circuit Γ (i.e., the circuit satisfying $\mathcal{C} = \mathrm{Int}^{\mathrm{V}}(\Gamma)$, which exists by Fact 2) is vacant in ω and is contained in $\mathbb{T} \setminus \mathbb{T}^0$. Furthermore, we have $\partial \mathrm{Int}^{\mathrm{V}}(\Gamma) \subset V(\omega, \Gamma)$. Indeed, this follows as Γ is vacant in ω and, by the definition of $B(\omega)$, no vertex of $\partial \mathrm{Int}^{\mathrm{V}}(\Gamma)$ belongs to a trivial loop surrounding a hexagon in $\mathbb{T}^0$. Thus, denoting by $\mathcal{G}$ the set of circuits $\gamma \subset \mathbb{T} \setminus \mathbb{T}^0$ having $u \in \mathrm{Int}^{\mathrm{V}}(\gamma)$, Lemma 5 implies that

$$\begin{aligned}
&\mathbb{P}\binom{u \text{ is not ground-connected}}{\text{to the boundary of } H} \\
&\le \sum_{\gamma\in\mathcal{G}} \mathbb{P}\big(\gamma \text{ vacant and } \partial\mathrm{Int}^{\mathrm{V}}(\gamma) \subset V(\omega,\gamma)\big) \\
&\le \sum_{\gamma\in\mathcal{G}} (cn\cdot\min\{x^6,1\})^{-|\partial\mathrm{Int}^{\mathrm{V}}(\gamma)|/15} \\
&\le \sum_{k\ge 1} D^k (cn\cdot\min\{x^6,1\})^{-k/15} \le C(n\cdot\min\{x^6,1\})^{-c},
\end{aligned}$$

where in the third inequality we used the facts that the length of a circuit γ such that $|\partial\mathrm{Int}^{\mathrm{V}}(\gamma)| = k$ is at most $3k$, and that the number of circuits of length at most $3k$ surrounding u is bounded by D^k for some sufficiently large constant D.

References

1. Aizenman, M.: Absence of an intermediate phase for a general class of one-component ferromagnetic models. Phys. Rev. Lett. **54**(8), 839 (1985)
2. Aizenman, M.: Rigorous studies of critical behavior. II. In: Statistical Physics and Dynamical Systems (Köszeg, 1984). Progress in Physics, vol. 10, pp. 453–481. Birkhäuser, Boston (1985)
3. Aizenman, M.: On the slow decay of O(2) correlations in the absence of topological excitations: remark on the Patrascioiu-Seiler model. J. Statist. Phys. **77**(1–2), 351–359 (1994)
4. Aizenman, M., Barsky, D.J., Fernández, R.: The phase transition in a general class of Ising-type models is sharp. J. Statist. Phys. **47**(3–4), 343–374 (1987)
5. Aizenman, M., Bricmont, J., Lebowitz, J.: Percolation of the minority spins in high-dimensional Ising models. J. Stat. Phys. **49**(3), 859–865 (1987)
6. Aizenman, M., Duminil-Copin, H., Sidoravicius, V.: Random currents and continuity of Ising model's spontaneous magnetization. Commun. Math. Phys. **334**(2), 719–742 (2015)
7. Andrews, G.E., Baxter, R.J., Forrester, P.J.: Eight-vertex sos model and generalized Rogers-Ramanujan-type identities. J. Stat. Phys. **35**(3), 193–266 (1984)
8. Balaban, T.: A low temperature expansion for classical N-vector models. I. A renormalization group flow. Commun. Math. Phys. **167**(1), 103–154 (1995)
9. Balister, P.N., Bollobás, B.: Counting regions with bounded surface area. Commun. Math. Phys. **273**(2), 305–315 (2007)
10. Bauerschmidt, R.: Ferromagnetic spin systems (2016). Lecture notes: http://www.statslab.cam.ac.uk/~rb812/doc/spin.pdf
11. Baxter, R.: Colorings of a hexagonal lattice. J. Math. Phys. **11**(3), 784–789 (1970)
12. Baxter, R.J.: Hard hexagons: exact solution. J. Phys. A **13**(3), L61–L70 (1980)
13. Baxter, R.J.: Exactly Solved Models in Statistical Mechanics. Academic Press Inc. [Harcourt Brace Jovanovich Publishers], London (1989). Reprint of the 1982 original
14. Beffara, V., Gayet, D.: Percolation without FKG. Preprint arXiv:1710.10644 (2017)
15. Benassi, C., Lees, B., Ueltschi, D.: Correlation inequalities for classical and quantum XY models. Preprint arXiv:1611.06019 (2016)

16. Benoist, S., Hongler, C.: The scaling limit of critical Ising interfaces is CLE(3). arXiv preprint arXiv:1604.06975 (2016)
17. Berezinskii, V.: Destruction of long-range order in one-dimensional and two-dimensional systems possessing a continuous symmetry group. II. Quantum systems. Sov. J. Exp. Theor. Phys. **34**, 610 (1972)
18. Berlin, T.H., Kac, M.: The spherical model of a ferromagnet. Phys. Rev. **86**(6), 821 (1952)
19. Biskup, M.: Reflection positivity and phase transitions in lattice spin models. In: Methods of Contemporary Mathematical Statistical Physics, pp. 1–86 (2009)
20. Bollobás, B.: The Art of Mathematics. Cambridge University Press, New York (2006)
21. Bonato, C., Perez, J.F., Klein, A.: The Mermin-Wagner phenomenon and cluster properties of one-and two-dimensional systems. J. Stat. Phys. **29**(2), 159–175 (1982)
22. Bricmont, J., Fontaine, J., Landau, L.: On the uniqueness of the equilibrium state for plane rotators. Commun. Math. Phys. **56**(3), 281–296 (1977)
23. Brydges, D., Fröhlich, J., Spencer, T.: The random walk representation of classical spin systems and correlation inequalities. Commun. Math. Phys. **83**(1), 123–150 (1982)
24. Camia, F., Newman, C.: Critical percolation exploration path and SLE_6: a proof of convergence. Probab. Theory Relat. Fields **139**(3–4), 473–519 (2007)
25. Camia, F., Newman, C.M.: Continuum nonsimple loops and 2D critical percolation. J. Stat. Phys. **116**(1), 157–173 (2004)
26. Camia, F., Newman, C.M.: The full scaling limit of two-dimensional critical percolation. Preprint arXiv:math/0504036 (2005)
27. Camia, F., Newman, C.M.: Two-dimensional critical percolation: the full scaling limit. Commun. Math. Phys. **268**(1), 1–38 (2006)
28. Cardy, J.: Conformal field theory and statistical mechanics. In: Exact Methods in Low-Dimensional Statistical Physics and Quantum Computing, pp. 65–98 (2008)
29. Chayes, L., Pryadko, L.P., Shtengel, K.: Intersecting loop models on $\mathbb{Z}^d$: rigorous results. Nucl. Phys. B **570**(3), 590–614 (2000)
30. Chelkak, D., Duminil-Copin, H., Hongler, C., Kemppainen, A., Smirnov, S.: Convergence of Ising interfaces to Schramm's SLE curves. C. R. Math. Acad. Sci. Paris **352**(2), 157–161 (2014)
31. Chelkak, D., Hongler, C., Izyurov, K.: Conformal invariance of spin correlations in the planar Ising model. Ann. Math. (2) **181**(3), 1087–1138 (2015)
32. Chelkak, D., Smirnov, S.: Universality in the 2D Ising model and conformal invariance of fermionic observables. Invent. Math. **189**(3), 515–580 (2012)
33. Cohen-Alloro, O., Peled, R.: Rarity of extremal edges in random surfaces and other theoretical applications of cluster algorithms. Preprint arXiv:1711.00259 (2017)
34. Crawford, N., Glazman, A., Harel, M., Peled, R.: Macroscopic loops in the loop $O(n)$ model via the XOR trick (in preparation)
35. Dimock, J.: The renormalization group according to Balaban I: Small fields. Rev. Math. Phys. **25**(07), 1330,010 (2013)
36. Dobrushin, R., Shlosman, S.: Absence of breakdown of continuous symmetry in two-dimensional models of statistical physics. Commun. Math. Phys. **42**(1), 31–40 (1975)
37. Domany, E., Mukamel, D., Nienhuis, B., Schwimmer, A.: Duality relations and equivalences for models with O(n) and cubic symmetry. Nuclear Phys. B **190**(2), 279–287 (1981)

38. Duminil-Copin, H.: Parafermionic observables and their applications to planar statistical physics models. Ensaios Matematicos **25**, 1–371 (2013)
39. Duminil-Copin, H., Glazman, A., Peled, R., Spinka, Y.: Macroscopic loops in the loop $O(n)$ model at Nienhuis' critical point. Preprint arXiv:1707.09335 (2017). J. Eur. Math. Soc. (2017, to appear)
40. Duminil-Copin, H., Kozma, G., Yadin, A.: Supercritical self-avoiding walks are space-filling. Ann. Inst. H. Poincaré Probab. Stat. **50**(2), 315–326 (2014)
41. Duminil-Copin, H., Peled, R., Samotij, W., Spinka, Y.: Exponential decay of loop lengths in the loop $O(n)$ model with large n. Commun. Math. Phys. **349**, 777–817 (2016)
42. Duminil-Copin, H., Raoufi, A., Tassion, V.: Sharp phase transition for the random-cluster and Potts models via decision trees. Preprint arXiv:1705.03104 (2017). Ann. Math. (2017, published)
43. Duminil-Copin, H., Sidoravicius, V., Tassion, V.: Continuity of the phase transition for planar Potts models with $1 \leq q \leq 4$. Commun. Math. Phys. **349**(1), 47–107 (2017)
44. Duminil-Copin, H., Smirnov, S.: The connective constant of the honeycomb lattice equals $\sqrt{2+\sqrt{2}}$. Ann. Math. (2) **175**(3), 1653–1665 (2012)
45. Duminil-Copin, H., Tassion, V.: A new proof of the sharpness of the phase transition for Bernoulli percolation and the Ising model. Commun. Math. Phys. **343**(2), 725–745 (2016)
46. Dyson, F.J., Lieb, E.H., Simon, B.: Phase transitions in quantum spin systems with isotropic and nonisotropic interactions. In: Nachtergaele, B., Solovej, J.P., Yngvason, J. (eds.) Statistical Mechanics, pp. 163–211. Springer, Heidelberg (1978)
47. Edwards, R.G., Sokal, A.D.: Generalization of the Fortuin-Kasteleyn-Swendsen-Wang representation and Monte Carlo algorithm. Phys. Rev. D (3) **38**(6), 2009–2012 (1988)
48. Fernández, R., Fröhlich, J., Sokal, A.D.: Random Walks, Critical Phenomena, and Triviality in Quantum Field Theory. Springer, Heidelberg (2013)
49. Fisher, M.E.: Critical temperatures of anisotropic Ising lattices. II. General upper bounds. Phys. Rev. **162**(2), 480 (1967)
50. Friedli, S., Velenik, Y.: Statistical mechanics of lattice systems: a concrete mathematical introduction. Cambridge University Press (2017, to appear). http://www.unige.ch/math/folks/velenik/smbook
51. Fröhlich, J., Israel, R., Lieb, E.H., Simon, B.: Phase transitions and reflection positivity. I. General theory and long range lattice models. In: Nachtergaele, B., Solovej, J.P., Yngvason, J. (eds.) Statistical Mechanics, pp. 213–246. Springer, Heidelberg (1978)
52. Fröhlich, J., Israel, R.B., Lieb, E.H., Simon, B.: Phase transitions and reflection positivity. II. Lattice systems with short-range and Coulomb interactions. In: Nachtergaele, B., Solovej, J.P., Yngvason, J. (eds.) Statistical Mechanics, pp. 247–297. Springer, Heidelberg (1980)
53. Fröhlich, J., Simon, B., Spencer, T.: Infrared bounds, phase transitions and continuous symmetry breaking. Commun. Math. Phys. **50**(1), 79–95 (1976)
54. Fröhlich, J., Spencer, T.: The Kosterlitz-Thouless transition in two-dimensional Abelian spin systems and the Coulomb gas. Commun. Math. Phys. **81**(4), 527–602 (1981). http://projecteuclid.org/getRecord?id=euclid.cmp/1103920388
55. Fröhlich, J., Spencer, T.: Massless phases and symmetry restoration in abelian gauge theories and spin systems. Commun. Math. Phys. **83**(3), 411–454 (1982)

56. Gagnebin, M., Miłoś, P., Peled, R.: In preparation
57. Gagnebin, M., Velenik, Y.: Upper bound on the decay of correlations in a general class of O(N)-symmetric models. Commun. Math. Phys. **332**(3), 1235–1255 (2014)
58. Georgii, H.O., Higuchi, Y.: Percolation and number of phases in the two-dimensional Ising model. J. Math. Phys. **41**(3), 1153–1169 (2000)
59. Ginibre, J.: Simple proof and generalization of Griffiths' second inequality. Phys. Rev. Lett. **23**(15), 828 (1969)
60. Ginibre, J.: General formulation of Griffiths' inequalities. Commun. Math. Phys. **16**, 310–328 (1970)
61. Glazman, A., Manolescu, I.: Exponential decay in the loop $O(n)$ model: $n > 1$, $x < \frac{1}{\sqrt{3}} + \varepsilon(n)$. arXiv preprint arXiv:1810.11302 (2018)
62. Glazman, A., Manolescu, I.: Uniform Lipschitz functions on the triangular lattice have logarithmic variations. arXiv preprint arXiv:1810.05592 (2018)
63. Griffiths, R.: Correlation in Ising ferromagnets I, II. J. Math. Phys. **8**, 478–489 (1967)
64. Griffiths, R.B.: Correlations in Ising ferromagnets. III. Commun. Math. Phys. **6**(2), 121–127 (1967)
65. Grimmett, G.: The Random-Cluster Model. Grundlehren der Mathematischen Wissenschaften [Fundamental Principles of Mathematical Sciences], vol. 333. Springer, Berlin (2006)
66. Häggström, O., Jonasson, J., et al.: Uniqueness and non-uniqueness in percolation theory. Probab. Surv. **3**, 289–344 (2006)
67. Hasenbusch, M.: The two-dimensional XY model at the transition temperature: a high-precision Monte Carlo study. J. Phys. A Math. Gen. **38**(26), 5869 (2005)
68. Herring, C., Kittel, C.: On the theory of spin waves in ferromagnetic media. Phys. Rev. **81**(5), 869 (1951)
69. Hohenberg, P.: Existence of long-range order in one and two dimensions. Phys. Rev. **158**(2), 383 (1967)
70. Hongler, C., Kytölä, K.: Ising interfaces and free boundary conditions. J. Amer. Math. Soc. **26**(4), 1107–1189 (2013)
71. Hurst, C., Sherman, S.: Griffiths' theorems for the ferromagnetic Heisenberg model. Phys. Rev. Lett. **22**(25), 1357 (1969)
72. Ioffe, D., Shlosman, S., Velenik, Y.: 2D models of statistical physics with continuous symmetry: the case of singular interactions. Commun. Math. Phys. **226**(2), 433–454 (2002)
73. Ito, K.: Clustering in low-dimensional SO(N)-invariant statistical models with long-range interactions. J. Stat. Phys. **29**(4), 747–760 (1982)
74. Izyurov, K.: Smirnov's observable for free boundary conditions, interfaces and crossing probabilities. Commun. Math. Phys. **337**(1), 225–252 (2015)
75. Jasnow, D., Fisher, M.E.: Broken symmetry and decay of order in restricted dimensionality. Phys. Rev. Lett. **23**(6), 286 (1969)
76. Kac, M., Thompson, C.J.: Spherical model and the infinite spin dimensionality limit. Phys. Norveg. **5**(3–4), 163–168 (1971)
77. Kager, W., Nienhuis, B.: A guide to stochastic Löwner evolution and its applications. J. Stat. Phys. **115**(5–6), 1149–1229 (2004)
78. Kennedy, T., King, C.: Spontaneous symmetry breakdown in the Abelian Higgs model. Commun. Math. Phys. **104**(2), 327–347 (1986)
79. Kharash, V., Peled, R.: The Fröhlich-Spencer proof of the Berezinskii-Kosterlitz-Thouless transition. Preprint arXiv:1711.04720 (2017)

80. Komura, Y., Okabe, Y.: Large-scale Monte Carlo simulation of two-dimensional classical XY model using multiple GPUs. J. Phys. Soc. Jpn. **81**(11), 113,001 (2012)
81. Kosterlitz, J.M., Thouless, D.J.: Long range order and metastability in two dimensional solids and superfluids. (Application of dislocation theory). J. Phys. C Solid State Phys. **5**, L124–L126 (1972)
82. Kosterlitz, J.M., Thouless, D.J.: Ordering, metastability and phase transitions in two-dimensional systems. J. Phys. C Solid State Phys. **6**(7), 1181–1203 (1973)
83. Kramers, H.A., Wannier, G.H.: Statistics of the two-dimensional ferromagnet. I. Phys. Rev. **2**(60), 252–262 (1941)
84. Kupiainen, A.J.: On the $1/n$ expansion. Commun. Math. Phys. **73**(3), 273–294 (1980). http://projecteuclid.org/getRecord?id=euclid.cmp/1103907876
85. Lebowitz, J.L., Mazel, A.E.: Improved Peierls argument for high-dimensional Ising models. J. Stat. Phys. **90**(3–4), 1051–1059 (1998)
86. McBryan, O.A., Spencer, T.: On the decay of correlations in SO(n)-symmetric ferromagnets. Commun. Math. Phys. **53**(3), 299–302 (1977)
87. McCoy, B., Wu, T.: The Two-Dimensional Ising Model. Harvard University Press, Cambridge (1973)
88. Mermin, N.D.: Absence of ordering in certain classical systems. J. Math. Phys. **8**(5), 1061–1064 (1967)
89. Mermin, N.D., Wagner, H.: Absence of ferromagnetism or antiferromagnetism in one-or two-dimensional isotropic Heisenberg models. Phys. Rev. Lett. **17**(22), 1133 (1966)
90. Messager, A., Miracle-Sole, S., Ruiz, J.: Upper bounds on the decay of correlations in SO(N)-symmetric spin systems with long range interactions. Ann. Inst. H. Poincaré Sect. A (NS) **40**(1), 85–96 (1984)
91. Miłoś, P., Peled, R.: Delocalization of two-dimensional random surfaces with hard-core constraints. Commun. Math. Phys. **340**(1), 1–46 (2015)
92. Naddaf, A.: On the decay of correlations in non-analytic SO(n)-symmetric models. Commun. Math. Phys. **184**(2), 387–395 (1997)
93. Nienhuis, B.: Exact critical point and critical exponents of O(n) models in two dimensions. Phys. Rev. Lett. **49**(15), 1062–1065 (1982)
94. Nienhuis, B.: Locus of the tricritical transition in a two-dimensional q-state Potts model. Phys. A Stat. Mech. Its Appl. **177**(1–3), 109–113 (1991)
95. Onsager, L.: Crystal statistics. I. A two-dimensional model with an order-disorder transition. Phys. Rev. (2) **65**, 117–149 (1944)
96. Patrascioiu, A., Seiler, E.: Phase structure of two-dimensional spin models and percolation. J. Stat. Phys. **69**(3–4), 573–595 (1992)
97. Peierls, R.: On Ising's model of ferromagnetism. Math. Proc. Camb. Philos. Soc. **32**, 477–481 (1936)
98. Peled, R.: High-dimensional Lipschitz functions are typically flat. Ann. Probab. **45**(3), 1351–1447 (2017)
99. Pfister, C.E.: On the symmetry of the Gibbs states in two dimensional lattice systems. Commun. Math. Phys. **79**(2), 181–188 (1981)
100. Pinson, H.: Rotational invariance of the 2D spin-spin correlation function. Commun. Math. Phys. **314**(3), 807–816 (2012)
101. Polyakov, A.M.: Interaction of Goldstone particles in two dimensions. Applications to ferromagnets and massive Yang-Mills fields. Phys. Lett. B **59**(1), 79–81 (1975)
102. Richthammer, T.: Translation-invariance of two-dimensional Gibbsian point processes. Commun. Math. Phys. **274**(1), 81–122 (2007)

103. Sakai, A.: Lace expansion for the Ising model. Commun. Math. Phys. **272**(2), 283–344 (2007)
104. Schramm, O.: Scaling limits of loop-erased random walks and uniform spanning trees. Israel J. Math. **118**, 221–288 (2000)
105. Sheffield, S.: Exploration trees and conformal loop ensembles. Duke Math. J. **147**(1), 79–129 (2009)
106. Shlosman, S.B.: Absence of continuous symmetry breaking in two-dimensional models of statistical physics. Theor. Math. Phys. **33**(1), 897–902 (1977)
107. Shlosman, S.B.: Decrease of correlations in two-dimensional models with continuous symmetry group. Theor. Math. Phys. **37**(3), 1118–1120 (1978)
108. Simon, B.: Mean field upper bound on the transition temperature in multicomponent ferromagnets. J. Stat. Phys. **22**(4), 491–493 (1980)
109. Simon, B., Sokal, A.D.: Rigorous entropy-energy arguments. J. Stat. Phys. **25**(4), 679–694 (1981)
110. Smirnov, S.: Critical percolation in the plane: conformal invariance, Cardy's formula, scaling limits. C. R. Acad. Sci. Paris Sér. I Math. **333**(3), 239–244 (2001)
111. Smirnov, S.: Towards conformal invariance of 2D lattice models. In: International Congress of Mathematicians, vol. II, pp. 1421–1451. European Mathematical Society, Zürich (2006)
112. Smirnov, S.: Conformal invariance in random cluster models. I. Holomorphic fermions in the Ising model. Ann. Math. (2) **172**(2), 1435–1467 (2010)
113. Stanley, H.E.: Spherical model as the limit of infinite spin dimensionality. Phys. Rev. **176**, 718–722 (1968)
114. Sylvester, G.S.: The Ginibre inequality. Commun. Math. Phys. **73**(2), 105–114 (1980)
115. Symanzik, K.: Euclidean quantum field theory. In: Proceedings of the 45th International School of Physics 'Enrico Fermi': Local Quantum Theory. Acaedmic Press, New York, London (1969)
116. Taggi, L.: Shifted critical threshold in the loop $O(n)$ model at arbitrary small n. Preprint arXiv:1806.09360 (2018)
117. Tassion, V.: Crossing probabilities for Voronoi percolation. Ann. Probab. **44**(5), 3385–3398 (2016)
118. Timár, Á.: Boundary-connectivity via graph theory. Proc. Amer. Math. Soc. **141**(2), 475–480 (2013)
119. Ueltschi, D.: Quantum spin systems and phase transitions (2013). Lecture notes: http://www.ueltschi.org/publications.php
120. Wolff, U.: Collective Monte Carlo updating for spin systems. Phys. Rev. Lett. **62**(4), 361 (1989)

Nature vs. Nurture in Discrete Spin Dynamics

Daniel L. Stein[1,2,3](✉)

[1] Department of Physics and Courant Institute of Mathematical Sciences, New York University, New York, NY 10012, USA
daniel.stein@nyu.edu
[2] NYU-ECNU Institutes of Physics and Mathematical Sciences, NYU Shanghai, 3663 Zhongshan Road North, Shanghai 200062, China
[3] Santa Fe Institute, 1399 Hyde Park Rd., Santa Fe, NM 87501, USA

Abstract. The problem of predictability, or "nature vs. nurture", in both ordered and disordered Ising systems following a deep quench from infinite to zero temperature is reviewed. Two questions are addressed. The first deals with the nature of the final state: for an infinite system, does every spin flip infinitely often, or does every spin flip only finitely many times, or do some spins flip infinitely often and others finitely often? Once this question is determined, the evolution of the system from its initial state can be studied with attention to the issue of how much information contained in the final state depends on that contained in the initial state, and how much depends on the detailed history of the system. This problem has been addressed both analytically and numerically in several papers, and their main methods, results, and conclusions will be reviewed. The discussion closes with some open problems that remain to be addressed.

Keywords: Nature vs. nurture · Discrete spin dynamics · Deep quench · Persistence · Aging · Chaotic time dependence · Damage spreading · Spin glasses · Random ferromagnets

1 Introduction

It is a great pleasure to contribute to this volume in honor of Chuck Newman's 70th birthday. Chuck has made so many fundamental contributions to probability theory, mathematical statistical mechanics, percolation theory, and many related fields that it was difficult to choose which topic to write about. The problem was resolved by settling on a problem of longstanding interest to Chuck, and one in which he remains heavily involved: the nonequilibrium dynamics of interacting spin systems, in particular following a *deep quench*, in which a system at high temperature is rapidly cooled to low temperature, after which it evolves according to equilibrium dynamics.

V. Sidoravicius (Ed.): *Sojourns in Probability Theory and Statistical Physics - I*, PROMS 298, pp. 321–338, 2019.
https://doi.org/10.1007/978-981-15-0294-1_11

Our initial foray into this problem appeared in [24], although we had earlier looked at the closely related problems of persistence [25] and (somewhat later) aging [16]. Our initial interest was in determining how and under what conditions equilibration occurs (or doesn't occur) in an infinite system following a deep quench. One has to specify first exactly what one means by "equilibration in an infinite system". Our proposal was that one can sensibly talk about a thermodynamic system settling down to an equilibrium state in the sense of *local equilibration*: that is, fix a region of diameter L and ask whether after a finite time $\tau(L)$ domain walls cease to sweep across the region. If the answer is yes, i.e., if every finite L corresponds to a $\tau(L) < \infty$ after which no spins inside the region ever flip again, then the infinite system locally equilibrates. (It's not only perfectly fine, but also expected in most cases, that $\tau(L) \to \infty$ as $L \to \infty$.)

The main thrust of [24] then went off in a different direction, but it set the seeds for further examination of the problem of local equilibration, its presence or absence in given systems, and its consequences. A paper with Seema Nanda soon followed that focused strictly on zero temperature [21] ([24] studied positive temperature only), with numerous results that began the sorting of systems in which local equilibration occurred and those in which it did not. At zero temperature, a new set of issues and problems arises, in particular the possibility of trapping in a metastable state [26]; nevertheless, if every spin flips only a finite number of times, then local equilibration has occurred, regardless of whether the final state is metastable or a global minimum.

2 Local Equilibration, Weak Nonequilibration, and Chaotic Time Dependence

The first question to be resolved is then whether local equilibration occurs for a given system. When it does not occur, then there are further questions to be resolved. There are two possibilities when local *non*equilibration (LNE) manifests itself: if one averages over all dynamical realizations, does the *dynamically averaged* configuration have a limit—i.e., is there a limiting distribution of configurations? Or does even the distribution not settle down? We refer to the first possibility as "weak LNE", while the second is referred to as *chaotic time dependence* (CTD) [24]. Weak LNE implies a complete lack of predictability, while CTD implies that some amount of predictability remains [21,24].

These claims are justified as follows. Consider for specificity the infinite uniform ferromagnet and consider a fixed finite region centered at the origin. If weak LNE occurs, then (because of global spin-flip symmetry) after some time roughly half the dynamical realizations are in the all-plus state and half are in the all-minus state in that region. This then remains true for all subsequent times, although in any particular dynamical realization, the system never settles down into either — the spin configuration in the finite region continues to alternate (on increasingly long timescales) between the all-plus and all-minus states.

On the other hand, if CTD occurs, then the dynamical averaging fails to fully mix the two possible outcomes at any time; a time-independent distribution is never achieved. In that case, given the initial configuration, one retains some predictive power at arbitrarily large times. These will presumably reflect fluctuations favoring one phase or the other (in that particular region) in the initial state. Compare this to weak LNE, which requires stronger mixing that destroys information contained in these initial fluctuations. Because at zero temperature the configuration evolves according to a deterministic set of rules (given the initial configuration and a realization of the dynamics), the lack of a limit of the averaged configuration in CTD corresponds to the usual notions of deterministic chaos, hence the name "chaotic time dependence". It is amusing that in this context, chaotic time dependence implies *greater* predictability than its nonchaotic alternative.

These considerations motivate the following specific question, first proposed in [24]: given a typical initial configuration, which then evolves under a specified dynamics, how much can one predict about the state of the system at later times? We have colloquially referred to this as a "nature vs. nurture" problem, with "nature" representing the influence of the initial configuration and "nurture" representing the influence of the random dynamics.

The study of nature vs. nurture therefore provides interesting information on some central dynamical issues concerning different classes of models. This problem is related to the general area of phase ordering kinetics [3]. In particular, Krapivsky, Redner and collaborators [29,30] investigated both the $2D$ and $3D$ Ising models with zero temperature Glauber dynamics to understand the time scales and final states of the dynamics. Derrida, Bray and Godrèche introduced the *persistence exponent* [12], which characterizes the power law decay of the fraction of spins that are unchanged from their initial value as a function of time after a quench. This exponent was measured for the zero temperature $2D$ Ising model by Stauffer [32] and calculated exactly for $1D$ by Derrida, Hakim and Pasquier [13].

3 Uniform vs. Disordered Systems

From here on, we restrict our attention to Ising systems at zero temperature and in the absence of any external field, where the bulk (but not all) of our studies have focused. In every case, we start with an infinite temperature spin configuration; i.e., the starting configuration is a realization of a Bernoulli process in which each spin is chosen independently of the others following the flip of a fair coin. The ensuing evolution of the system is governed by zero-temperature Glauber dynamics, in which each spin has an attached Poisson clock with rate 1. When a spin's clock rings, it looks at its neighbors and computes the energy change ΔE associated with flipping (with all other spins remaining fixed). If the energy decreases ($\Delta E < 0$) as a result of the flip, the flip is carried out. If the energy increases ($\Delta E > 0$), the flip is not accepted. If the energy remains the same ($\Delta E = 0$), a flip is carried out with probability $1/2$. This last "tie-breaking" rule can occur only for models (such as the homogenous ferromagnet, or the $\pm J$ spin glass) in which a zero-energy flip can occur; for models with continuous coupling distributions, there is zero probability of a tie. (This is also the case for uniform systems with an odd number of neighbors, such as the $2D$ uniform ferromagnet on the honeycomb lattice.)

For a uniform system, there are two sources of randomness in the problem: the choice of realization of the initial spin configuration, and the realization of the dynamics (i.e., the order in which Poisson clocks for different spins ring, and for models where zero-energy flips are possible, the results of the tie-breaking coin flips). For disordered systems, a third source of randomness enters in the choice of realization of the couplings.

As indicated in the Introduction, the first question that must be resolved is whether the system under study equilibrates locally. This is generally a difficult problem, and our knowledge remains incomplete. Nevertheless, we have obtained results on several systems that include some of the most studied in statistical mechanics.

The simplest is the $1D$ Ising chain. For the uniform ferromagnetic chain, it was proved in [21] (although the result was known earlier [1,5]) that there is no local equilibration: every spin flips infinitely often. In general dimension, the Hamiltonian is

$$\mathcal{H} = -\sum_{\langle x,y \rangle} \sigma_x \sigma_y \,, \tag{1}$$

where $\sigma_x = \pm 1$ is the spin at site x and $\langle \cdot \rangle$ denotes a sum over nearest neighbor spins only.

The proof that every spin flips infinitely often is straightforward and will be informally summarized here. (For the remainder of the paper, most results will be quoted, with references to the original papers for the detailed proofs.) To begin, it is easy to see that the only fixed points of the dynamics are $\sigma_x = +1$ for all x or $\sigma_x = -1$ for all x. Let ω denote a realization of the dynamics and σ^0 a realization of the initial conditions. Then, because both are i.i.d., their joint distribution $P_{\sigma^0,\omega}$ is translation-invariant and translation-ergodic. Because the events that the system lands in the all-plus or the all-minus final states are each translation-invariant, and because both the initial condition and the dynamics are invariant under a global spin flip, each of these events has $P_{\sigma^0,\omega}$-probability zero.

Now let A_x^+ be the event that σ_x flips only finitely many times and its final state is $+1$; similarly for A_x^-. By the same reasoning as above, we must have $P_{\sigma^0,\omega}(A_x^+) = P_{\sigma^0,\omega}(A_x^-) = p$, with $0 \le p \le 1/2$. If $p > 0$, then there must exist a pair of spins at two distinct sites x and x' with opposite final states. This implies a domain wall somewhere between the two that never moves past either one. But it is easy to see that, with positive probability in the dynamics, a sequence of clock rings exists that moves the domain wall outside of this confined range, leading to a contradiction.

We will call systems in which every spin flips infinitely often as being in class $\mathcal{I}$; if every spin flips only finitely often, we refer to it as being in class $\mathcal{F}$. There are also systems in which some spins flip infinitely often and others only finitely often; we refer to these as being in class $\mathcal{M}$. In [21] it was proved that the $2D$

uniform ferromagnet also belongs to class $\mathcal{I}$; the proof is similar in spirit to the $1D$ case but is slightly more elaborate.

What about the uniform ferromagnet in higher than two dimensions? This remains an open problem. Rather old numerical studies [32] indicate that above five dimensions these models are no longer in class $\mathcal{I}$, but the situation remains unclear. The possible presence of dynamical fixed points with complicated geometries in three dimensions and above have so far prevented further analytical progress.

Before leaving the subject of the uniform ferromagnet, it's worth noting that Chuck and collaborators have looked at other mathematically interesting situations, including quasi-$2D$ slabs of varying thicknesses and boundary conditions (which they showed are either type $\mathcal{F}$ or $\mathcal{M}$ depending on thickness and boundary condition) [7], and on $\mathbb{Z}^2$ with a single fixed spin (class $\mathcal{I}$, modulo the frozen spin) [8].

Our discussion has focused so far on uniform ferromagnets. What about disordered systems, in particular, random-bond Ising models? We consider two important classes: random ferromagnets, in which the couplings are i.i.d. non-negative random variables, and spin glasses, where couplings can be positive or negative. (Minor point: in the latter case, if the coupling distribution is asymmetrically distributed about zero, one will have a ferromagnet if the average ratio of positive to negative couplings is sufficiently high. But it turns out this will have no effect on determining whether equilibration occurs.) The Hamiltonian is now

$$\mathcal{H} = -\sum_{\langle x,y\rangle} J_{xy}\sigma_x\sigma_y\,, \tag{2}$$

where the J_{xy} are independent random variables chosen from a common probability distribution.

If one is interested in thermodynamic behavior — e.g., presence or absence of a phase transition, or the number of pure states at a fixed positive temperature and dimension — then central limit theorem-type considerations lead one to expect that the form of the coupling distribution is unimportant, as long as certain features of the distribution, such as mean and variance, are unchanged. So, for example, one expects the same thermodynamic behavior when the coupling distribution is the normal distribution $N(0,1)$ or ± 1, the latter referring to a distribution where each coupling is assigned the value ± 1 independently with the flip of a fair coin.

However, from the point of view of dynamics, particularly with regard to the question of whether local equilibration occurs, the differences between distributions becomes important. In particular, Theorem 3 in [21] shows that, under very mild conditions (existence of a finite mean), *any* model with Hamiltonian (2) and with i.i.d. couplings chosen from a continuous distribution will belong to class $\mathcal{F}$.

In fact the proof is more general, showing that in any discrete-spin model there are only finitely many flips of any spin resulting in a *nonzero* energy change. This immediately implies that models belonging to class $\mathcal{I}$ (or $\mathcal{M}$), such as the

$1D$ and $2D$ uniform ferromagnets, do so exclusively because of an infinite number of zero-energy flips at each site (or a subset of sites). Consequently, models with continuous coupling distributions all belong to class $\mathcal{F}$, because the chance of a tie at any site has zero probability; but the same conclusion applies, say, to the above-mentioned $2D$ uniform ferromagnet on a honeycomb lattice, where the chance of a tie is also zero.

Consequently, the question of the ultimate dynamical fate of a disordered system with continuous coupling distribution is answered in all dimensions.

The restriction to finite mean was needed for technical reasons in the proof, but may not be necessary. Also in [21] it was proved that the so-called *highly disordered* models [2,22,23] in any dimension belong to class $\mathcal{F}$. These are models in which the coupling magnitudes are sufficiently "stretched out" so that the magnitude of any coupling is at least twice the value of the next smaller magnitude and no more than half that of the next larger magnitude. (For a formal definition, see [21].) Of course, this needs to be done in a size-dependent manner, rescaling the values of the couplings as one increases the volume under consideration [22,23]. It turns out that $1D$ chains with continuous coupling distributions, although not satisfying the above criterion, still belong to this class. That is, $1D$ chains with *any* continuous coupling distribution, regardless of whether the mean is finite, will fall into class $\mathcal{F}$.

As a final remark on continuous coupling distributions in general, an important point is that it doesn't matter (for the purposes of whether all spins eventually fixate) whether one is talking about a random ferromagnet or a spin glass—the signs of the coupling do not enter into these considerations, only the coupling magnitudes.

We close this section by mentioning a result for a model with discrete disorder: the $\pm J$ spin glass in $2D$. It was shown in [17] that this model belongs to class $\mathcal{M}$. We are unaware of any other results on this class of models in terms of their final evolutionary states.

4 Nature vs. Nurture

Given the classification of systems in the previous section, we turn now to the main focus of this review—namely, the extent to which information contained in a configuration at time t can be inferred from knowledge of the initial state of the system. We first need to quantify this concept, and the approach used rests on the idea of determining what proportion of the information contained in the state at time t depends on the initial condition and what proportion on the realization of the dynamics.

If the system is type $\mathcal{I}$, one might expect that the information contained in the initial state will decay to zero as $t \to \infty$. We will discuss below how this is indeed the case in the $1D$ and $2D$ uniform ferromagnets, where the decay takes the form of a power law. This leads to a new exponent characterizing the power-law decay, which we have denoted the *heritability exponent*. We will return to the heritability exponent in Sect. 4.2.

We turn next to the (perhaps) simpler case of type-$\mathcal{F}$ systems. Because these by definition equilibrate locally, we compare the final state with the initial state, averaged over many dynamical trials, to determine to what degree initial information has been retained. In [21] we introduced a type of dynamical order parameter, denoted q_D (here "D" refers to dynamical, not dimension).

Let σ^0 denote the realization of the initial condition and σ^t the configuration at time t later. This of course depends on the dynamical realization ω, but to keep the notation from becoming too unwieldy, we suppress the dependence of σ^t on σ^0 and ω. Let $\langle\sigma_x\rangle_t$ denote the state of σ_x averaged over all dynamical realizations up to time t for a *fixed* σ^0. We then study the resulting quantity averaged over all initial configurations and (if the system is disordered) coupling realizations.

Denoting the latter averages (with respect to the joint distribution $P_{\mathcal{J},\sigma^0} = P_{\mathcal{J}} \times P_{\sigma^0}$) by $\mathbf{E}_{\mathcal{J},\sigma^0}$, we define $q_D = \lim_{t\to\infty} q_t$ (providing the limit exists), where

$$q_t = \lim_{L\to\infty} |\Lambda_L|^{-1} \sum_{x\in\Lambda_L} (\langle\sigma_x\rangle_t)^2 = \mathbf{E}_{\mathcal{J},\sigma^0}(\langle\sigma_x\rangle_t^2) \tag{3}$$

and Λ_L is a d-dimensional cube of side L centered at the origin. The equivalence of the two formulas for q_t in (3) follows from translation-ergodicity [21].

The order parameter q_D measures the extent to which σ^∞ is determined by σ^0 rather than by ω; it is a dynamical analog to the usual Edwards-Anderson order parameter.

4.1 1D and Highly Disordered Models

The dynamical order parameter can be exactly calculated for the 1D Ising chain. For the uniform ferromagnet, σ_x^∞ does not exist. But it does when the couplings are drawn from a continuous distribution (the details of the distribution are unimportant, as long as its support is continuous), and moreover for these models $q_D = 1/2$. The result and proof appear in [21], but the argument underlying the result is easy to state informally. The basic idea is to note that every spin lives in the domain of influence of a "bully bond". This is a coupling whose magnitude is larger than the couplings to either side; such a coupling must be satisfied in any fixed point of the 1-spin-flip dynamics (and more generally, in any ground state). As one moves along the chain starting from either side of the bully bond, couplings decrease in magnitude until arriving at a local minimum: i.e., a coupling whose magnitude is *smaller* than the couplings to either side. (Of course, the bond whose coupling value is a local minimum could be adjacent to the bully bond.)

The domain of influence of the bully bond is then the set of spins that live on the sites of the bonds to either of its sides until one reaches the nearest local minimum bonds to its right and left. For the local minimum bond to the right of the bully bond, the spin on its lefthand site belongs to the domain of influence of the bully bond; for the local minimum bond to the left of the bully bond, the spin on its righthand site belongs to the domain of influence of the bully bond. (The remaining two spins on the two local minimum bonds belong to other domains of influence.) In this way the entire chain is partitioned into domains of influence, each "governed" by a single bully bond.

Regardless of the dynamical realization, it must be the case that once the bully bond is satisfied, the final state of every spin in its domain of influence is completely determined. Now consider the initial state σ^0. In a.e. realization of the initial state, half the bully bonds will be satisfied and half will be unsatisfied. For those that are satisfied, the final state of every spin in its domain of influence is completely determined by the initial state; i.e., a.e. realization of the dynamics will result in the same final state. For those that are unsatisfied, the final state is determined by which Poisson clock, of the two spins on the sites connected to the bully bond, rings first. Since these two events have equal probability, and since the two possible outcomes give equal and opposite contributions to the twin overlap, all spins in such domains of influence contribute zero to q_D. Consequently, $q_D = 1/2$ for the continuously disordered $1D$ chain: in σ^∞ half the final states of the spins are completely determined by the initial configuration, and half are completely determined by the dynamics.

For highly disordered models in any dimension, q_D is also $1/2$. The main idea behind the proof is similar to that above, but more involved. Here, one can define *influence clusters* of similarly defined bully bonds, but one needs to show as well that these influence clusters do not percolate. Details can be found in [21].

Before leaving this section, it is worth noting that in models with continuous disorder on the Euclidean lattice $\mathbb{Z}^d$ with $1 < d < \infty$, an argument similarly based on bully bonds can be used to show that $0 < q_D < 1$. In any dimension, the density of bully bonds is strictly positive (though decreasing as dimension increases). If $\rho_b(d) > 0$ is the density of bully bonds in d dimensions, then considerations similar to those above provide a lower bound of $(1/2)\rho_b(d)$ and an upper bound of $1 - (1/2)\rho_b(d)$ for q_D.

Needless to say, these are poor upper and lower bounds; the main point is that $0 < q_D < 1$ for disordered models in any finite dimension. We return to this discussion is Sect. 6.

4.2 Heritability, Damage Spreading, and Persistence

We turn now to type-$\mathcal{I}$ systems, in which σ_x^∞ does not exist for any x, and therefore q_D is not defined. However, q_t remains perfectly well-defined for all finite t, and one can study this quantity to determine how the initial information contained in σ^0 changes with time.

If one is studying the system numerically (necessary in most cases), then one can model q_t in the following way: prepare two Ising systems with the same initial configuration, and then allow them to evolve independently using zero-temperature Glauber dynamics. One then computes the spin overlap between these "twin" copies, chooses another initial condition, and so on, eventually computing a twin overlap over many different initial conditions. This overlap as a function of time, which we refer to as the "heritability", is essentially the same as q_t, and for the $2D$ uniform ferromagnet was found to decay as a power law in time [34]. We denote the exponent θ_h associated with the power-law decay of heritability as the "heritability exponent".

Heritability is the opposite of "damage spreading" [6,19,31], given that the latter involves starting with two slightly different initial configurations and letting them evolve with the *same* dynamical realization. In damage spreading studies one is interested in the extent of the spread of the initial difference throughout the system as time proceeds.

The nature vs. nurture question is also related to persistence [12], which studies the fraction of spins that have not flipped up to time t. This was found to decay as a power law in a number of systems, in particular uniform ferromagnets and Potts models in low dimensions, and the associated decay exponent θ_p is known as the "persistence exponent". Although the concepts are related, heritability is not the same as persistence, which asks which spins have not flipped up to a time t. In contrast, heritability asks to what extent the *information* contained in the initial state persists up to time t. A spin may have flipped multiple times during this interval but its final state might still be predictable knowing the initial condition.

5 The $1D$ and $2D$ Uniform Ferromagnets

Both the persistence and heritability exponents can be computed exactly in the $1D$ uniform Ising ferromagnet. It was shown in [13,14] that $\theta_p = 3/8$ for this system. Correspondingly, it can be shown that $\theta_h = 1/2$, as discussed in [34], by using the mapping to the voter model and coalescing random walks (see, e.g., [14,16]).

The $2D$ uniform Ising ferromagnet on the square lattice is considerably more difficult, and requires numerical study. The (finite-volume) absorbing states include not only the uniform plus and minus states, but also "striped" configurations which appear in roughly 1/3 of the runs [29]. A striped state has one (or more) vertical or horizontal stripes (but not both) whose boundaries constitute domain walls separating regions of antiparallel spin orientation.

An initial study of the nature vs. nurture problem was reported in [9], in which evidence was found for chaotic time dependence. However, the problem was not fully analyzed and solved until almost a decade later, when Ye et al. [34] showed that a power-law decay of initial information did occur and computed the heritability exponent.

Ye et al. did twin studies on 21 $L \times L$ squares, from $L = 10$ to $L = 500$. For each size 30,000 runs on independent pairs of twins were taken out to times such that (almost) all of the samples landed in an absorbing state. (For each initial condition only two dynamical trajectories were computed, one for each twin.) Because these runs were done on finite lattices, the authors studied $q_t(L) = \frac{1}{N}\sum_{i=1}^{N}\sigma_i^1(t)\sigma_i^2(t)$, where $\sigma_i^1(t)$ denotes the state of the ith spin at time t in twin 1, and similarly for $\sigma_i^2(t)$.

For each pair of twins $q_t(L)$ was computed, and averaging over all runs gave the average $\overline{q_t(L)}$. Ye et al. studied both the size dependence of the final overlap, $\overline{q_\infty(L)} = \lim_{t\to\infty} \overline{q_t(L)}$ and the time dependence of the infinite volume limit $\overline{q_t} = \lim_{L\to\infty} \overline{q_t(L)}$. It was also shown analytically, and confirmed numerically, that the behavior of $\overline{q_\infty(L)}$ and $\overline{q_t}$ are connected by a finite size scaling ansatz [34].

Consider first the behavior of $\overline{q_t(L)}$ vs. t for several L, shown in Fig. 1. For short and intermediate times, $\overline{q_t(L)}$ appears to follow a single curve for all L, until an L-dependent time scale when $\overline{q_t(L)}$ separates from the main curve and a plateau is reached. Ye et al. made the natural assumption that the single curve represented the infinite volume behavior $\overline{q_t}$ to good approximation. The long-time behavior of $\overline{q_t}$ is well-described by a power law of the form $\overline{q_t} = dt^{-\theta_h}$, with $d = 0.62(3)$ and $\theta_h = 0.225(6)$ computed at the two largest sizes studied ($L = 400$ and $L = 500$). The error bars are obtained by the bootstrap method. Using other large sizes also, the heritability exponent describing the decay of the overlap with time was found to be $\theta_h = 0.22 \pm 0.02$.

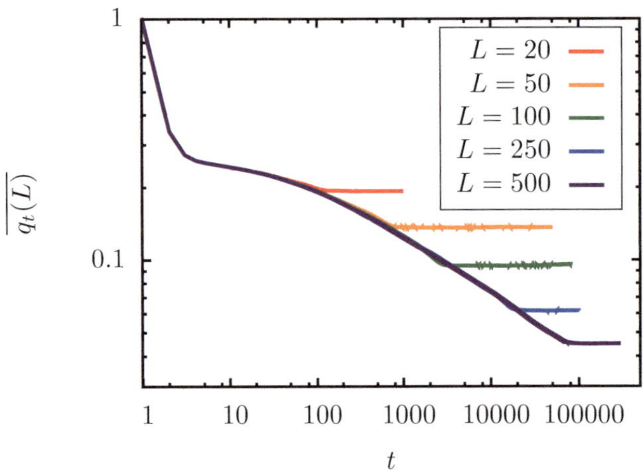

Fig. 1. $\overline{q_t(L)}$ vs. t for several L and for quenches to $T = 0$. The plateau value decreases from small to large L. From [34].

Next consider $\overline{q_\infty(L)}$, the finite size behavior of the absorbing value of $\overline{q_t(L)}$, shown in Fig. 2. Ye et al. found that the data were well fit by a power law of the form $\overline{q_\infty(L)} = aL^{-b}$. Their best estimate of b, taking into account both statistical and possible systematic errors, was 0.46 ± 0.02.

Summarizing, the main results are:

1. Heritability exponent: $\overline{q_t} \sim dt^{-\theta_h}$ and $\theta_h = 0.22 \pm 0.02$.
2. Size dependence: For finite L and $T = 0$, $\overline{q_\infty(L)} \sim aL^{-b}$ with $b = 0.46 \pm 0.02$.
3. Finite size scaling considerations suggest that $b = 2\theta_h$, consistent with the numerically determined values and with the exact $1D$ values.

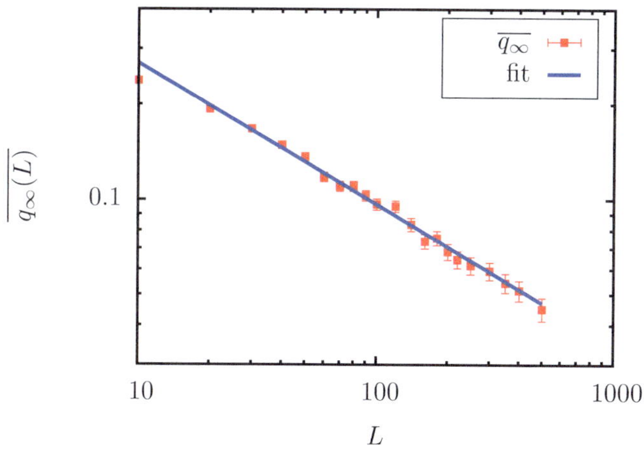

Fig. 2. $\overline{q_\infty(L)}$ vs. L for quenches to $T = 0$. The solid line is the best power law fit for sizes 20 to 500, and corresponds to $\overline{q_\infty(L)} \sim L^{-0.46}$. From [34].

6 Random Ferromagnets and Spin Glasses in Finite Dimensions

In [35], the nature vs. nurture question was studied in random Ising ferromagnets and Edwards–Anderson (EA) spin glasses [15] in dimensions greater than one. Both use the Hamiltonian (2), the difference being in the form of the probability distribution from which the couplings J_{xy} are chosen. In both cases the J_{xy} are i.i.d. random variables, but in the ferromagnetic case the couplings are chosen from a continuous distribution supported on nonnegative real numbers, and in the spin glass the support of the coupling distribution lies on both positive and negative real numbers. Specifically, in [35] the coupling distribution for the EA spin glass in d dimensions was taken to be a normal distribution with mean zero and variance one, and for the random ferromagnet the distribution was taken to be a one-sided Gaussian, in which each bond is chosen as the absolute value of a standard Gaussian random variable (again with mean zero and variance one).

In Sect. 3, it was observed that both of these models belong to class $\mathcal{F}$ in all dimensions; therefore, the proper quantity to study is the dynamical order parameter q_D. We showed in Sect. 4.1 that in any finite dimension, q_D lies strictly between 0 and 1. We also know that the random ferromagnet and spin glass are identical (up to a trivial gauge transformation) in an infinite $1D$ chain, and that $q_D = 1/2$. As also discussed in Sect. 4.1, a (poor) lower bound for q_D is provided by the density of bully bonds $\rho_b(d)$, which goes to 0 as $d \to \infty$. A central question posed in [35] is then: does $q_D(d) \to 0$ as dimension $d \to \infty$?

Figure 3 shows numerically derived values for both systems in 2, 3, and 4 dimensions.

The results for the random ferromagnet are quite close to those of the Edwards–Anderson model, especially in $d = 2$. This suggests the possibility that

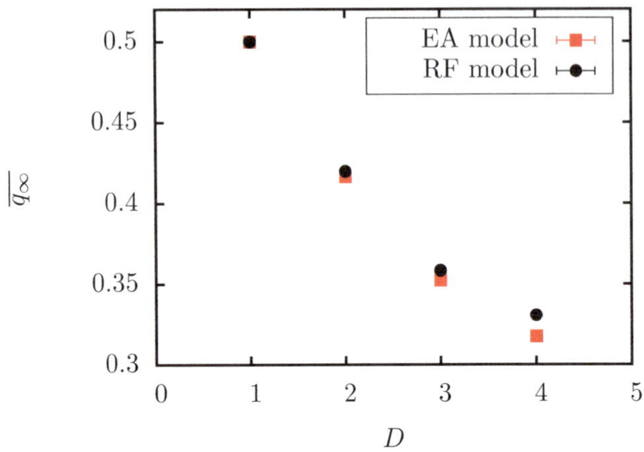

Fig. 3. The average twin overlap in the absorbing state for the largest systems size $\overline{q_\infty(N_{\max})}$ vs. dimension d for the Edwards–Anderson spin glass (red squares) and random ferromagnet (black circles). From [35].

frustration, present in the spin glass but not the random ferromagnet, plays little or no role in the nature vs. nurture problem for low dimensionality. This could change, however, as dimension increases; indeed, Fig. 3 indicates that as dimension increases the difference between $\overline{q_\infty(N_{\max})}$ for the spin glass and random ferromagnet likewise increases, with q_D(random ferromagnet)$>$ q_D(spin glass).

While q_D monotonically decreases with dimension up to 4, limits on what can be done numerically at this time preclude going to higher dimensions, and the question of the behavior of $q_D(d)$ as $d \to \infty$ remains open. Although the results from [35] are consistent with the conjecture that $q_D(d) \to 0$ as $d \to \infty$, they don't preclude other possibilities. One might then take a look at the behavior of the random Curie–Weiss ferromagnet and the Sherrington–Kirkpatrick (SK) [27] infinite-range spin glass to infer the high-dimensional behavior of the finite-dimensional random ferromagnet and spin glass. As we will see in the next section, however, the mean-field results are surprising, and so we will return to this question there.

Before leaving this section, it is worth noting that several other features of the nature vs. nurture problem besides q_D were studied in [35]. These include convergence rates of overlaps as the number of spins N increases and the mean survival time $\tau(N)$ (i.e., the average number of spin flips per spin as a function of system size and dimension). The reader is referred to [35] for a discussion of these quantities. One further study is especially interesting and will be briefly mentioned here, namely the spatial structure of the overlap in the final state. It is natural to ask whether "like spins" (i.e., spins whose final state is the same) percolate in the twin samples. Figure 4 shows the overlap configuration for a typical pair of final states for the two-dimensional EA spin glass with $L = 100$. From the figure it is indeed seen that like spins (shown in red) are well above the percolation threshold. (Note that there are no isolated singleton unlike spins since those must be eliminated by the zero-temperature dynamics.)

Fig. 4. (color online). The overlap configuration for a typical pair of final states for the two-dimensional Edwards–Anderson model with $L = 100$. Like spins percolate and are shown in red while unlike spins are shown in black.

7 Mean-Field Models

We begin by examining the SK model, which is the spin glass whose N spins lie on the nodes of the complete graph. Its Hamiltonian is

$$\mathcal{H} = -\frac{1}{\sqrt{N}} \sum_{i<j} J_{ij}\sigma_i\sigma_j \,. \tag{4}$$

The couplings are again chosen from a normal distribution with mean zero and variance one, and the rescaling factor $N^{-1/2}$ ensures a sensible thermodynamic limit of the energy and free energy per spin.

Figure 5 is a plot of $\overline{q_\infty(N)}$ as a function of N. While it is clear that $\overline{q_\infty(N)}$ is decreasing with N, it again is not obvious whether $\overline{q_\infty(N \to \infty)}$ is zero or greater than zero.

The best fit implies that $\overline{q_\infty(N \to \infty)} \to 0$, although slowly. However, a different fit—not as good, but still reasonable—implies a nonzero limit. For details, see [35]. Nevertheless, a heuristic argument presented in [35] suggests that it is most reasonable to expect that indeed $\overline{q_\infty(N \to \infty)} \to 0$. The main

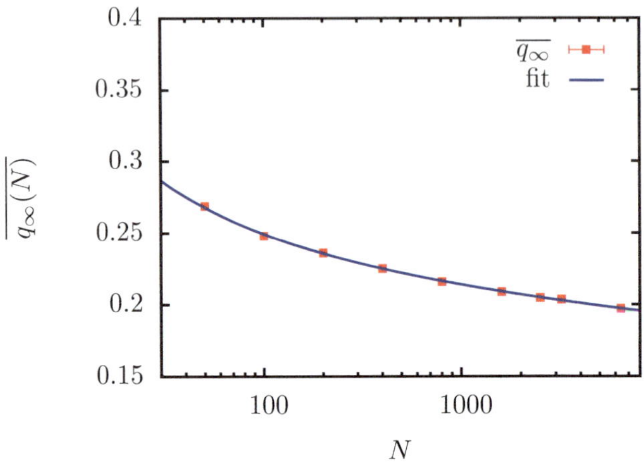

Fig. 5. Simulation results for $\overline{q_\infty(N)}$ vs.!N for the SK model. The curve is the highest quality fit: $\overline{q_\infty(N)} = \frac{a}{(\log N)^{1/3}} + \frac{b}{N}$, with a and b both close to 0.4. From [35].

idea is the following: it is known that the SK model has exponentially many (in N) one-spin-flip metastable states [4], and moreover numerics indicate that the number of spin flips grows linearly with N [35]. Given the $O(N)$ distance traveled by the SK model on the state space hypercube to find an absorbing state, and given that these states have been shown to be uncorrelated [4], the decay of $q_\infty \to 0$ as $N \to \infty$ appears to be its most likely behavior.

Again, it is worth noting that several other dynamical behaviors of the SK model were studied, including the median time for a system of N spins to reach the absorbing state, the fraction of active spins (those that have not finished flipping) as a function of time and system size, and the energy per spin as a function of time. The reader is referred to [35] for a discussion of these behaviors.

The evidence for $\overline{q_\infty(N \to \infty)} \to 0$ as $N \to \infty$ for the SK model may be taken to imply that $q_D(d) \to 0$ for the EA spin glass as $d \to \infty$. This is reasonable, and even likely to be correct, but a cautionary note should be added: the behavior of the Curie–Weiss random ferromagnet is completely different — in fact, for this model $\overline{q_\infty(N \to \infty)} \to 1$ as $N \to \infty$ [35]! While presumably the information in the initial state is completely lost in the SK model for large systems at long times, for the random ferromagnet it's completely retained. The profound difference between nature and nurture for these models demonstrates that frustration plays a centrally important role in infinite dimensions, and so it may play an important role in high but finite dimensions as well.

It is easy to understand why the Curie–Weiss model behaves this way. Consider first the uniform case, where all couplings have equal magnitude. A typical initial condition will have an excess (of order $\sqrt{N}$) of spins in one state (say the plus state) over the other, so every spin feels the same positive internal field, and this can only increase with time. Given the usual Glauber dynamics, it's clear that the final state will then be all plus, so the initial condition completely determines the final configuration. The only initial conditions in which this will not be true is that for which $-1 \leq \sum_{i=1}^{N} \sigma_i \leq 1$. But the contribution to q_∞ from such configurations goes to zero (as $N^{-1/2}$) as $N \to \infty$.

One expects exactly the same behavior for the random Curie–Weiss model, but the proof is considerably more difficult. As before, a typical initial spin configuration will have $O(\sqrt{N})$ excess of plus or minus spins, but now the internal fields acting on the spins will vary, and if the initial excess of spins is positive, the internal fields on some spins could be negative. One therefore has to study the *distribution* of the internal fields at each site, which at time zero has positive mean. The main technical issue is to show that the fraction of sites with positive internal field increases steadily with time.

A proof covering the case of the randomly diluted Curie–Weiss ferromagnet, where the J_{ij}'s are chosen from Ber(p) for some fixed $p \in (0, 1)$, showing that again $\overline{q_\infty(N \to \infty)} \to 1$ as $N \to \infty$, appeared in [18]. The proof demonstrates that after a time of order $N^{1/2+\epsilon}$, where $\epsilon > 0$ is independent of N, every site has positive internal field with probability going to one as $N \to \infty$. At this point, the dynamics monotonically leads to absorption into the all-plus state, so that $q_\infty \to 1$ as $N \to \infty$.

The proof further demonstrates that the final state of the system is one of the two uniform states. One possibility is then that the random Curie-Weiss ferromagnet possesses many metastable states (like finite-dimensional random ferromagnets and spin glasses [24] or the SK model [4]), but that the dynamics somehow avoids them. More likely, though, is the possibility that the random Curie-Weiss model possesses no metastable states (in the sense that their number falls to zero as $N \to \infty$), or else too few for the system to find; i.e., the union of their basins of attraction has zero measure in the $N \to \infty$ limit. This problem was studied in Wang *et al.* [33] and Song *et al.* [28] and it was found that such metastable states do exist in these models, but their numbers are small and grow slowly with N.

These results lead to a problem in interpreting the finite-dimensional random ferromagnetic model: either $q_D(d)$ reaches a minimum at some finite d and then increases to 1 as $d \to \infty$, or else $q_D(d)$ falls to zero as conjectured and the infinite-dimensional limit is singular for the nature vs. nurture problem in the random ferromagnet as $d \to \infty$. (There are other possibilities, of course, but these are the most plausible alternatives.) In [35] a heuristic argument is presented in support of the latter alternative, but the problem remains open in the absence of a more detailed argument (or preferably, a proof).

Other models were studied as well in [35], in particular the random energy model (REM) of Derrida [10,11], where it was again found that $q_\infty \to 1$. The argument can again be found in [35] and will not be repeated here. What is important to note is that all of these studies indicate strongly that two conditions appear to be necessary (though perhaps not sufficient) in order for $q_\infty < 1$. The first is the presence of a large number of uncorrelated metastable states, so that the system has many possible final states whose overlap is small. This condition is satisfied for the EA model and random ferromagnet in all finite dimensions (although the metastable states retain some correlations, which presumably decrease to zero as $d \to \infty$), as well as by the SK model and the REM. The second, equally important, condition is that a finite system undergoes at least $O(N)$ spin flips before reaching the absorbing state, and correspondingly for an infinite system, the average number of spin flips per site is strictly positive. This is the case for the EA model and random ferromagnet in all finite dimensions, as well as for the SK model and both mean-field ferromagnets. Of the mean-field models studied, only the SK model satisfies both conditions.

In contrast, the random Curie–Weiss ferromagnet and REM have $q_\infty \to 1$, but for very different reasons. In the random Curie–Weiss ferromagnet, there are $O(N)$ steps in the random walk on the configuration space hypercube (with 2^N vertices) but (presumably) no metastable states to trap the walk before reaching the absorbing uniform final state consistent with the initial configuration. The REM, on the other hand, has many metastable states, but its random walk travels only $O(\log N)$ steps before being trapped in a metastable state [20, 35]; hence the overlap with the initial state approaches 1 as $N \to \infty$.

8 Conclusion

This relatively brief review has touched on some of the central topics in the nature vs. nurture problem, but omitted many interesting issues and results which can be further pursued in the papers listed in the bibliography. There are numerous outstanding questions and open problems, but perhaps the most fundamental ones are the following.

1. Determine the dynamical class ($\mathcal{I}$, $\mathcal{F}$, or $\mathcal{M}$) of the uniform Ising ferromagnet in $d \geq 3$. For those dimensions belonging to class $\mathcal{I}$ (or $\mathcal{M}$), determine the heritability exponent and study its behavior as a function of dimension. How does it relate to the persistence exponent, particularly as dimension increases?
2. What is the behavior of $q_D(d)$ for the random ferromagnet and EA spin glass as a function of dimension? Does $q_D(d) \to 0$ as $d \to \infty$? The case of the random ferromagnet is particularly interesting: does it begin to increase at some finite d, or is the dynamical behavior singular at $d = \infty$? Answering this question would shed light on the equally important question of the role of frustration in finite dimensions: is it as important as it appears to be in infinite dimensions?
3. Prove (or disprove) that $\overline{q_\infty(N \to \infty)} \to 0$ as $N \to \infty$ for the SK model.
4. A particularly difficult problem is disordered systems with discrete coupling distributions, most notably the $\pm J$ spin glass. This was shown in [17] to be in class $\mathcal{M}$ in two dimensions. What is its dynamical behavior in higher dimensions? The nature vs. nurture question has not been studied at all in this model, nor more generally has a serious attempt been made to understand how to approach this problem in the context of class-$\mathcal{M}$ models.

If nothing else, a major aim of this review is to provide the reader with the sense that the nature vs. nurture approach to dynamics constitutes a set of deep problems and rich phenomena whose explication can provide significant illumination on the dynamical behavior of both ordered and disordered statistical mechanical systems.

Acknowledgments. I thank my collaborators on various aspects of this work—Reza Gheissari, Jon Machta, Chuck Newman, Paolo Maurilo de Oliveira, Vladas Sidoravicius, Eric Song, Lily Wang, and Jing Ye—for a productive and enjoyable collaboration. And of course, happy birthday (again) to Chuck!

References

1. Arratia, R.: Site recurrence for annihilating random walks on $\mathbb{Z}^d$. Ann. Prob. **11**, 706–713 (1983)
2. Banavar, J.R., Cieplak, M., Maritan, A.: Optimal paths and domain walls in the strong disorder limit. Phys. Rev. Lett. **72**, 2320–2323 (1994)
3. Bray, A.J.: Theory of phase-ordering kinetics. Adv. Phys. **43**, 357–459 (1994)
4. See, for example, Bray, A., Moore, M.A.: Metastable states in spin glasses. J. Phys. C (Sol. St. Phys.) 13, L469–L476 (1980)
5. Cox, J.T., Griffeath, D.: Diffusive clustering in the two-dimensional voter model. Ann. Prob. **14**, 347–370 (1986)
6. Creutz, M.: Deterministic Ising dynamics. Ann. Phys. **167**, 62–72 (1986)
7. Damron, M., Kogan, H., Newman, C., Sidoravicius, V.: Fixation for coarsening dynamics in $2D$ slabs. Electron. J. Probab. **18**(105), 20 (2013)
8. Damron, M., Kogan, H., Newman, C., Sidoravicius, V.: Coarsening with a frozen vertex. Electron. J. Probab. **21**(9), 4 (2016)
9. de Oliveira, P.M.C., Newman, C.M., Sidoravicius, V., Stein, D.L.: Ising ferromagnet: zero-temperature dynamical evolution. J. Phys. A **39**, 6841–6849 (2006)
10. Derrida, B.: Random-energy model: limit of a family of disordered models. Phys. Rev. Lett. **45**, 79–82 (1980)
11. Derrida, B.: Random-energy model: an exactly solvable model of disordered systems. Phys. Rev. B **24**, 2613–2626 (1981)
12. Derrida, B., Bray, A.J., Godreche, C.: Non-trivial exponents in the zero temperature dynamics of the $1D$ Ising and Potts models. J. Phys. A: Math. Gen. **27**, L357–L361 (1994)
13. Derrida, B., Hakim, V., Pasquier, V.: Exact first-passage exponents of $1D$ domain growth: relation to a reaction-diffusion model. Phys. Rev. Lett. **75**, 751–754 (1995)
14. Derrida, B., Hakim, V., Pasquier, V.: Exact exponent for the number of persistent spins in the zero-temperature dynamics of the one-dimensional Potts model. J. Stat. Phys. **85**, 763–797 (1996)
15. Edwards, S., Anderson, P.W.: Theory of spin glasses. J. Phys. F **5**, 965–974 (1975)
16. Fontes, L.R., Isopi, M., Newman, C.M., Stein, D.L.: Aging in $1D$ discrete spin models and equivalent systems. Phys. Rev. Lett. **87**, 110201 (2001)
17. Gandolfi, A., Newman, C.M., Stein, D.L.: Zero-temperature dynamics of $\pm J$ spin glasses and related models. Commun. Math. Phys. **214**, 373–387 (2000)
18. Gheissari, R., Newman, C.M., Stein, D.L.: Zero-temperature dynamics in the dilute Curie–Weiss model. J. Stat. Phys. **172**, 1009–1028 (2018)

19. Grassberger, P.: Damage spreading and critical exponents for Ising dynamics. Phys. A **214**, 547–559 (1995)
20. Kauffman, S., Levin, S.A.: Towards a general theory of adaptive walks on rugged landscapes. J. Theor. Biol. **1281**, 11–45 (1987)
21. Nanda, S., Newman, C.M., Stein, D.L.: Dynamics of Ising spin systems at zero temperature. Amer. Math. Soc. Transl. **2**, 183–194 (2000). In On Dobrushin's Way (from Probability Theory to Statistical Physics), R. Minlos, S. Shlosman and Y. Suhov, eds
22. Newman, C.M., Stein, D.L.: Spin-glass model with dimension-dependent ground state multiplicity. Phys. Rev. Lett. **72**, 2286–2289 (1994)
23. Newman, C.M., Stein, D.L.: Ground state structure in a highly disordered spin glass model. J. Stat. Phys. **82**, 1113–1132 (1996)
24. Newman, C.M., Stein, D.L.: Equilibrium pure states and nonequilibrium chaos. J. Stat. Phys. **94**, 709–722 (1999)
25. Newman, C.M., Stein, D.L.: Blocking and persistence in the zero-temperature dynamics of homogeneous and disordered Ising models. Phys. Rev. Lett. **82**, 3944–3947 (1999)
26. Newman, C.M., Stein, D.L.: Metastable states in spin glasses and disordered ferromagnets. Phys. Rev. E **60**, 5244–5260 (1999)
27. Sherrington, D., Kirkpatrick, S.: Solvable model of a spin glass. Phys. Rev. Lett. **35**, 1792–1796 (1975)
28. Song, Y., Gheissari, R., Newman, C., Stein, D.L.: Searching for local minima in a random landscape. In preparation
29. Spirin, V., Krapivsky, P.L., Redner, S.: Fate of zero-temperature Ising ferromagnets. Phys. Rev. E **63**, 036118 (2000)
30. Spirin, V., Krapivsky, P.L., Redner, S.: Freezing in Ising ferromagnets. Phys. Rev. E **65**, 016119 (2001)
31. Stanley, H.E., Stauffer, D., Kertesz, J., Hermann, H.: Dynamics of spreading phenomena in two-dimensional Ising models. Phys. Rev. Lett. **59**, 2326–2328 (1987)
32. Stauffer, D.: Ising spinodal decomposition at $T = 0$ in one to five dimensions. J. Phys. A: Math. Gen. **27**, 5029–5032 (1994)
33. Wang, L., Gheissari, R., Newman, C., Stein, D.L.: Nature vs. nurture: dynamical evolution in disordered Ising ferromagnets. In: Gayrard, V., Arguin, L.-P., Kistler, N., Kourkova, I. (eds.) Statistical Mechanics of Classical and Disordered Systems. Springer, New York (2019). to appear
34. Ye, J., Machta, J., Newman, C., Stein, D.L.: Nature vs. nurture: predictability in zero-temperature Ising dynamics. Phys. Rev. E **88**, 040101 (2013)
35. Ye, J., Gheissari, R., Machta, J., Newman, C., Stein, D.L.: Zero-temperature dynamics in the dilute Curie-Weiss model. J. Stat. Phys. **172**, 1009–1028 (2018)

MIX
Papier aus verantwortungsvollen Quellen
Paper from responsible sources
FSC® C105338

If you have any concerns about our products,
you can contact us on
ProductSafety@springernature.com

In case Publisher is established outside the EU,
the EU authorized representative is:
Springer Nature Customer Service Center GmbH
Europaplatz 3, 69115 Heidelberg, Germany

Printed by Libri Plureos GmbH
in Hamburg, Germany